From a human viewpoint, the cold waters of deep sea and polar marine regions present an inhospitable and harsh environment for life. The study of organisms which have exploited this seemingly hostile ecological niche has revealed intriguing physiological adaptations. This volume summarises the most recent information on the adaptations exhibited by representatives of the main groups of polar animals, including marine invertebrates, fish, mammals and birds, at scales ranging from the whole organism through to the underlying molecular adaptational mechanisms. Unique in its breadth of coverage, this volume will provide an important resource for all those concerned specifically with how animals have adapted to life in extreme conditions, as well as being of general interest to all marine biologists.

SOCIETY FOR EXPERIMENTAL BIOLOGY
SEMINAR SERIES: 66

COLD OCEAN PHYSIOLOGY

SOCIETY FOR EXPERIMENTAL BIOLOGY SEMINAR SERIES

A series of multi-author volumes developed from seminars held by the Society for Experimental Biology. Each volume serves not only as an introductory review of a specific topics, but also introduces the reader to experimental evidence to support the theories and principles discussed, and points the way to new research.

6. Neurones without impulses: their significance for vertebrate and invertebrate systems. *Edited by A. Roberts and B. M. H. Bush*
8. Stomatal physiology. *Edited by P. C. G. Jarvis and T. A. Mansfield*
16. Gills. *Edited by D. F. Houlihan, J. C. Rankin and T. J. Shuttleworth*
31. Plant canopies: their growth, form and function. *Edited by G. Russell, B. Marshall and P. G. Jarvis*
33. Neurohormones in invertebrates. *Edited by M. Thorndyke and G. Goldsworthy*
34. Acid toxicity and aquatic animals. *Edited by R. Morris, E. W. Taylor, D. J. A. Brown and J. A. Brown*
35. Division and segregation of organelles. *Edited by S. A. Boffey and D. Lloyd*
36. Biomechanics in evolution. *Edited by J. M. V. Rayner and R. J. Wootton*
37. Techniques in comparative respiratory physiology: An experimental approach. *Edited by C. R. Bridges and P. J. Butler*
38. Herbicides and plant metabolism. *Edited by A. D. Dodge*
39. Plants under stress. *Edited by H. C. Jones, T. J. Flowers and M. B. Jones*
40. *In situ* hybridisation: application to developmental biology and medicine. *Edited by N. Harris and D. G. Wilkinson*
41. Physiological strategies for gas exchange and metabolism. *Edited by A. J. Woakes, M. K. Grieshaber and C. R. Bridges*
42. Compartmentation of plant metabolism in non-photosynthesis tissues. *Edited by M. J. Emes*
43. Plant growth: interactions with nutrition and environment. *Edited by J. R. Porter and D. W. Lawlor*
44. Feeding and texture of foods. *Edited by J. F. V. Vincent*
45. Endocytosis, exocytosis and vesicle traffic in plants. *Edited by G. R. Hawes, J. O. D. Coleman and D. E. Evans*
46. Calcium, oxygen radicals and cellular damage. *Edited by C. J. Duncan*
47. Fruit and seed production: aspects of development, environmental physiology and ecology. *Edited by C. Marshall and J. Grace*
48. Perspectives in plant cell recognition. *Edited by J. A. Callow and J. R. Green*
49. Inducible plant proteins: their biochemistry and molecular biology. *Edited by J. L. Wray*
50. Plant organelles: compartmentation of metabolism in photosynthetic cells. *Edited by A. K. Tobin*
51. Oxygen transport in biological systems: modelling of pathways from environment to cell. *Edited by S. Eggington and H. F. Ross*
52. New insights in vertebrate kidney function. *Edited by J. A. Brown, R. J. Balment and J. C. Rankin*
53. Post-transitional modifications in plants. *Edited by N. H. Battey, H. G. Dickinson and A. M. Hetherington*
54. Biomechanics and cells. *Edited by F. / J. Lyall and A. J. El Haj*
55. Molecular and cellular aspects of plant reproduction. *Edited by R. Scott and A. D. Stead*
56. Amino acids and their derivatives in plants. *Edited by R. M. Wallsgrove*
57. Toxicology of aquatic pollution: physiological, cellular and molecular approaches. *Edited by E. W. Taylor*
58. Gene expression and adaptation in aquatic organisms. *Edited by S. Ennion and G. Goldspink*
59. Animals and temperature: phenotype and evolutionary adaptation. *Edited by I. A. Johnston and A. F. Bennett*
60. Molecular physiology of growth. *Edited by P. T. Loughna and J. M. Pell*
61. Global warming: implications for freshwater and marine fish. *Edited by C. M. Wood and G. McDonald*
62. Fish stress and health in aquaculture. *Edited by G. K. Iwama, A. D. Pickering, J. Sumpter and C. Schreck*
63. Scaling-up: from cell to landscape. *Edited by P. van Gardingen, P. Curran and G. Foody*
64. Plants and UV-B: responses to environmental change. *Edited by P. Lumsden*
65. Recent advances in arthropod endocrinology. *Edited by G. Coast and S. Webster*

COLD OCEAN PHYSIOLOGY

Edited by

H.O. Pörtner

Biologie I/Ökophysiologie
Alfred-Wegener-Institut für Polar- und Meeresforschung
Bremerhaven, Fed. Rep. Germany

and R.C. Playle

Department of Biology
Wilfrid Laurier University
Waterloo, Ontario, Canada

CAMBRIDGE
UNIVERSITY PRESS

PUBLISHED BY THE PRESS SYNDICATE OF THE UNIVERSITY OF CAMBRIDGE
The Pitt Building, Trumpington Street, Cambridge CB2 1RP, United Kingdom

CAMBRIDGE UNIVERSITY PRESS
The Edinburgh Building, Cambridge CB2 2RU, United Kingdom
40 West 20th Street, New York, NY 10011-4211, USA
10 Stamford Road, Oakleigh, Melbourne 3166, Australia

First published 1998

Printed in the United Kingdom at the University Press, Cambridge

Typeset in 10/12pt Monophoto Times NR [SE]

A catalogue record for this book is available from the British Library

Library of Congress cataloguing in publication data

Cold ocean physiology / edited by H.O. Pörtner and R. Playle.
p. cm. – (Society for Experimental Biology seminar series ; 65)
Includes index.
ISBN 0 521 58078 1 (hardback)
1. Marine animals—Adaptation—Polar regions. 2. Deep-sea animals—Adaptation—Polar regions. 3. Cold adaptation. I. Pörtner, H. O. (Hans-Otto), 1955– II. Playle, R. (Richard), 1956– . III. Series: Seminar series (Society for Experimental Biology (Great Britain)) ; 66.
QL121.C65 1998
571.4′641177–dc21 97-22063 CIP

Contents

PART V: Applied approaches

Contributors

ABELE-OESCHGER, D.
Biologie I/Ökophysiologie, Alfred-Wegener-Institut für Polar- und Meeresforschung, Columbusstraße, D-27515 Bremerhaven, Fed. Rep. Germany

ACIERNO, R.
Laboratorio di Fisiologia Generale, Dipartimento di Biologia, Università di Lecce, via Prov. Ie, Lecce-Monteroni, I-73100 Lecce, Italia

AGNISOLA, C.
Dipartimento di Fisiologia Generale et Ambientale, Università di Napoli 'Federico II', I-80127 Napoli, Italia

AXELSSON, M.
Dept. of Zoophysiology, University of Göteborg, Medicinaregatan 18, S-41390 Göteborg, Sweden

BEVAN, R.M.
Research Department, Wildfowl and Wetlands Trust, Slimbridge, Gloucester GL2 7BT, UK

BISHOP, C.M.
The University of Birmingham, School of Biological Sciences, Edgbaston, Birmingham B15 2TT, UK

BOUTILIER, R.G.
Department of Zoology, Downing Street, Cambridge CB2 3EJ, UK

BUCHHOLZ, F.
Biologische Anstalt Helgoland, D-27498 Helgoland, Fed. Rep. Germany

BURLANDO, B.
Dipartimento die Scienze e Tecnologie Avanzate, Universita di Torino, Corso Borsalino 54, I-15100 Alessandria, Italia

BUTLER, P.J.
The University of Birmingham, School of Biological Sciences, Edgbaston, Birmingham B15 2TT, UK

CLARKE, A.
British Antarctic Survey, High Cross, Madingley Road, Cambridge CB3 0ET, UK

DAVIES, P.L.
Department of Biochemistry, Queen's University, Kingston, Ontario, K7L 3N6, Canada

D'AVINO, R.
Institute of Protein Biochemistry and Enzymology, CNR, Via Marconi 10, I-80125 Napoli Italia

DAVISON, W.
Department of Zoology, University of Canterbury, Private Bag 4800, Christchurch, New Zealand

DI PRISCO, G.
Institute of Protein Biochemistry and Enzymology, CNR, Via Marconi 10, I-80125 Napoli Italia

EGGINTON, S.
Physiology, University of Birmingham, Medical School, Vincent Drive, Birmingham B15 2TT, UK

EWART, K.V.
NRC Institute for Marine Biosciences, 1411 Oxford Street, Halifax, Nova Scotia, B3H 3Z1, Canada

FLETCHER, G.L.
Ocean Sciences Center, Memorial University of Newfoundland, St. John's, Newfoundland, A1C 5S7, Canada

GODDARD, S.V.
Ocean Sciences Center, Memorial University of Newfoundland, St. John's, Newfoundland, A1C 5S7, Canada

GONG, Z.
Department of Zoology, National University of Singapore, Singapore 0511

GUDERLEY, H.
Department of Biology, Universite Laval, Quebec City, G1K 7P4, Canada

HARDEWIG, I.
Biologie I/Ökophysiologie, Alfred-Wegener-Institut für Polar- und Meeresforschung, Columbusstraße, D-27515 Bremerhaven, Fed. Rep. Germany

HEW, C.L.
Department of Clinical Biochemistry, Banting Institute 100, College Street, Toronto, Ontario, M5G 1L5, Canada

HOCHACHKA, P.W.
Department of Zoology, University of British Columbia, Vancouver, BC, V6T 1Z4, Canada

MAFFIA, M.
Laboratorio di Fisiologia Generale, Dipartimento di Biologia, Università di Lecce, via Prov. Ie, Lecce-Monteroni, I-73100 Lecce, Italia

MOTTISHAW, P.D.
Department of Zoology, University of British Columbia, Vancouver, BC, V6T 1Z4, Canada

NILSSON, S.
Department of Zoophysiology, University of Göteborg, Medicinaregatan 18, S-41390 Göteborg, Sweden

PECK, L.S.
British Antarctic Survey, High Cross, Madingley Road, Cambridge CB3 0ET, UK

PÖRTNER, H.O.
Biologie I/Ökophysiologie, Alfred-Wegener-Institut für Polar- und Meeresforschung, Columbusstraße, D-27515 Bremerhaven, Fed. Rep. Germany

SARTORIS, F.J.
Biologie I/Ökophysiologie, Alfred-Wegener-Institut für Polar- und Meeresforschung, Columbusstraße, D-27515 Bremerhaven, Fed. Rep. Germany

SIDELL, B.D.
School of Marine Sciences, University of Maine, 5741 Libby Hall, Orono, Maine 04469-5741, USA

SOMERO, G.N.
Hopkins Marine Station, Stanford University, Ocean Blvd., Pacific Grove, CA 93950-3094, USA

STORELLI, C.
Laboratorio di Fisiologia Generale, Dipartimento di Biologia, Università di Lecce, via Prov. Ie, Lecce-Monteroni, I-73100 Lecce, Italia

TAMBURRINI, M.
Institute of Protein Biochemistry and Enzymology, CNR, Via Marconi 10, I-80125 Napoli, Italia

TOTA, B.
Statione Zoologica A. Dohrn, Villa Comunale, I- 80121 Napoli, Italia

VAN DIJK, P.L.M.
Biologie I/Ökophysiologie, Alfred-Wegener-Institut für Polar- und Meeresforschung, Columbusstraße, D-27515 Bremerhaven, Fed. Rep. Germany

VAYDA, M.E.
Department of Biochemistry, Microbiology and Molecular Biology, University of Maine, 5735 Hitchner Hall, Orono, Maine 04469-5735, USA

VETTER, R.A.H.
Biologische Anstalt Helgoland, D-27498 Helgoland, Fed. Rep. Germany

VIARENGO, A.
Instituto di Fisiologia Generale, Palazzo delle Scienze, Corso Europa 26, I-16132 Genova, Italia

WÖHRMANN, A.P.A.
Institut für Polarökologie, Universität Kiel, Wischhofstr. 1–3, Geb. 12, D-24148 Kiel, Fed. Rep. Germany

Preface

The last two decades have seen a drastic increase in our knowledge of adaptational mechanisms in cold ocean environments. General hypotheses have been developed addressing the potentials and limitations of life under these conditions which from an anthropocentric understanding have always been seen as harsh and demanding. Some of the early hypotheses have recently been questioned. For example, the concept of metabolic cold adaptation, which postulated a rise in energy expenditure to compensate for the thermodynamic effect of low temperature is currently being modified by several laboratories and the mechanistic and molecular background of changes in metabolic rate is being studied. The volume reviews both the general concepts and the detailed mechanisms of adaptation. Presentations cover a wide range of topics on cold ocean physiology, from molecular to whole animal, including invertebrates, fish, marine mammals, and birds.

The volume opens with a general outline of animal physiology in the cold addressing the physical challenges and physiological constraints that exposure to low temperature may imply (A. Clarke). Mechanisms of cold compensation are briefly summarised, setting the stage for the following, more detailed treatments. The general conclusion of this overview is that constraints in the physiological design required in the cold may preclude lifestyles requiring high rates of aerobic power output. The following papers describe the compensatory mechanisms in detail which would enable marine ectotherms to thrive in the cold. Differentiation between polar and deep-sea environments is required, since an investigation of cold adaptation in deep-sea organisms requires the consideration of depth-related variables like hydrostatic pressure, light intensity and oxygen levels. This comparison allows evaluation of the overall relevance of low temperature for the evolutionary adaptation to cold ocean environments (G.N. Somero). None the less, temperature is the focus of the subsequent chapters. This section considers the metabolic capacity of pathways and organelles like mitochondria as well as enzyme characteristics in general and enzymes in specific metabolic pathways. Again, factors like food availability and growth rate under cold ocean conditions may influence considerably the picture of compensatory cold adaptation (H. Guderley). The maintenance of functional integrity in the cold requires an adjustment of low and high critical temperatures, which are characterised by a transition to anaerobic energy metabolism, suggesting that maintenance of oxygen supply by ventilation and circulation is a key question during cold adaptation. The patterns of ionic and alphastat acid–base regulation during a temperature decrease are interpreted as attempts to maintain the energy content of the adenylate system and to balance the dissipative ion fluxes predominating in the cold (H.O. Pörtner, I. Hardewig, F.J. Sartoris and P.L.M. van Dijk). Oxygen supply problems are less on the level of haemoglobin and myoglobin functional properties, owing to increased oxygen solubility in the cold and to reduced oxygen requirements by the animal. Myoglobin function is low or absent in cold ocean fish as suggested by multiple losses of myoglobin expression and its function being restricted to the heart ventricle of some species (B.D. Sidell and M.E. Vayda). Haemoglobin (Hb) function is reduced in benthic notothenioids, but emphasized in active pelagic notothenioids as demonstrated by the expression of multiple, functionally distinct Hbs. Haemoglobin diversity can be used as a marker for the phylogenetic history of the Antarctic and non-Antarctic members of this group (G. di Prisco, M. Tamburrini and R. D'Avino).

Functional integrity is also provided by the adaptive modification of membrane lipids and proteins through the maintenance of membrane fluidity and an increase in enzyme activity, largely via a selection of enzyme isoforms active at low temperature. Adjustment of the lipid microenvironment plays a key role in maintaining the activity of membrane transporters (C. Storelli, R. Acierno and M. Maffia). A decrease in activation energy, conserved or decreasing Michaelis constants or an increase in total enzyme activity as well as a modulation of effector mediated control represent adaptations observed in cytosolic or mitochondrial matrix enzymes important in the regulation of metabolic flux. Temperature dependent flexibility is less evident in the free hydrolases and proteases of the digestive tract (R.A.H. Vetter and F. Buchholz). An adaptive response is also discernible in the activity of antioxidant enzymes and the tissue concentrations of scavengers of reactive oxygen species (oxygen radicals). The use of branched-chain instead of polyunsaturated fatty acids in the homeoviscous response of membranes may reduce the potential susceptibility to oxidative damage (A. Viarengo, D. Abele-Oeschger and B. Burlando).

The third section investigates innovative adaptations like antifreeze proteins which allow hypoosmotic fish to exploit the cold ocean environment. The presence of various types of antifreeze peptides and glycopeptides in extracellular fluids and in external epithelia of sub-Arctic species depends on seasonal factors and not only supports freeze avoidance but appears to contribute to cold adaptation of membrane transport (G.L. Fletcher, S.V. Goddard, P.L. Davies, Z. Gong, K.V. Ewart and C.L. Hew). Patterns of antifreeze composition and plasma contents have also been investigated in Antarctic species in relation to their mode of life and geographical as well as depth-related distribution (A.P.A. Wöhrmann).

Integrated views of organismic functions are presented in the next section starting with an analysis of the susceptibility of Antarctic fishes, mostly notothenioids, to stress. These studies revealed differences in the hormonal response to stress but similar recovery times as in temperate species suggesting effective rate compensation in the cold (S. Egginton and W. Davison). The maintenance of efficient circulatory function reflects all processes of cold adaptation at the molecular, membrane, cellular and systemic level. Choosing cold adapted myosin isoforms, homeoviscous adaptation and mitochondrial proliferation all contribute to maintain function in the cold but also impose constraints to myocardial performance linked to distinct differences in mechanical and hormonal control of the (notothenioid) system from temperate zone fish (M. Axelsson, C. Agnisola, S. Nilsson and B. Tota). Components of metabolic capacity are also analysed, focusing on the metabolic rate increment in response to feeding (SDA=specific dynamic action) and on ammonia excretion. Specific dynamic action peaks are smaller in cold ectotherms and of longer duration, possibly because they are limited by less aerobic capacity in cold adapted versus temperate zone animals (L.S. Peck). This section concludes with chapters on specific adaptations in birds and mammals specialised to living in productive but cold oceans. Diving apnea, bradycardia, tissue hypoperfusion and hypometabolism of hypoperfused tissues are general features of the diving response in phocids and otariids whereas increases in spleen weight and whole body oxygen carrying capacity correlate with the higher diving capacity in phocids (P.W. Hochachka and P.D. Mottishaw). Birds breeding in polar areas mostly take advantage of being able to forage during long summer days in productive oceans, but most species leave to avoid the polar winter. Several physiological and behavioural mechanisms are used to combat the low temperature and to store sufficient food reserves to supplement long periods of food deprivation during incubation, the migration period or the polar winter (R.M. Bevan, C.M.Bishop and P.J.Butler).

The volume concludes with a perspective of how physiological studies can be important in more applied fields like fisheries. Using behavioural, physiological and molecular genetic techniques has improved our understanding of the ecology of Atlantic cod. The main conclusion is that temperature gradients are most important in determining cod distribution and that the existence of geographically and genetically isolated populations needs to be considered in fisheries management (R.G. Boutilier).

This seminar series volume is the result of a joint Society for Experimental Biology (SEB) and Canadian Society of Zoologists (CSZ) symposium called 'Cold Ocean Physiology'. The symposium was held May 10, 1996, at the Annual CSZ meeting held that year at Memorial University, St. John's, Newfoundland, Canada. We thank the following sponsors for their financial support: Elsevier and John Wiley publishers, the SEB Respiration group, the CSZ, and the Comparative Physiology and Chemistry section of the CSZ as well as the sponsors at Memorial University. We also thank the local organising committee of the CSZ May 1996 meeting in St. John's, especially Derek Burton, Memorial University for their support and for providing such a stimulating environment. Finally, we thank the contributors to that symposium and those who joined us and contributed to this volume on cold ocean physiology.

Hans O. Pörtner
Biologie I/Ökophysiologie
Alfred-Wegener-Institut
Columbusstraße
D-27515 Bremerhaven
Fed. Rep. Germany

Richard C. Playle
Department of Biology
Wilfrid Laurier University
Waterloo, Ontario
N2L 3C5
Canada

PART I General concepts

A. CLARKE

Temperature and energetics: an introduction to cold ocean physiology

Cold has long been regarded as inimicable to life. From the speculations of the earliest geographers and explorers to the azoic hypothesis of Edward Forbes, it was widely believed that the frozen polar regions and the cold, dark depths of the ocean would be essentially devoid of life. Even today, the richness of some polar marine invertebrate assemblages comes as a surprise to many. The tropics are still regarded as a less demanding environment, at least for marine organisms, and many palaeoecologists continue to refer to a cooling of the climate as 'deterioration' and a warming as 'amelioration'.

The first physiologists to study polar marine organisms naturally turned their attention to the key areas of how teleost fish avoided freezing when living in waters significantly colder than the equilibrium freezing point of their body fluids, and how marine ectotherms managed to sustain metabolic activity at low temperatures (Scholander *et al.*, 1953; Wohlschlag, 1960). In the marine environment, the avoidance of freezing is a peculiarly polar problem, and the essential mechanism was first elucidated in Antarctic fish by DeVries (DeVries & Wohlschlag, 1969; DeVries, 1971). Subsequent work on Arctic fish showed that most northern taxa utilised antifreeze proteins in contrast to the antifreeze glycoproteins of Antarctic notothenioids and the true cods (gadoids), thereby revealing an intriguing case of parallel evolution (Scott, Fletcher & Davies, 1986; Eastman, 1993).

In the terrestrial environment, exposure to freezing temperatures is a widespread environmental challenge. In contrast, for marine organisms freezing is essentially a problem only for polar teleost fish, intertidal organisms and those high latitude benthic invertebrates which become encased in anchor-ice. In the wider context of adaptation to temperature in general, freezing resistance therefore represents something of a special case in the marine environment. In this review, a general picture of temperature adaptation will be considered, attempting to place the polar regions in the context of the marine environment overall.

Temperature adaptation in marine organisms has been reviewed many times (Clarke, 1983, 1991; Hochachka & Somero, 1973, 1984; Cossins &

Bowler, 1987; Johnston, 1990). Instead of covering this ground again, an attempt will be made to identify the major constraints to physiological function in cold water, for it is through these constraints that temperature has its greatest impact on ecology and lifestyle.

The physical properties of cold seawater

Just as the ecology of any organism can be understood only in relation to those other organisms which share its environment, the physiology of that organism can only be understood in relation to its physical environment. The equations of state for seawater all contain a temperature term, and thus a change in environmental temperature brings with it concomitant changes in all other physical properties of seawater. Seawater at 0 °C and seawater at 30 °C vary in more than just temperature, and are quite different environments from the point of view of an organism attempting to make a living there (Clarke, 1983). In Table 1, a number of important environmental variables are compared for polar and tropical seawater. All show significant variation with temperature, and in all cases this variation is non-linear (although for some variables this non-linearity is not important over the range of temperatures of interest to physiologists).

The ecological and physiological implications of some of this variability have long been recognised. Thus the increased solubility of oxygen in cold seawater means that more oxygen can be carried in physical solution in the blood of polar ectotherms. As a result, there is a well-described tendency for a reduced haematocrit in polar fish (Kunzmann, 1991), which reaches its ultimate expression in the notothenioid ice-fish (Channichthyidae) (Macdonald & Wells, 1991). In icefish the genes for haemoglobin are not expressed (Cocca *et al.*, 1995), and there are few circulating erythocytes. Oxygen is carried to the tissues in physical solution and there are parallel adjustments in blood volume, heart size and circulatory architecture (Eastman, 1993). Interestingly, although the genes for myoglobin are also generally not expressed, a number of icefish have recently been shown to express myoglobin genes in heart tissue (Sidell and Vayda, this volume).

Although many species of ice-fish are sluggish and appear to have a rather sedentary lifestyle, it is likely that this reflects, at least in part, the ancestral niche of notothenioids as a taxon. Notothenioids evolved from a demersal form, which had secondarily lost a functional swimbladder (Eastman, 1993). In consequence, most notothenioids remain benthic or demersal in habit, although a number of species have subsequently evolved a semi-pelagic, fully pelagic or cryo-pelagic way of life, and these include at least one ice-fish (*Champsocephalus gunnari*). This species is partly demersal, but is also capable of moving into the water column to take planktonic prey such as

Table 1. *A comparison of the physical properties of seawater at 0 °C and 30 °C (based on Clarke, 1983)*

Variable	Value in seawater at poles (0 °C)	Value in seawater at tropics (30 °C)
Thermodynamic temperature (K)	273.15	303.15
Acidity ($\log_{10}[H^+]=pH$) (NB data for pure water)[a]	7.47	6.92
Dynamic viscosity (mPa s)[b]	1.8823	0.8582
Density (g ml^{-1})[c]	1.0273	1.0210
Oxygen content (mol l^{-1})[d]	8.05	4.35
Total carbon dioxide content (mol l^{-1})[e]	644.9	255.1
Apparent solubility product for calcite ($K^1_{calc} \times 10^6$)[f]	0.711	0.522

Notes:
[a] Data for pure water, since the pH of seawater is affected by many factors in addition to temperature.
[b] Data for salinity 34 psu (from Table 25 in Riley & Skirrow, 1975).
[c] Data for salinity 34 psu (from Table 2.1 in Riley & Chester, 1971).
[d] Data for salinity 35 psu, relative to atmosphere of 20.95% 0_2 at pressure of 101.325 kPa and 100% relative humidity (from Table 6 in Riley & Skirrow, 1975).
[e] Data for acidified seawater of chlorinity 19‰, atmospheric pressure of 101.325 kPa and 100% relative humidity (from Table 6.4 in Riley & Chester, 1971).
[f] Data for seawater of chlorinity 19‰ (Edmund & Gieskes, 1970); data are approximate because of the problem of interference by other ionic species, especially Mg^{2+}.

euphausiids. It also lives in warmer waters where the benefits of enhanced oxygen solubility are not as marked as they are closer to the Antarctic continent. The haemoglobin-less condition does not therefore necessarily bring with it a concomitant inactive lifestyle, nor a limitation to the coldest waters for *Champsocephalus esox* is found north of the polar front (Gon & Heemstra, 1990).

Other aspects of the variation of seawater physics with temperature have been less well explored. For example, the viscosity of seawater is of fundamental significance, particularly to small organisms living at large Reynolds numbers. Recently, Podolsky and Emlet (1993) have shown that relatively small temperature-related changes in viscosity can have significant effects on the energetics of echinoderm larvae. Nothing, however, appears to be known about the impact of temperature-mediated changes in viscosity on the energetics of sessile benthos that rely on ciliary mechanisms for feeding, or on the energetics of cold-water protozooplankton.

Temperature and physiology

Since all processes involving a change in free energy will be affected by temperature, it is to be expected that, in the absence of any evolutionary compensation, all physiological and biochemical processes will also show marked differences in rate between tropical and polar species. Thus for a typical enzyme-mediated reaction, a significant contribution to the free energy of activation for the reaction comes from the enthalpy of the reactants. At any given temperature, only a small fraction of the total population of molecules has sufficient energy to react, and this proportion can be estimated from statistical mechanics (specifically the Maxwell–Boltzmann distribution law). As temperature decreases, the proportion of molecules with sufficient kinetic energy to react also decreases, and the reaction rate slows accordingly; this relationship is effectively exponential.

Almost a century of physiological work has shown that, within an individual system (such as an individual organism, or isolated subsystem), this thermal dependency typically falls within narrow bounds, equivalent roughly to an increase in rate of $\times$ 2 to $\times$ 3 for each rise in temperature of 10 K (that is Q_{10}=2–3). It is striking that the temperature dependence of isolated systems, such as an enzyme reaction *in vitro,* and that of many processes in whole organisms are broadly the same. This suggests that evolution has resulted in an integrated physiology where most processes have a broadly similar relationship to temperature.

Extrapolation of the typical temperature dependency of physiological rates to the marine ecosystem as a whole would predict that physiological processes in tropical marine ectotherms at 30 °C will proceed at between 27 (Q_{10}=3) and 8 (Q_{10}=2) times the rate in polar species at 0 °C. Clearly, for some processes this would pose severe ecological difficulties: consider the fate of a fish whose escape response was curtailed by a factor as low as $\times$ 8 when attempting to escape a predatory seal or seabird whose locomotor ability was maintained at a high level by endothermy.

Compensation

The simple observation that a rich and diverse marine flora and fauna can be found from the tropics to the poles indicates that some form of compensation has evolved to offset these direct temperature effects. Even if not

Table 2. *Kinetic parameters of enzymes isolated from three allopatric species of Pacific barracuda* Sphyraena

	Species of *Sphyraena*		
	argentea	*lucasana*	*ensis*
Kinetic parameters measured at 25°C			
K_m (mM)	0.34±0.03	0.26±0.02	0.20±0.02
k_{cat} (s^{-1})	893±54	730±37	658±19
Representative environmental temperature			
	18	23	26
Kinetic parameters measured at representative environmental temperature			
K_m (mM)	0.24	0.24	0.24
k_{cat}	667	682	700

Note:
Kinetic parameters are for isolated M_4 LDH operating in the direction of pyruvate reduction; K_m is for pyruvate (mM). Data are presented for both 25 °C and a temperature representative of the environment in which the fish lives (K_m, Michaelis–Menten constant; k_{cat}, turnover number).
Source: From Graves & Somero (1982).

'perfect' in theoretical terms, this compensation is clearly highly effective in ecological and evolutionary terms. A single, classical, example which demonstrates such compensation is the activity of M_4-LDH in related species of the Pacific barracuda *Sphyraena* (Graves & Somero, 1982).

The enzyme LDH catalyses the conversion of lactate to pyruvate (and vice versa) and is thereby critical in regulating the flux of the products of glycolysis into the Krebs cycle. Graves and Somero (1982) examined the thermodynamic properties of M_4-LDH isolated from four species of *Sphyraena* ranging from tropical to temperate waters in the eastern Pacific. Although enzymes isolated from each species showed typical thermal dependency in kinetic parameters, there was much less variation in these parameters when measured at the temperatures at which the fish lived (Table 2). This study exemplifies two important general conclusions. The first is that the enzymes isolated from organisms living at different temperatures themselves differ in some way, and the second is that these differences mean that the enzymes have broadly comparable activities at the temperatures at which the organisms actually live.

There are two important consequences which follow from this simple

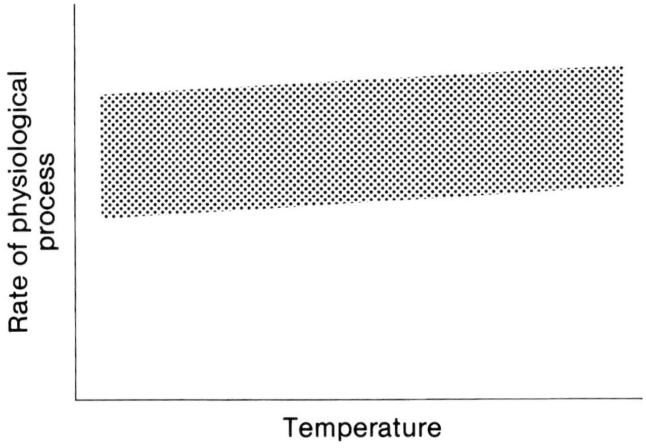

Fig. 1. A conceptual model of compensation as homeostasis; evolution has modified cellular or organismal physiology such that the same process has the capacity to operate at broadly the same speed in organisms living at different temperatures. Note that this compensation plot has both bandwidth and slope. The bandwidth indicates that, at any given environmental temperature, there may be a range of rates in different organisms, perhaps associated with ecology or phylogenetic history. Perfect compensation across the ecological temperature range would be indicated by a slope of zero (that is, the same rate at all temperatures). The slight slope shown here indicates that underlying physical or physiological constraints may mean that compensation is not perfect. (Reproduced from Clarke, 1991.)

observation. The most important point is that it is unwise to extrapolate from experimental data obtained from one organism (say a eurythermal temperate marine ectotherm) to other organisms living elsewhere (say a polar or tropical species). Unfortunately, this is still done to predict the physiology of polar or tropical species, or to speculate on the likely effects of global warming.

The second consequence is a straightforward definition of temperature compensation, which is:

> *the maintenance of physiological rate in the face of temperature change*

This definition of compensation is shown diagrammatically in Fig.1, and it applies equally to other forms of environmental challenge (such as pressure) as well as to temperature.

In the past few decades, many physiological processes in marine ectotherms have been studied to determine the extent to which such compensation has evolved, and the mechanisms by which it has been achieved. It is now clear that some processes have evolved almost perfect compensation (for example, microtubule assembly: Detrich & Overton, 1986; Detrich *et al.*, 1989), others have achieved only partial compensation (for example, burst swimming performance in teleost fish: Johnston, Johnson & Battram, 1991) and some almost none at all (for example, in the speed of nervous conduction: Macdonald, 1981; Montgomery & Macdonald, 1984; or embryonic development: Bosch *et al.*, 1987; Clarke, 1992). This poses the evolutionary question of whether there are particular constraints to temperature compensation, and what these might be.

Constraints to temperature adaptation

Viewed in evolutionary terms, an organism takes up material from its environment and manipulates this in such a way as to maximise its fitness (that is, to maximise the relative representation of its genes in the next generation). Since this process is fuelled by the oxidation of foodstuffs or reserves, the rate and efficiency of energetic pathways are critical to fitness.

An organism, however, contains a vast number of enzymes involved in a complex web of metabolic interactions. This complexity poses a problem in metabolic control for the organism, and an intellectual challenge to any physiologist attempting to understand how that organism can respond in an integrated way to a change in temperature. Traditionally, physiologists interested in temperature compensation have looked in detail at selected aspects of cellular or organismal operation. In this chapter, a different approach will be taken in that an attempt will be made to look at temperature compensation as a whole organism phenomenon, and to try to identify those aspects of physiology which appear to act as a constraint upon compensation. Rather than simply cataloguing known examples of temperature compensation, work will be within the conceptual framework first outlined by Hochachka and Somero (1973, 1984).

A conceptual framework for temperature compensation

Cellular physiology, indeed all life, depends absolutely on enzyme-mediated catalysis, and this catalysis is affected significantly by temperature. As discussed above, a typical enzyme reaction in a tropical marine invertebrate or fish will proceed at between about 10 and 30 times the rate in a polar relative, unless the organism does something about it. It is traditional amongst physiologists to concentrate on how organisms adjust to cope with the rate-depressing effect of low temperature rather than regarding a temperature-mediated increase in rate as a physiological problem. This reflects the long-held bias towards the tropics being 'amenable' or 'easy' places to live, and the poles as 'challenging' or 'difficult'.

Table 3. *The major categories of compensatory response to a lowered cellular temperature*

Response	Description
Quantitative strategy	Increase the number of enzyme molecules to offset the reduced catalytic efficiency of an individual enzyme molecule at lower temperatures.
Qualitative strategy	A change in the type of enzyme(s) present. This could include a shift in the balance of allozymes with differing kinetic properties.
Modulation strategy	Modulation in the activity of pre-existing enzymes. This could include changes in enzyme–substrate binding kinetics with temperature, changes in cofactor binding kinetics, or modification of the immediate enzyme environment (membrane or cytosol).

Source: From Hochachka & Somero (1973).

Hochachka and Somero (1973) have classified the range of possible responses to the rate-depressing effects of low temperature into three broad categories. These are the quantitative strategy (more enzymes), the qualitative strategy (different enzymes) or modulation of the enzyme environment (Table 3). These strategies are not, of course, mutually exclusive.

Modulation of the enzyme environment

Modification of the enzyme environment offers the opportunity to modulate the effect of temperature on many enzyme systems at once. It would thus appear, *a priori*, to be an effective way of simultaneously offsetting the impact of temperature and of maintaining the overall balance of metabolic pathways.

Many enzymes are membrane bound and it is now well established that the primary response of biological membranes to a change in temperature is an alteration in the fatty acid composition of the constituent phospholipids. There may also be changes in phospholipid head-group composition and phospholipid/cholesterol ratio. These changes serve to maintain the physical state of the membrane and the process is often referred to as homeoviscous adaptation (see Cossins, 1994 for a recent discussion). Most work in this area has been concerned with seasonal adjustments in eurythermal organisms. Unfortunately, most analyses of the lipid composition of polar organisms are not very helpful in attempting to understand evolutionary adjustments to membrane composition at low temperatures for they involve whole organism or whole tissue extracts which mix fatty acids from membranes with those from storage lipids. Nevertheless what evidence there is points to homeostatic changes in membrane composition having occurred during evolutionary adaptation to low temperatures.

For cytosolic enzymes, most attention has been directed at the pH of the intracellular environment. The ionic dissociation of pure water is temperature dependent, with $\Delta pH/\Delta T$ being about 0.017. It is now well established that the pH of blood in marine ectotherms shows a similar temperature dependency, though offset to higher pH values to produce a higher relative alkalinity compared with pure water. This pH is actively regulated and it has been proposed that pH is adjusted to maintain the dissociation state of changed groups, and in particular the imidazole groups of histidyl residues (the alphastat hypothesis: Reeves, 1972, 1977). Regulation of alphastat and regulation of a constant relative alkalinity are different processes, but the two hypotheses would be difficult to distinguish by experiment.

There is evidence for the regulation of net protein change over a wide range of temperature in some organisms (for example the intracellular environment of skeletal muscle in the turtle *Pseudemys scripta*: Malan, Wilson & Reeves, 1976). There is also widespread evidence for the maintenance of the net alkalinity of ectotherm blood across a range of environmental temperatures (Reeves, 1977; Clarke, 1987). This would suggest that a significant degree of compensation for temperature is achieved by a homeostatic adjustment of the intracellular ionic environment (see Pörtner *et al.*, this volume), but firm evidence for this as a universal primary mechanism for temperature compensation is still lacking.

Changes in the number or type of enzymes

Changes in either the number or the type of enzymes (or perhaps both for these are not mutually exclusive processes) would appear to present a number of evolutionary difficulties. There is a limit to the solvent capacity of the cell and hence it seems intuitively unlikely that temperature compensation could be achieved by a simultaneous increase in the concentration of all cellular enzymes (to say nothing of substrates, cofactors, products and transcribed messages) without severe consequences for intracellular viscosity, diffusion, transport, and cellular energetics. In contrast, it would seem likely that, over evolutionary time, the process of natural selection will have driven changes in the nature of many enzymes and proteins. The key question here is the strength of selection, and this is likely to vary from enzyme to enzyme depending on its function and contribution to the overall fitness of the

Table 4. *A simple classification of the major cellular energetic processes*

Process	Description
Gene expression	The regulated transcription and translation of genetic information.
Catabolism	The regulated oxidation of reserves or exogenous foodstuffs to provide a supply of high-energy phosphate, reducing power and a suite of intermediary compounds[a], together with the production of CO_2, heat and nitrogenous waste (typically ammonia and/or urea).
Work	Biosynthesis of macromolecules for turnover, growth, and mechanical work (including locomotor activity)
Homeostasis	Chemical work associated with the active regulation of the intracellular environment within boundary conditions set by physiological viability. This includes ion exchange, macromolecular turnover, and removal of waste products. In endotherms it also involves heat production (heat produced by ectotherms is largely lost to the environment).

Note:
[a] The major intermediary compounds are DHAP (dihydroxyacetone phosphate), PGA (phosphoglyceric acid), G6P (glucose-6-phosphate), R5P (ribulose-5-phosphate), E4P (erythrose-4-phosphate), PEP (phosphenol pyruvate), pyruvate, acetyl CoA, succinyl CoA and oxaloacetate. The major high energy phosphate is ATP (adenosine triphosphate) and the major sources of reducing power are the nicotinamide adenine dinucleotide (NAD) and its phosphate (NADP).

organism. If we accept that most organisms are, at least for parts of their life-history, resource limited (although ecologists are still arguing over this) this would suggest that selection would be strongest, and hence evolutionary adjustments most likely, in those enzymes with the greatest regulatory influence over metabolic flux and efficiency.

Despite its overall complexity, cellular metabolism can be broken down into a small number of key processes, thereby providing a useful framework for any discussion of the impact of low temperature on physiology. These key processes are gene expression, catabolism (energy production), work (energy utilisation) and homeostasis (Table 4). The latter is, of course, a component of energy utilisation but it is convenient to distinguish it for two reasons. The first is that it encompasses most of those processes necessary to keep the organism alive in the absence of any other physiological processes such as growth, locomotion or reproduction; it is therefore the major component of basal (maintenance) metabolic rate. The second is that homeostatic processes set the internal milieu within which all other processes must take place, and thereby establish a conceptual link to the strategies of temperature compensation discussed above. It is worth noting that roughly three-quarters of the enthalpy content of the chemical bonds in foodstuffs is dissipated as heat during catabolism; in ectotherms this heat is usually lost to the environment (Hochachka & Somero, 1984), whereas in endotherms a significant fraction of this heat is used to maintain the organism's internal temperature around a set point.

To what extent has temperature compensation been achieved in polar marine organisms?

A full discussion of the evidence for temperature compensation, and the mechanisms involved, in polar marine organisms is beyond the scope of this chapter for it would require a major review. Instead an attempt will be made to identify the main themes before moving to a discussion of constraints and ecological consequences.

The first point to be made may seem trivial, but it is important. The polar regions support a rich and diverse marine flora and fauna. Whatever limitations there may be to temperature compensation at low seawater temperatures, they have not prevented ecological success for a wide range of polar marine organisms.

Despite the fundamental importance of gene expression in cellular physiology and ecology, very little is known about temperature compensation in this process. Although there are a number of well-documented examples of specific genes being switched on or off by changes in temperature, essentially nothing is known about the impact of temperature on the processes involved in routine transcription, message transport and processing, or translation. As with all other enzyme-mediated processes, these activities will slow down at lower temperatures but it is not known what organisms do about this, or even whether it is ever an energetic or evolutionary problem.

Catabolic processes have the main function of generating ATP, reduced pyridine nucleotides and intermediary compounds for biosynthesis. The thermodynamic efficiency of ATP synthesis will itself be temperature dependent but the effect is small (Gnaiger, 1987) and hence likely to be of limited ecological impact. Of much greater significance is the impact of temperature on the maximum flux possible through critical pathways. Although a number of studies have revealed an increase at low temperatures in the activity of enzymes associated with aerobic metabolism in eurythermal organisms

(Hazel & Prosser, 1974; Sidell, 1983), it is not clear to what extent this reflects a change in absolute enzyme concentration. One study where an increase in enzyme concentration has been demonstrated unequivocally is that of cytochrome oxidase in the skeletal muscle of green sunfish by Sidell (1977, 1983). There are as yet no unequivocal demonstrations of an increase in the absolute concentration of specific enzymes in marine ectotherms that have adapted over evolutionary time to cold waters. There are, however, clear demonstrations of an increase in mitochondrial density in polar fish (Johnston, Camm & White, 1988), and in the surface cristae density of those mitochondria (Archer & Johnston, 1991), which imply an associated increase in mitochondrial enzyme concentration.

A number of detailed studies have demonstrated kinetic differences in enzyme variants from organisms living at different temperatures. These include both a shift in the proportion of allozymes with different kinetic properties along a latitudinal cline (and, by inference, driven by selection for thermal properties along the associated temperature cline), and kinetic differences in enzymes isolated from related species inhabiting different thermal environments (for example, isocitrate dehydrogenase in the actinian *Metridium senile*: Hoffmann, 1981*a*, *b*; 1983). The most thorough study of a clinal variation in allozyme frequency associated with temperature within a single marine species is that of M_4-LDH in the killifish *Fundulus heteroclitus* (Powers & Place, 1978).

Many early studies of the kinetic properties of enzymes isolated from organisms living at different temperatures suffered from the use of inappropriate buffer conditions which made it impossible to distinguish the effects of temperature on enzyme activity from those of pH (Yancey & Somero, 1978; Clarke, 1983; Hochachka & Somero, 1984). These difficulties are now well recognised, and three general conclusions have emerged from detailed studies of enzyme homologues isolated from organisms living at different temperatures, namely:

(i) Enzymes from organisms with lower body temperature tend to have lower activation energies ($\Delta G^{\ddagger}$, and hence E_a).

(ii) Enzymes from organisms with lower body temperature tend to have lower thermal stabilities (when all measurements are made at the same incubation temperature), although thermal denaturation of these enzymes generally occurs only at temperatures well above those at which the organisms live.

(iii) There is a large shift in the relative contributions of enthalpy ($\Delta H^{\ddagger}$) and entropy ($\Delta S^{\ddagger}$) to the overall free energy of activation ($\Delta G^{\ddagger}$), with a greater enthalpic contribution at high temperatures and a greater entropic contribution at lower temperatures.

The reduced activation energies at lower temperatures would seem an intuitively reasonable evolutionary response to the decreased thermal energy at lower temperatures, but data collected to date suggest that changes in $\Delta G^{\ddagger}$ with habitat temperature are typically quite small and contribute relatively little to the observed increase in activity in enzymes isolated from cold water species (Hochachka & Somero, 1984; Moerland, 1995). The critical change is the shift in the balance between the enthalpic and entropic contribution to the overall free energy change. Thus, at lower cell temperatures, where there is less thermal energy, a greater contribution is made by entropic processes. These involve constraints to the conformational changes associated with enzyme activity, and are linked intimately to the greater flexibility and reduced thermal stability of enzymes from organisms with lower body temperatures (Somero, 1978; Moerland, 1995).

The two most detailed studies of the relationship between the entropic and enthalpic contribution to the free energy of activation both involve contractile protein systems (myofibrillar ATPase: Johnston & Goldspink, 1975; skeletal muscle actin: Swezey & Somero, 1982). Other studies, however, suggest that the conclusions may be general for cytosolic proteins (see summary by Hochachka & Somero, 1984). These results would thus tend to support a picture of a generally high degree of compensation for temperature in cellular enzymes. In those cases (very few) where enzyme homologues have been sequenced, changes in amino acid composition have been identified: an excellent example is shown by a comparison of tubulins in polar fish compared with warmer water fish (Detrich & Overton, 1986; Detrich *et al.*, 1989).

The evidence discussed above might suggest the general result that natural selection has been able to modify the structure and/or environment of most enzymes to achieve a significant degree of compensation for temperature. A recent study of mitochondrial activity has, however suggested the possibility that there may be powerful constraints on the overall rates of ATP production that are possible at low temperatures, implying that, for some physiological processes, significant evolutionary temperature compensation may not have been achieved. This is discussed in the next section.

Mitochondria: a critical constraint?

The relationship between mitochondrial activity and temperature is absolutely fundamental to overall temperature adaptation, for mitochondria are the site of aerobic ATP generation. Although mitochondria are complex organelles whose overall function depends on a diverse suite of enzymes, diffusion processes and membrane-associated activities, for the purposes of ATP production they may conveniently be regarded as discrete entities. In

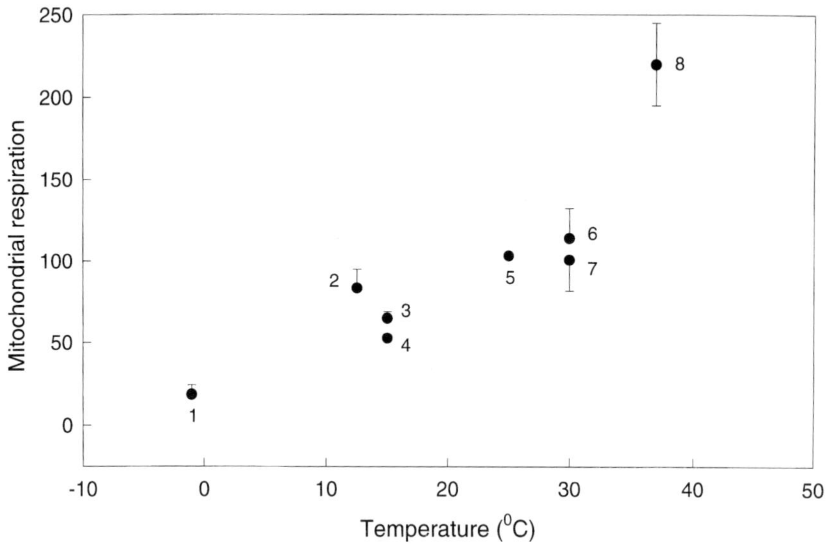

Fig. 2. State 3 respiration rate (ng-atom oxygen consumed per mg mitochondrial protein per minute) with pyruvate as substrate of mitochondria isolated from red muscle of fish, as a function of the temperature at which the fish usually live. The species are: 1. *Notothenia coriiceps*, 2. *Myoxocephalus scorpius* (short-horned sculpin, warm acclimated (summer) data), 3. *Oncorhynchus mykiss* (rainbow trout), 4. *Carassius carassius* (common carp), 5. *Katsuwonus pelamis* (skipjack tuna), 6. *Oreochromis andersoni*, 7. *Oreochromis niloticus* and 8. *Oreochromis alcalicus grahami*. Redrawn from Johnston *et al.* (1994) with extra data from H. Guderley (personal communication). If temperature compensation were perfect, the slope of the relationship would be zero (horizontal). (See also Fig. 2 in Guderley, this volume.)

the context of ecological physiology, the critical factor is the maximum rate at which ATP can be regenerated, and this can be assayed conveniently by the maximum rate at which oxygen is consumed by mitochondria oxidising pyruvate as fuel. Such measurements do, however, necessitate the isolation of the mitochondria from the internal cellular milieu within which they operate.

A recent study suggests that mitochondria isolated from the red muscle of fish living at different temperatures show only limited temperature compensation (Johnston *et al.*, 1994; Guderley, this volume; Fig. 2). If evolution had resulted in perfect temperature compensation, then maximum respiration rates *in vitro* under non-limiting conditions of oxygen, ADP and pyruvate would be expected to be independent of temperature, that is, the slope of Fig. 2 would be zero (compare the theoretical prediction in Fig. 1).

In fact, the relationship is equivalent to a Q_{10} of 1.6 over the range 0 to 30 °C (although the underlying relationship is probably strongly curvilinear: see Guderley, this volume). These data suggest that the degree of compensation for temperature in fish red muscle mitochondria is low, despite evolutionary adjustments to cristae surface density (Archer & Johnston, 1991) and presumably membrane lipid composition (demonstrated by Wodtke, 1981 following thermal acclimation in *Cyprinus carpio*, but not so far examined in fish adapted evolutionary to different temperatures).

This is an important result, for it suggests a major constraint of cellular temperature on energetics at low temperature. Since the volume-specific power generation of aerobic muscle is related to the proportion of the cell devoted to mitochondria (Pennycuick, 1992), the low rate of *per mitochondrion* ATP regeneration in polar fish means that very high mitochondrial volume densities are necessary to achieve even moderate muscle power outputs. In the relatively slow-moving *Pleurogramma antarcticum*, myofibrils in the red muscle are arranged in columns just one fibril thick, surrounded by mitochondria which constitute 56% of the overall fibre volume (Johnston *et al.*, 1988). This is a considerably higher mitochondrial volume density than in much more active fish from warmer waters, and is similar to values typical of hummingbirds and some insect flight muscles. It would seem to be an inescapable conclusion that the highest rates of muscle power output are achievable only at relatively high cell temperatures, whether those temperatures are achieved by endothermy (hummingbirds), heat conservation (tuna and some sharks) or pre-flight muscle warming (many insects). The low temperature of polar waters would thus preclude lifestyles requiring the highest rates of aerobic power output.

Whereas aerobic (red) muscles power cruising locomotion, burst swimming in fish is powered by anaerobic white muscle; intermediate levels of activity involve the recruitment of a varying proportion of white muscle. Detailed studies of whole muscle mechanics have shown that, in polar fish, maximum stress shows perfect compensation for temperature and the maximum shortening speed is similar to tropical species, whereas deactivation rates show a limited compensation (Johnston *et al.*, 1991*b*). The impact of low temperature on prey capture or predator avoidance in polar fish would therefore primarily be through recovery rates (that is, the rate at which mitochondrial activity or gluconeogenesis could clear accumulated products of anaerobic utilisation of substrates), although any direct comparison with warmer water fish would be confounded by differences in the use of short-term high-energy phosphate stores and in circulatory physiology.

The obvious evolutionary question posed by these data is why natural selection has apparently not resulted in better low temperature mitochondria in the way it has produced, for example, fully functional low temperature microtubules. The answer may be, in part, because rate compensation in mitochondria depends not only on adjustments to features subject to natural selection (enzyme structure and function, membrane composition, organelle structure) but also to physical processes, such as diffusion, which are not. It is also possible that the real constraints to aerobic function at low temperatures lie elsewhere in the organism, and that the apparently poor performance of mitochondria from cold-water species simply reflects an adjustment of mitochondrial activity to constraints somewhere else.

One caveat on this general conclusion is that it is based on a study of one type of mitochondrion (red muscle) from one type of organism (fish). Mitochondrial characteristics can vary from tissue to tissue within an organism, indicating a significant degree of evolutionary flexibility. Differences between within-species and between-species comparisons have long beset comparative studies in ecology and physiology (Harvey & Pagel, 1991); they do, however, warn against an overly simplistic interpretation of the data in Fig. 2.

There are two other important physiological considerations related to the relationship between mitochondrial ATP output and temperature shown in Fig. 2. The first is that mitochondria can only realise their maximal power output if they are not limited by the supply of oxygen or substrate. This means that there may be a requirement for large-scale tissue reorganisation at low temperatures to maintain substrate and oxygen delivery. The second is that the limitations to the rate at which individual mitochondria can deliver ATP at low temperatures may have major consequences for overall organismal energetics.

Higher levels of integration

Not all enzymes, and hence not all metabolic pathways, will be affected equally by a lowering of cell temperature. In order to avoid a potentially catastrophic metabolic imbalance during a change of temperature, the relative balance of the various pathways must be regulated. This need not mean that the balance must remain unaltered, for in both *Fundulus heteroclitus* (Moerland & Sidell, 1981) and *Morone saxatilus* (Stone & Sidell, 1981; Jones & Sidell, 1982) there is an increase in the activity of the pentose shunt relative to glycolysis following acclimation to low temperatures. Studies of fish adapted evolutionarily to living at low temperatures have indicated subtle shifts in preferred metabolic substrates (Sidell & Crockett, 1987; Crockett & Sidell, 1990). Although the adaptive significance of these shifts is not immediately clear (if indeed there is one), they do suggest a regulation of metabolic balance during acclimation and evolutionary adaptation to low temperature.

Structural organisation

The rate of supply of oxygen and metabolites to tissues depends on the respiratory and circulatory systems. Delivery to individual cells, or organelles within cells, however depends on diffusion, and diffusion is slowed significantly at low temperatures (Sidell & Hazel, 1987). Compensation for reduced diffusion rates can be achieved by shortening diffusion distances (Tyler & Sidell, 1984), and this is one benefit of the high mitochondrial packing densities within red muscle of polar fish. Recent studies using sophisticated models of oxygen delivery have suggested that diffusion of oxygen may be critical, and anatomical studies have shown that the capillary structure is modified in fish living at low temperatures to offset the limitations of reduced diffusion distances (Egginton & Sidell, 1989; Egginton & Rankin, 1991). Oxygen is highly soluble in lipid, and there are indications that the lipid droplets characteristic of polar fish may enhance oxygen flow through the cell to organelles (Sidell, 1990).

Power generation: a major ecological constraint in polar ectotherms?

There are two relationships which are fundamental to the overall energetics of a marine ectotherm. The first is the three-way relationship between protein structure, protein function and protein stability, and the second is the relationship between mitochondrial ATP generation and temperature (assuming that the pattern exhibited by fish red muscle mitochondria is a general one).

Proteins, energetics and temperature

The basic cost of existence, the energy which is expended in the absence of growth, reproduction, locomotor or other overt costs, may be categorised as basal or maintenance metabolism (see Clarke, 1983, 1991, and references cited therein). Although easy to define in theory, basal metabolism *sensu stricto* is difficult to measure in practice; the usual approach is to measure the respiration rate of a quiescent organism following short-term deprivation of food and at a season when neither growth nor reproduction is proceeding actively (this is usually called *standard* or *resting* metabolism) or to determine the relationship between activity and metabolic rate and extrapolate this to estimate the metabolic rate of the organism at zero activity.

The literature contains a great many studies of the respiration rate of fish, and those data which may be taken as approximating to resting metabolism show a positive monotonic relationship with temperature (Clarke, data in preparation), similar to the pattern long known for marine invertebrates (for

Table 5. *Resting oxygen consumption (Rest $\dot{M}o_2$) of six species of sedentary fish*

Species	Body mass (g)	Temperature (°C)	Rest $\dot{M}o_2$ Mean	SD	n
Notothenia neglecta	22–63	0	40.6	9.4	9
Agonus cataphractus	8–18	4	65.6	18.8	6
Myoxocephalus scorpius	27–164	7	75.0	31.3	9
Paracirrhites forsteri	57–187	24	115.6	28.1	9
Paracirrhites arcatus	13–36	24	106.3	25.0	6
Neocirrhites armatus	6–8	25	137.5	31.3	4

Note:
Data (mean ± SD) are given for the temperature at which the fish normally live and have been corrected to a common mass of 50 g using a mass exponent of 0.82. Units are μmol/h (1 μmol oxygen=2 μg-atoms oxygen=32 μg). The relationship between rest $\dot{M}o_2$ and temperature is positive and highly significant (ANOVA, $F=44.0$, $P=0.03$).
Source: From Johnston *et al.* (1991*a*).

review see Clarke, 1983, 1991). The problem with simply compiling literature data is that these data were collected by many different investigators using different techniques on fish of very different lifestyles, ecology and phylogenetic history.

An alternative approach is to undertake a detailed investigation of a small number of species selected carefully to minimise variation in lifestyle or ecology. The results of a study of resting metabolic rate in sedentary fish from tropical, temperate and polar waters are given in Table 5. This study also revealed a positive relationship between resting metabolic rate and environmental temperature. We may therefore take this as a general result, whilst acknowledging that ecology and life-style also play a role: at any given temperature, active fish are likely to have a greater resting oxygen consumption than sedentary fish (Morris & North, 1984; Macdonald, Montgomery & Wells, 1987).

For many years, it was thought that oxygen consumption should show compensation in a manner directly analogous to that of enzyme activity or locomotion. Indeed, it was a direct comparison of the relationships of locomotor ability and oxygen consumption to temperature that led Krogh (1916) to propose that an elevated respiration rate should be expected in organisms that had become adapted over evolutionary time to live in cold waters. This elevation was relative to the metabolic rate that would be predicted for polar organisms on the basis of extrapolating the rates observed in temperate or tropical species to polar temperatures. This suggestion, enshrined in the concept of metabolic cold adaptation (MCA), has had enormous influence in the field of polar physiology.

Krogh advanced his suggestion before the development of modern physiology (indeed before it had been established that enzymes were proteins), and the suggestion that respiration should show some form of compensation for temperature was intuitively reasonable. It is now clear, however, that to expect respiration to respond to temperature in this way is to misunderstand the nature of respiratory demand. Oxygen acts as the terminal electron acceptor in the regeneration of ATP, and hence the rate of oxygen consumption is a measure of the rate at which reserves are being consumed to fuel that ATP regeneration. ATP regeneration does not, however, proceed regardless; it is under complex and subtle feedback control. The demand for ATP, and hence oxygen, is driven by the requirement for that ATP to fuel work. Since the costs of that work, be it chemical or mechanical, must be met from food or previously stored reserves, selection will act to reduce those costs as far as is consonant with survival or fitness (Clarke, 1993*b*). An organism will only increase the rate at which ATP is synthesised (and hence oxygen is consumed) if the benefits obtained outweigh the costs involved in obtaining food or utilising reserves. Metabolic cold adaptation, in the sense of 'maintaining metabolism' at some arbitrary or fixed level is thus a meaningless concept energetically.

It is quite possible that there may be extra costs associated with living in cold water, for example, antifreeze production in polar teleost fish (DeVries, 1983), but the extent of these costs are currently unknown (although they are not believed to be great). Early studies of the metabolism of polar marine ectotherms (principally fish) provided data in support of a relative elevation of respiration rate at low temperatures (for review see Clarke 1980, 1983). It was Holeton (1973, 1974) who first questioned both the concept of metabolic cold adaptation and the experimental evidence in its support. Subsequent work has largely validated Holeton's criticisms, and the concept of metabolic cold adaptation (*sensu* Krogh) has now largely been abandoned (Clarke, 1991). In its place is the empirical observation of a positive relationship between resting metabolic rate and environmental temperature (Table 5).

This result implies that basal processes consume less ATP at polar temperatures than at tropical temperatures; data from marine invertebrates suggests that this difference falls in the range ×7 to ×26, depending on which relationship is used (Clarke, 1991). Evidence from isolated cells in culture (discussed by Wieser, 1994) and the mussel *Mytilus edulis* (Hawkins, 1991) suggests that protein turnover costs form a substantial fraction of resting (and therefore, by influence, basal) metabolic costs. Furthermore, Smith & Haschemeyer (1980) and Smith *et al.* (1980) showed that overall rates of protein synthesis were much lower in polar fish than in tropical fish.

Whilst other contributions to basal metabolic costs, and especially those associated with ionic regulation, are important, it would appear that differences in protein turnover contribute substantially to the differences in resting metabolic rate with temperature. These differences in protein turnover would appear to be the result of the interaction between protein function, protein structure (and hence stability) and temperature. Perhaps counterintuitively, although proteins from polar ectotherms are less thermally stable than homologues from tropical species, turnover requirements are less at the low temperatures at which these polar fish live (Clarke, 1993*b*). One intriguing footnote of potential energetic importance is that there is a small but growing body of evidence that the costs of synthesising protein (in terms of moles ATP required per mole of peptide bond synthesised) may be significantly greater at lower temperatures and/or lower rates of synthesis (Pannevis & Houlihan, 1992; Wieser, 1994; Whiteley, Taylor & El Haj, 1996). The most reasonable interpretation of this relationship is that the overall costs of protein synthesis comprises two components. The first is a fixed cost (more or less irrespective of the rate), and the second is a cost which is a function of the rate of synthesis. The latter is the cost of peptide bond synthesis, but the nature of the fixed cost is not at all clear and may involve subtle trade-offs (Wieser, 1994).

Overall, however, the lower resting metabolic rates in polar fish compared with tropical fish would appear to confer a distinct energetic advantage. Basal costs are, in ecological terms, wasted in that they represent utilisation of resources which could otherwise be used for processes such as growth, reproduction or locomotion (Clarke, 1983, 1991). Reduced basal costs therefore mean that a greater proportion of the energy intake can be directed at processes which enhance fitness or survival. This theoretical energetic advantage in polar marine ectotherms can, however, only be utilised if sufficient food can be obtained, this food can be processed sufficiently rapidly, and regenerated ATP can be supplied fast enough.

The generation of metabolic power at low temperatures: evidence for constraint from the costs of feeding

The question as to whether heterotrophic organisms are ever food limited in the wild has long been debated by ecologists. Whilst there is ample evidence that some species in some places at some times are undoubtedly resource limited, there is no consensus on whether this is a general result. There is, however, no evidence that polar regions are different from anywhere else in the marine environment in this respect, and so in the absence of evidence to the contrary it may be assumed that the reduced basal metabolic requirements of polar marine ectotherms probably mean that resource limitation overall is no more severe than elsewhere. It may, however, be very severe seasonally (Clarke, 1988) although not for all species (Barnes & Clarke, 1994, 1995), and overall growth rates are generally slow (Brey & Clarke, 1993).

When an organism consumes a meal, there is a concomitant increase in oxygen consumption (and hence heat production) and nitrogen excretion. This is often termed the specific dynamic action of feeding (SDA) or the heat increment of feeding (HIF). It is believed that the major components of the SDA/HIF are the costs of digestion, peristalsis and absorbtion, together with the costs of growth (and especially protein synthesis) following absorption of a meal (Parry, 1983; Jobling, 1981, 1983; Wieser, 1994).

Johnston & Battram (1993) first demonstrated that the peak (absolute) increase in respiration rate following a meal (the metabolic scope of feeding) was lower in polar fish compared with tropical or temperate fish. A reduced metabolic scope of feeding has now been confirmed for another species of Antarctic fish, *Harpagifer antarcticus* (Boyce & Clarke, 1997), the scavenging amphipod *Waldeckia obesa* (Chapelle, Peck & Clarke, 1994), the suspension feeding brachiopod *Liothyrella uva* (Peck, 1996), the grazing limpet *Nacella concinna* (Peck, personal communication) and the carnivorous nemertean *Parborlasia corrugatus* (Clarke & Prothero-Thomas, 1997) (see also Peck, this volume).

The result of this reduced metabolic scope of feeding is a greatly extended SDA response (Johnston & Battram, 1993; Peck, 1996). Whilst current evidence does not suggest enhanced food processing costs at low temperatures (Johnston & Battram, 1993), the evidence for a possible increase in protein synthesis costs at low rates of synthesis (Wieser, 1994; Whiteley *et al.*, 1996) suggests that this may be a possibility.

It is also possible that the combination of a reduced metabolic scope of feeding and a lengthened SDA response may limit the rate at which food can be processed (and hence energy acquired) at low temperatures. This, in turn, may limit the rate at which polar organisms can grow or synthesise reproductive tissues when these processes are fuelled directly from food intake, limitations which are of profound ecological importance.

Final remarks

Our views of the energetics of cold-water marine ectotherms have changed significantly over the past half century. The original suggestion that evolutionary adaptation to polar waters required an elevation of routine metabolic rate was intuitively reasonable and provided a logical explanation for the widely observed slow rates of growth (Krogh, 1916; Scholander *et al.*, 1953; Wohlschlag, 1960, 1964; Dunbar, 1968). This view was first queried (Holeton, 1973, 1974) and then replaced by the observation that metabolic rates were actually generally low (summarised in Clarke, 1980, 1983); such low metabolic rates would suggest higher growth efficiencies, and reduced costs of existence (Clarke, 1983, 1991). At the same time, an extensive series of studies have revealed extensive temperature compensation at the molecular level in some, but by no means all, cellular processes (Somero, 1978; Sidell, 1983; Macdonald *et al.*, 1987; Johnston, 1989; Detrich *et al.*, 1989).

More recently, it has become clear that, although a low resting metabolic rate may provide some energetic advantages (for example, in reducing the need for overwintering energy reserves), it may necessarily be associated with other features which may be regarded as constraints. In particular, it may involve a concomitant reduction in maximum absolute aerobic scope, thereby ruling out energy-intensive lifestyles. It may also limit the rate at which food can be processed, which itself may act, in turn, as a constraint on energetics.

There is also the growing body of data suggesting that the overall costs of protein synthesis may be greater at lower rates of synthesis and/or low temperature (Pannevis & Houlihan, 1992; Wieser, 1994; Whiteley *et al.*, 1996). This would mean that, despite the lower resting metabolic rates, growth efficiencies in polar marine ectotherms would actually be low. What data exist, however, suggest that growth efficiencies are higher in cold water (Clarke, 1991). It is clear that to understand better the energetics of cold water marine ectotherms, it is necessary to clarify the relationship between resting (basal) metabolic rate and maximum aerobic scope, and that between the rate and cost of protein synthesis.

References

Archer, S.D. & Johnston, I.A. (1991). Density of cristae and distribution of mitochondria in the slow muscles of Antarctic fish. *Physiological Zoology*, **64**, 242–58.

Barnes, D.K.A. & Clarke, A. (1994). Seasonal variation in the feeding activity of four species of Antarctic bryozoan in relation to environmental factors. *Journal of Experimental Marine Biology and Ecology*, **181**, 117–31.

Barnes, D.K.A. & Clarke, A. (1995). Seasonality of feeding activity in Antarctic suspension feeders. *Polar Biology*, **15**, 335–40.

Bosch, I., Beauchamp, K.A., Steele, M.E. & Pearse, J.S. (1987). Development, metamorphosis and seasonal abundance of embryos and larvae of the Antarctic sea urchin *Sterechinus neumayeri*. *Biological Bulletin*, **173**, 126–35.

Boyce S.J. & Clarke A. (1997). Effects of body size and ration on specific dynamic action in the Antarctic plunderfish *Harpagifer antarcticus* Nybelin, 1947. *Physiological Zoology*, **70**, in press.

Brey, T. & Clarke, A. (1993). Population dynamics of marine benthic invertebrates in Antarctic and subantarctic environments: are there unique adaptations? *Antarctic Science*, **5**, 253–66.

Chapelle, G., Peck, L.S. & Clarke, A. (1994). Effects of feeding and starvation on the metabolic rate of necrophagous Antarctic amphipod *Waldeckia obesa* (Chevreux, 1905). *Journal of Experimental Marine Biology and Ecology*, **183**, 63–76.

Clarke, A. (1980). A reappraisal of the concept of metabolic cold adaptation in polar marine invertebrates. *Biological Journal of the Linnean Society*, **14**, 77–92.

Clarke, A. (1983). Life in cold water: the physiological ecology of polar marine ectotherms. *Oceanography and Marine Biology: An Annual Review*, **21**, 341–453.

Clarke, A. (1987). The adaptation of aquatic animals to low temperature. In *The Effects of Low Temperatures on Biological Systems*, ed. B.W.W. Grout & G.J. Morris, pp. 315–48. London: Edward Arnold.

Clarke, A. (1988). Seasonality in the Antarctic marine ecosystem. *Comparative Biochemistry and Physiology*, **90B**, 461–73.

Clarke, A. (1991). What is cold adaptation and how should we measure it? *American Zoologist*, **31**, 81–92.

Clarke, A. (1992). Reproduction in the cold: Thorson revisited. *Invertebrate Reproduction and Development*, **22**, 175–84.

Clarke, A. (1993*a*). Temperature and extinction in the sea: a physiologist's view. *Paleobiology*, **19**, 499–518.

Clarke, A. (1993*b*). Seasonal acclimatisation and latitudinal compensation in metabolism: do they exist? *Functional Ecology*, **7**, 139–49.

Clarke, A. & Prothero-Thomas, E. (1997). The influence of feeding on oxygen consumption and nitrogen excretion in the Antarctic nemertean *Parborlasia corrugatus*. *Physiological Zoology*, **70**, in press.

Cocca, E., Ratnayakelecamwasam, M., Parker, S.K., Camardella, L., Ciaramella, M., di Prisco, G. & Detrich, H.W. (1995). Genomic remnants of alpha-globin genes in the hemoglobinless Antarctic icefishes. *Proceedings of the National Academy of Sciences, USA*, **92**, 1817–21.

Cossins, A.R. (1994). Homeoviscous adaptation of biological membranes and its functional significance. In *Temperature Adaptation of Biological Membranes*, ed. A.R. Cossins, pp. 63–76. London: Portland Press.

Cossins, A.R. & Bowler, K. (1987). *Temperature Biology of Animals*. London: Chapman & Hall.

Crockett, E.L. & Sidell, B.D. (1990). Some pathways of energy metabolism are cold adapted in Antarctic fishes. *Physiological Zoology*, **63**, 472–88.

Detrich, H.W., III & Overton, S.A. (1986). Heterogeneity and structure of brain tubulins from cold-adapted fishes. *Journal of Biological Chemistry*, **261**, 10922–30.

Detrich, H.W., III, Johnson, K.A. & Marchese-Ragona, S.P. (1989). Polymerization of Antarctic fish tubulins at low temperature: energetic aspects. *Biochemistry*, **28**, 10085–93.

DeVries, A.L. (1971). Glycoproteins as biological antifreeze agents in Antarctic fishes. *Science*, **172**, 1152–5.

DeVries, A.L. (1983). Antifreeze peptides and glycopeptides in cold-water fishes. *Annual Review of Physiology*, **45**, 245–60.

DeVries, A.L. & Wohlschlag, D.E. (1969). Freezing resistance in some Antarctic fishes. *Science*, **163**, 1073–5.

Dunbar, M.J. (1968). *Ecological Development in Polar Regions. A Study in Evolution*. Englewood Cliffs, New Jersey: Prentice-Hall, 119pp.

Eastman, J.T. (1993). *Antarctic Fish Biology: Evolution in a Unique Environment*. New York: Academic Press.

Edmund, J.M. & Gieskes, J.A.T.M. (1970). On the calculation of the degree of saturation of seawater with respect to calcium carbonate under *in situ* conditions. *Geochimica Cosmochimica Acta*, **34**, 1261–91.

Egginton, S. & Rankin, J.C. (1991). The vascular supply to skeletal muscles in fishes with and without respiratory pigments. *International Journal of Microcirculation – Clinical and Experimental*, **10**, 396.

Egginton, S. & Sidell, B.D. (1989). Thermal acclimation induces adaptive changes in subcellular structure of fish skeletal muscle. *American Journal of Physiology*, **256**, R1–9.

Gnaiger, E. (1987). Optimum efficiencies of energy transformation in anoxic metabolism; the strategies of power and economy. In *Evolutionary Physiological Ecology*, ed. P. Calow, pp. 7–36. Cambridge University Press.

Gon, O. & Heemstra, P.C. (eds.) (1990). *Fishes of the Southern Ocean*. J.L.B. Smith Institute of Ichthyology, South Africa.

Graves, J.E. & Somero, G.N. (1982). Electrophoretic and functional enzyme evolution in four species of eastern Pacific barracudas from different thermal environments. *Evolution*, **36**, 97–106.

Harvey, P.H. & Pagel, M.D. (1991). *The Comparative Method in Evolutionary Biology*. Oxford University Press, 239 pp.

Hawkins, A.J.S. (1991). Protein turnover: a functional appraisal. *Functional Ecology*, **5**, 222–33.

Hazel, J.R. & Prosser, C.L. (1974). Molecular mechanisms of temperature compensation in poikilotherms. *Physiological Reviews*, **54**, 620–77.

Hochachka, P.W. & Somero, G.N. (1973). *Strategies of Biochemical Adaptation*. Philadelphia: W.B. Saunders, 357pp.

Hochachka, P.W. & Somero, G.N. (1984). *Biochemical Adaptation*. Princeton: Princeton University Press, 537pp.

Hoffman, R.J. (1981*a*). Evolutionary genetics of *Metridium senile*. I. Kinetic differences in phosphoglucose isomerase allozymes. *Biochemical Genetics*, **19**, 129–44.

Hoffman, R.J. (1981*b*). Evolutionary genetics of *Metridium senile*. II. Geographic patterns of allozyme variation. *Biochemical Genetics*, **19**, 145–54.

Hoffman, R.J. (1983). Temperature modulation of the kinetics of phosphoglucose isomerase genetic variants from the sea anemone *Metridium senile*. *Journal of Experimental Zoology*, **227**, 361–70.

Holeton, G.F. (1973). Respiration of Arctic char (*Salvelinus alpinus*) from a high arctic lake. *Journal of the Fisheries Research Board of Canada*, **30**, 717–23.

Holeton, G.F. (1974). Metabolic cold adaptation of polar fish: fact or artefact? *Physiological Zoology*, **47**, 137–52.

Jobling, M. (1981). The influences of feeding on the metabolic rate of fishes: a short review. *Journal of Fish Biology*, **18**, 385–400.

Jobling, M. (1983). Towards an explanation of specific dynamic action (SDA). *Journal of Fish Biology*, **23**, 549–55.

Johnston, I.A. (1989). Antarctic fish muscles – structure, function and physiology. *Antarctic Science*, **1**, 97–108.

Johnston, I.A. (1990). Cold adaptation in marine organisms. *Philosophical Transactions of the Royal Society of London, Series B*, **326**, 655–67.

Johnston, I.A. & Battram, J. (1993). Feeding energetics and metabolism in demersal fish species from Antarctic, temperate and tropical environments. *Marine Biology*, **115**, 7–14.

Johnston, I.A. & Goldspink, G. (1975). Thermodynamic activation parameters of fish myofibrillar ATPase enzyme and evolutionary adaptations to temperature. *Nature*, **257**, 620–2.

Johnston, I.A., Camm, J.P. & White, M.G. (1988). Specialisations of swimming muscles in the pelagic Antarctic fish, *Pleurogramme antarcticum*. *Marine Biology*, **100**, 3–12.

Johnston, I.A., Clarke, A. & Ward, P. (1991*a*). Temperature and metabolic rate in sedentary fish from the Antarctic, North Sea and Indo-West Pacific Ocean. *Marine Biology*, **109**, 191–5.

Johnston, I.A., Johnson, T.P. & Battram, J.C. (1991*b*). Low temperature limits burst swimming performance in Antarctic fish. In *Biology of Antarctic Fishes*, ed. G. di Prisco, B. Maresca & B. Tota, Berlin: Springer-Verlag.

Johnston, I.A., Guderley, H., Franklin, C.E., Crockford, T. and Kamunde, C. (1994). Are mitochondria subject to evolutionary temperature adaptation? *Journal of Experimental Biology*, **195**, 293–306.

Jones, P.L. & Sidell, B.D. (1982). Metabolic responses of striped bass (*Morone saxatilis*) to temperature acclimation. II. Alterations in metabolic carbon sources and distributions of fiber types in locomotory muscle. *Journal of Experimental Zoology*, **219**, 163–71.

Krogh, A. (1916). *The Respiratory Exchange of Animals and Man*. London: Longmans.

Kunzmann, A. (1991). Blood physiology and ecological consequences in Weddell Sea fishes (Antarctica). *Berichte zur Polarforschung*, **91**, 71pp.

Macdonald, J.A. (1981). Temperature compensation in the peripheral nervous system: Antarctic vs temperate poikilotherms. *Journal of Comparative Physiology*, **142**, 411–18.

Macdonald, J.A. & Wells, R.M.G. (1991). Viscosity of body fluids from Antarctic notothenioid fish. In *Biology of Antarctic Fishes*, ed. G. di Prisco, B. Maresca & B. Tota, pp. 163–78. Berlin: Springer-Verlag.

Macdonald, J.A., Montgomery, J.C. & Wells, R.M.G. (1987). Comparative physiology of Antarctic fishes. *Advances in Marine Biology*, **24**, 321–88.

Malan, A., Wilson, T.L. & Reeves, R.B. (1976). Intracellular pH in cold-blooded vertebrates as a function of body temperature. *Respiration Physiology*, **28**, 29–47.

Moerland, T.S. (1995). Temperature: enzyme and organelle. In *Biochemistry and Molecular Biology of Fishes, volume 5*, ed. P.W. Hochachka & T.P. Mommsen, pp. 57–71. Amsterdam: Elsevier.

Moerland, T.S. & Sidell, B.D. (1981). Characterisation of metabolic carbon flow in hepatocytes isolated from thermally acclimated killifish *Fundulus heteroclitus*. *Physiological Zoology*, **54**, 379–89.

Montgomery, J.C. & Macdonald, J.A. (1984). Performance of motor systems in antarctic fishes. *Journal of Comparative Physiology*, **117B**, 181–91.

Morris, D.J. & North, A.W. (1984). Oxygen consumption of five species of fish from South Georgia. *Journal of Experimental Marine Biology and Ecology*, **265**, 75–86.

Pannevis, M.C. & Houlihan, D.F. (1992). The energetic cost of protein synthesis in isolated hepatocytes of rainbow trout (*Oncorhynchus mykiss*). *Journal of Comparative Physiology*, **162B**, 393–400.

Parry, G. (1983). The influence of the costs of growth on ectotherm metabolism. *Journal of Theoretical Biology*, **107**, 453–77.

Peck, L.S. (1996). Metabolism and feeding in the Antarctic brachiopod *Liothyrella uva*: a low energy lifestyle species with restricted metabolic scope. *Proceedings of the Royal Society of London, Series B*, **236**, 223–8.

Pennycuick, C.J. (1992). *Newton Rules Biology*. Oxford: Oxford University Press, 111pp.

Place, A.R. & Powers, D.A. (1984). Kinetic characterisation of the lactate dehydrogenase (LDH-B_4) allozymes of *Fundulus heteroclitus*. *Journal of Biological Chemistry*, **259**, 1309–15.

Podolsky, R.D. & Emlet, R.B. (1993). Separating the effects of temperature and viscosity on swimming and water-movement by sand dollar larvae (*Dendraster excentricus*). *Journal of Experimental Marine Biology and Ecology*, **176**, 207–21.

Powers, D.A. & Place, A.R. (1978). Biochemical genetics of *Fundulus heteroclitus* (L.) I. Temporal and spatial variation in gene frequencies of Ldh-B, Mdh-A, Gpi-B, and Pgm-A. *Biochemical Genetics*, **16**, 593–607.

Reeves, R.B. (1972). An imidazole alphastat hypothesis for vertebrate acid–base regulation: tissue carbon dioxide content and body temperature in bullfrogs. *Respiration Physiology*, **28**, 49–63.

Reeves, R.B. (1977). The interaction of acid–base balance and body temperature in ectotherms. *Annual Review of Physiology*, **39**, 559–86.

Riley, J.P. & Chester, R. (1971). *Introduction to Marine Chemistry*. London: Academic Press.

Riley, J.P. & Skirrow, G. (1975). *Chemical Oceanography*. 2nd edn. London: Academic Press.

Scholander, P.F., Flagg, W., Walters, V. & Irving, L. (1953). Climatic adaptation in arctic and tropical poikilotherms. *Physiological Zoology*, **26**, 67–92.

Scott, G.K., Fletcher, G.L. & Davies, P.L. (1986). Fish antifreeze proteins: recent gene evolution. *Canadian Journal of Fisheries and Aquatic Sciences*, **43**, 1028–34.

Sidell, B.D. (1977). Turnover of cytochrome c in skeletal muscle of green sunfish (*Lepomis cyanellus*, R.) during thermal acclimation. *Journal of Experimental Zoology*, **199**, 233–50.

Sidell, B.D. (1983). Cellular acclimation to environmental change by quantitative alterations in enzymes and organelles. In *Cellular Acclimatisation to Environmental Change*, ed. A.R. Cossins & P. Sheterline, pp. 103–20. Cambridge: Society for Experimental Biology, Seminar Series 17, Cambridge University Press.

Sidell, B.D. (1990). Diffusion and ultrastructural adaptations of ectotherms. In *Microcompartmentation*, ed. D.P. Jones, Boca Raton, Florida: CRC Press.

Sidell, B.D. & Crockett, L. (1987). Characterisation of energy metabolism in Antarctic fishes. *Antarctic Journal of the United States*, **22**, 213–14.

Sidell, B.D. & Hazel, J.R. (1987). Temperature affects the diffusion of small molecules through the cytosol of fish muscle. *Journal of Experimental Biology*, **129**, 191–203.

Smith, M.A.K. & Haschemeyer, A.E.V. (1980). Protein metabolism and cold adaptation in Antarctic fishes. *Physiological Zoology*, **53**, 373–82.

Smith, M.A.K., Mathews, R.W., Hudson, A.P. & Haschemeyer, A.E.V. (1980). Protein metabolism of tropical reef and pelagic fish. *Comparative Biochemistry and Physiology*, **65B**, 415–18.

Somero, G.N. (1978). Temperature adaptation of enzymes: biological optimisation through structure–function compromises. *Annual Review of Ecology and Systematics*, **9**, 1–29.

Stone, B.B. & Sidell, B.D. (1981). Metabolic responses of striped bass (*Morone saxatilis*) to temperature acclimation. I. Alterations in carbon sources for hepatic energy metabolism. *Journal of Experimental Zoology*, **218**, 371–9.

Swezey, R.R. & Somero, G.N. (1982). Polymerization thermodynamics and structural stabilities of skeletal muscle actins from vertebrates adapted to different temperatures and hydrostatic pressures. *Biochemistry*, **21**, 4496–503.

Tyler, S. & Sidell, B.D. (1984). Changes in mitochondrial distribution and diffusion distances in muscle goldfish upon acclimation to warm and cold temperatures. *Journal of Experimental Zoology*, **232**, 1–9.

Whiteley, N.M., Taylor, E.W. & El Haj, A.J. (1996). A comparison of the metabolic cost of protein synthesis in stenothermal and eurythermal isopod crustaceans. *American Journal of Physiology*, **271**, R1295–303.

Wieser, W. (1994). Cost of growth in cells and organisms: general rules and comparative aspects. *Biological Reviews*, **68**, 1–33.

Wodtke, E. (1981). Temperature adaptation of biological membranes. Compensation of the molar activity of cytochrome c oxidase in the mitochondrial energy-transducing membrane during thermal acclimation of the carp (*Cyprinus carpio* L.). *Biochimica et Biophysica Acta*, **640**, 710–20.

Wohlschlag, D.E. (1960). Metabolism of antarctic fish and the phenomenon of cold adaptation. *Ecology*, **41**, 287–92.

Wohlschlag, D.E. (1964). Respiratory metabolism and ecological characteristics of some fishes in McMurdo Sound, Antarctica. In *Biology of the Antarctic Seas*, ed. M.O. Lee, pp. 33–62. Washington: American Geophysical Union.

Yancey, P.H. & Somero, G.N. (1978). Temperature dependence of intracellular pH: its role in the conservation of pyruvate apparent K_m values of vertebrate lactate dehydrogenases. *Journal of Comparative Physiology*, **125**, 129–34.

PART II Compensatory adaptations in cold ocean environments

G.N. SOMERO

Adaptation to cold and depth: contrasts between polar and deep-sea animals

Two regions of the world ocean are characterised by extremely cold temperatures: the polar oceans and the deep sea. In each of these regions, water temperatures usually are close to 0 °C. In the polar oceans, especially in the waters surrounding Antarctica, temperatures may rise seasonally by no more than a few hundredths of a degree centigrade above the freezing point of sea water, −1.86 °C (Eastman, 1993). Deep-sea temperatures generally are also very stable, with the notable exception of waters found at the deep-sea hydrothermal vents, where animals may encounter temperatures between approximately 2 °C and at least 40 °C (Fustec, Desbruyeres & Juniper, 1987; Johnson, Childress & Beehler, 1988; Dahlhoff *et al.*, 1991). Thus, ectothermic species living in polar and typical deep-sea waters are characterised by some of the lowest and most stable body temperatures found in any species, in any environment. One would predict that many physiological and biochemical systems of shallow-living polar species and deep-sea species would manifest similar adaptations to low and stable temperatures.

Despite facing similar thermal conditions, deep-sea species and shallow-living polar species encounter other environmental differences that may lead to divergent patterns of adaptation. Hydrostatic pressure is one important difference between shallow and deep cold water habitats. Pressure rises by approximately 0.1 MPa (~1 atm) with each 10 m increase in depth, so deep-living organisms encounter pressures of up to 100 MPa (=~1000 atm) in the deepest trenches of the ocean. Because most physiological and biochemical systems are sensitive to pressure (Siebenaller, 1991; Somero, 1993), adaptations to high pressure, as well as to low temperature, would be expected to characterise diverse physiological and biochemical traits of deep-sea species. Cold-adapted shallow-water organisms would not be predicted to exhibit resistance to pressure unless adaptations to low temperature happen to be pre-adaptive for life at high pressure, a point to be considered below.

A second important difference between shallow polar waters and the deep sea is the availability of light. Although photoperiod varies enormously throughout the year in high latitude waters, sunlight reaches the ocean surface for at least several months each year. During the summer, there may be continuous light in high latitude seas. In contrast, the deep sea is continuously dark. Darkness not only precludes sunlight-driven primary production, but may render difficult the location of what food is available. Because darkness may have a strong selective effect on animal activity and metabolic rate (Childress & Somero, 1979; Childress, 1995; Childress & Thuesen, 1995), differences in light availability may favour different evolutionary trajectories in cold-adapted organisms from shallow polar waters and the deep sea.

This review compares and contrasts several physiological and biochemical systems of cold-adapted, shallow-living polar animals with those of deep-sea species. The analysis attempts to distinguish between adaptations that appear to be related strictly to facilitating life at low temperatures, that is, mechanisms of cold adaptation broadly defined, and other adaptations that are reflections of life under the conditions of high pressure and reduced light in the deep sea. The following questions serve as central themes. First, are cold-adapted physiological and biochemical systems of shallow-living polar species pre-adapted for function at depth, or must adaptations to elevated pressure be superimposed on these cold-adapted systems to allow function in the deep sea? Secondly, when adaptation to high pressure is necessary, do these adaptations to pressure entail trade-offs such that systems adapted to high pressures no longer retain the full complement of properties associated with adaptation to cold? Thirdly, in the context of rates of physiological and biochemical function, how do low temperatures, high pressures, and reduced availability of light and food interact to establish the metabolic rates of deep-sea species? Is temperature a dominant factor in establishing the metabolic rates in cold-adapted species, whatever their depths of occurrence? Conversely, is temperature of only minor significance among the broad suite of factors that select for metabolic rates? Finally, what insights into adaptation to temperature, pressure, and light availability can be gleaned from studies of hydrothermal vent fauna that live in the presence of high and extremely variable temperatures, high pressures, and an abundant food supply relative to that of the typical deep sea? To what degree do the hydrothermal vent animals serve as exceptions that prove the rules that have been proposed to explain the unique physiological and biochemical characteristics of species from the typical cold deep sea?

Metabolic rates: effects of temperature and depth

An appropriate starting point for analysing the physiological and biochemical properties of ectothermic species from cold ocean habitats is an examination of their rates of metabolism. A question that often has dominated, and often confused, discussions about adaptation to low temperature concerns the phenomenon of metabolic cold adaptation (Clarke, 1991). Much work and a great deal of speculation have surrounded the question of whether cold-adapted ectotherms like polar fishes display metabolic (respiratory) rates that are higher than the rates that would be predicted from extrapolating the metabolic rates of tropical and temperate species down to the low environmental (body) temperatures of cold-adapted species. Clarke (1991) has provided a cogent summary of studies of metabolic cold adaptation at the level of whole organism oxygen consumption. As Clarke shows, examination of this issue is fraught with experimental difficulties, not the least of which is finding the appropriate study species at different latitudes to allow meaningful interspecific comparisons. The frequent absence of closely related species at low and high latitudes (Eastman, 1993) renders comparative work on metabolic rates in warm- and cold-adapted species very difficult. Some of the conclusions about metabolic cold adaptation definitely have involved 'apples to oranges' comparisons in which species with different phylogenies, ecological characteristics, and feeding and locomotory habits have been compared. It has been extremely difficult to study the effects of temperature in strict isolation from interspecific differences in other traits which may play vastly stronger roles than the one played by temperature in governing metabolic rates. Methodological problems also cloud the picture of metabolic cold adaptation (Holeton, 1974; Clarke, 1991). For example, unless one carefully controls for excitement level of the specimens and for the effects of specific dynamic action, a strong contribution to metabolic rate from these sources can create serious artifacts in studies of temperature compensation.

One of the few sets of studies in which ecologically and taxonomically similar species from different latitudes have been compared, using a common methodology, is the work of Joseph Torres and collaborators (Torres, Belman & Childress, 1979; Donnelly & Torres, 1988; Torres & Somero, 1988*a,b*). Because these studies examined not only temperature effects, but also depth-related influences on metabolism, analysis of their findings is a good starting point for distinguishing the effects of differences in temperature from effects due to depth-related variables such as hydrostatic pressure, light intensity, and, where relevant, oxygen concentration. Torres and colleagues have been careful to select, whenever possible, confamilial and congeneric species, in order to compare taxonomically and ecologically similar species to the fullest extent allowed by the vagaries of collecting specimens in the deep ocean. Note that, in their analyses and in the studies of Childress and colleagues discussed later (Childress, 1995), minimal depth of occurrence (MDO) is used to define how deep a species occurs. Use of MDO is based on the fact that vertically migrating species that feed in shallow water at night and return to depth during the day have higher locomotory capacities and metabolic rates than species that co-occur with migrators at depth, but do not undergo the vertical displacements noted for migrators (Childress & Nygaard, 1973). The period spent feeding in shallow water is marked by more intense predator–prey interactions, that is, by a requirement for more robust locomotory ability, than the time spent at depth (Childress & Somero, 1979; Childress, 1995). Use of MDO therefore helps to factor out some of the interspecific variation in respiration rate that results from inherent differences in activity levels among species.

For pelagic fishes belonging to several families, a decrease in respiration rate with increasing MDO was found in all water columns studied: temperate coastal waters off California (Torres *et al.*, 1979), tropical waters in the Gulf of Mexico (Donnelly & Torres, 1988), and the extremely cold waters of Antarctica (Torres & Somero, 1988*a,b*) (Fig. 1). The decrease in respiration rate with increasing MDO shown in Fig. 1 is due, in part, to depth-related changes in ambient temperature for all but the Antarctic fishes. In the coastal California data, species with MDOs of 100 m or less were measured at 10 °C, and deeper-occurring species were measured at 5 °C. For the Gulf of Mexico specimens, respiration rate data are for temperatures of 20 °C (100 m and shallower), 17 °C (100–200 m), 14 °C (200–300 m), 10–12 °C (300–600 m), 7 °C (600–850 m), and 5 °C (850 m and deeper). For the Antarctic species, all rates are for 0.5 °C.

It is apparent that complete metabolic compensation to temperature, that is, similar rates of respiration by these three groups of fishes at their respective habitat temperatures, is not present. At normal habitat temperatures, the Gulf of Mexico fishes have much higher respiration rates than fishes from the other two regions, and this difference persists for fishes from all depths. The respiration rates of the California fishes are slightly higher than those of the Antarctic fishes at MDOs above approximately 100 m, but at greater depths the rates of the Antarctic species are slightly higher. Using a Q_{10} of 2.0 to extrapolate the respiration rates of the California fishes from 10 °C and 5 °C to 0.5 °C, the measurement temperature used to study the Antarctic species, Torres and Somero (1988*a,b*) showed that the predicted rates of respiration of the California fishes were lower than those of the Antarctic fishes regardless of MDO. None the less, the whole organism respiration rates shown in Fig. 1 show that temperature compensation is far from complete, especially when the tropical species are compared with the more cold-adapted species.

The extent to which interspecific differences in activity levels and other metabolic-rate-determining factors preclude estimating the precise extent of whole organism metabolic cold adaptation in the studies of Torres and colleagues is not clear. It is pertinent in this context to review data from studies with a more reductionist focus, in which estimates of metabolic potential

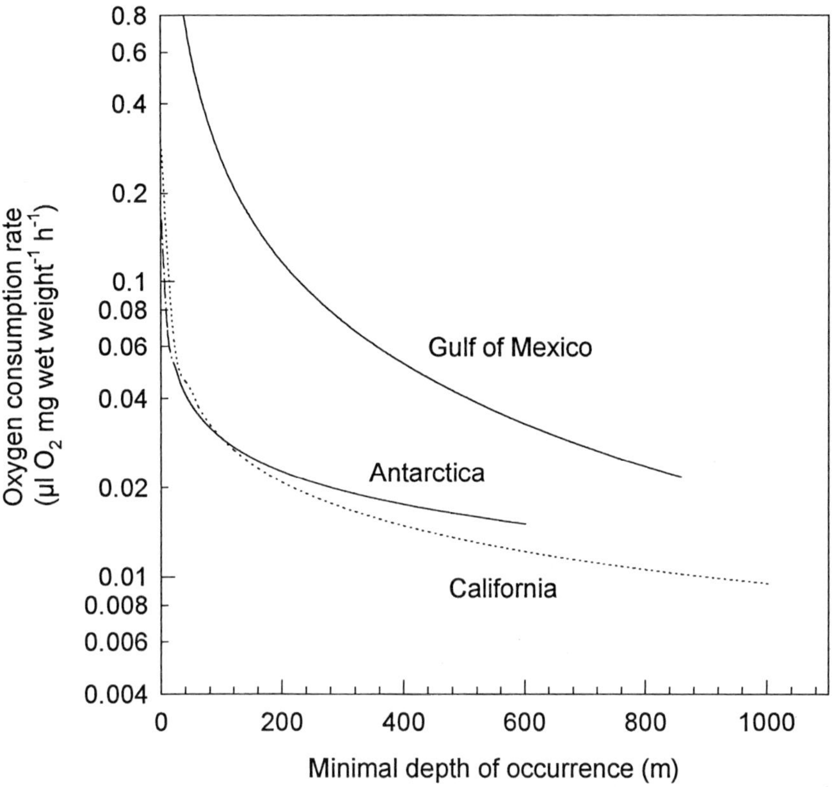

Fig. 1. Rates of oxygen consumption as a function of minimal depth of occurrence (MDO) for pelagic fishes from three different regions: coastal California (Torres *et al.*, 1979), Antarctica (Torres & Somero, 1988*a,b*), and the Gulf of Mexico (Donnelly & Torres, 1988). The curves illustrated are based on the regression equations expressing respiration rate versus MDO given in the above citations. See text for details of temperatures of measurement (Antarctica=0.5 °C; California=5 °C and 10 °C; Gulf of Mexico=5–20 °C.)

have been made using isolated tissues and enzymatic activities. A study of gill and brain tissue metabolism *in vitro* reported a high degree of metabolic compensation to temperature in Antarctic fishes (Somero, Giese & Wohlschlag, 1968). Measurements of the activities of enzymes of ATP-generating pathways in Antarctic and non-Antarctic fishes have likewise provided evidence for cold adaptation (Crockett & Sidell, 1990; Kawall & Somero, 1996). In an attempt to eliminate confounding effects of phylogeny and differences in locomotory activity among species, Kawall and Somero

(1996) compared activities of enzymes in brain of Antarctic, temperate and tropical species. Unlike skeletal muscle, whose metabolic activity varies enormously among species (Childress & Somero, 1979), brain is likely to function at similar rates in different fishes, except for possible differences related to temperature compensation. A high degree of metabolic compensation to temperature was found for enzymes of the glycolytic pathway and citric acid cycle in brain. These data suggest that, if the caveats of Holeton (1974), Clarke (1991) and others are heeded and metabolic rate comparisons are made with an 'apples to apples' logic, a substantial degree of metabolic temperature compensation will be discovered.

Even though the absolute rates of whole organism respiration differ among the three groups of fishes shown in Fig. 1, there is, for all three groups, a fairly similar decrease in respiration rate with increasing MDO. What depth-related factors can account for the fall in respiration rate with increasing MDO? Because the Antarctic fishes live in an essentially isothermal water column (Torres & Somero, 1988*a,b*), none of the decrease in respiration rate with increasing MDO can be attributed to direct effects of temperature. For the coastal California fishes too, temperature cannot be the primary cause of the fall in respiration rate with increasing MDO, even though deeper-living species do encounter colder temperatures than shallower-living ones. The almost 25-fold decrease in respiration rate that is associated with a 5 °C decrease in temperature is far too great to be explained by Q_{10} effects (Torres *et al.*, 1979). For the Gulf of Mexico fishes, however, much of the decrease in respiration rate with depth has been attributed to the steep temperature gradient in these waters (Donnelly & Torres, 1988). In this water column, surface temperatures reach 27 °C, while at 1000 m, temperatures are near 5 °C. Because of the very high Q_{10} values found for the Gulf of Mexico fishes (an average of 3.22 for all species studied), this steep decline in temperature may account for most of the fall in respiration rate with increasing MDO. However, the findings that different fishes exhibited different Q_{10} values and that Q_{10} varied substantially with temperature make it difficult to draw firm, general conclusions about the role of temperature in explaining the pattern shown for Gulf of Mexico fishes in Fig. 1.

For at least the Antarctic and coastal California water columns, then, depth-related variables other than temperature must be invoked to explain the decreased respiration rate with depth. Before considering these causal relationships, it is important to stress that, although pelagic fishes and crustaceans (Childress, 1995) may exhibit decreases in respiration rate with depth, this pattern is not found in all taxa. A number of invertebrate taxa, including chaetognaths (Thuesen & Childress, 1993*a*), pelagic worms (Thuesen & Childress, 1993*b*), and jellyfishes (Childress & Thuesen, 1995) failed to show depth-related changes in metabolism. The lack of a decrease in respiration with increasing MDO in these taxa indicates that different types of animals may be influenced differently by depth-related variables, for example, changes in light levels. By distinguishing what is different between taxa that display depth-related decreases in respiration and taxa that do not, one might be able to determine the ultimate causes of the declines in respiration with depth shown in Fig. 1.

Among the potentially important depth-related variables that could influence metabolic rate is the increase in hydrostatic pressure with depth. Because all the data shown in Fig. 1 were obtained at 0.1 MPa (1 atm), they provide no information about possible pressure effects on rates of respiration. Although the effects of hydrostatic pressure on respiration are less well understood than those of temperature, where pressure effects on respiration rate have been measured for pelagic fishes, these effects have been found to be too small to explain depth-related changes in respiration rate (for review, see Childress, 1995). The low respiration rates for deep-sea fishes measured using *in situ* respirometry are further support for the hypothesis that pressure effects cannot account for any significant degree of MDO-related decreases in metabolism (Smith, 1978; Smith & Brown, 1983).

Dissolved oxygen concentrations might seem a possible cause for reduced metabolic rate at depth, at least in water columns that have a highly developed oxygen minimum zone. However, the fact that the decrease in respiration rate with increasing MDO is similar for water columns having (California) or lacking (Antarctica and Gulf of Mexico) strongly developed oxygen minimum zones shows that oxygen concentration is not a determinant of depth-related changes in metabolic rate (Fig. 1; Torres *et al.*, 1979; Donnelly & Torres, 1988; Childress, 1995).

The most persuasive explanation for the decrease in respiration rate with increasing MDO for pelagic fishes and crustaceans emphasises the effects of life in darkness. Childress (1995) has proposed that the principal ultimate cause of reduced respiratory rates in deep-living pelagic species that depend on visual cues for predation (and escape therefrom) is relaxed selection for powerful swimming ability under conditions of darkness (Childress, 1995). Childress (1995) buttresses this hypothesis by contrasting pelagic species that rely on vision, that is, species for which light availability will help to determine activity levels, with those that do not, and with benthic species. Pelagic species that rely on visual detection of predators and prey exhibit a rapid decrease in respiration rate with increasing MDO, from the surface through the bottom of the euphotic zone (approximately 100 m, although this depth will vary with season, rates of primary production, ice cover, and other factors that attenuate light penetration in different water columns). At depths below the euphotic zone, the decrease in respiration rate with increasing MDO is smaller than at shallower depths (Fig. 1). Species that are not visual predators, for example, jellyfish and chaetognaths, do not exhibit the MDO-related decreases in metabolism found for visual predators like pelagic fishes and crustaceans (Childress, 1995; Thuesen & Childress, 1995). For benthic species too, there is no decrease in respiration rate with increasing MDO, other than decreases due to low temperature (Childress & Mickel, 1985; Childress, 1995). Benthic invertebrates exhibit no apparent temperature compensation of metabolism.

If darkness allows reduced expenditure of energy in locomotion by visually orientated pelagic species living at depth, then these species should exhibit substantial MDO-related reductions in physiological systems associated with locomotory activity, compared to species that are not reliant on vision. Furthermore, within visually orientated species living at depth, systems associated with locomotion should be reduced disproportionately relative to systems whose function is not expected to be reduced with increasing MDO. These predictions have been tested by using the activities of metabolic enzymes involved in ATP generation or ATP consumption as indices of physiological activity in different taxa and in different tissues of shallow- and deep-living species (Childress & Somero, 1979; Sullivan & Somero, 1980; Torres & Somero, 1988*a,b*; Gibbs & Somero, 1990a; Thuesen & Childress, 1993*a,b*; Childress & Thuesen, 1995).

For pelagic fishes, the predicted relationship between MDO and enzymatic power in locomotory (skeletal) muscle is observed (Fig. 2*a,b*). The enzyme showing the greatest decrease in activity with increasing MDO is lactate dehydrogenase (LDH), an enzyme whose activity is an index of a muscle's capacity for supporting ATP generation by anaerobic glycolysis during bursts of high-speed swimming (Childress & Somero, 1979; Somero & Childress, 1980). The decrease in LDH activity in locomotory muscle, expressed as units of enzymatic activity per gram of tissue, between highly active, shallow-living species like tunas and very sluggish deep-sea fishes is about two orders of magnitude. Note that all of the data given in Fig. 2*a* were obtained at a common measurement temperature, 10 °C. Thus, at normal body temperatures of the different species, even greater depth-related decreases in LDH activity (differences of over three orders of magnitude) would exist than those shown in Fig. 2*a*. Data were also gathered at a common measurement pressure, 0.1 MPa. Because the maximal velocity of the LDH reaction is relatively insensitive to pressure and is less pressure-sensitive for LDHs of deep-living fishes than for shallow species (Siebenaller & Somero, 1979), the trends found at 0.1 MPa are likely to reflect differences in LDH activity at *in situ* pressures.

These enormous differences among fishes in capacity for anaerobic ATP generation during vigorous swimming show unequivocally that deep-living fishes have greatly reduced locomotory capacities compared to shallow-

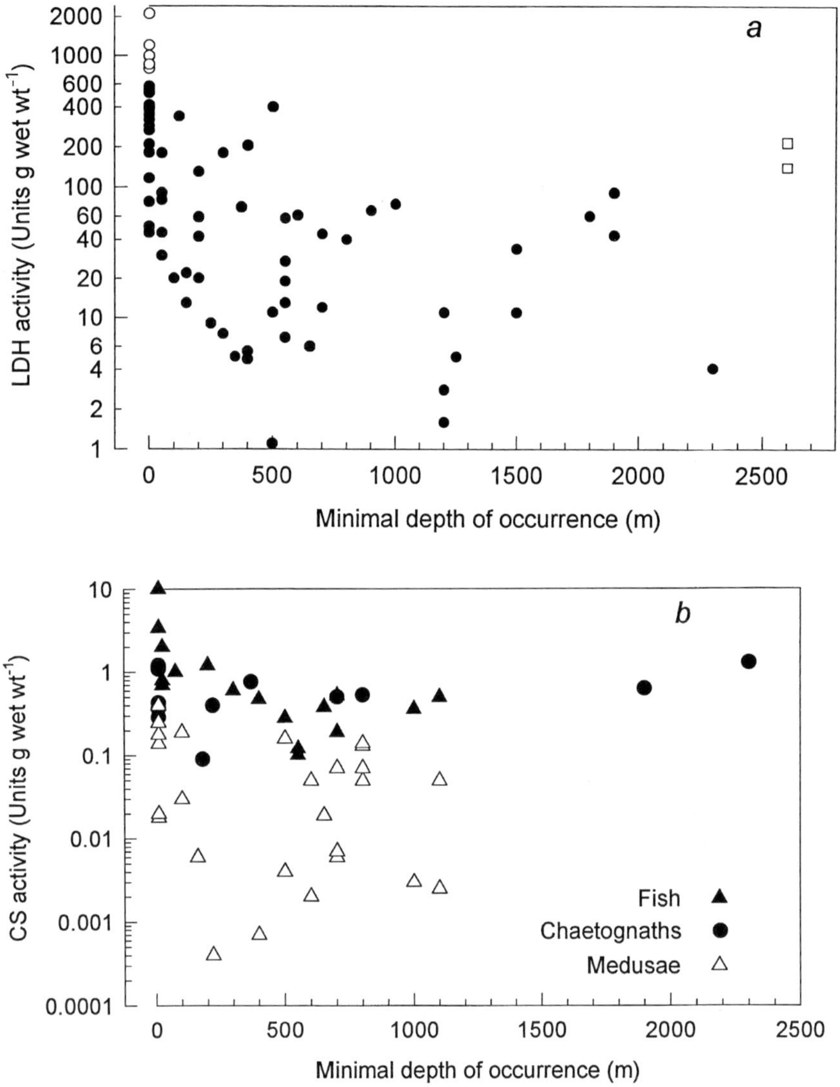

Fig. 2. The relationship between weight-specific enzymatic activity and minimal depth of occurrence for enzymes of several marine taxa. *a*. Lactate dehydrogenase (LDH) activity in locomotory muscle of marine fishes. Each symbol represents a single species, for which one to several specimens were analysed. Open circles represent tunas; open squares represent hydrothermal vent fishes; all other species are represented by filled circles. All measurements were at 10 °C and 0.1 MPa. Data are from Childress and Somero (1979), Sullivan and Somero (1980), Hand and Somero (1983) and

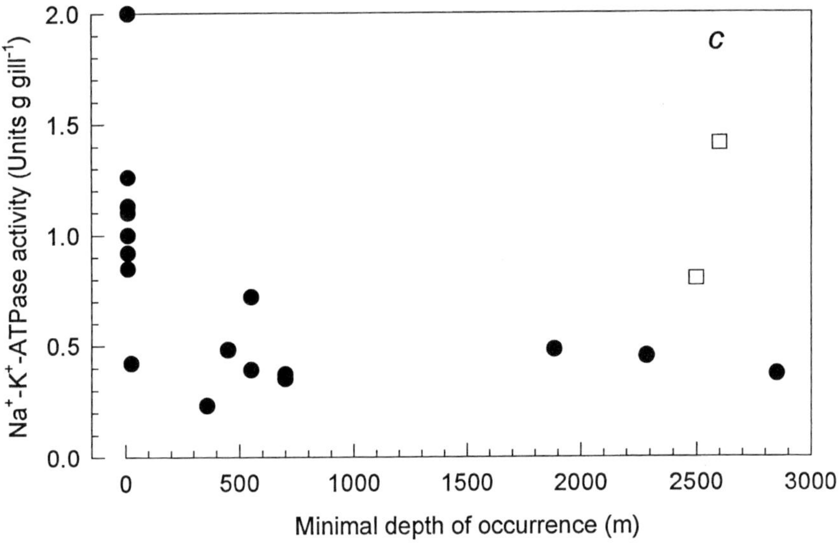

unpublished data of the author. *b*. Citrate synthase (CS) activity in fish muscle, chaetognaths (whole body) and medusae (whole body). Data from Thuesen and Childress (1993*a*, 1995). *c*. Na^+-K^+-ATPase in gills of marine teleosts. Each symbol represents data from one to several specimens of a single species. Open squares represent hydrothermal vent fishes. All other species are represented by filled circles. (Data from Gibbs & Somero, 1990*a*.)

living species. The variations in LDH activity among shallow-living species also are reflections of the different locomotory habits of the species compared. Actively swimming pelagic species have much higher LDH activities than benthic fishes found at similar depths (Sullivan & Somero, 1980).

The data for citrate synthase (CS), an enzyme associated with aerobic production of ATP through the Krebs cycle, also illustrate a fall in ATP-generating potential in locomotory muscle with increasing MDO (Fig. 2*b*), although the decrease is not as great as that found for LDH. This difference between enzymes associated with anaerobic (LDH) and aerobic (CS) ATP generation is consistent with a much larger decrease in capacity for high-speed swimming than for slow, routine locomotion with increasing MDO.

In contrast to the MDO-related patterns found for enzymes of pelagic fishes, CS activities in chaetognaths (Thuesen & Childress, 1993*a*) and medusae (Thuesen & Childress, 1995) fail to show any decrease with depth (Fig. 2*b*), in accord with the depth independence of their metabolic rates. These species are not active visual predators, and the role of darkness in selection for metabolic rate is proposed to be insignificant (Childress, 1995).

Reductions in activity of the ion-transporting enzyme Na^+-K^+-adenosine triphosphatase (Na^+-K^+-ATPase) in gill tissue of teleosts as a function of MDO also are substantial and are fully consistent with the predictions of reduced locomotory activity in deep-living pelagic fishes (Fig. 2*c*; Gibbs & Somero, 1990*a*). Activity of the gill Na^+-K^+-ATPase reflects Na^+ and K^+ flux rates at the gill surface. These flux rates depend, in part, on the amount of blood flowing through the gills and the amount of water flowing over the gills. In fishes with reduced locomotory activity, the needs for oxygen, which will determine the amount of blood flow through the gills, and the exchange of ions with seawater, which will be influenced by the amount of water flowing over the gills as well as by blood flow through the gills, may be greatly reduced, thus permitting decreased levels of ion pumping enzymes like the Na^+-K^+-ATPase. Because the costs of ion exchange and osmotic regulation may account for a substantial fraction of total metabolism (Gibbs & Somero, 1990*a*), reductions in ion flux at the gills associated with reduced locomotory activity may provide a substantial energy saving in deep-sea fishes.

Further support for the hypothesis that reduced locomotory activity is responsible for much of the decrease in respiration rate with increasing MDO also comes from the finding that activities of ATP generating enzymes do not decrease with depth in all tissues. In marine fishes, activities of enzymes in brain do not decrease with depth of occurrence (Sullivan & Somero, 1980). Thus, for an organ whose function is not likely to decrease with increasing MDO, there is no indication of any depth-related fall in metabolic rate. Enzymatic activities in hearts of pelagic fishes also did not show the sharp decrease with MDO noted for locomotory muscle enzymes (Childress & Somero, 1979). Thus, reduction of metabolic rate with increasing MDO is a specific adaptation restricted to tissues that are linked either directly (skeletal muscle) or indirectly (gill ion regulatory systems) to the level of locomotory function.

Further insights into the ultimate and proximate causes of reduced metabolic rates in deep-sea pelagic fishes come from studies of enzymatic activities of hydrothermal vent fishes (Fig. 2*a,c*). Activities of skeletal muscle LDH (Fig. 2*a*) and gill Na^+-K^+-ATPase (Fig. 2*c*) in two species of hydrothermal vent fishes fall into the same range as activities found in shallow-living species (Hand & Somero, 1983; Gibbs & Somero, 1990*a*; Somero, 1992). At the hydrothermal vents, requirements for robust locomotory capacity may be due to several characteristics of this environment that distinguish it from the typical deep sea. The need to escape from abundant and highly active predators, for example, carnivorous brachyuran crabs may select for strong swimming ability. At many vent sites, extremely hot waters rising at high velocity from smoker chimneys may select for an ability to accelerate rapidly to avoid lethal temperatures (Hand & Somero, 1983). Therefore, high capacities for ATP generation in muscle, as indicated by activity of LDH, and the need to support a high rate of ion exchange in gills, as shown by the high levels of Na^+-K^+-ATPase, can be understood. The abundant food supply at the vents may provide sufficient energy to support high rates of protein synthesis and metabolism, rates that might be similar to those found in food-rich surface waters. These data show especially clearly that neither high pressure nor low temperature can account for depth-related decreases in metabolic rate. Rather, when there is a selective advantage in possessing high locomotory capacities and metabolic rates, and when abundant food is present in the environment, deep-sea species may attain metabolic potentials similar to those of shallow-living species. Thus, hydrothermal vent animals are, in many ways, exceptions that prove the rules developed about organisms from the typical deep sea.

Protein adaptation to cold temperatures and high pressures

In addition to physiological and biochemical adaptations that influence rates of energy metabolism ('capacity' adaptations), other types of adaptation help to establish the environmental tolerance ranges of an organism ('tolerance' or 'resistance' adaptations). Because temperature and pressure have such marked effects on protein function and structure (Siebenaller, 1991; Somero, 1993, 1995), homologous proteins of species adapted to different temperature and pressure conditions commonly exhibit adaptive variations that influence the range of temperature or pressure within which protein function can be sustained. By comparing the structural and functional properties of sets of homologous proteins from animals of cold–shallow waters, cold–deep waters, and deep and thermally variable vent waters, it is possible to obtain answers to questions concerning (i) possible pre-adaptation of biochemical systems of cold-adapted, shallow-living species for function at depth, (ii) the interactions between adaptive changes made to pressure and temperature, and (iii) the unique characteristics of proteins of vent organisms which face high pressures and widely varying temperatures.

Actin

Actin plays a broad role in contractile processes in muscle and other cell types. One important capacity of actin, which must be maintained at physiological temperatures and pressures, is the ability to polymerise reversibly. Protein assembly processes like reversible polymerisation of actin generally are extremely sensitive to both temperature and pressure (Swezey & Somero, 1982, 1985). Comparisons of orthologous homologues of actin from vertebrates adapted to a wide range of temperatures and pressures have demonstrated important differences between species in the energy and volume changes that accompany polymerisation. Assembly of actin from a deep-sea fish, *Coryphaenoides armatus*, which occurs to depths of approximately 5 km, was accompanied by a volume change of only 9 cm^3/mol, while assembly of actin of cold-adapted, shallower-occurring fishes, including a congener, *Coryphaenoides acrolepis*, occurred with volume increases of approximately 60 cm^3/mol (Swezey & Somero, 1985). Because the size of the volume changes accompanying a process determines the sensitivity of the process to pressure, the low polymerisation volume of *C. armatus* can be interpreted as pressure adaptive. The polymerised state of actin thus was much more resistant to disruption by elevated pressure for the actin of *C. armatus* than for other actin homologues (Swezey & Somero, 1985).

Enthalpy and entropy changes accompanying actin polymerisation also differed among species. The increases in enthalpy and entropy during polymerisation both rose with adaptation temperature. Polymerisation of actin from a thermophilic desert reptile (*Dipsosaurus dorsalis*) occurred with enthalpy and entropy increases of 14.57 kcal/mole and 80.2 entropy units, respectively, while assembly of actin from two Antarctic fishes, *Pagothenia borchgrevinki* and *Gymnodraco acuticeps*, assembled with enthalpy and entropy increases of only approximately 2 kcal/mole and 33 entropy units, respectively (Swezey & Somero, 1985). Actin of the deepest-living fish examined in this study, *C. armatus*, had the lowest enthalpy and entropy changes of polymerisation, 0.67 kcal/mol and 29 entropy units, respectively. These data indicate that temperature plays a significant role in selecting for the energy changes that accompany actin polymerisation. The lower enthalpy and entropy changes associated with actin polymerisation in cold-adapted species are consistent with reduced reliance on hydrophobic interactions, whose formation is associated with large enthalpy requirements which are paid for by concomitant rises in entropy. The low enthalpy, entropy, and, in particular, volume changes that accompany actin self-assembly in cold-adapted, shallow-living species suggest that, for this protein, adaptation to low temperature may pre-adapt the system for function at high pressures. None the less, even lower values for the enthalpy, entropy and volume changes accompanying actin polymerisation were found for *C. armatus* than for highly cold-adapted Antarctic fishes, suggesting that adaptation to high pressure requires additional changes in this protein above and beyond the adaptations required for effective polymerisation at low temperatures.

Further evidence for the occurrence of specific pressure-adaptive differences in actin structure is provided by the variation in thermal stability of the actin monomers (globular (G) actin) of species adapted to different temperatures and pressures (Swezey & Somero, 1982). Actin, like virtually all proteins so examined (Somero, 1995), exhibits a regular increase in thermal stability as a function of adaptation to high temperatures. Thus, actin from the desert iguana, *D. dorsalis*, has a higher thermal stability than actins from a bird and a mammal. The latter actin homologues are more thermally stable than actins from temperate fishes which, in turn, are more stable than actins from Antarctic fishes. Despite this strong correlation between actin thermal stability and adaptation temperature, actins from three deep-sea fishes were found to have thermal stabilities greater than those of shallow water fishes, birds, and mammals. This discovery suggests that adaptation to high pressure may necessitate increases in protein structural stability. This increased rigidity of structure may offset perturbation of the proteins by pressure but, as discussed below, it may come at the cost of loss of certain functional advantages associated with proteins of cold-adapted, shallow-living ectotherms.

Dehydrogenase enzymes

Three families of dehydrogenases comprise the most thoroughly studied enzymes from species adapted to different temperatures and pressures. The A-isozyme of lactate dehydrogenase (LDH-A) has been compared in several cold-adapted fishes with depths of occurrence up to approximately 5 km (Siebenaller & Somero, 1978, 1979; Hennessey & Siebenaller, 1985, 1987; Yancey & Siebenaller, 1987; Dahlhoff, Schneidemann & Somero, 1990; Siebenaller, 1991). The cytosolic isoforms of malate dehydrogenase (cMDH) have been studied in shallow- and deep-living fishes (Siebenaller, 1984) and in several taxa of invertebrates, including hydrothermal vent species (Dahlhoff & Somero, 1991). Glyceraldehyde-3-phosphate dehydrogenase has been studied in a limited range of fishes (Siebenaller, 1984). Studies of these three sets of homologous proteins have yielded several insights into adaptations of enzymes to temperature and pressure, and into the interacting effects of these two physical variables on protein evolution.

First, for all three sets of dehydrogenases, the homologues of cold-adapted, shallow-living species are not pre-adapted for function at elevated pressures. Perturbation of substrate or cofactor binding by elevated pressure, indexed by measuring the effect of pressure on the Michaelis–Menten constant (K_m), was significantly greater for homologues of cold-adapted, shallow-living fishes and invertebrates than for homologues from deep-sea species, including hydrothermal vent species (Siebenaller & Somero, 1978, 1979; Siebenaller, 1984; Dahlhoff & Somero, 1991). Figure 3 shows comparisons of cMDHs from several species of deep- and shallow-living marine invertebrates, including clams, mussels, polychaete worms, and a vestimentiferan tube worm (Dahlhoff & Somero, 1991). Differences between shallow- and deep-living species were observed across all taxa. These results, like those

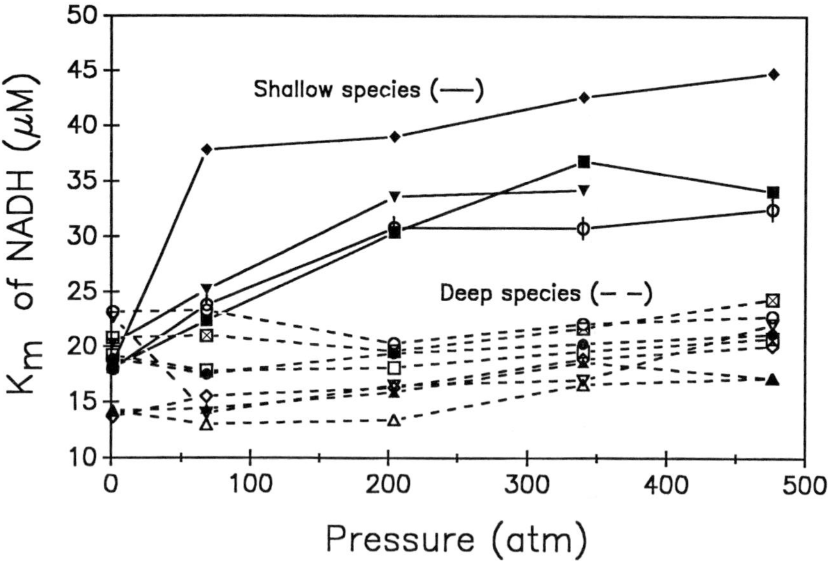

Fig. 3. The effect of pressure on the Michaelis–Menten constant (K_m) of NADH for cytosolic malate dehydrogenases (cMDHs) of shallow- and deep-sea invertebrates. Shallow species (symbols connected by solid lines): *Mytilus galloprovincialis* (filled square), *Calyptogena elongata* (clam; filled inverted triangle), *Cancer antennarius* (crab; filled diamond), *Chaetopterus variopedatus* (polychaete; circle with slash). Deep-sea (symbols connected by dashed lines): *Calyptogena magnifica* (clam; filled circle), *Alvinella pompejana* (vent polychaete; open triangle), *Alvinella caudata* (vent polychaete; filled triangle), *Bathymodiolus thermophilus* (vent mussel; open square), Florida escarpment mussel (square with x), *Riftia pachyptila* (vent tube worm; open diamond), *Calyptogena phaseoliformes* (clam; open circle), and *Bythograea thermydron* (vent crab; open inverted triangle). Data are from Dahlhoff and Somero (1991). (Figure is modified after Somero, 1993.)

from studies of fishes, indicate that colonisation of the deep sea by species originating in cold, shallow waters entails the development of resistance to pressure by proteins already adapted for function at the low temperatures of the deep sea.

A second finding from the comparative analyses of dehydrogenases is that pressures as low as 5–10 MPa (50–100 atm) appear sufficient to favour selection for increased resistance to pressure in these enzymes (Fig. 3) (Siebenaller & Somero, 1978; Siebenaller, 1984, 1991; Dahlhoff & Somero, 1991). Therefore, species moving downwards in the marine water column may encounter pressure-based barriers to colonisation at depths of only 500–1000 m. However, other classes of proteins may have different pressure thresholds, that is, pressures adequate to favour selection for adaptive change. For actin (Swezey & Somero, 1985) and for the Na^+-K^+-ATPase (Gibbs & Somero, 1989), pressures near 20 MPa appear necessary to perturb the systems sufficiently to favour adaptive change.

A third finding from comparative studies of dehydrogenase homologues is that, in common with actin, adaptation to pressure by these enzymes involves acquiring resistance to denaturation by pressure. Hennessey and Siebenaller (1985, 1987) compared the resistance to denaturation by high pressure of LDH-A homologues from six macrourid fishes with different depths of occurrence. There was a significant increase in resistance to pressure denaturation with rising depth of maximal abundance (Hennessey & Siebenaller, 1985). Another indication of the enhanced structural stabilities of the LDH-As from deep-sea fishes was the lower rate of proteolytic degradation of the LDH-A homologues from deep-living fishes relative to homologues from cold-adapted, shallow-living species (Hennessey & Siebenaller, 1987). These data, like those obtained in study of actin homologues, suggest that proteins of deep-sea species may have enhanced structural stabilities relative to homologous proteins from cold-adapted organisms. However, not all proteins of deep-sea species have high structural stabilities; eye lens proteins of deep-sea fishes had thermal stabilities similar to those of cold-adapted Antarctic fishes (McFall-Ngai & Horwitz, 1990).

A fourth conclusion from studies of enzymes of deep-sea fishes is that the catalytic efficiencies of these enzymes may be lower than one would predict for a cold-adapted enzyme. Catalytic efficiency, as indexed by the catalytic rate constant (k_{cat}), a measure of how rapidly an enzyme can convert substrate to product, is generally highest for homologues from cold-adapted species and lowest for high-body-temperature organisms like mammals and birds (Somero, 1995). The LDH-A homologues of deep-sea fishes were found to have k_{cat} values 30–40% lower than the homologues of cold-adapted shallow-living fishes, including an Antarctic species (Somero & Siebenaller, 1979). The divergence of the LDH-As of deep-sea fishes from the common trend found between adaptation temperature and k_{cat} may be a consequence of the more rigid structure in these high-pressure-adapted proteins, a rigidity that may increase the energy costs of the conformational changes associated with catalysis (Somero & Siebenaller, 1979). It should be emphasised that the lower k_{cat}s found for LDH-As of deep-living fishes cannot explain a significant amount of the decrease in LDH activity with increasing MDO (Fig. 2). The activity of LDH in muscle differs by almost two orders of magnitude between the most active shallow-living species and the most sluggish deep-sea species. The approximately 30–40% difference in k_{cat} between LDH-As of deep-sea species and shallow cold-adapted species thus represents only a very minor contribution to depth-related decreases in enzymatic activity.

A fifth conclusion from comparative studies of dehydrogenases concerns the question about whether the pressure-adapted proteins of deep-sea species are pre-adapted for function at the elevated temperatures found in the vent ecosystem. To address this question, comparisons were made of pressure and temperature effects on K_m values for enzymes from species found in the typical, cold deep sea and from the hydrothermal vents. For LDH-A homologues of the deep-sea rattail fish *C. armatus*, which occurs only in cold deep-sea habitats, and the hydrothermal vent fish *Bythites hollisi*, which occurs at the vents but not in the heated vent effluents, cofactor binding was resistant to pressure at 5 °C, but not at higher temperatures (Dahlhoff *et al.*, 1990). In contrast, a hydrothermal vent zoarcid fish, *Thermarces andersoni*, which occurs in the heated waters at the vent sites, has an LDH-A which exhibits stable cofactor binding over a temperature range of 5 to 20 °C and at pressures between 0.1 MPa and approximately 35 MPa.

Similar differences in temperature and pressure effects on the K_m of NADH were found for cMDHs from several invertebrate taxa from cold–shallow, cold–deep, and hydrothermal vent habitats (Dahlhoff & Somero, 1991). The cMDHs of hydrothermal vent species displayed the ability to retain cofactor binding ability at high pressure throughout the range of body temperatures predicted to occur for these species, whereas cMDHs of invertebrates from cold deep-sea habitats exhibited strong thermal perturbation of K_m. Therefore, much as proteins of cold-adapted, shallow-living species are not pre-adapted for function at high pressures, the proteins of species from the typical cold deep sea are not pre-adapted for function at the elevated temperatures found at hydrothermal vents. However, the enzymes of hydrothermal vent animals do appear well adapted for function at low temperatures, a trait that may be critical for survival during life history stages which may be spent outside of the warm vent waters.

In summary, comparisons of families of homologous proteins from animals adapted to different ranges of temperature and pressure show that both of these environmental variables have marked effects on protein evolution. Adaptive differences in protein structural stability and in the temperature and pressure ranges over which critically important kinetic parameters like K_m values are retained within physiologically appropriate ranges may be of major importance in establishing environmental tolerance ranges.

Membrane adaptations to temperature and pressure

The effects of temperature and pressure on membrane-localised systems that include both protein and lipid components are generally strong, and can lead to significant effects on a variety of physiological processes, including transport and signal transduction (Macdonald, 1984). The sensitivities of membrane lipoprotein systems to temperature and pressure derive not only from the types of protein perturbations discussed earlier, but also from the sensitivities of the physical states of lipid bilayers to temperature and pressure. Moreover, the effects of temperature and pressure on membrane-localised processes can be strongly synergistic. Decreases in temperature and increases in pressure both lead to decreases in membrane 'fluidity' (increases in membrane static order) (Hazel & Williams, 1990). For a wide variety of membrane lipid-based processes, an increase in pressure of 100 MPa is equivalent to a temperature decrease of 15–20 °C (Macdonald, 1984). Thus, in the deepest regions of the ocean, the effective temperature for membrane-based systems is approximately −13 to −18 °C. One would predict that, for membrane-based systems, there must be extensive adaptation to offset the effects of high pressure as well as low temperature, and that the synergistic effects of increased pressure and reduced temperature would favour membranes in deep-sea organisms that are distinctly different from those of shallow-living polar organisms. In the terminology of the membrane biochemist, one would predict extensive shifts in membrane lipid composition to effect 'homeoviscous adaptation' of membrane structure and function, that is, the retention of the appropriate membrane fluidity or static order at ambient temperature and pressure (Hazel & Williams, 1990; Hazel, 1995).

This prediction has been verified through study of membranes from fishes adapted to different depths. Cossins and Macdonald (1984, 1986) examined the lipid composition and fluidity of membrane fractions prepared from fishes collected over a wide range of depths. In membrane fractions enriched in mitochondrial membranes, they found a regular increase with depth in the weight percent of unsaturated fatty acids in ethanolamine-containing phospholipids, a compositional change consistent with homeoviscous adaptation.

Further insights into membrane adaptations to pressure have come from studies of membrane-localised enzymes. For the Na^+-K^+-ATPase of teleost gills, a lipoprotein enzyme, the inhibition of maximal velocity by pressure was much less for the enzymes of cold-adapted, deep-sea species than for shallow-living, cold-adapted species, including Antarctic fishes (Fig. 4; Gibbs & Somero, 1989). As in the case of dehydrogenases, adaptation to low temperature did not fully pre-adapt the Na^+-K^+-ATPases for function at elevated pressures. Correlated with the reduced sensitivity to pressure of the Na-K^+-ATPases of deep-living fishes was a higher membrane fluidity, as measured using fluorescence polarisation (Gibbs & Somero, 1990*b*). Another difference noted between the Na^+-K^+-ATPases of shallow-living, cold-adapted fishes and that of the deepest-living species studied, the rattail *C. armatus*, was that the effect of varying pressure on membrane fluidity was significantly less for the latter species. It appears that adaptation to great

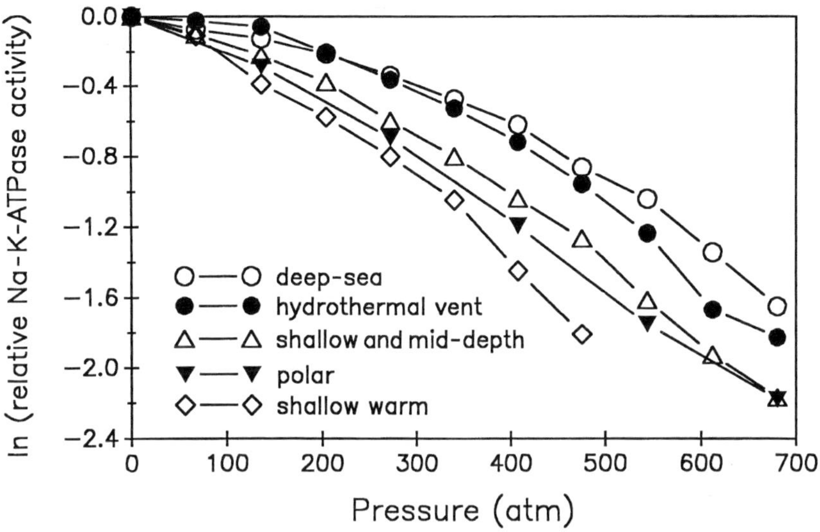

Fig. 4. The effect of pressure on the maximal velocity of the Na^+-K^+-ATPase from gills of fishes from different depths. Data are plotted as natural logarithm of ratio of Na^+-K^+-ATPase activity at higher pressures relative to 1 atm value. For species studied see Gibbs and Somero (1989). (Figure modified after Gibbs & Somero, 1989.)

depth may involve the incorporation into membranes of lipids which not only maintain appropriate fluidity at high pressure and low temperature, but which have fluidities that are resistant to changes in pressure. Both traits would be advantageous to species that not only live under high pressure, but which may encounter substantial changes in pressure during vertical excursions in the water column.

In view of sensitivities of both proteins and lipids to pressure and temperature, it is appropriate to ask whether the adaptations that facilitate function at high pressures and low temperatures reside strictly in the lipid moiety of a lipoprotein or involve as well changes in the protein component. This question was addressed by conducting 'mix and match' experiments in which the lipids surrounding the Na^+-K^+-ATPase were removed and replaced with lipids from a Na-K^+-ATPase from a differently adapted species (Gibbs & Somero, 1990*b*). These experiments showed that both the lipid and protein components of the enzyme were adapted to high pressure: the addition of lipids from the enzyme of *C. armatus* to the Na^+-K^+-ATPase proteins of shallow-living fishes did not make these proteins as pressure-resistant as the enzyme of the deep-sea rattail. Conversely, lipids isolated from the Na^+-K^+-ATPase of shallow-living fishes did not make the protein isolated from the enzyme of *C. armatus* as pressure-sensitive as the enzymes from the shallow-living species. These results suggest that neither the protein nor the lipid moiety of the Na^+-K^+-ATPases of shallow-living, cold-adapted species is pre-adapted for function under conditions typical of the deep sea.

Studies of a membrane-localised process, mitochondrial respiration, in animals from the hydrothermal vents have provided evidence for a substantial degree of temperature adaptation in these species (Dahlhoff *et al.*, 1991). The temperatures at which mitochondrial respiration exhibited thermal inactivation, as indexed by changes in the sign of the slope of Arrhenius plots (Arrhenius break temperatures), were significantly higher for vent invertebrates found in the warmest sites (two Alvinellid polychaetes, a crab, and the tube worm *Riftia pachyptila*) than for species found at cold sites peripheral to the warmest vent regions (the clam *Calyptogena magnifica* and the mussel *Bathymodiolus thermophilus*). The interspecific differences in Arrhenius break temperatures suggest that under conditions of high pressure, which would favour the generation of membrane bilayers with high inherent fluidity, there can be superimposed a large component of homeoviscous adaptation related to the ambient temperatures experienced by the vent fauna. Thus, temperature and pressure can independently govern the composition of membranes. It is not known whether hydrothermal vent species modify the fluidity of their membranes in response to changes in temperature, as might result from alterations in vent water flow or from transitions between life stages that occur in different thermal regimes. One might expect that the ability to acclimatise to temperature would be substantially greater in hydrothermal vent species than in typical deep-sea animals found only in stable thermal regimes. The latter species are likely to share with cold-adapted polar animals a high degree of stenothermy and limited ability to acclimatize to warm temperatures.

Conclusions

Comparisons of animals from cold and shallow polar waters, the typical cold deep sea, and the thermally variable deep-sea hydrothermal vents have yielded a substantial number of insights into the types of mechanisms that help to establish tolerance of different temperature and pressure regimes, and that set metabolic rates at levels appropriate for the physical and biological characteristics of a given environment. In examining these mechanisms in a variety of marine taxa, this review has emphasised two primary points, both having to do with the complexity of adaptational processes, and of their investigation. One point is that adaptations of biochemical systems to temperature and pressure occur by separate yet interacting mechanisms. Adaptation to low temperatures does not appear to pre-adapt animals for life at depth. At pressures as low as 50–100 MPa, biochemical systems of cold-adapted ectotherms may be sufficiently perturbed by rising pressure to necessitate adaptive change. The structural adaptations that facilitate function under high pressure may offset some of the effects of adaptation to low temperature, as seen for catalytic rate constants of enzyme homologues and structural stabilities of actins and dehydrogenases. The amino acid sequence changes that are involved in adapting proteins for function at high pressure remain to be elucidated. It will be interesting to discover whether the similar patterns noted among diverse taxa in adaptation of protein function are due to common types of changes in amino acid sequence. Likewise, it will be interesting to learn how the proteins of hydrothermal vent species living at high temperatures and high pressures differ structurally from homologues of other species.

A second emphasis of this review has been on the complex set of environmental factors that are ultimate causes of rates of metabolism. Although temperature has often been singled out as a major, if not a primary, cause of metabolic rate (Clarke, 1991), this factor may be of limited significance in many circumstances. Even though there is evidence for some degree of cold adaptation of metabolism in polar fishes, at least in the case of some organs (Somero *et al.*, 1968; Torres & Somero, 1988*a,b*; Crockett & Sidell, 1990; Kawall & Somero, 1996), the extent of this compensation to temperature at the whole organism level is not complete, at least when polar and tropical species are compared (Fig. 1). The absence of a dominant role of temperature in establishing the metabolic rates of ectotherms under some circumstances is perhaps illustrated best by the decrease in respiration rate of Antarctic fishes with increasing MDO. The large decrease in respiration rate through this essentially isothermal water column is thought to be due to a variety of environmental factors, notably depth-related changes in light intensity, that far outweigh temperature as ultimate causes of metabolic rate. Thus, in accord with the conclusions of studies of temperature and pressure effects on isolated proteins and lipoprotein systems, analysis of the factors determining respiration rate shows the need for maintaining a broad focus when examining how specific environmental variables, notably temperature, affect the function and evolution of marine organisms. Animals from the coldest regions of the oceans provide excellent study systems for addressing issues of environmental adaptation, in part because they allow the experimentalist to 'dissect out' the influences of diverse physical and biological variables on evolutionary processes.

Acknowledgement

Portions of this research were supported by National Science Foundation grant IBN 92–06660.

References

Childress, J.J. (1995). Are there physiological and biochemical adaptations of metabolism in deep-sea animals? *Trends in Ecology and Evolution*, **10**, 30–6.

Childress, J.J. & Mickel, T.J. (1985). Metabolic rates of animals from the hydrothermal vents and other deep-sea habitats. *Biological Society of Washington Bulletin*, **6**, 249–60.

Childress, J.J. & Nygaard, M.H. (1973). The chemical composition of midwater fishes as a function of depth of occurrence off Southern California. *Deep-Sea Research,* **20**, 1093–109.

Childress J.J. & Somero, G.N. (1979). Depth related enzymic activities in muscle, brain and heart of deep-living pelagic marine teleosts. *Marine Biology*, **52**, 273–83.

Childress, J.J. & Thuesen, E.V. (1995). Metabolic potentials of deep-sea fishes: A comparative approach. In *Biochemistry and Molecular Biology of Fishes,* Vol. 5, ed. P.W. Hochachka & T. Mommsen, pp. 175–96. London: Elsevier Science BV.

Clarke, A. (1991). What is cold adaptation and how should we measure it? *American Zoologist*, **31**, 81–92.

Cossins, A.R. & Macdonald, A.G. (1984). Homeoviscous theory under pressure: II. The molecular order of membranes from deep-sea fish. *Biochimica Biophysica Acta*, **776**, 144–50.

Cossins, A.R. & Macdonald, A.G. (1986). Homeoviscous adaptation under pressure:III. The fatty acid composition of liver mitochondria phospholipids of deep-sea fish. *Biochimica Biophysica Acta*, **860**, 325–35.

Crockett, E.L. & Sidell, B.D. (1990). Some pathways of energy metabolism are cold adapted in Antarctic fishes. *Physiological Zoology*, **63**, 472–88.

Dahlhoff, E. & Somero, G.N. (1991). Pressure and temperature adaptation of cytosolic malate dehydrogenases of shallow- and deep-living marine invertebrates: evidence for high body temperatures in hydrothermal vent animals. *Journal of Experimental Biology*, **159**, 473–87.

Dahlhoff, E., O'Brien, J., Somero, G.N. & Vetter, R.D. (1991). Temperature effects on mitochondria from hydrothermal vent invertebrates: evidence for adaptation to elevated and variable habitat temperatures. *Physiological Zoology*, **64**, 1490–508.

Dahlhoff, E., Schneidemann, & Somero, G.N. (1990). Pressure-temperature interactions on M_4-lactate dehydrogenases from hydrothermal vent fishes: evidence for adaptation to elevated temperatures by the zoarcid *Thermarces andersoni* but not by the bythitid *Bythites hollisi*. *Biological Bulletin (Woods Hole)*, **179**, 134–9.

Donnelly, J. & Torres, J.J. (1988). Oxygen consumption of midwater fishes and crustaceans from the eastern Gulf of Mexico. *Marine Biology*, **97**, 483–94.

Eastman, J. (1993). *Antarctic Fish Biology*. San Diego: Academic Press.

Fustec, A., Desbruyeres, D. & Juniper, S.K. (1987). Deep-sea hydrothermal vent communities at 13° N on the East Pacific Rise: microdistribution and temporal variations. *Biological Oceanography*, **4**, 121–64.

Gibbs, A. & Somero, G.N. (1989). Pressure adaptation of Na^+/K^+-ATPase in gills of marine teleosts. *Journal of Experimental Biology*, **143**, 475–92.

Gibbs, A. & Somero, G.N. (1990*a*). Na^+-K^+-adenosine triphosphatase activities in gills of marine teleost fishes: changes with depth, size and locomotory activity level. *Marine Biology*, **106**, 315–21.

Gibbs, A. & Somero, G.N. (1990*b*). Pressure adaptation of teleost gill Na^+/K^+-adenosine triphosphatase: role of the lipid and protein moieties. *Journal of Comparative Physiology*, **160B**, 431–9.

Hand, S.C. & Somero, G.N. (1983). Energy metabolism pathways of hydrothermal vent animals: adaptations to a food-rich and sulfide-rich deep-sea environment. *Biological Bulletin (Woods Hole)*, **165**, 167–81.

Hazel, J.R. (1995). Thermal adaptation in biological membranes: is homeoviscous adaptation the explanation? *Annual Review of Physiology*, **57**, 19–42.

Hazel, J.R. & Williams, E.E. (1990). The role of alterations in membrane lipid composition in enabling physiological adaptations of organisms to their physical environment. *Progress in Lipid Research*, **29**, 167–227.

Hennessey, J.P. Jr. & Siebenaller, J.F. (1985). Pressure inactivation of tetrameric lactate dehydrogenase homologues of confamilial deep-living fishes. *Journal of Comparative Physiology*, **155B**, 647–52.

Hennessey, J.P. Jr. & Siebenaller, J.F. (1987). Pressure-adaptive differences in proteolytic inactivation of M_4-lactate dehydrogenase homologues from marine fishes. *Journal of Experimental Zoology*, **241**, 9–15.

Holeton, G.F. (1974). Metabolic cold adaptation of polar fish: fact or artefact. *Physiological Zoology*, **47**, 137–52.

Johnson, K.S., Childress, J.J. & Beehler, C.L. (1988). Short-term temperature variability in the Rose Garden hydrothermal vent field: an unstable deep-sea environment. *Deep-Sea Research*, **35**, 1711–21.

Kawall, H.G. & Somero, G.N. (1996). Temperature compensation of enzymatic activities in brain of Antarctic fishes: Evidence for metabolic cold adaptation. *Antarctic Journal of the United States*, in press.

Macdonald, A.G. (1984). The effect of pressure on the molecular structure and physiological functions of cell membranes. *Philosophical Transactions of the Royal Society of London, Series B*, **304**, 47–68.

McFall-Ngai, M.J. & Horwitz, J. (1990). A comparative study of the thermal stability of the vertebrate eye lens: Antarctic ice fish to the desert iguana. *Experimental Eye Research*, **50**, 703–9.

Siebenaller, J.F. (1984). Pressure-adaptive differences in NAD-dependent dehydrogenases of congeneric marine fishes living at different depths. *Journal of Comparative Physiology*, **154B**, 443–8.

Siebenaller, J.F. (1991). Pressure as an environmental variable: magnitude and mechanisms of perturbation. In *Biochemistry and Molecular Biology of Fishes*, vol. 1, ed. P.W. Hochachka & T. Mommsen, pp. 323–43. London: Elsevier Science Publishers B.V.

Siebenaller, J.F. & Somero, G.N. (1978). Pressure adaptive differences in lactate dehydrogenases of congeneric fishes living at different depths. *Science*, **201**, 255–7.

Siebenaller, J.F. & Somero, G.N. (1979). Pressure-adaptive differences in the binding and catalytic properties of muscle-type (M_4) lactate dehydrogenases of shallow- and deep-living marine fishes. *Journal of Comparative Physiology*, **129B**, 295–300.

Smith, K.L. Jr. (1978). Metabolism of the abyssopelagic rattail *Coryphaenoides armatus* measured *in situ*. *Nature*, **274**, 362–4.

Smith, K.L. Jr. & Brown, N.O. (1983). Oxygen consumption of pelagic juveniles and demersal adults of the deep-sea fish *Sebastolobus altivelis*, measured at depth. *Marine Biology*, **76**, 325–32.

Somero, G.N. (1992). Biochemical ecology of deep-sea animals. *Experientia*, **48**, 537–43.

Somero, G.N. (1993). Adaptation to high hydrostatic pressure. *Annual Review of Physiology*, **54**, 557–77.

Somero, G.N. (1995). Proteins and temperature. *Annual Review of Physiology*, **57**, 43–68.

Somero, G.N. & Childress, J.J. (1980). A violation of the metabolism-size scaling paradigm: activities of glycolytic enzymes in muscle increase in larger size fishes. *Physiological Zoology*, **53**, 322–37.

Somero, G.N. & Siebenaller, J.F. (1979). Inefficient lactate dehydrogenases of deep-sea fishes. *Nature*, **282**, 100–2.

Somero, G.N., Giese, A.C. & Wohlschlag, D.E. (1968). Cold adaptation of the Antarctic fish *Trematomus bernacchii*. *Comparative Biochemistry and Physiology*, **26**, 223–33.

Sullivan, K.M. & Somero, G.N. (1980). Enzyme activities of fish skeletal muscle and brain as influenced by depth of occurrence and habits of feeding and locomotion. *Marine Biology*, **60**, 91–9.

Swezey, R.R. & Somero, G.N. (1982). Polymerization thermodynamics and structural stabilities of skeletal muscle actins from vertebrates adapted to different temperatures and pressures. *Biochemistry*, **21**, 4496–503.

Swezey, R.R. & Somero, G.N. (1985). Pressure effects on actin self-assembly: Interspecific differences in the equilibrium and kinetics of the G to F transformation. *Biochemistry*, **24**, 852–60.

Thuesen, E.V. & Childress, J.J. (1993*a*). Enzymatic activities and metabolic rates of pelagic chaetognaths: lack of depth-related declines. *Limnology and Oceanography*, **38**, 935–48.

Thuesen, E.V. & Childress, J.J. (1993*b*). Metabolic rates, enzyme activities and chemical compositions of some deep-sea pelagic worms, particularly *Nectonemertes mirabilis* (Nerertea; Hoplonemertinea) and *Poeobius meseres* (Annelida; Polychaeta). *Deep-Sea Research I*, **40**, 937–51.

Thuesen, E.V. & Childress, J.J. (1995). Respiratory rates and metabolic enzyme activities of oceanic California medusae in relation to body size and habitat depth. *Biological Bulletin (Woods Hole)*, **187**, 84–98.

Torres, J.J. & Somero, G.N. (1988*a*). Metabolism, enzymic activities and cold adaptation in Antarctic mesopelagic fishes. *Marine Biology*, **98**, 169–80.

Torres, J.J. & Somero, G.N. (1988*b*). Vertical distribution and metabolism in Antarctic mesopelagic fishes. *Comparative Biochemistry and Physiology*, **90B**, 521–8.

Torres J.J., Belman, B.W. & Childress, J.J. (1979). Oxygen consumption rates of midwater fishes as a function of depth of occurrence. *Deep-Sea Research*, **26A**, 185–97.

Yancey, P.H. & Siebenaller, J.F. (1987). Coenzyme binding ability of homologs of M_4-lactate dehydrogenase in temperature adaptation. *Biochimica et Biophysica Acta*, **924**, 483–91.

H. GUDERLEY

Temperature and growth rates as modulators of the metabolic capacities of fish muscle

Much as the external morphology of fish is indicative of their swimming style, the metabolic capacities of their muscle reflect this locomotor style. The overall capacity for ATP generation, in both the glycolytic and aerobic pathways is much higher in the muscle of the fast and continuously swimming tunas than in intermittently active flatfish or the slow-moving cod. Abyssal fish take this pattern to an extreme in maintaining low levels of protein and enzymes in their musculature (Siebenaller & Yancey, 1984). On the other hand, for any given fish, muscle metabolic capacities change with the abiotic and biotic factors, principally temperature, food availability and reproductive activity, to which fish are exposed. In this chapter, the impact of temperature, food availability and growth rates upon muscle metabolic capacities in fish will be examined and potential interactions among these factors addressed.

Temperature is a major factor determining physiological performance of fishes. For most fish, thermal changes in the habitat modify body temperature due to rapid thermal equilibration across the gills. Thermal fluctuations in any living system affect reaction rates as well as the non-covalent interactions which stabilise biochemical structures. In response to either short- or long-term thermal changes, individual organisms often adjust their physiological rates as well as their exact biochemical composition. Therefore, temperature effects on muscle metabolic capacities must be considered both on the level of long-term adjustments which may imply genetic modifications and on the level of short-term phenotypic adjustments. Muscle metabolic capacities may also be modified by indirect thermal effects upon food availability and the ensuing impact upon growth and energetic status.

Is there evidence for thermal compensation of metabolic processes?

When physiological processes in ectotherms living in cold oceans are considered, two major questions arise. How have these organisms adjusted their physiology for life at cold temperatures? Do physiological processes occur at rates which are higher than those which would characterise animals from warmer habitats exposed to such low temperatures? The second aspect is commonly known as thermal compensation. Biologists have long been fascinated by the possibility that evolution could overcome such fundamental constraints as that posed by the Arrhenius principle.

Numerous studies have examined whether basal metabolic rates have undergone thermal compensation during the evolution of organisms which inhabit cold oceans. For any ectothermal species, an increase in temperature within the range of tolerated temperatures generally accelerates basal metabolic rates. Whereas early studies of this question over-estimated the degree of thermal compensation due to experimental problems, more recent studies taking locomotor style, allometry and depth of occurrence into account indicate that significant thermal compensation of metabolic rates has occurred (Johnston, Clarke & Ward, 1991; Somero, 1991). However, this compensation is far from that required to bridge the performance gap brought about by a 25 °C change in temperature.

In cold oceans, thermal compensation of fish swimming performance would presumably be particularly advantageous given the abundance of warm-blooded predators such as marine mammals and birds. For most species of fish, sustained swimming increases with temperature to an optimum and then decreases. Burst swimming seems less affected by temperature (for a review, see Guderley & Blier, 1988), although a growing literature on fast start performance suggests a greater thermal sensitivity than that previously found with swimming tunnels (Beddow, Van Leeuwen & Johnston, 1995; Johnson & Bennett, 1995). The optimal temperature for sustained and burst swimming varies with the thermal habitat and with acclimation status. Antarctic fish swim best around 0 °C. In contrast, maximum aerobic performance of a temperate zone fish such as the largemouth bass, *Micropterus salmoides* occurs at 25 to 30 °C, and the fish become torpid below 7 °C (Beamish, 1970; Lemons & Crawshaw, 1985). Thermal acclimation shifts the thermal optimum of sustained swimming for salmon (*Oncorhynchus nerka*), goldfish and striped bass (*Morone saxatilis*) (Brett, 1967; Fry & Hart, 1948; Sisson & Sidell, 1987). The thermal optimum of fast-start swimming is changed by thermal acclimation of sculpin (*Myoxocephalus scorpius*), goldfish and killifish (*Fundulus heteroclitus*) (Beddow *et al.*, 1995; Johnson & Bennett, 1995), whereas trout (*Oncorhynchus mykiss*) only show minor acclimatory shifts of fast start performance (Johnson, Bennett & McLister, 1996). Thermal acclimation of sculpin leads to near-perfect compensation of power production per contraction cycle in isolated fast muscle fibres stimulated to contract at the sarcomere length at which they contract *in vivo* (Johnston, Davidson & Goldspink, 1995). Thus, the thermal optimum for sustained and burst swimming reflects a fish's habitat both on a seasonal and evolutionary perspective.

Modifications of the contractile and metabolic components of muscle are likely to underlie short- and long-term changes in the thermal optimum of swimming. The considerable literature concerning thermal adaptation of contractile properties is beyond the scope of this chapter, but overall, it is clear that the contractile properties of muscle are adjusted to the species' thermal habitat and may be modified by thermal acclimation. Marked shifts in the thermal sensitivity of myosin ATPase activity are apparent in comparisons of this protein from Antarctic and tropical species. The increased activity in cold-adapted fishes seems to come at a cost, as the thermal stability of the enzyme from Antarctic fishes is markedly lower than that in tropical fishes (Johnston *et al.*, 1975). Thermal acclimation changes the contractile properties of isolated fibres from carp (*Cyprinus carpio*) and sculpin as well as the activities and properties of the myosin ATPase from carp (Johnston, Sidell & Driedzic, 1985; Crockford & Johnston 1990; Gerlach *et al.*, 1990; Hwang *et al.*, 1991; Beddow & Johnston, 1995). In sculpin, the changes in contractile properties seem due to changes in the myosin light chains as myosin ATPase activity remains unaltered (Ball & Johnston, 1996). A relationship between the contractile properties of isolated fibres and the types of myosin light chains present also occurs in the tilapia, *Oreochromis andersoni* (Crockford, Johnston & McAndrew, 1995). None the less, despite the thermally induced changes in contractile proteins and properties of isolated fibres in cyprinids and sculpins, thermal acclimation of trout, chain pickerel, striped bass and killifish leads to no, or only to limited, modifications of contractile properties and myosin ATPase activities (for a review see Guderley & Blier, 1988; also Johnson & Bennett, 1995; Johnson *et al.*, 1996).

Thermal adjustments of muscle metabolism: glycolytic elements

The metabolic elements of muscle are located primarily in the cytoplasm, home to glycolysis, and the mitochondrion, the seat of oxidative phosphorylation and the Krebs cycle. Glycolytic enzymes may be associated with each other or with the contractile filaments, but our current understanding of their response to temperature is largely based on the behaviour of the isolated enzymes. Given the extensive studies of Somero and coworkers, the thermal sensitivity of the kinetics of lactate dehydrogenase (LDH) is better understood than that of any other enzyme. Both the maximal capacities and the affinity for pyruvate have been adjusted during evolution to different thermal habitats. Maximal catalytic capacities are higher in enzymes from organisms living in cold habitats, whereas their thermal stabilities are lower. Whereas substrate affinities generally decrease slightly with temperature, they have been adjusted so that the apparent K_m values at habitat temperatures are highly similar when organisms from different thermal habitats are compared (for review see Somero, 1995). Thus, during specialisation of LDH for different thermal habitats, biochemical adjustments have allowed conservation of rates of pyruvate reduction, of pyruvate affinities and of k_{cat}/K_m ratios at habitat temperatures.

Unfortunately, much less is known about the thermal sensitivity of the enzymes, such as phosphofructokinase (PFK), which are central in glycolytic regulation. As is the case for LDH, histidine protonation is central in the kinetic properties of PFK and modulates the interactions among PFK dimers. As virtually all modulators of PFK activity act by modifying these subunit interactions, they provide the basis for the regulation of PFK activity (Bock & Frieden, 1974; Tellam & Frieden, 1981; Hand & Somero, 1983). Since intracellular pH changes with temperature in a fashion which is fairly similar to the thermal changes in the pK of histidine (Cameron, 1984), the regulatory finesse of PFK may be preserved over a considerable thermal range. None the less, some evidence suggests that the regulatory properties of PFK may change with temperature and with thermal acclimation. For example, exhaustive exercise leads to an enhanced binding of PFK to structural proteins in warm-acclimated but not in cold-acclimated goldfish (Huber & Guderley, 1993). This change seems to be related to a decrease of the exercise-induced inhibition of PFK in the cold-acclimated fish, as it does not reflect different degrees of lactate accumulation after exhaustion in the two acclimation groups. In muscle of many lower vertebrates and invertebrates, pyruvate kinase (PK) assumes a regulatory role. The kinetic properties of muscle PK from krill, *Meganyctiphanes norvegica*, change with thermal acclimation. The PEP affinity is higher, whereas the activation energy is lower, in winter- than in summer-acclimatised krill (R.A.H. Vetter & F. Buchholz, personal communication). Such a change would enhance activity at low temperatures, if PEP concentrations remain stable.

Phosphorylation of glycolytic enzymes is important in the control of metabolic depression, which in some species, occurs in response to exposure to low temperatures. Numerous invertebrates, including mussels and whelks, undergo metabolic depression in response to anoxia. As part of this response, muscle pyruvate kinase (PK) is inhibited by phosphorylation primarily by a reduction in its affinity for phosphoenolpyruvate (PEP) (Plaxton & Storey, 1984). Cold temperatures accentuate the effect of phosphorylation by strengthening the ATP inhibition and decreasing the PEP affinity (Michaelidis & Storey, 1990).

During cold acclimation of fishes, the levels of glycolytic enzymes increase in white muscle of trout (*Oncorhynchus mykiss*) and lake whitefish (*Coregonus clupeaformis*) (Blier & Guderley, 1988; Guderley & Gawlicka, 1992) and decrease in red and white muscle of goldfish and striped bass (Sidell, 1980; Sidell & Moerland, 1989). The specific response seems to depend upon whether thermal acclimation changes the mitochondrial volume fraction. When mitochondrial indicators do not rise (as is the case for the above salmonids), glycolytic enzymes may increase with cold acclimation whereas when the mitochondrial volume fraction increases (as is the case for striped bass and goldfish), the glycolytic enzymes tend to decrease.

Thermal adjustments of muscle metabolism: mitochondrial processes

As the centres of oxidative phosphorylation, mitochondria are crucial sites of ATP provision during sustained swimming. Given the adjustments of the thermal sensitivity of sustained swimming during specialisation for life in cold oceans, it would seem likely that the thermal sensitivity of mitochondria in oxidative fibres was changed. Qualitative adjustments in mitochondrial capacities are suggested by the differences in mitochondrial membrane phospholipid composition of species living at different temperatures (Hazel & Williams, 1990). The muscles of fish living in cold oceans tend to have higher oxidative capacities than their counterparts with similar locomotor styles in temperate waters. Citrate synthase and cytochrome oxidase activities per wet muscle mass are 1.5 to 5 times higher at 1 °C in Antarctic than in temperate zone species (Crockett & Sidell, 1990). The mitochondrial volume fractions in the muscles of Antarctic species are markedly higher than those in temperate zone species, with values reaching 60% in some Antarctic species (Johnston, Camm & White, 1988). The ultrastructural data suggest that, if mitochondrial properties were modified during specialisation for cold temperatures, these changes were not sufficient to overcome the constraints of low temperatures, making increases in mitochondrial abundance necessary.

The increase in muscle aerobic capacity with cold acclimation of many fish species also suggests that some constraints limit mitochondrial activity at low temperatures. This increase has been observed in the form of enhanced mitochondrial volume fractions in red and white muscle or as higher activities of mitochondrial enzymes. The central role of the enhanced aerobic capacity in the overall response to thermal acclimation is particularly well documented for striped bass (Jones & Sidell, 1982; Egginton & Sidell, 1989), goldfish (Sidell, 1980) and carp (Johnston *et al.*, 1985; Rome, Loughna & Goldspink, 1985) where responses on many levels of organisation have been examined. Cold acclimation enhances muscle aerobic capacity for other temperate zone fish (for a review, see Guderley, 1990), indicating that muscle mitochondrial abundance is a prime target during cold acclimation.

The functional limitations on mitochondrial activity at low temperatures could reflect insufficient (i) maximal oxidative capacities, (ii) surface for exchange between the myofibrillar and mitochondrial compartments (Egginton & Sidell, 1989) or (iii) regulatory sensitivity. The first two possibilities are supported by the following studies. ADP-stimulated rates of oxygen uptake at 0 °C are higher in skinned muscle fibre segments from Antarctic than temperate species (Johnston, 1987). However, when the rates of oxygen uptake are normalised for the mitochondrial surface area in these fibres, the rates are more similar among species (Johnston, 1987). Thus, expressing oxygen uptake per unit mitochondria removes the evidence for significant cold-adaptation of mitochondrial oxygen uptake in muscles from Antarctic fish. That the rates of oxygen uptake versus habitat temperature fall upon one line suggests only limited adjustments of the thermal sensitivity of mitochondrial respiration (Johnston, 1987). Mitochondrial cristae density is unchanged by thermal acclimation in crucian carp (*Carassius carassius*) and striped bass (Johnston & Maitland, 1980; Egginton & Sidell, 1989). These results imply that, to overcome the effects of low temperature upon mitochondrial respiration, an increase in mitochondrial number is required. An increase in mitochondrial number would also amplify the change in oxygen uptake for a given change in ADP concentration (Dudley, Tullson & Terjung, 1987).

Have maximal mitochondrial capacities specialised for life in cold oceans?

Interspecific comparisons were used to evaluate the hypothesis that muscle mitochondria from fish living at cold temperatures do not demonstrate thermal compensation of their oxidative capacities. To this end, the thermal sensitivity of maximal rates of oxygen uptake by mitochondria isolated from red muscle of fish inhabiting a variety of thermal habitats were characterised (Blier & Guderley, 1993*a*; Johnston *et al.*, 1994). Two points are critical in such a comparison. First, the substrates used to assess oxidative capacities must be physiologically relevant and lead to maximal rates and secondly the quality of the mitochondrial preparations examined must be similar. In the teleosts, maximal rates of oxygen uptake by muscle mitochondria are stimulated by pyruvate at all temperatures (Moyes, Buck & Hochachka, 1988; Moyes *et al.*, 1992; Johnston *et al.*, 1994; St. Pierre & H. Guderley, unpublished observations). As pyruvate can originate from glycogen or from amino acids, both of which are important fuels for muscle, pyruvate is a physiologically significant substrate. As differences in isolation and assay conditions can modify the measured properties of mitochondria through effects upon the homogeneity, yield and respiratory coupling of the preparation, the same methodology was applied to all the species. The coupling of mitochondrial preparations is important as uncoupled mitochondria may provide erroneously high state 3 rates (state 3 rates refer to the oxygen consumption obtained for mitochondria provided with saturating levels of carbon substrate and stimulated with saturating levels of ADP). High state 4 rates (those obtained once all the ADP has been phosphorylated) may reflect contamination by ATPases (generally myofibrillar in muscle) or damaged mitochondrial membranes. By using the same methodology for the different species, it was aimed to minimise variation in the contamination of the mitochondrial fraction by other subcellular fractions. This is important since oxygen uptake rates are expressed per mg protein in the 'mitochondrial' fraction.

Mitochondria were prepared from the oxidative fibres of teleosts living at the thermal extremes inhabited by fish, including the Lake Magadi tilapia, *Oreochromis alcalcus grahami*, which has one of the highest temperature tolerances of any fish, the tilapias, *Oreochromis niloticus* and *O. andersoni* which live at warm temperatures (20–30 °C), the sculpin *M. scorpius* and the rainbow trout *O. mykiss* which live in temperate waters (1–18 °C) and the Antarctic nototheniid, *Notothenia coriiceps*, which requires antifreeze proteins to avoid freezing (Blier & Guderley, 1993*a*; Johnston *et al.*, 1994). The data obtained can be compared with results obtained for muscle mitochondria from carp, *C. carpio*, tuna, *Katsuwonus pelamis* and mako shark *Isurus oxyrinchus*, since similar isolation and assay methods were used (Moyes *et al.*, 1988, 1992; Ballantyne, Chamberlin & Singer, 1992).

For mitochondria isolated from each species, rates of pyruvate oxidation rise with temperature and tend to increase with habitat temperature (Fig. 1). Extrapolation of the rate–temperature curve for the Lake Magadi tilapia suggests that these mitochondria would stop oxidising pyruvate at temperatures above 0 °C, where the mitochondria from *N. coriiceps* are optimally active. Thus, some thermal compensation of maximal mitochondrial capacities seems to have occurred in the Antarctic species. However, an exponential relationship relates the maximal mitochondrial capacities at habitat temperatures (Fig. 2). The rates for the Lake Magadi tilapia (*O. a. grahami*) at their habitat temperature are markedly higher than those of the Antarctic species at their habitat temperature. Mitochondria from *N. coriiceps* oxidise pyruvate at 15.5 nmole O mg protein^{-1} min^{-1} at -1.2 °C, whereas those from Lake Magadi tilapia show a rate of 275 nmole O mg protein^{-1} min^{-1} at 42 °C. More thermal compensation of maximal mitochondrial capacities seems apparent for species living at intermediate temperatures (12 to 30 °C). For example, at 15 and 30 °C, respectively, rates of 110 nmole O mg protein^{-1}

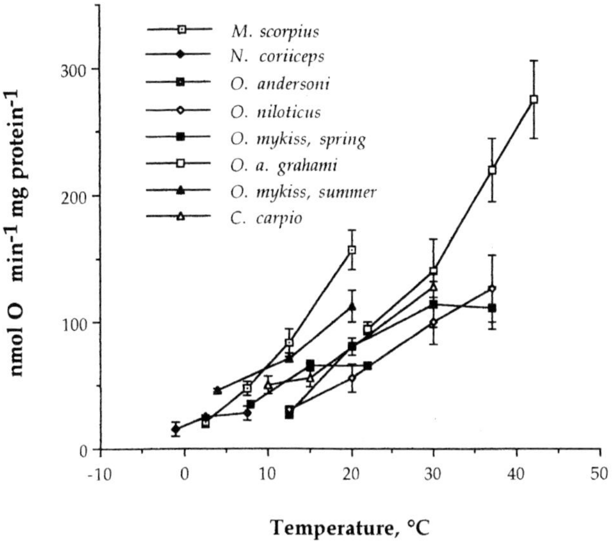

Fig. 1. Maximal rates of pyruvate oxidation by mitochondria isolated from oxidative muscle of teleost fishes living at a variety of habitat temperatures. Rates are expressed as nmol O min^{-1} mg $protein^{-1}$; mean±SE; n=5 to 10 and were obtained in the studies cited in the text.

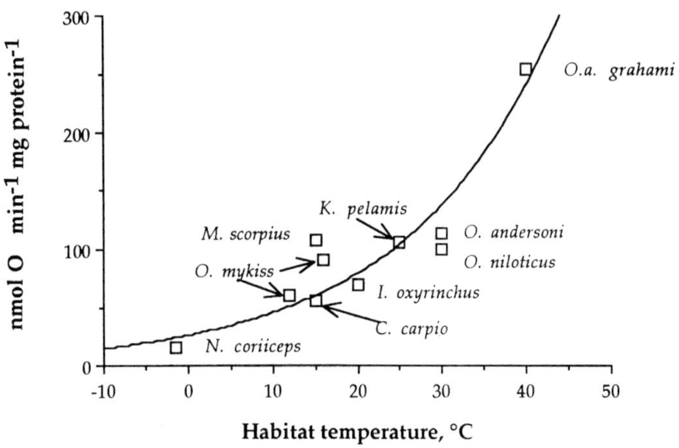

Fig. 2. Rates of pyruvate oxidation at habitat temperatures for mitochondria isolated from the oxidative muscle of fishes living at a variety of temperatures. Rates were obtained from the data provided in the studies cited in the text. The relationship among the points is $y=26.1\times10^{(0.024\,x)}$.

min^{-1} are obtained for mitochondria from *M. scorpius* and *O. andersoni*. The tuna and mako shark values fall near these rates. Whereas mitochondria from the Antarctic nototheniid, functioning at their habitat temperature, cannot attain the rates which characterise the Lake Magadi tilapia at 42 °C, oxygen uptake rates of mitochondria from cold-adapted and temperate species seem capable of a certain degree of thermal compensation.

Possibly the limited thermal compensation of mitochondrial capacity in the Antarctic species reflects constraints associated with enhancing capacity at cold temperatures. In other biochemical systems, increases in capacity can lead to trade-offs such as loss of thermal stability in the case of myosin ATPase (Johnston *et al.*, 1975) and lactate dehydrogenase (Somero, 1995) or contractile failure at higher temperatures in the case of isolated muscle fibres from cold-acclimated sculpin (Beddow & Johnston, 1995). Perhaps enhancement of maximal capacities would have decreased regulatory sensitivity. The high mitochondrial volume fractions present in the muscles of Antarctic fish (Johnston *et al.*, 1988) limit the volume available for other cellular elements, so it seems logical that they present a response to some constraint.

Thermal sensitivity of mitochondrial regulatory properties: general considerations

The thermal sensitivity of mitochondrial regulatory properties is poorly understood. Oxidative phosphorylation requires numerous processes, including nucleotide and substrate transport into the mitochondrion, substrate oxidation and the production of NADH, electron transport and the creation of a proton gradient, and ATP production by the H^+-ATP synthase. Respiratory rates may be limited by each of these processes, but the regulatory roles vary with the rates and conditions under which mitochondria are respiring (Brand & Murphy, 1987). During state 4, proton leak is the major determinant of oxygen uptake (Brand *et al.*, 1994), whereas in mitochondria oxidising succinate in the presence of ADP, control is distributed almost equally among the adenine nucleotide translocator, the dicarboxylate carrier and cytochrome C oxidase (Groen *et al.*, 1982). Thus, mitochondrial affinities for carbon substrates and ADP may be important determinants of respiratory rates during ATP synthesis.

The thermal sensitivity of rates of adenine nucleotide exchange, pyruvate transport and activation by Ca^{2+} have been described for mammalian mitochondria. Adenine nucleotide exchange and pyruvate transport by mammalian mitochondria slow markedly at low temperatures (<15 °C) (Halestrap, 1975; Klingenberg, Grebe & Appel, 1982). The calcium activation of maximal respiration rates by rat liver mitochondria is much weaker at low (25–28 °C) than at high (32–34 °C) temperatures (Moreno-Sanchez &

Torres-Marquez, 1991). From the perspective of thermal specialisation, these results show that thermal dependence of potential regulatory processes increases as temperatures diverge from the normal set-point.

Thermal sensitivity of mitochondrial regulatory properties from ectotherms

Few studies have examined the thermal sensitivity of mitochondrial regulatory properties from non-mammalian species. The oxygen affinity of isolated mitochondria ($K_{m,app}$=0.9 μM) from goldfish red muscle is independent of temperature (Bouwer & Van den Thillart, 1984). This suggests that the mitochondrial affinity for oxygen is less of a determinant of mitochondrial activity than thermally induced changes in oxygen delivery. Thermal influences upon the affinity of mitochondria for their carbon substrates are minimised when pH is allowed to fluctuate with temperature in a physiological fashion. Physiological shifts in pH and PCO_2 with temperature minimise thermally induced shifts in succinate affinity of mitochondria from iguana liver (Yacoe, 1986). The pyruvate affinity of mitochondria isolated from *O. mykiss* red muscle remains approximately 45 μM between 8 and 22 °C, when pH covaries with temperature (Blier & Guderley, 1993*a*). The pyruvate affinity of pyruvate dehydrogenase isolated from *O. mykiss* red muscle shows a similar response, suggesting that the thermal sensitivity of pyruvate dehydrogenase drives the pyruvate affinity of the isolated mitochondria (Blier & Guderley, 1993*a*). This pyruvate affinity lies in the range of pyruvate concentrations in rested muscle (0.02–0.14 mM) from a variety of fish species (Hochachka & Somero, 1984).

The thermal sensitivity of the ADP affinity of isolated mitochondria from ectothermal muscle presents a more complex picture. For mitochondria isolated from *O. mykiss* red muscle, the $K_{m,app}$ for ADP decreases from a value of 11 μM at 8 °C to values of 4 and 6 μM at 15 and 22 °C, even when pH covaries with temperature (Blier & Guderley, 1993*b*). On the other hand, the $K_{m,app}$ for ADP of isolated mitochondria from scallop (*Euvola ziczac*) adductor muscle and sculpin red muscle rises slightly with increases in temperature. At 18 °C, scallop muscle mitochondria have a $K_{m,app}$ of 12 μM, at 25 °C the value is 13 μM and at 32 °C it rises to 19 μM (Guderley, unpublished observations). For the sculpin mitochondria, the $K_{m,app}$ values are 37 μM at 2.5 and 7.5 °C and rise slightly to 42 and 50 μM at 12.5 and 20 °C (Guderley & Johnston, 1996). These values are similar to the ADP $K_{m,app}$ values for oxidative phosphorylation (20 – 60 μM) of isolated mammalian heart mitochondria (Harris & Das, 1991).

To evaluate the functional significance of thermally induced shifts in ADP affinity, one must know the free ADP concentrations to which the mitochondria are exposed. Since much of the ADP in the cell is protein-bound, free ADP concentrations must be estimated by calculations which assume equilibrium of the creatine phosphokinase reaction and use measured values of ATP, creatine phosphate, creatine and intracellular pH. Another approach is to follow NMR spectra and to estimate the metabolite concentration responsible for the signals with independent enzymatic analyses in tissue extracts. Such estimates of cytosolic ADP concentrations in muscle of carp, tilapia and goldfish in a variety of physiological states range from 0.02 to 0.12 $mmol\,l^{-1}$ (Van Waarde *et al.*, 1990). As these concentrations lie above the $K_{m,app}$ values measured for trout mitochondria, it is possible that trout mitochondria are exposed to ADP levels well above $K_{m,app}$ values and that the loss of affinity at low temperatures does not drastically modify rates of ATP production. On the other hand, the $K_{m,app}$ values for the sculpin muscle mitochondria lie in the range of these estimates of free ADP concentrations.

Adenylate concentrations in muscle are such that mitochondria would rarely encounter the low ATP and high ADP levels used to obtain maximal state 3 rates. *In vivo*, the ADP used for mitochondrial ATP synthesis may be produced near the adenylate transporter, such as by the creatine phosphokinase in mammalian heart mitochondria (Gellerich & Saks, 1982). Direct transfer of ADP from CPK to the adenylate transporter would limit competition by high ATP levels upon mitochondrial ADP uptake. However, it is not clear whether mitochondria from highly oxidative tissues consistently have a mechanism for preferentially providing ADP to the adenylate transporter. For example, preferential coupling between mitochondrial arginine phosphokinase and the adenylate translocase is not found in heart mitochondria from the horseshoe crab, *Limulus polyphemus* (Doumen & Ellington, 1990). Mitochondria are frequently associated with other ADP-generating enzymes, such as hexokinase and adenylate kinase, which have been shown to facilitate adenylate traffic to and from isolated mitochondria. The addition of adenylate kinase to isolated rabbit heart mitochondria enhances the transport of ATP equivalents from mitochondria to hexokinase (Dzeja *et al.*, 1985). Adenylate kinase preferentially provides ADP for oxidative phosphorylation by isolated rat liver mitochondria (Gellerich, Khuchua & Kuznetsov, 1993). Substantial levels of adenylate kinase are present in preparations of trout muscle mitochondria (Blier & Guderley, 1993*b*). *In vivo*, these enzymes may facilitate the access of ADP to the adenylate transporter and circumvent competition by the high ATP levels typical of non-fatigued muscle. A complete understanding of the thermal sensitivity of the ADP control of mitochondria will require knowledge of the thermal sensitivity of the physiological ADP delivery mechanism as well as of thermal shifts in the ADP concentrations to which the mitochondria are exposed.

Thermal adjustments of muscle mitochondria during thermal acclimation

After cold acclimation, fish oxidative fibres show an increased mitochondrial volume density, a greater specific activity of mitochondrial enzymes as well as considerable restructuring of membrane lipid composition (Wodtke, 1981*a*; Van den Thillart & de Bruin, 1981). Membrane restructuring most commonly involves changes in the unsaturation of the acyl chains of membrane phospholipids as well as changes in the phospholipid head groups (Hazel & Williams, 1990). These changes maintain membrane fluidity and permeability at the different acclimation temperatures. Thermal acclimation has been shown to shift the transbilayer distribution of phosphatidylethanolamine in mitoplasts prepared from trout liver (Miranda & Hazel, 1996). Despite the thorough understanding of how mitochondrial abundance and membranes change with thermal acclimation, little is known of the functional impact of thermal acclimation on mitochondria.

In goldfish, thermal acclimation shifts the temperature at which the Arrhenius curve for maximal oxygen consumption shows a marked shift in slope (Arrhenius break temperature) (Van den Thillart & Modderkolk, 1978). The mitochondrial respiration rate per g muscle increases with cold acclimation, but the rise in cytochrome C oxidase (CCO) activity is greater than that of mitochondrial respiration (Van den Thillart & Modderkolk, 1978), suggesting that the rise in oxidation rate per g tissue is due mainly to the increased mitochondrial abundance. On the other hand, the rise in CCO activity during cold acclimation of carp has been explained by changes in the lipid composition of the mitochondrial membrane (Wodtke, 1981*b*). By contrast, hepatic and cardiac mitochondria from the sea bass, *Dicentrarchus labrax*, decrease maximal rates of glutamate oxidation at 20 °C with 12 weeks of cold acclimation (Trigari *et al.*, 1992). In the sea bass, no changes in membrane lipid unsaturation accompany cold acclimation.

The maximal capacities of red muscle mitochondria from trout and sculpin change with thermal acclimation and acclimatisation. To examine the first possibility, *M. scorpius* were acclimated for 6 weeks to 5 and 15 °C, whereas for the second, *O. mykiss* were sampled after acclimatisation to summer (16 °C) and fall (12 °C) conditions. The maximal oxidative capacities of sculpin mitochondria increase considerably with cold acclimation (Fig. 3) (Guderley & Johnston, 1996). This shift leads to virtually complete compensation of the rates at acclimation temperatures. Maximal capacities for pyruvate oxidation are considerably higher in fall- than in summer-acclimatised trout muscle mitochondria (Fig. 3) (St. Pierre & H. Guderley, unpublished observations), again leading to similar rates at habitat temperatures. Thus, the maximal oxidative capacities of mitochondria increase with cold

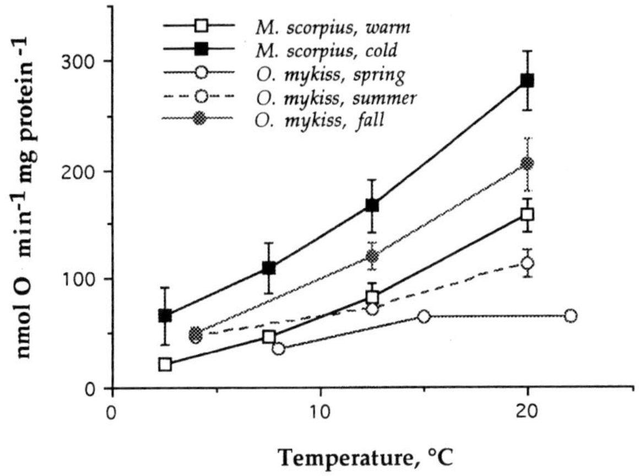

Fig. 3. The effect of thermal acclimation and acclimatisation upon the maximal rates of pyruvate oxidation for mitochondria isolated from the oxidative muscle of the sculpin, *Myoxocephalus scorpius*, and trout, *Oncorhynchus mykiss*. Filled symbols represent cold-acclimated or -acclimatised fish. Rates are expressed as mean±SE; n=6 to 9 and were obtained in the studies cited in the text.

acclimation, both in sculpin and trout. The increases in muscle aerobic capacity frequently observed during cold acclimation therefore seem likely to reflect modifications in both the properties and numbers of mitochondria present.

The maximal rates for the cold-acclimated sculpins and fall-acclimatised trout show a much greater thermal compensation of the pyruvate oxidation rates than was apparent from interspecific comparisons (Fig. 4). At 20 °C, the maximal capacities of mitochondria from cold-acclimated sculpins are similar to those of the mitochondria from the Lake Magadi tilapia at 42 °C. The pyruvate oxidation rates for mitochondria from cold-acclimated sculpins at their acclimation temperature are considerably higher than those of *N. coriiceps* at 2.5 °C.

To ascertain whether or not the regulatory properties of muscle mitochondria change with thermal acclimation, the apparent ADP affinity of muscle mitochondria isolated from cold- and warm-acclimated sculpins was compared. In marked contrast to the thermal independence of the ADP affinity of mitochondria from warm-acclimated sculpins, mitochondria isolated from cold-acclimated sculpins lose their ADP affinity at high temperatures. The ADP $K_{m,app}$ values obtained at high temperatures for these mitochondria were twice as high as those obtained at 2.5 and 7.5 °C (Fig. 5). Cold acclimation leads to a loss of the thermal independence of the ADP affinity of sculpin muscle mitochondria.

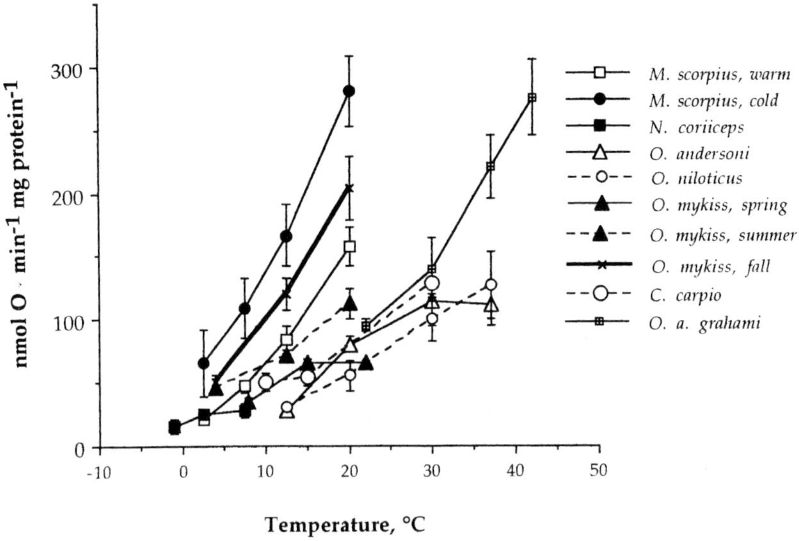

Fig. 4. Maximal rates of pyruvate oxidation by isolated mitochondria from oxidative muscle: comparison of rates from cold-acclimated and -acclimatised fish with those from fish inhabiting various thermal habitats. Cold-acclimated sculpins were acclimated to 5 °C whereas warm-acclimated sculpins were acclimated to 15 °C. *N. coriiceps* were held at 0–1 °C, *Oreochromis andersoni* and *O. niloticus* were held at 28 °C, *Oncorhynchus mykiss* sampled in the spring were at 12 °C, summer *O. mykiss* were at 16 °C whereas as fall *O. mykiss* were at 8 °C, *Cyprinus carpio* were at 15 °C and *Oreochromis alcalicus grahami* were held at 35 °C. Values are shown as mean ± SE; n = 5 to 10 and come from the studies cited in the text.

Evolutionary constraints on mitochondrial plasticity

In considering why specialisation for life at cold temperatures did not exploit the mechanisms which are available during cold acclimation, possible costs of increases in capacity must be considered. For sculpin mitochondria, a loss of mitochondrial ADP affinity at high temperatures seems to be the price of the high catalytic capacity achieved during cold acclimation. For organisms which have specialised for life in cold oceans by establishing a narrow thermal optimum, such a loss of ADP affinity may occur at lower temperatures and prevent their exploitation of the enhanced capacity which obviously can be attained within the basic design of the fish muscle mitochondrion. An alternative interpretation is that the maximal oxidative capacities were not the level for which the system was specialised; that selection did not act upon maximal oxidative capacities. For example, the

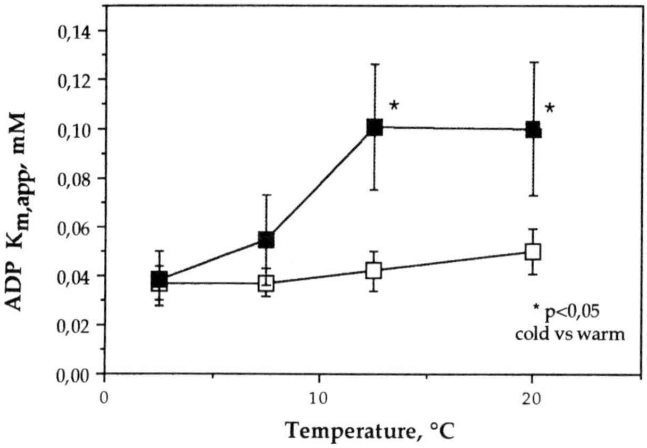

Fig. 5. Effect of thermal acclimation of sculpins upon the apparent ADP affinity of mitochondria isolated from oxidative muscle. The ADP affinities were obtained for mitochondria oxidising pyruvate and are expressed as mean±SE n=6 to 7. Filled symbols are for cold-acclimated sculpins and empty symbols are for warm-acclimated scuplins. (Data are from Guderley & Johnston, 1996.)

mitochondrial surface area available for exchanges between the cytoplasm and the mitochondrial matrix may become limiting at low temperatures, favouring increases in the mitochondrial volume fraction. In this context, shifts in mitochondrial clustering and intracellular lipid content which may maximise rates of mitochondrial ATP generation are apparent in the more active Antarctic fish (Londraville & Sidell, 1990). Evaluation of the availability of substrates, adenylates and oxygen to mitochondria within the muscle of fish living at cold temperatures should help situate the constraints which limit mitochondrial activity.

Thermal effects mediated by food availability and growth rates

Direct effects of temperature upon the metabolic capacities of fish muscle are numerous, ranging from short-term perturbations during acute changes, to medium range adjustments during seasonal acclimatisation and long-term specialisation for specific thermal habitats, be they of constant or fluctuating temperatures. As presented in the preceding section, these adjustments bring modifications in the contractile and metabolic capacities to the extent to which they are feasible. However, indirect thermal effects such as those which occur through shifts in food availability and growth rates cannot be underestimated as means by which temperature modifies tissue and organismal metabolic capacities. Effectively, by changing growth rates of prey organisms, temperature can change food availability, as well as directly modifying rates of assimilation and growth of ectothermal organisms (Guderley, Lavoie & Dubois, 1994). In the natural environment, food tends to be distributed in a heterogeneous manner. When such patchiness becomes pronounced, fish will undergo periods of feeding interspersed with periods of food deprivation. Starvation and refeeding may be the predominant lifestyle of many fishes, particularly in temperate zones, where patchiness in food availability has a strong seasonal component. In the natural habitat of temperate zone fishes, thermal changes are inescapably linked to changes in food availability and thus have both direct and indirect effects on muscle metabolic capacities. In polar environments, food availability will also vary seasonally due to changes in primary production.

Food limitation and muscle metabolic capacities

Food availability has long been recognised as a determinant of muscle metabolic capacities, particularly from the angle of food limitation or starvation. In contrast with mammals, fish and other ectotherms can survive extensive periods without food, particularly in conjunction with overwintering. During such extensive fasts, mobilisation of muscle protein becomes significant and muscle water content rises. In fish, the order of reserve mobilisation varies among species. In eels and salmonids, liver glycogen supplies are maintained until late during starvation (Mommsen, French & Hochachka, 1980; Moon, 1983), whereas in gadoids, liver glycogen depletion starts during the mobilisation of hepatic lipids (Beardall & Johnston, 1985; Black & Love, 1986). Generally, lipids are mobilised before proteins. During starvation, rates of protein synthesis decrease, particularly in the fast glycolytic muscle, whereas the rates of protein degradation rise (Loughna & Goldspink, 1984; Lowery & Somero, 1990). The oxidative fibres undergo less marked changes of biochemical composition and protein synthetic rates reflecting their essential role in continuous swimming (Loughna & Goldspink, 1984). Although the muscle mass will atrophy during starvation, muscle hydratation rises to higher levels during starvation of fish than of terrestrial animals, presumably as high tissue water contents pose no particular disadvantage (i.e. no net mass to carry) to aquatic organisms.

When food deprivation is coupled to migration, as is the case during the reproductive migration of the various species of Pacific salmon, several biochemical transitions facilitate the mobilisation of muscle protein and its transformation into easily oxidised metabolic intermediates. While muscle protein synthesis rates are presumably decreasing, the levels of muscle proteases and transaminases required for the formation of glucogenic amino acids from muscle protein rise (Mommsen *et al.*, 1980). Hepatic capacities for gluconeogenesis also increase, facilitating formation of glucose or glycogen from the amino acids liberated from the muscle. The dramatic changes in migrating salmon are exceptional, given that salmon die at the end of their reproductive migration. More commonly, food limitation is seasonal, leading fish to decrease their metabolic rates and subsist until food becomes available again. Mobilisation of muscle protein follows a similar but more moderate pattern as that found in salmon (Moon & Johnston, 1980). When food limitation coincides with winter temperatures, certain species further suppress metabolic rates and enter into dormancy (Lemons & Crawshaw, 1985) to extend the period during which their reserves will last.

Positive growth and muscle metabolic capacities

While the impact of starvation upon muscle metabolic capacities is well known in fish, the relationship between growth rate and tissue metabolic capacities has only been addressed recently. Growth rate is a more complex variable than food limitation since it expresses a change in status. Increased growth is achieved by higher ration levels, but feeding hierarchies can make individuals in a group fed a given ration achieve rather different levels of feeding and growth (Van Duren & Glass, 1992). Increases in food availability modify metabolic capacities of fish tissues and in particular raise levels of glycolytic enzymes in white muscle (Sullivan & Somero, 1983; Kiessling *et al.*, 1989; Yang & Somero, 1993) and the rates of protein synthesis in many tissues (Houlihan *et al.*, 1988; Von der Decken & Lied, 1992*a*). Refeeding after starvation leads to rapid compensatory growth (Beardall & Johnston, 1985; Black & Love, 1986; Mendez & Wieser, 1993). Growth rate and tissue activities of cytochrome C oxidase (CCO) are positively correlated in largemouth bass, *Micropterus salmoides* (Goolish & Adelman, 1987). In white muscle of the saithe, *Pollachius virens*, the activities of LDH, citrate synthase and CCO are positively correlated with growth rates (Mathers, Houlihan & Cunningham, 1992).

Response of muscle metabolic capacities to changes in growth rate were studied in the Atlantic cod, *Gadus morhua*, to examine whether the aerobic capacity of cod white muscle could be limiting growth rates. This question was stimulated by the observations of positive correlations between the levels of CCO in white muscle and growth rates in largemouth bass and saithe (Goolish & Adelman, 1987; Mathers *et al.*, 1992) and the suggestion that tissue aerobic capacities rise to permit higher growth rates. The fast growth rates of the cod coupled with the low aerobic capacity of white muscle and suggestions that the maximal aerobic capacity of cod is used during recuperation from exhaustive exercise or during the post-prandial increase in metabolic rate (Soofiani & Priede, 1985) seemed to make cod an ideal species in which to demonstrate such a limitation.

Our studies were carried out during four seasons with cod acclimated to three temperatures (4, 10 and 13 °C), fed at a series of ration levels and tagged to allow determination of individual growth rates. Acclimation temperature has little direct impact upon muscle metabolic capacities, but the cod grow more rapidly at 10 °C than at 4 or 13 °C. The activities of glycolytic enzymes in white muscle are strongly correlated with growth rates at all acclimation temperatures and seasons while the activities of mitochondrial enzymes increase little with growth rates (Pelletier, Guderley & Dutil, 1993*a*,*b*). The season of study affected the levels of some mitochondrial enzymes. In subsequent experiments in which the cod were individually housed to eliminate feeding hierarchies and to limit locomotor activity, the positive correlations between growth rate and mitochondrial enzyme activities in white muscle disappeared, while those between glycolytic enzyme activities and growth rate remained (Pelletier *et al.*, 1994). Thus, the levels of glycolytic enzymes in cod white muscle are more consistently correlated with growth rates than are those of mitochondrial enzymes. Therefore, increases in the aerobic capacity of cod muscle are not required for enhanced growth rates.

Our data and those of others do not support the concept that the aerobic capacity of cod muscle may be limiting growth rates. In their examination of the questions raised by Soofiani & Priede (1985), Tang *et al.* (1994) conclusively demonstrated that cod reach their maximal rates of oxygen uptake during sustained swimming and that respiration rates during recovery from exhaustion never exceed those attained during sustained swimming. Their data suggest that the aerobic capacity of white muscle does not limit maximal organismal respiration rates, even though white muscle may well possess the bulk of the mitochondria present in cod. The changes in glycolytic capacity of white muscle with growth do not indicate that the glycolytic capacity of muscle could be limiting growth, rather the increased glycolytic capacity is probably the result of increased protein synthesis in muscle.

Coordinated biochemical responses of fish muscle to growth and starvation

The positive correlation between growth rates and the glycolytic enzyme activities in cod white muscle suggests that faster growing cod would have a higher capacity for burst swimming. For increases in glycolytic enzyme activities to modify locomotor capacities, substrate levels, contractile elements and tissue buffer capacity should vary with the levels of glycolytic enzymes, so as to prevent these factors from limiting burst locomotion. In cod, muscle glycogen levels rise early during refeeding, preceding increases in muscle protein levels (Black & Love, 1986). Faster growth rates should be correlated with a greater muscle buffer capacity as well as higher muscle protein levels, particularly in the sarcoplasmic fraction which contains the glycolytic enzymes. To examine this question, cod was sampled in late spring (0 °C), one group placed at a high ration, the other under starvation (both at 7.8 °C) and at the end of our laboratory feeding period, wild cod were sampled again (fall, 0 °C) to compare their status with that of the laboratory cod. It was found that LDH activity, sarcoplasmic protein concentration and buffer capacity in white muscle respond in concert to feeding ration and to natural changes in food availability and environmental conditions (Fig. 6) (Guderley, Dutil & Pelletier, 1996). These results support the hypothesis that shifts in growth rate lead to coordinated changes in the glycolytic apparatus.

Whereas the levels of myofibrillar and total protein decreased with starvation, the myofibrillar proteins did not follow the seasonal changes in the glycolytic apparatus and the sarcoplasmic proteins, as myofibrillar protein levels were higher in spring than in fall cod (Fig. 6). Since the starved, fed and fall cod were sampled at a similar date, a seasonal cycle may be the source of this difference. Similarly, winter-acclimatised cod from Scotland have somewhat higher total muscle protein concentrations than fall-acclimatised cod (Foster, Hall & Houlihan, 1993*a*). Our data indicate that the levels of sarcoplasmic and myofibrillar proteins are regulated independently in cod white muscle.

While starvation leads to catabolism of muscle protein, muscle atrophy is less than would be expected since muscle water contents rise considerably during starvation. The retention of myofibrillar and other structural proteins may allow the fish to respond rapidly to increases in food availability, by replacing the excess intracellular water with glycogen and sarcoplasmic proteins. The extent of protein mobilisation may vary among the regions of the musculature (Kiessling, Hung & Storebakken, 1993). During the spawning migration of sockeye salmon, *Oncorhynchus nerka*, the soluble protein fraction declines more than the insoluble protein fraction (Mommsen *et al.*, 1980). Androgen stimulation of chum salmon, *Oncorhynchus keta*, leads to mobilisation of sarcoplasmic proteins while leaving myofibrillar proteins unchanged (Ando, Hatano & Zama, 1986). Changes in the feeding frequency of cod modify sarcoplasmic proteins more than myofibrillar proteins, with the myosin heavy chain content changing little with the feeding regime (von der Decken & Lied, 1992*a,b*). In our study, the concentrations of sarcoplasmic proteins differed more than those of the myofibrillar proteins, when the starved, fed, spring and fall cod were compared. The spring cod had lower levels of sarcoplasmic proteins than fall cod (Fig. 6). In general, these studies suggest that food limitation leads to preferential mobilisation of

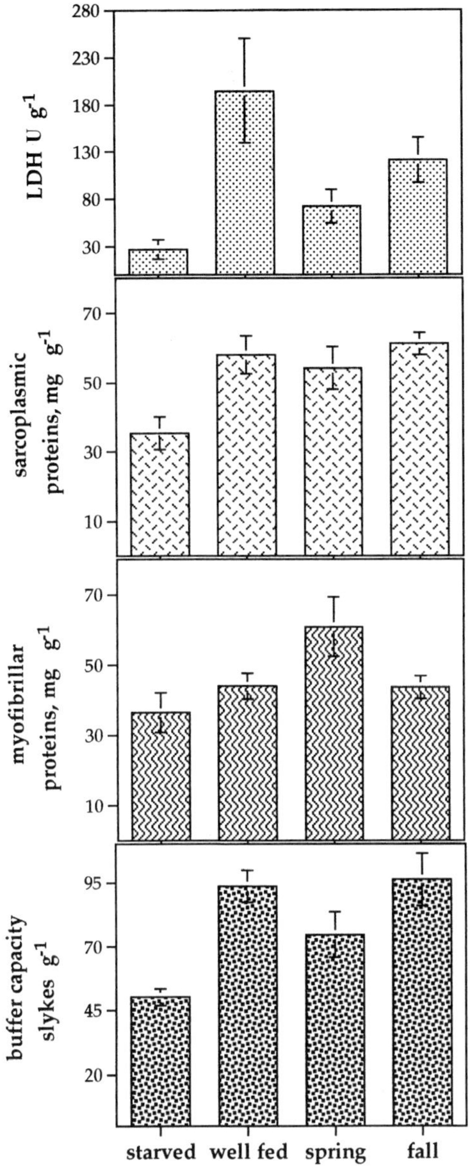

Fig. 6. Changes in the lactate dehydrogenase activity, sarcoplasmic and myofibrillar protein concentrations and buffer capacity of white muscle of cod, *Gadus morhua* submitted to starvation, satiation feeding, or sampled in the spring or fall. The starved and satiation fed cod were maintained at 7.8 °C whereas the wild cod were caught at temperatures near 0 °C. Buffering capacities are the amount of base (NaOH) required to change the pH of an extract of 1 g of muscle by 1 pH unit and are likely to overestimate the actual buffering capacity (Pörtner, 1990). (Data are from Guderley *et al.*, 1996.)

sarcoplasmic proteins, possibly because accumulation of excess water around the structural proteins maintains the overall form of the fish.

Muscle metabolic capacities in fish are markedly modified by food availability and growth rates. The metabolic capacities of muscle will track changes in food availability with a delay which depends upon the processes (all temperature dependent) intervening between food assimilation and muscle growth. During starvation and refeeding, sequential changes occur in biochemical composition. Refeeding leads to rapid increases in muscle and liver glycogen which overshoot control fed levels and precede increases in muscle protein synthesis (Beardall & Johnston, 1985; Black & Love, 1986; Mendez & Wieser, 1993). The intestinal absorptive and aerobic capacity would need to rise before muscle and liver glycogen levels could increase. The rapid turnover of the intestinal epithelium makes it probable that intestinal aerobic capacities reflect capacities of food assimilation and more recent growth rates than muscle glycolytic capacities. Accordingly, during refeeding of cod after starvation, CCO levels in the intestine and stomach rise more rapidly than those in white muscle. On the other hand, increases in RNA levels in muscle, intestine and stomach during refeeding follow a more similar time course (Foster, Houlihan & Hall, 1993*b*). Therefore, different physiological and biochemical parameters are likely to reflect growth rates and food availability on different time scales.

Application of the correlation between individual growth rates and the activity of LDH in white muscle closely predicts the growth rates of cod held in the laboratory for 10 weeks (Guderley *et al.*, 1996). The application of this relation to the spring and fall cod suggests that the former were food restricted whereas the fall cod were growing as fast as the fed cod. During late summer and early fall, saithe captured in the open sea also have positive growth rates according to the levels of several biochemical indicators (Mathers *et al.*, 1992). Enzyme levels in field-caught short-spine thornyhead, *Sebastolobus alascanus*, are more similar to those of fasted than fed laboratory fish (Yang & Somero, 1993). The physiological condition of fish in their natural habitats is of increasing concern, particularly in light of the declines in the abundance of species which were once the object of major fisheries. Poor physiological condition of fish due to a decreased food abundance or lower mean temperatures could have accelerated this stock depletion.

Conclusions

In summary, muscle metabolic components have undergone considerable modifications during specialisation for life at cold temperatures. Individual enzymes show more complete compensation for the effect of low temperatures than mitochondria, particularly on the level of maximal capacities.

Whereas interspecific comparisons suggest that mitochondrial substrate affinities are fairly independent of temperature, maximal oxidative capacities show less evidence of thermal compensation. On the other hand, muscle mitochondria can undergo considerable modification of their catalytic and regulatory properties during thermal acclimation. For sculpin and trout mitochondria, the increases in capacity lead to complete equivalence of the rates at acclimation temperature. However, for sculpin mitochondria this increase in capacity is offset by a loss of ADP affinity at high temperatures. Such trade-offs may have prevented increases in maximal mitochondrial capacity during evolution of life in cold oceans.

Cold temperatures can set the metabolic capacities of muscle directly and indirectly by modifying food availability and growth rates. Muscle metabolic capacities change considerably in response to shifts in food availability: not only does food limitation decrease levels of muscle proteins and metabolic capacities, but also positive growth stimulates protein deposition and increases enzymatic capacities. Whereas the increases in metabolic capacity with growth rate should logically reach a maximum specific to each species, it is intriguing that in our experiments with cod, a 'saturation' of the response of glycolytic enzyme activities to growth rate was never observed (Pelletier *et al.*, 1993*a*, 1994). Experimentally, growth rates are determined as the difference in mass or length of fish between the beginning and the end of an experiment. During this time, the energetic status of the fish also changes. Since protein deposition in muscle rises with energetic condition, muscle metabolic capacities are likely to improve given the increase in energetic condition brought about by growth, rather than growth *per se*. Experiments in which changes in condition are separated from growth rates are required to resolve this question.

In temperate zones, seasonal decreases in temperature tend to decrease productivity and lower food availability. Whereas maintenance requirements also fall, lowered assimiliation efficiencies lead to decreases in growth rate. The critical role of temperature in setting growth rates is illustrated by the suggestion that the high growth rates of the endothermic tuna may largely be due to the increased temperature at which digestion and food assimilation occur and that the impressive metabolic capacities of tuna muscles reflect their high growth rates rather than their locomotor requirements (Brill, 1996). Most fish do not have the vascular anatomy permitting increases in the temperature of the digestive tract and allowing accelerated digestion and food assimilation. Therefore, changes in thermal habitat are the only fashion by which most fish species can shift their temperature to enhance rates of physiological processes. Thus, the thermal constraint remains central in setting the metabolic capacities of fish muscle, both through its direct effects as well as its indirect effects on food availability and growth.

Acknowledgements

The author's research is supported by grants from NSERC Canada and from DFO Canada. I wish to gratefully acknowledge my collaborators, in particular Pierre Blier, Julie St. Pierre, Ian Johnston, Dany Pelletier and Jean-Denis Dutil, for their influence upon my understanding of the subject of this chapter. I also thank Ralf Vetter for allowing me to present his data on pyruvate kinase from krill.

References

Ando, S., Hatano, M. & Zama, K. (1986). Protein degradation and protease activity of chum salmon (*Oncorhynchus keta*) muscle during spawning migration. *Fish Physiology and Biochemistry*, **1**, 17–26.

Ball, D. & Johnston I.A. (1996). Molecular mechanisms underlying the plasticity of muscle contractile properties with temperature acclimation in the marine fish *Myoxocephalus scorpius*. *Journal of Experimental Biology*, **199**, 1363–73.

Ballantyne, J.S., Chamberlin, M.E. & Singer, T.D. (1992). Oxidative metabolism in thermogenic tissues of the swordfish and mako shark. *Journal of Experimental Zoology*, **261**, 110–14.

Beamish, F.W.H. (1970). Oxygen consumption of largemouth bass, *Micropterus salmoides*, in relation to swimming speed and temperature. *Canadian Journal of Zoology*, **48**, 1221–8.

Beardall, C.H. & Johnston, I.A. (1985). The ultrastructure of myotomal muscles of the saithe *Pollachius virens* L. following starvation and refeeding. *European Journal of Cell Biology*, **39**, 105–11.

Beddow, T.A. & Johnston, I.A. (1995). Plasticity of muscle contractile properties following temperature acclimation in a marine fish, *Myoxocephalus scorpius*. *Journal of Experimental Biology*, **198**, 193–201.

Beddow, T.A., Van Leeuwen, J.L. & Johnston, I.A. (1995). Swimming kinematics of fast-starts are altered by temperature acclimation in the marine fish, *Myoxocephalus scorpius*. *Journal of Experimental Biology*, **198**, 203–8.

Black, D. & Love, R.M. (1986). The sequential mobilization and restoration of energy reserves in tissues of Atlantic cod during starvation and refeeding. *Journal of Comparative Physiology*, **156B**, 469–79.

Blier, P.U. & Guderley, H.E. (1988). Metabolic responses to cold acclimation in the swimming musculature of lake whitefish, *Coregonus clupeaformis*. *Journal of Experimental Zoology*, **246**, 244–52.

Blier, P.U. & Guderley, H.E. (1993*a*). Effects of pH and temperature on the kinetics of pyruvate oxidation by muscle mitochondria from rainbow trout (*Oncorhynchus mykiss*). *Physiological Zoology*, **66**, 474–89.

Blier, P.U. & Guderley, H.E. (1993*b*). Mitochondrial activity in rainbow trout red muscle: the effect of temperature on the ADP-dependence of ATP synthesis. *Journal of experimental Biology*, **176**, 145–57.

Bock, P.E. & Frieden, C. (1974). pH induced cold inhibition of rabbit skeletal muscle phosphofructokinase. *Biochemistry*, **13**, 4191–6.

Bouwer, S. & Van Den Thillart, G. (1984). Oxygen affinity of mitochondrial state III respiration of goldfish red muscle: the influence of temperature and O_2 diffusion on K_m values. *Molecular Physiology*, **6**, 291–306.

Brand, M.D. & Murphy, M.P. (1987). Control of electron flux through the respiratory chain in mitochondria and cells. *Biological Reviews*, **62**, 141–93.

Brand, M.D., Chien, L.-F., Ainscow, E.K., Rolfe, D.F.S. & Porter, R.K. (1994). The causes and functions of mitochondrial proton leak. *Biochimica et Biophysica Acta*, **1187**, 132–9.

Brett, J. R. (1967). Swimming performance of sockeye salmon in relation to fatigue time and temperature. *Journal of the Fisheries Research Board of Canada*, **24**, 1731–41.

Brill, R. (1996). Selective advantages conferred by the high performance physiology of tunas, billfishes, and dolphin fish. *Comparative Biochemistry and Physiology*, **113A**, 3–17.

Cameron, J.N. (1984). Acid–base status of fish at different temperatures. *American Journal of Physiology*, **246**, R452–9.

Crockett, E.L. & Sidell, B.D. (1990). Some pathways of energy metabolism are cold adapted in Antarctic fishes. *Physiological Zoology*, **63**, 472–88.

Crockford, T. & Johnston, I.A. (1990). Temperature acclimation and the expression of contractile protein isoforms in the skeletal muscles of the common carp (*Cyprinus carpio* L.). *Journal of Comparative Physiology*, **160B**, 23–30.

Crockford, T., Johnston, I.A. & McAndrew, B.J. (1995). Functionally significant alleiic variation in myosin light chain composition in a tropical cichlid. *Journal of Experimental Biology*, **198**, 2501–8.

Doumen, C. & Ellington, W.R. (1990). Mitochondrial arginine kinase from the heart of the horseshoe crab, *Limulus polyphemus*. II Catalytic properties and studies of potential functional coupling with oxidative phosphorylation. *Journal of Comparative Physiology*, **160B**, 459–68.

Dudley, G.A., Tullson, P.C. & Terjung, R.L. (1987). Influence of mitochondrial content on the sensitivity of respiratory control. *Journal of Biological Chemistry*, **262**, 9109–14.

Dzeja, P., Kalvenas, A., Toleikis, A. & Praskevicius, A. (1985). The effect of adenylate kinase activity on the rate and efficiency of energy transport from mitochondria to hexokinase. *Biochemistry International*, **10**, 259–65.

Egginton, S. & Sidell, B.D. (1989). Thermal acclimation induces adaptive changes in subcellular structure of fish skeletal muscle. *American Journal of Physiology*, **256**, R1–9.

Foster, A.R., Hall, S.J. & Houlihan, D.F. (1993*a*). The effects of seasonal acclimatization on correlates of growth rate in juvenile cod, *Gadus morhua*. *Journal of Fish Biology*, **42**, 461–4.

Foster, A.R., Houlihan, D.F. & Hall, S.J. (1993*b*). Effects of nutritional regime on correlates of growth rate in juvenile Atlantic cod (*Gadus morhua*): comparison of morphological and biochemical measurements. *Canadian Journal of Fisheries and Aquatic Sciences*, **50**, 502–12.

Fry, F.E.J. & Hart, J.S. (1948). Cruising speed of goldfish in relation to water temperature. *Journal of the Fisheries Research Board of Canada*, **7**, 169–75.

Gellerich, F. & Saks, V.A. (1982). Control of heart mitochondrial oxygen consumption by creatine kinase: the importance of enzyme localization. *Biochemical and Biophysical Research Communications*, **105**, 1473–81.

Gellerich, F.N., Khuchua, Z.A. & Kuznetsov, A.V. (1993). Influence of the mitochondrial outer membrane and the binding of creatine kinase to the mitochondrial inner membrane on the compartmentation of adenine nucleotides in the intermembrane space of rat heart mitochondria. *Biochimica et Biophysica Acta*, **1140**, 327–34.

Gerlach, G., Turay, L., Mailik, K., Lida, J., Scutt, A. & Goldspink, G. (1990). The mechanisms of seasonal temperature acclimation in carp: a combined physiological and molecular biology approach. *American Journal of Physiology*, **259**, R237–44.

Goolish, E.A. & Adelman, I.R. (1987). Tissue-specific cytochrome oxidase activity in largemouth bass: the metabolic costs of feeding and growth. *Physiological Zoology*, **60**, 454–64.

Groen, A.K., Wanders, R.J.A., Westerhoff, H.S., Van Der Meer, R., & Tager, J.M. (1982). Quantification of the contribution of various steps to the control of mitochondrial respiration. *Journal of Biological Chemistry*, **257**, 2754–7.

Guderley, H. (1990). Functional significance of metabolic responses to thermal acclimation in fish muscle. *American Journal of Physiology*, **259**, R245–52.

Guderley, H. & Blier, P.U. (1988). Thermal acclimation in fish: conservative and labile properties of swimming muscle. *Canadian Journal of Zoology*, **66**, 1105–15.

Guderley, H. & Gawlicka, A. (1992). Qualitative modification of muscle organization with thermal acclimation of rainbow trout, *Oncorhynchus mykiss*. *Fish Physiology and Biochemistry*, **10**,123–32.

Guderley, H., Lavoie, B.A. & Dubois, N. (1994). The interaction among age, thermal acclimation and growth rate in determining muscle metabolic capacities and tissue masses in the threespine stickleback, *Gasterosteus aculeatus*. *Fish Physiology and Biochemistry*, **13**, 419–31.

Guderley, H.E., Dutil, J-D. & Pelletier, D. (1996). The physiological status of Atlantic cod, *Gadus morhua*, in the wild and the laboratory: estimates of growth rates under field conditions. *Canadian Journal of Fisheries and Aquatic Sciences*, **53**, 550–7.

Guderley, H. & Johnston, I.A. (1996). Plasticity of fish muscle mitochondria with thermal acclimation. *Journal of Experimental Biology*, **199**, 1311–17.

Halestrap, A. P. (1975). The mitochondrial pyruvate carrier. Kinetics and specificity for substrates and inhibitors. *Biochemical Journal*, **148**, 85–96.

Hand, S.C. & Somero G.N. (1983). Phosphofructokinase of the hibernator *Citellus beecheyi*: temperature and pH regulation of activity via influences on the tetramer-dimer equilibrium. *Physiological Zoology*, **56**, 380–8.

Harris, D.A. & Das, A.M. (1991). Control of mitochondrial ATP synthesis in the heart. *Biochemical Journal*, **280**, 561–73.

Hazel, J.R. & Williams, E.E. (1990). The role of alterations in membrane lipid composition in enabling physiological adaptation of organisms to their physical environment. *Progress in Lipid Research*, **29**, 167–227.

Hochachka, P.W. & Somero, G.N. (1984). *Biochemical Adaptation*. Princeton Univ. Press, Princeton, New Jersey, USA.

Houlihan, D.F., Hall, S.J., Gray, C. & Noble, B.S. (1988). Growth rates and protein turnover in Atlantic cod, *Gadus morhua*. *Canadian Journal of Fisheries and Aquatic Sciences*, **45**, 951–64.

Huber, M. & Guderley H. (1993). The effect of thermal acclimation and exercise upon the binding of glycolytic enzymes in muscle of the goldfish *Carassius auratus*. *Journal of Experimental Biology*, **175**, 195–209.

Hwang, G.-C., Ochiai, Y., Watabe, S. & Hashimoto, K. (1991). Changes of carp myosin subfragment-1 induced by temperature acclimation. *Journal of Comparative Physiology*, **157B**, 241–52.

Johnson, T.P. & Bennett, A.F. (1995). The thermal acclimation of burst escape performance in fish: an integrated study of molecular and cellular physiology and organismal performance. *Journal of Experimental Biology*, **198**, 2165–75.

Johnson, T.P., Bennett, A.F. & McLister, J.D. (1996). Thermal dependence and acclimation of fast start locomotion and its physiological basis in rainbow trout (*Oncorhynchus mykiss*). *Physiological Zoology*, **69**, 276–92.

Johnston, I.A. (1987). Respiratory characteristics of muscle fibres in a fish (*Chaenocephalus aceratus*) that lacks haeme pigments. *Journal of Experimental Biology*, **133**, 415–28.

Johnston, I.A., Davidson, W. & Goldspink, G. (1975). Adaptations in magnesium-activated myofibrillar ATPase activity induced by environmental temperature. *FEBS Letters*, **50**, 293–5.

Johnston, I.A. & Maitland, B. (1980). Temperature acclimation in crucian carp (*Carassius carassius* L.): morphometric analysis of muscle fibre ultrastructure. *Journal of Fish Biology*, **17**, 113–25.

Johnston, I.A., Sidell, B.D. & Driedzic, W.R. (1985). Force-velocity characteristics and metabolism of carp muscle fibres following temperature acclimation. *Journal of Experimental Biology*, **119**, 239–49.

Johnston, I.A., Camm, J-P. & White, M. (1988). Specialisations of swimming muscles in the pelagic Antarctic fish *Pleuragramma antarcticum*. *Marine Biology*, **100**, 3–12.

Johnston, I.A., Clarke, A. & Ward, P. (1991). Temperature and metabolic rate in sedentary fish from the Antarctic, North Sea and Indo-West Pacific Ocean. *Marine Biology*, **109**, 191–5.

Johnston, I.A., Guderley, H.E., Franklin, C.E., Crockford, T. & Kamunde, C. (1994). Are mitochondria subject to evolutionary temperature adaptation? *Journal of Experimental Biology*, **195**, 293–306.

Johnston, I.A., van Leeuwen, J.L., Davies, M.L.F. & Beddow, T. (1995). How fish power predation fast-starts. *Journal of Experimental Biology*, **198**, 1851–61.

Jones, P.L. & Sidell, B.D. (1982). Metabolic responses of striped bass (*Morone saxatilis*) to temperature acclimation. II. Alterations in metabolic carbon sources and distributions of fiber types in locomotory muscle. *Journal of Experimental Zoology*, **219**, 163–71.

Kiessling, A., Storebakken,T., Asgard,T., Anderson, I.L. & Kiessling, K.H. (1989). Physiological changes in muscle of rainbow trout fed different ration levels. *Aquaculture*, **79,** 293–301.

Kiessling, A., Hung, S.S.O. & Storebakken, T. (1993). Differences in protein mobilization between ventral and dorsal parts of white epaxial muscle from fed, fasted and refed white sturgeon (*Acipenser transmontanus*) *Journal of Fish Biology*, **43**, 401–8.

Klingenberg, M., Grebe, K. & Appel, M. (1982). Temperature dependence of ADP/ATP translocation in mitochondria. *European Journal of Biochemistry*, **126**, 263–9.

Lemons, D.E. & Crawshaw, L.I. (1985). Behavioural and metabolic adjustments to low temperatures in the largemouth bass (*Micropterus salmoides*). *Physiological Zoology*, **58**, 175–80.

Londraville, R.L. & Sidell, B.D. (1990). Ultrastructure of aerobic muscle in Antarctic fishes may contribute to the maintenance of diffusive fluxes. *Journal of Experimental Biology*, **150**, 205–20.

Loughna, P.T. & Goldspink, G. (1984). The effects of starvation upon protein turnover in red and white myotomal muscle of rainbow trout, *Salmo gairdneri* Richardson. *Journal of Fish Biology*, **25**, 223–30.

Lowery, M.S. & Somero, G.N. (1990). Starvation effects on protein synthesis in red and white muscle of the barred sand bass, *Paralabrax nebulifer*. *Physiological Zoology*, **63**, 630–48.

Mathers, E.M., Houlihan, D.F. & Cunningham, M.J. (1992). Nucleic acid concentrations and enzyme activities as correlates of growth rate of the saithe *Pollachius virens*: growth rate estimates of open-sea fish. *Marine Biology*, **112**, 363–9.

Mendez G. & Wieser, W. (1993). Metabolic responses to food deprivation and refeeding in juveniles of *Rutilus rutilus* (Teleostei: Cyprinidae). *Environmental Biology of Fishes*, **36**, 73–81.

Michaelidis, B. & Storey, K.B. (1990). Interactions of temperature and pH on the regulatory properties of pyruvate kinase from organs of a marine mollusc. *Journal of Experimental Marine Biology and Ecology*, **140**, 187–96.

Miranda E.J. & Hazel, J.R. (1996). Temperature-induced changes in the transbilayer distribution of phosphatidylethanolamine in mitoplasts of rainbow trout (*Oncorhynchus mykiss*) liver. *Journal of Experimental Zoology*, **274**, 23–32.

Mommsen, T.P., French, C.J. & Hochachka, P.W. (1980). Sites and patterns of protein and amino acid utilization during the spawning migration of salmon. *Canadian Journal of Zoology*, **58**, 1785–99.

Moon, T.W. & Johnston, I.A. (1980). Starvation and the activities of glycolytic and gluconeogenic enzymes in skeletal muscles and liver of the plaice, *Pleuronectes platessa*. *Journal of Comparative Physiology*, **136B**, 31–8.

Moon, T.W. (1983). Metabolic reserves and enzyme activities with food deprivation in immature American eels *Anguilla rostrata* (LeSueur). *Canadian Journal of Zoology*, **61**, 802–11.

Moreno-Sanchez, R. & Torres-Marquez, M.E. (1991). Control of oxidative phosphorylation in mitochondria, cells and tissues. *International Journal of Biochemistry*, **23**, 1162–74.

Moyes, C.D., Buck, L.T. & Hochachka, P.W. (1988). Temperature effects on pH of mitochondria isolated from carp red muscle. *American Journal of Physiology* **254**, R611–15.

Moyes, C.D., Mathieu-Costello, O.A., Brill, R.W. & Hochachka, P.W. (1992). Mitochondrial metabolism of cardiac and skeletal muscles from a fast (*Katsuwonus pelamis*) and a slow (*Cyprinus carpio*) fish. *Canadian Journal of Zoology*, **70**, 1246–53.

Pelletier, D., Guderley, H. & Dutil, J-D. (1993*a*). Effects of growth rate, temperature, season and body size on glycolytic enzyme activities in the white muscle of Atlantic cod (*Gadus morhua*). *Journal of Experimental Zoology*, **265**, 477–87.

Pelletier, D., Guderley, H. & Dutil, J-D. (1993*b*). Does the aerobic capacity of fish muscle change with growth rates? *Fish Physiology and Biochemistry* **12**, 83–93.

Pelletier, D, Dutil, J-D., Blier, P. & Guderley, H. (1994). Relation between growth rate and metabolic organization of white muscle, liver and digestive tract in cod, *Gadus morhua*. *Journal of Comparative Physiology*, **164B**, 179–90.

Plaxton, W.C. & Storey, K.B. (1984) Phosphorylation *in vivo* of red muscle pyruvate kinase from the channeled whelk, *Busycotypus canaliculatum*, in response to anoxic stress. *European Journal of Biochemistry*, **143**, 267–72.

Pörtner, H-O. (1990). Determination of intracellular buffer values after metabolic inhibition by fluoride and nitrilotriacetic acid. *Respiration Physiology*, **81**, 275–88.

Rome, L.C., Loughna, P.T. & Goldspink, G. (1985). Temperature acclimation: improved sustained swimming performance in carp at low temperatures. *Science*, **228**, 194–6.

Sidell, B. D. (1980). Response of goldfish (*Carassius auratus* L.) muscle to acclimation temperature: alterations in biochemistry and proportions of different fibre types. *Physiological Zoology*, **53**, 98–107.

Sidell, B.D. & Moerland, T.S. (1989). Effects of temperature on muscular function and locomotory function in teleost fish. *Advances in Comparative and Environmental Physiology*, **5**, 115–56.

Siebenaller, J.F. & Yancey, P.H. (1984). Protein composition of white skeletal muscle from mesopelagic fishes having different water and protein contents. *Marine Biology*, **57**, 181–91.

Sisson III, J.E. & Sidell, B.D. (1987). Effect of thermal acclimation on muscle fiber recruitment of swimming striped bass (*Morone saxatilis*). *Physiological Zoology*, **60**, 310–20.

Somero, G.N. (1991). Biochemical mechanisms of cold adaptation and stenothermality in Antarctic fish. In *Biology of Antarctic Fish*, ed. G. di Prisco, B. Maresca, & B. Tota, pp. 232–47. Berlin: Springer Verlag, .

Somero, G.N. (1995). Proteins and temperature. *Annual Review of Physiology*, **57**, 43–68.

Soofiani, N.M. & Priede, I.G. (1985). Aerobic metabolic scope and swimming performance in juvenile cod, *Gadus morhua*. *Journal of Fish Biology*, **26**, 127–38.

Sullivan, K.M. & Somero, G.N. (1983). Size- and diet-related variation in enzymic activity and tissue composition in the sablefish, *Anoplopoma fimbria*. *Biological Bulletin*, **164**, 315–26.

Tang, Y., Nelson, J.A., Reidy, S.P., Kerr, S.R. & Boutilier, R.G. (1994). A reappraisal of activity metabolism in Atlantic cod (*Gadus morhua*). *Journal of Fish Biology*, **44**, 1–10.

Tellam, R. & Frieden, C. (1981). The purification and properties of frog skeletal muscle phosphofructokinase. *Comparative Biochemistry and Physiology*, **69B**, 517–22.

Trigari, G., Pirini, M., Ventralla, V., Pagliarani, A., Trombetti, F. & Borgatti, A.R. (1992). Lipid composition and mitochondrial respiration in warm- and cold-adapted sea bass. *Lipids*, **27**, 371–7.

Van Den Thillart, G. & Modderkolk, J. (1978). The effect of temperature on state III respiration and on the unsaturation of membrane lipids of goldfish mitochondria. *Biochimica et Biophysica Acta*, **510**, 38–51.

Van den Thillart, G. & De Bruin, G. (1981). Influence of environmental temperature on mitochondrial membranes. *Biochimica et Biophysica Acta*, **640**, 439–47.

Van Duren, L.A. & Glass, C.W. (1992). Choosing where to feed: the influence of competition on feeding behaviour of cod, *Gadus morhua* L. *Journal of Fish Biology*, **41**, 463–71.

Van Waarde, A., Van den Thillart, G., Erkelens C., Addink, A. & Lugtenburg, J. (1990). Functional coupling of glycolysis and phosphocreatine utilization in anoxic fish muscle. *Journal of Biological Chemistry*, **265**, 914–23.

Von der Decken, A. & Lied, E. (1992*a*). Dietary protein levels affect growth and protein metabolism in trunk muscle of cod, *Gadus morhua*. *Journal of Comparative Physiology*, **162B**, 351–7.

Von der Decken, A. & Lied, E. (1992*b*). Feeding frequency and ration size alter the metabolism of white trunk muscle in cod (*Gadus morhua*). *Journal of Animal Physiology and Animal Nutrition*, **67**, 215–24.

Wodtke, E. (1981*a*). Temperature adaptation of biological membranes. The effects of acclimation temperature on the unsaturation of the main neutral and charged phospholipids in mitochondrial membranes of the carp (*Cyprinus carpio*). *Biochimica et Biophysica Acta*, **640**, 698–709.

Wodtke, E. (1981*b*). Temperature adaptation of biological membranes. Compensation of the molar activity of cytochrome c oxidase in the mitochondrial energy-transducing membrane during thermal acclimation of the carp (*Cyprinus carpio* L.). *Biochimica et Biophysica Acta*, **640**, 710–20.

Yacoe, M.E. (1986). Effects of temperature, pH and CO_2 tension on the metabolism of isolated hepatic mitochondria of the desert iguana, *Dipsosaurus dorsalis*. *Physiological Zoology*, **59**, 263–72.

Yang, T-H. & Somero, G.N. (1993). Effects of feeding and food deprivation on oxygen consumption, muscle protein concentration and activities of energy metabolism enzymes in muscle and brain of shallow-living (*Scorpaena guttata*) and deep-living (*Sebastolobus alascanus*) scorpaenid fishes. *Journal of Experimental Biology*, **181**, 213–32.

H.O. PÖRTNER, I. HARDEWIG, F.J. SARTORIS and P.L.M. VAN DIJK

Energetic aspects of cold adaptation: critical temperatures in metabolic, ionic and acid–base regulation?

Temperature is considered to be one of the most important abiotic factors shaping marine ecosystems due to its major impact on all biological processes. Therefore, low or high temperature extremes characterise the limits of geographical distribution of many species, and global change has already caused a change in the distribution of species (Southward, Hawkins & Burrows, 1995). An investigation of marine ectotherms surviving in seasonally or permanently cold ocean environments and their comparison with ectotherms from temperate and warm waters should help to reveal those biochemical or physiological mechanisms which determine geographical distribution limits. These studies should also reveal which molecular, cellular and systemic functions have been shifted to levels compatible with the steady-state maintenance of all life-sustaining processes in the cold. In permanent cold, the latter must also include growth and reproduction, whereas during seasonal cold exposure these processes may be suspended.

Adaptation to cold started from life forms that evolved in warm waters (e.g. Arntz, Brey & Gallardo, 1994; Thiel, Pörtner & Arntz, 1996). Therefore, the characteristics of the adaptational process must be seen in the light of this evolutionary trend. Cold adaptation then becomes a special physiological feature rather than a basic ability of all life forms. If life conquered the cold after having evolved in warm waters, the question arises about what were the limiting factors in this adaptational process and how would these limiting factors affect the whole organism, thereby preventing an easy access to cold ocean environments. Animals from latitudes outside the polar regions are therefore included in our analysis, in order to elaborate the general validity of those adaptational strategies. As a first step, critical thresholds need to be defined beyond which steady-state function is no longer possible. Furthermore, physiological or biochemical characteristics or processes have to be identified which are responsible for limiting survival. In a logical second step, key processes of physiological and biochemical adjustment should be identified which support seasonal and permanent life in the cold, and which allow for a shift in tolerance and distributional limits under different and changing temperature regimes. Certainly, research has not yet provided final answers to these questions and the present study is intended to summarise current knowledge and stimulate further research in this direction.

Critical temperatures

As a first step towards a deeper understanding of the integrated regulatory cascades leading to adaptation in cold ocean environments, the mechanisms and limits of adaptation to cold were investigated in temperate zone animals (Zielinski & Pörtner, 1996; Sommer, Klein & Pörtner, 1997). The marine sipunculid worm, *Sipunculus nudus*, is a suitable invertebrate model for an animal without a circulatory system and, generally, with a simple organisational level at the low end of the animal kingdom. In its natural environment in sandy sediments of the intertidal zone around Brittany, France, this worm is subjected to regular fluctuations of environmental parameters like oxygen and CO_2 levels as well as temperature, and modifications of cellular set points are required to adjust to these fluctuations. Long-term exposure to unfavourable conditions (low oxygen and high CO_2 levels) is tolerated, based on an adaptive drop in metabolic rate (Hardewig *et al.*, 1991; Pörtner, Reipschläger & Heisler, 1997). However, fluctuations of abiotic factors will only be tolerated within certain limits. Limiting temperature thresholds may be reached during seasonal fluctuations. Extremely low temperatures including winter frost are very rare in the worm's natural environment, and if low temperature is a limiting factor these animals should show stress effects during cold exposure.

In a study with cannulated animals dwelling in their natural burrows the correlated changes in ventilatory activity, gas exchange and the mode of energy production were investigated, using anaerobic metabolites as stress indicators since they will indicate insufficiency of aerobic ATP production and transition to a time-limited situation (Zielinski & Pörtner, 1996). Ventilation decreased sharply below 4 °C, and blood gas values as well as tissue metabolite levels indicate that hypoxia developed owing to insufficient oxygen supply. Succinate and volatile fatty acids like acetate and propionate accumulated in the body wall musculature and in the coelomic fluid. These metabolites are formed in the mitochondria, emphasising that insufficient oxygen supply elicits anaerobic metabolism. Obviously, a low critical temperature exists in *S. nudus* (between 4 and 0 °C), that is characterised by a failure of ventilation and the transition to anaerobic metabolism which finally causes death of the animals.

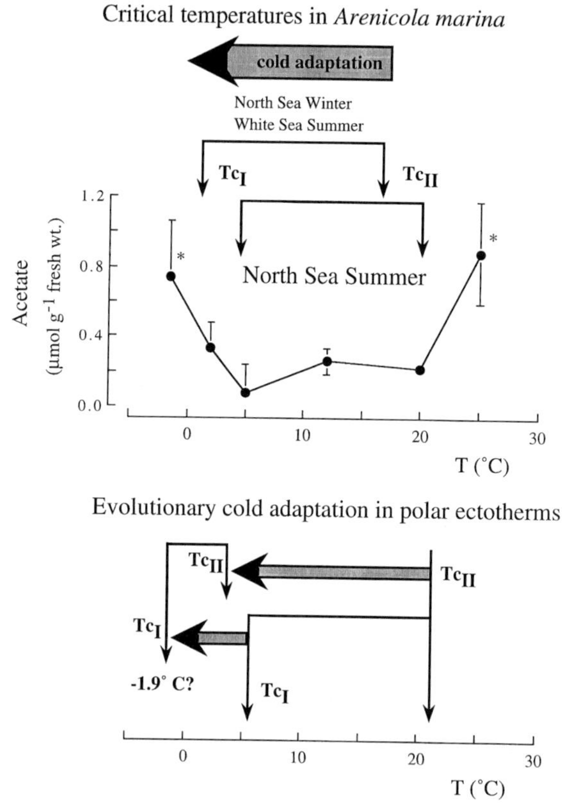

Fig. 1. Low (Tc$_I$) and high (Tc$_{II}$) critical temperatures in the lugworm, *Arenicola marina,* are characterized by the accumulation of anaerobic end products, especially acetate. Both Tc$_I$ and Tc$_{II}$ shift to lower values during adaptation to cold. They are found at lower values in animals collected at higher latitudes, e.g. the Russian White Sea, indicating cold adaptation (based on data by Sommer *et al.*, 1997, *=significantly different from controls, $P<0.05$). In polar ectotherms, Tc$_I$ falls to below freezing associated with a large reduction in the distance between Tc$_I$ and Tc$_{II}$ (see text, supported by findings of H.O. Pörtner, L. Peck, S. Zielinski & L.Z. Conway, unpublished data, in the Antarctic bivalve *Limopsis marionensis*).

Work with *Arenicola marina*, another animal model well studied in hypoxia research, provided a comparison of low and high critical temperatures in populations of the same species in a latitudinal gradient (Sommer *et al.*, 1997, Fig. 1). *A. marina* is found from the Mediterranean to the North Sea, and its distribution ranges via Norway to the Russian White Sea. Although, in contrast to *S. nudus*, this species possesses a well-developed circulatory system, temperature stress also leads to insufficient oxygen supply and a transition to anaerobic metabolism. The comparison of populations from different latitudinal areas revealed that critical thresholds as defined above are set to low and high values depending on the latitudinal and seasonal temperature regime. When the animals experienced temperatures beyond low and high critical thresholds while dwelling in their natural burrows, they accumulated largely acetate, but also propionate in their body fluids. It appears that both low and high critical temperatures are characterised by the transition to anaerobic mitochondrial metabolism. Both temperature thresholds are set to lower values in White Sea than in North Sea animals (Fig. 1). During winter adaptation of North Sea animals, the low critical temperature was also found to decrease, but a value 1–2 °C below ambient was not eliminated when animals collected at 2 °C in January were kept at 2 °C in the laboratory. Consequently, animals leave the intertidal zone and migrate to areas below the low water line to avoid lethal cold exposure (Werner, 1956). Even animals exposed to seawater without sediment showed a transition to anaerobic metabolism in the cold, a phenomenon not evident in *S. nudus.* The comparison suggests that not only ventilatory mechanisms but also the performance of the circulatory system is prone to being disturbed by cold exposure, emphasising the threat of functional hypoxia. Further studies suggest that cold-induced anaerobiosis also develops in crustaceans such as the shore crab *Carcinus maenas* (DeWachter & Pörtner, 1997).

The complete set of mechanisms involved in cold adaptation is still unclear, but these data emphasise that mechanisms are required to eliminate the threat of functional hypoxia. North Sea animals appear unable to adjust to polar conditions and are less able, in general, than the White Sea animals to shift critical temperatures (Sommer, Hummel & Pörtner, 1996). White Sea animals adjust to Arctic water temperatures during winter and it is now known that these animals are genetically distinct from the North Sea population (Sommer *et al.*, 1996). The pattern of metabolite accumulation during temperature-induced anaerobiosis suggests that anaerobic mitochondrial metabolism is more pronounced in White Sea than in North Sea specimens (Sommer *et al.*, 1996, 1997).

Cold adaptation is therefore linked to the necessity to overcome the threat of cold-induced functional hypoxia. This process may contribute to the development of a high sensitivity to high temperatures (cf. Somero, 1991), i.e. a low upper Tc, Tc_{II}. Obviously, the distance between critical temperatures does not remain constant during evolutionary cold adaptation but rather falls when cellular functions in the cold are optimised (Fig. 1). It may very well be that the drop in the distance between the Tcs is obligatory for maintaining all life-sustaining functions in the extreme cold. The molecular mechanisms responsible for setting the Tcs are currently under investigation, especially those shifting the low Tc to below polar ambient temperatures. Adaptational changes would include, among others, a rise in aerobic capacity (combined with mitochondrial proliferation, as found by Egginton & Sidell, 1989 in fish muscle), improvement of muscle function and nervous conductivity by adjustments of ionic exchange mechanisms (see below), and adjustments of the metabolic machinery (for example, enzyme quantities and kinetic properties, see Vetter & Buchholz, this volume; Guderley, this volume). These changes overall can be summarised as 'metabolic cold adaptation' (Thiel *et al.*, 1996). This definition should be preferred over the historical definition, which is restricted to the view that the adaptation to low temperature may be associated with energy expenditures elevated above the decrease expected from the Q_{10} effect. This is a continuing area of discussion (see Clarke, this volume; Somero, this volume).

In this context, we hypothesise that mitochondrial proliferation as required during cold adaptation would inevitably lead to elevated metabolic rates, not just owing to the transient cost of mitochondrial synthesis, but also to the cost of mitochondrial maintenance which comprises the baseline 'idling' of mitochondrial oxygen consumption associated with the maintenance of ionic gradients and the compensation of H^+ leakage across mitochondrial membranes. In consequence, it will be more costly to maintain mitochondria in the same volume of tissue or animal than other cellular elements. In support of this hypothesis, we found consistently higher rates of oxygen consumption in populations of the same species (!), *Arenicola marina* from the sub-Arctic White Sea than from the more temperate North Sea, when the two populations were compared at identical temperatures over a wide temperature range (A. Sommer & H.O. Pörtner, unpublished observations). A comparison of fiddler crab populations (animals reared in the laboratory under the same conditions) from along the North American Atlantic coast yielded similar results (Vernberg & Costlow, cited in Cossins & Bowler, 1987), even suggesting genetic differences to develop in a latitudinal gradient. It would be difficult to draw this conclusion from traditional interspecies comparisons.

These considerations also fit the widely held principle that animals with a higher level of activity or cost of locomotion must exhibit higher standard or resting metabolic rates than more sluggish species in order to attain high rates of metabolism during exercise. Extreme differences in this respect are seen between fish and squid where, owing to the costly mode of swimming in squid, rates of resting metabolism are about ten times higher in squid than in equally active fish in order to allow for the extreme rates of oxygen consumption at high swimming speeds (cf. O'Dor, Pörtner & Shadwick, 1990). Further support of this concept arises from the finding that a seven times higher rate of oxygen consumption in an endotherm (the rat) compared to an ectotherm (a lizard) can be explained by the observation that overall mitochondrial density and mitochondrial leakiness for H^+ is higher in the rat than in the lizard (Brand *et al.*, 1991). More precisely, the level of standard metabolic rate is correlated with the total area of inner mitochondrial membrane and its degree of leakiness for H^+ according to body size, the level of endothermy (Brand, 1990; Brand *et al.*, 1992) and, most likely, the level and scope for activity (see above) of an animal. According to Brand (1990) a doubling of the number of mitochondria (of inner membrane surface area) without a concomitant rise in ATP demand will cause oxygen consumption to rise by 50 to 75%. The cost of maintaining the mitochondria (i.e. of compensating for the H^+ leak) is estimated to comprise up to 45% of the respiration rate of the individual cell or up to 70 % of the increment in respiration rate associated with an elevation of mitochondrial density! These numbers are vague estimates but qualify the price for maintaining a high aerobic capacity as needed for a maximisation of aerobic scope of activity (e.g. in squid) and immediately explain why an organism should strive to minimise the number of mitochondria and maximise 'fuel economy' in accordance with its mode of life and level of activity.

This discussion already suggests that with an obligatory mitochondrial proliferation in the cold, only a modification of inner mitochondrial membrane or a reduction of its surface area (both processes leading to a reduced leakiness for H^+) may offset some (but probably not all) of the metabolic rate increment following mitochondrial proliferation. This may be associated with a drop in mitochondrial oxidative capacity. Accordingly, mitochondria from fish living in permanent cold were modified to exhibit lower capacities of substrate oxidation and rates of oxygen consumption than mitochondria of temperate zone fish acclimated to cold water (Guderley & Johnston, 1996; Guderley, pers. comm.). To compensate for the Q_{10} effect the latter showed marked increases in oxidative capacity at low temperature (Guderley & Johnston, 1996), an adaptational change which was obviously reversed during evolutionary cold adaptation (Johnston *et al.*, 1994, Fig. 2), possibly associated with the adoption of a low activity mode of lifestyle (cf. Clarke, this volume; Thiel *et al.*, 1996).

In conclusion, mitochondrial proliferation leads to a higher cost and may contribute to an increment in whole animal standard metabolic rate in the cold which compensates for (some of) the Q_{10} effect (metabolic cold compensation). This metabolic rate increment is required to shift Tc_I to lower temperatures but will inevitably cause Tc_{II} to fall as well (Fig. 1). It may be balanced by a reduction in other processes like motor activity thus reducing the rise in SMR (Fig. 2). However, there may be further processes associated with a higher cost of maintenance than expected from the Q_{10} effect. One of

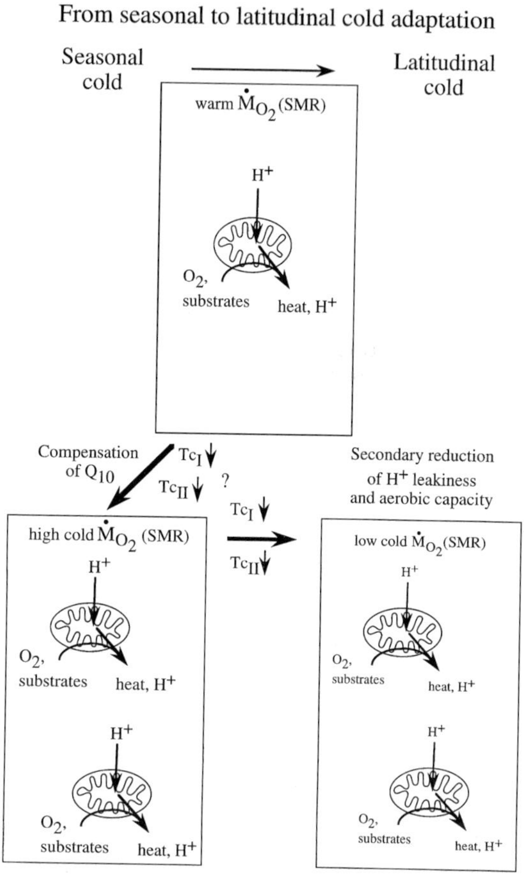

Fig. 2. Modelled depiction of the adjustment of mitochondrial density and of the H^+ leakiness of inner mitochondrial membranes during seasonal or latitudinal cold adaptation. The model assumes that an increase in mitochondrial density and aerobic capacity (possibly also via an increase in the surface area of cristae, Archer & Johnston, 1991) offsets the Q_{10} effect on standard metabolic rate (SMR, measured as the molar rate of oxygen consumption, $\dot{M}o_2$) and aerobic capacity, for example, during seasonal cold (if activity levels are to be maintained and seasonal dormancy does not occur). This process which very likely contributes to a shift of Tc_I and Tc_{II} to lower values (cf. Fig. 1), may partly be compensated during progressive evolutionary (latitudinal) cold adaptation at elevated levels of mitochondrial density when a reduction in aerobic capacity may occur to minimise futile proton cycling and associated energy dissipation. This is postulated to be achieved by a reduced H^+ leakiness of inner mitochondrial membranes (supported by a modification of the membrane or a reduction in surface area) and is interpreted to allow for even further reduction of the Tcs. Note that cold acclimated mitochondria ('seasonal cold') exhibit a higher oxidative capacity if compared to warm acclimated mitochondria at the same temperature (for further discussion and references, see text).

them may be especially important in marine fish, which are hypoosmotic relative to the ambient medium. In contrast to marine invertebrates, fishes maintain ion concentrations lower than ambient in the blood. Since the inward diffusion of ions along the osmotic gradient is only slightly altered by temperature, the metabolic costs for the elimination of surplus ions are maintained at low temperatures, whereas other metabolic processes such as respiration are more temperature-dependent. In the following section, the adaptive flexibility and cost of ion regulation will be dealt with in more detail.

Membrane transport mechanisms

Temperature effects on the cost of ion regulation

The preservation of ion balance despite changes in body temperature is crucial for ectothermal animals in order to maintain vital cellular functions. Ion homeostasis is influenced by active ion pumping, mostly via Na^+/K^+-ATPase, and opposing dissipative ion fluxes. Those dissipative fluxes, referred to as 'leaks', are caused by the diffusion of ions along the electrochemical gradients either non-specifically or mediated by specific pathways, such as cotransport, antiport, or propagation of electrical signals through voltage-gated channels.

Pump and leak processes are differentially affected by temperature changes. While Na^+/K^+-ATPase displays a Q_{10} of 2–4 (Ellory & Hall, 1987; Raynard & Cossins, 1991; Gibbs, 1995), leak processes are relatively temperature insensitive with Q_{10} values close to unity (Ellory & Hall, 1987; Raynard & Cossins, 1991). The disparity in Q_{10} values of the two opposing processes seems to widen at low temperature: generally, Q_{10} values for most physiological processes, including Na^+/K^+-ATPase activity, increase with decreasing temperature, whereas the Q_{10} for passive K^+ flux in red blood cells is smaller at low than at high temperatures (Hall & Willis, 1986). Accordingly, Raynard & Cossins (1991) found almost identical Q_{10} values for pumps and leaks in rainbow trout red cells between 20 and 10 °C. A disparity occurred only between 10 and 0 °C, where the Q_{10} for Na^+/K^+-ATPase increased to 3.3, while the Q_{10} for passive K^+ flux remained low at 1.4. This differential effect of temperature on ion movements may lead to an excess of dissipative fluxes over active ion pumping during cold exposure unless the organism is able to compensate for the difference in temperature sensitivity of these pathways. The necessity of compensatory mechanisms may be especially important during exposure to temperatures close to freezing, where the mismatch of Q_{10} values is particularly distinct.

Much literature is available on the compensatory mechanisms utilised to maintain ion homeostasis during seasonal acclimation to low temperatures

(Stuenkel & Hillegard, 1980; Raynard & Cossins, 1991; Rady, 1993; Ventrella *et al.*, 1993). However, data on the evolutionary adaptation to subzero temperatures in Antarctic species or polar species in general are scarce. According to the foregoing explanations Antarctic species are confronted with large dissipative ion fluxes, comparable to those in temperate species. This is particularly important in Antarctic teleosts, which are hypoosmotic to the ambient sea water, facing large inward fluxes of inorganic ions that have to be counterbalanced by appropriate rates of active ion pumping. Marine invertebrates, on the other hand, are isoosmotic to seawater and ion balance may therefore be less affected by temperature changes. However, gradients between haemolymph and ambient water are maintained for individual ions such as K^+ and Mg^{2+}. In addition, ion gradients (for Na^+, K^+, etc.) over cell membranes have to be maintained in all organisms despite temperature changes.

Generally, two strategies are possible to compensate for the differential effects of temperature on pumps and leaks: either the activity of ion pumps is upregulated during acclimation or adaptation to cold to match the relatively temperature insensitive dissipative fluxes, or mechanisms are employed to reduce those fluxes (Hochachka, 1988). A unifying trend is not yet visible, research has predominantly been carried out in fish, but much less is known about invertebrates.

Upregulation of pump activity in the cold

A compensatory increase in Na^+/K^+-ATPase activity during (seasonal) cold acclimation has been observed in different tissues of several eurythermal teleosts such as Atlantic cod *Gadus morhua*, trout *Oncorhynchus mykiss*, carp *Cyprinus carpio* or roach *Rutilus rutilus* (Raynard & Cossins, 1991; Schwarzbaum, Wieser & Niederstätter, 1991; Schwarzbaum, Niederstätter & Wieser, 1992*a*; Rady, 1993; Staurnes *et al.*, 1994, cf. Fig. 3). The increase in Na^+/K^+-ATPase activity can either be due to an enhanced number of carrier molecules or to an increased catalytic activity of individual transporters. In the fresh water roach *R. rutilus* , the increase of Na^+/K^+-ATPase activity in kidney and hepatocytes was at least partly due to an increase of the number of pumps as determined by ouabain binding studies (Schwarzbaum *et al.*, 1991, 1992*a*). While the binding sites in hepatocytes increased by a factor of 1.9 between fish acclimated to 20 and 5 °C, a 4.4-fold increase was observed in the kidney. Surprisingly, pump density decreased in gills at low acclimation temperatures despite a positive compensation of Na^+/K^+-ATPase activity (Schwarzbaum *et al.*, 1991). In trout erythrocytes, acclimation temperature had no influence on the number of ouabain binding sites, while total Na^+/K^+-ATPase activity was elevated in the cold (Raynard & Cossins,

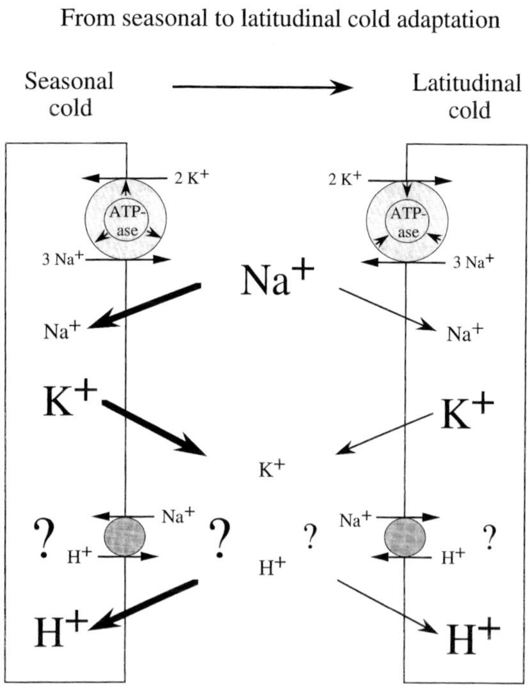

Fig. 3. Modelled depiction of the adjustment of cellular Na^+/K^+-ATPase levels during seasonal or latitudinal cold adaptation. Acclimatisation to seasonal cold is interpreted to reflect the short-term response to cold, whereas latitudinal cold adaptation reflects evolutionary adjustment to cold linked to a secondary reduction of the initial increment of Na^+/K^+-ATPase capacity. Downward arrows represent passive fluxes of the respective ions, the magnitude of dissipative flux being represented by line thickness. Question marks indicate that nothing is known about the potential adjustment of acid–base exchangers (represented by the Na^+/H^+-exchanger) during cold exposure. During seasonal cold the larger question marks indicate that an upregulation of the Na^+/H^+-exchanger is expected owing to large dissipative proton fluxes (see text).

1991). The difference in response between different cell types may correlate with the cellular protein turnover and biosynthetic capacity of the respective tissue. While highly active tissues such as kidney and liver produce more enzymes, tissues with a lower capacity of biosynthesis like red blood cells and gills may increase Na^+/K^+-ATPase activity by increasing the substrate turnover number of existing molecules. Changes in Na^+/K^+-ATPase turnover number may be induced by post-translational modification, such as enzyme phosphorylation, or by non-covalent interactions with allosteric modulators like cardiotenic steroids or membrane lipids (Blaustein & Hamlyn, 1985; Beguin *et al.*, 1994; Gibbs, 1995). The influence of membrane composition on the activity of membrane-bound enzymes has received considerable attention in the past. It has been proposed that desaturation of membrane lipids during cold adaptation, a phenomenon termed homeoviscous response, may lead to enhanced Na^+/K^+-ATPase activity (Cossins, Bowler & Prosser, 1981; Gibbs, 1995). In several cases, reductions of the level of membrane order following cold acclimation have been correlated with higher rates of Na^+/K^+-ATPase activity (Gibbs & Somero, 1989; Raynard & Cossins, 1991; Schwarzbaum, Wieser & Cossins, 1992*b*; Rady, 1993; see Storelli *et al.*, this volume). However, in the Arctic charr *Salvelinus alpinus* no compensation of Na^+/K^+-ATPase activity occurs during cold acclimation despite a large homeoviscous response (Schwarzbaum *et al.*, 1992*b*). Similar observations were made in trout erythrocytes where changes in membrane composition did not correlate with an increase in Na^+/K^+-ATPase activity in animals caught during the winter (Raynard & Cossins, 1991).

Whether or not cold adaptation also leads to the modulation of other ion transport activities is virtually unknown. In the context of the present study consideration of changes in transporters relevant in acid–base regulation is interesting. During seasonal cold an upregulation of the Na^+/H^+-exchanger is expected owing to a rise in dissipative proton fluxes (see below and Fig. 3).

Hochachka (1988) proposed that, in latitudinal cold adaptation, ion pump capacities are increased in polar compared to temperate zone fish to maintain the balance between pumps and leaks in the cold. A partial compensation may occur for Ca^{2+}-ATPase, since the sarcoplasmatic reticulum (SR) prepared from the fast muscle of the Antarctic *Notothenia rossii* accumulated calcium six times faster than SR prepared from a tropical fish at 0 °C. However, assayed at their respective environmental temperatures, calcium pumping rates were about five times faster in the warm acclimated species (McArdle & Johnston, 1980). Somero, Giese and Wohlschlag (1968) showed that gill filaments from an Antarctic species at 0 °C consume O_2 at the same rate as the goldfish gill at 10–15 °C, which may indicate cold adaptation of Na^+/K^+-ATPase activity. However, comparison of specific activities of Na^+/K^+-ATPase between polar, temperate and tropical species leads to a different picture (Fig. 4). The accumulated data suggest that evolutionary cold adaptation causes a reduction of Na^+/K^+-ATPase activity. In their study on pressure adaptation of Na^+/K^+-ATPase, Gibbs & Somero (1989) investigated Na^+/K^+-ATPase activity in the gills of 19 fish species of different geographical origin including several Antarctic teleosts. Unfortunately, the authors did not present the specific activities of Na^+/K^+-ATPase, but they did mention that Na^+/K^+-ATPase activity was barely measurable in Antarctic fish while activities in other investigated species

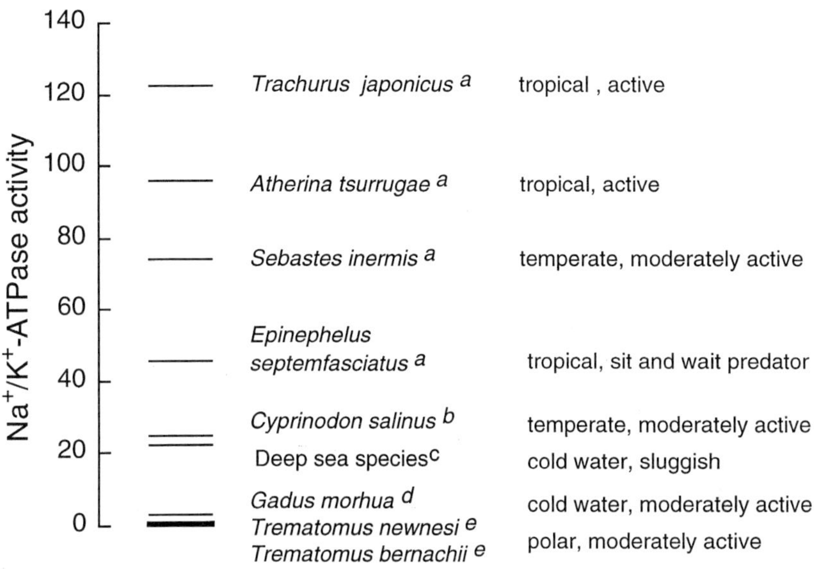

Fig. 4. Semi-quantitative depiction of branchial Na^+/K^+-ATPase activity (μmol P_i mg protein^{-1} h^{-1}) in fish species of different climatic zones and with different lifestyles. The data suggest that evolutionary adaptation to cold goes along with reduced Na^+/K^+-ATPase activities. Enzyme activities were determined at 37 °C or are extrapolated from lower assay temperatures using a Q_{10} of 2. Note that the method of analysis differs between studies and influences the absolute values of Na^+/K^+-ATPase activity. A trend is visible for the activity of Na^+/K^+-ATPase to decline in latitudinal cold. [a] Kamiya & Utida, 1969 (25 °C); [b] Stuenkel & Hillyard, 1980 (37 °C); [c] Gibbs & Somero, 1989 (10 °C); [d] Staurnes *et al.*, 1994 (37 °C); [e] Gonzales-Carbrera *et al.*, 1995 (37 °C).

ranged from 2–10 μmol P_i mg prot^{-1} h^{-1} at 10 °C. These data suggest no positive and possibly even a negative compensation of Na^+/K^+-ATPase activity in the gills of Antarctic teleosts. Accordingly, these animals appear to use a strategy of reduced ion leakage to maintain ion balance in the cold.

Reduction of ion leakage at low temperature

A reduction of ion leakage during cold acclimation has been demonstrated in the Arctic charr *Salvelinus alpinus*. Rb^{2+} efflux in kidney preparations was reduced by a factor of 2.3 at low temperatures (Schwarzbaum *et al.*, 1991). However, an adaptional decrease of membrane conductivity has not been observed in carp or trout erythrocytes (Bourne & Cossins, 1981; Raynard & Cossins, 1991). It is widely accepted that ions pass through membranes via water-filled pores and channels rather than through the lipid bilayer (Hochachka, 1986). A reduction of ion loss at low temperature may therefore be caused by decreasing channel density or by regulating the activity of existing channels. This phenomenon is termed channel arrest and has been suggested as an adaptive strategy during hypoxic or hypothermal exposure (Hochachka, 1986). Channel inhibition through Ca^{2+}, and modulation of channel activity by phosphorylation of the channel protein, are the most common features of channel regulation (Latorre *et al.*, 1989; Levitan, 1994). However, none of these mechanisms has yet been shown to become involved during cold acclimation. The role of homeoviscous adaptation in the down-regulation of dissipative fluxes during cold acclimation has been discussed (Cossins, Schwarzbaum & Wieser, 1995). In the Arctic charr *S. alpinus*, the reduction of passive K^+ fluxes is accompanied by a nearly perfect homeoviscous compensation (Schwarzbaum *et al.*, 1991). It seems unlikely, however, that membranes with a higher fluidity are less permeable to ions. Proton permeability of the inner mitochondrial membrane in rats is enhanced by the desaturation of membrane lipids (Brand *et al.*, 1992). Therefore, homeoviscous adaptation of the membranes may even lead to enhanced ion leakage in the cold, so the animal must compensate for ion leakage in a different manner.

A very interesting observation with respect to channel arrest during cold exposure was made by Rubinsky, Arav and Fletcher (1991). Using patch-clamp techniques they revealed that antifreeze proteins block ion channels in mammalian tissues and, therefore, prevent ion leakage. This indicates that antifreeze proteins in polar teleosts may serve not only as agents to avoid freezing, but may also have a function in ion regulatory processes.

Another possible mechanism to reduce passive ion fluxes is the reduction of ion gradients over cell membranes, which may be used during exposure to extreme cold. Burton (1986) reviewed the effects of temperature on the ion composition of the blood of several teleosts. The available data suggest that acclimation temperature has only minor effects on ion composition in most cases. Only acclimation to temperatures around 0 °C may lead to increased plasma osmolarities (Burton, 1986). Prosser, Mackay and Kato (1970) determined elevated concentrations of monovalent ions in the plasma of some cold stenothermal fish. Antarctic teleosts show plasma osmolarities about twice that of temperate fish (O'Grady & DeVries, 1982; Gonzalez-Cabrera *et al.*, 1995). This phenomenon agrees well with the lack of a positive compensation of Na^+/K^+-ATPase activity in the gills of Antarctic teleosts, and has been interpreted as an adaptive strategy to reduce the energetic costs of ion regulation (Prosser *et al.*, 1970). However, higher plasma osmolarity imposes higher Na^+ and K^+ gradients over the cell membranes in all tissues, since the ratios of intra- and extracellular concentrations of these ions remain constant (Prosser *et al.*, 1970). Thus, the energy conserved in the gills by reducing the gradient between plasma and ambient water may be used to counteract enhanced dissipative fluxes over the cell membranes in muscle and nerve tissues. This conclusion is supported by a decrease in muscle Na^+/K^+-ATPase activity in warm acclimated *Trematomus bernacchii* and *T. newnesi* where plasma osmolarity is reduced to values comparable to those in temperate species (Gonzalez-Carbrera *et al.*, 1995). Therefore, the role of hyperosmolarity of Antarctic teleosts in energy conservation remains questionable. Investigations by O'Grady & DeVries (1982) indicate that Antarctic fish are able to actively regulate plasma osmolarity despite changes in ambient salinity. Therefore, plasma osmolarity is not determined by the insufficient capacity of ion regulation but is actively adjusted by a shift of the set point of the ion pumps involved. Accordingly, elevated ion concentrations in the plasma may serve to decrease the freezing point of the plasma rather than represent a strategy to reduce the cost of ion regulation in the cold (O'Grady & DeVries, 1982).

The above survey suggests that two different strategies may be used to maintain ionic equilibrium during temperature changes. Eurythermal species like cod *Gadus morhua*, trout *Oncorhynchus mykiss* and roach *Rutilus rutilus* display a positive compensation of ATPase activity either by enhancing the number of enzyme molecules or by modulating the turnover number of existing proteins. The energetic costs of ion regulation in fish have been estimated to be 25–30% of the total metabolic rate, based on oxygen consumption measurements in euryhaline teleosts at different salinities (Rao, 1968; Febry & Lutz, 1987). A lower estimate of 10% is given by Gibbs & Somero (1989) based on maximum Na^+/K^+-ATPase activities in the gills of marine teleosts. The compensatory strategy used by eurythermal species implies that the energy requirement for ion regulation remains almost unchanged and comprises a larger fraction of energy turnover during cold acclimation when the overall metabolic rate is depressed (Fig. 3). Indeed, Rao (1968) determined that a larger percentage of the standard metabolic rate is attributed to ion regulatory processes in rainbow trout at 5 °C than at 15 °C.

As an alternative strategy, dissipative ion fluxes are reduced in the cold-stenothermal teleost *Salvelinus alpinus*. This energy saving strategy may also be used by Antarctic species through a reduction of osmotic gradients and through possible inhibition of ion leakage by antifreeze proteins. This strategy may be even more important when the mode of life allows for maximum reduction of ion exchange mechanisms, as has been discussed extensively in a comparison of lifestyles and energy savings in Antarctic vs. deep sea fish (for recent reviews, see Thiel *et al.*, 1996; Somero, this volume).

The previous discussion suggests that the change in the cost of ion regulation in polar fish may not be substantially different from the respective change expected for marine invertebrates. Actually, similar mechanisms may be involved in the reduction of dissipative ion fluxes. In general, the reduced cost of ion regulation at low temperature may help to shift the low critical temperature to below polar ambient temperature by reducing the energy requirements of the animals and the extent to which mitochondrial proliferation would otherwise be required. These cost reductions may also free enough energy for the purpose of growth and reproduction. Further research is needed to substantiate the different strategies discussed in this section.

Integrative signals of cold adaptation

Alphastat-regulation of pH

One important aspect of ionic regulation is the regulation of pH, which is known to be a parameter that integrates cellular functions (Busa & Nuccitelli, 1984). Passive proton distribution over the cell membrane is determined by the membrane potential and would lead to a ΔpH of about one pH unit between intra- and extracellular compartments (Thomas, 1984). A typical pH gradient of 0.4–0.6 pH units is maintained in most vertebrates and invertebrates, which is achieved by active proton extrusion from the cell through H^+-ATPases or by secondary active ion exchange, mostly via the Na^+/H^+-exchanger, which depends upon the Na-gradient established by Na^+/K^+-ATPase (see above, Fig. 3). Accordingly, an ATP cost of acid-base regulation has been quantified (A. Reipschläger & H.O. Pörtner, unpublished observations). The Na^+/H^+-exchanger displays an unusually high Q_{10} of 7.9 in trout erythrocytes (Cossins & Kilbey, 1990). In the cold, even a reduction of pH gradients between intra- and extracellular spaces has been observed (Heisler, 1986*b*; Sommer *et al.*, 1997), which would require enhanced rates of proton extrusion. Possibly, an up-regulation of Na^+/H^+-exchanger activity is required, associated with the larger deviation from thermodynamic equilibrium (see Fig. 3). However, these conclusions are valid only if the membrane potential remains unchanged during cold exposure.

A study of temperature effects on the regulation of pH may also give further insight into the regulatory signals causing shifts in critical temperatures. This question is not restricted to how the limiting temperatures are set but also to how the animals adjust to those between the critical temperatures. Each temperature change will not just mean a change in overall energy turnover according to Q_{10}, but also requires a readjustment of energy production and consumption, as discussed before. The temperature-dependent regulation of pH may relate to the fine tuning of these adjustments and support maintenance of steady-state function at different temperatures.

Reeves (1972) introduced the imidazole α-stat (=alphastat) hypothesis postulating that poikilotherms regulate the pH of their body fluids such that the degree of protonation (α) of imidazole groups is maintained despite changes in body temperature. α-stat pH regulation has been under reinvestigation during recent years, both concerning its existence or not in various tissues and animal species, and its potential importance for enzymatic regulation. Briefly, the alphastat process is proposed to play an important role in the maintenance of the structural integrity of proteins, which appears to be a prerequisite for the maintenance of cellular, especially enzyme functions (Hochachka & Somero, 1984). Since, on average, the pK of imidazole groups changes at -0.018 units $°C^{-1}$ a shift in intra- and extracellular pH with body temperature by $\Delta pH/\Delta T \sim -0.018\ °C^{-1}$, as observed in many poikilotherms, is sufficient to keep histidine protonation (α) constant (Fig. 5 A). However, the picture is not uniform and deviation from intracellular α-stat control has been observed in many species and tissues (Heisler, 1986*b*; Butler & Day, 1993; Whiteley & Taylor, 1993; Whiteley *et al.*, 1995*a*,*b*). $\Delta pH/\Delta T$ values range from $-0.003\ °C^{-1}$ in crayfish claw muscle (Whiteley *et al.*, 1995*a*) to $-0.031\ °C^{-1}$ in the red muscle of the dogfish *Scyliorhinus stellaris* (Heisler & Neumann, 1980). Some of these studies may not have considered the existence of critical temperatures beyond which the onset of anaerobic metabolism and a shift of the setpoints of pH regulation (Pörtner, 1993) may lead to the observed deviation from a linear pH/temperature relationship (Sommer *et al.*, 1997). Therefore, we are interested in the temperature range of α-stat control and in the mechanisms involved as well as in how these mechanisms are influenced by environmental change.

Animals acclimated to low temperatures during the winter season frequently exhibit relatively low pH values deviating from the alphastat pattern. Low intracellular pH has been proposed as a key mechanism eliciting metabolic depression, for example, in hibernating mammals (Malan, 1985; for further examples and recent review, see Hand & Hardewig, 1996). Accordingly, the shrimp *Palaemon elegans* tends to be inactive at temperatures below 10 °C in winter, metabolic depression being reflected by a drop in intracellular pH and an increase in the concentration of sugar phosphates (Thebault & Raffin, 1991, Fig. 6). Low pHi values were also reported by Whiteley *et al.* (1995*a*) for the crayfish *Austrapotamobius pallipes* in winter. Only tissues like abdominal muscle, which remain operative in winter, followed α-stat while less active tissues like claw muscle and hepatopancreas did not. The authors explained the relative acidosis in these tissues by low rates of protein synthesis and a lowered overall metabolic rate at low temperatures when crayfish are inactive. Based on these results they speculated that the relative acidosis observed in the haemolymph of *Glyptonotus antarcticus* and *A. pallipes* may be characteristic of crustaceans living at low temperatures when rates of protein synthesis and possibly catabolism are low (Whiteley *et al.*,

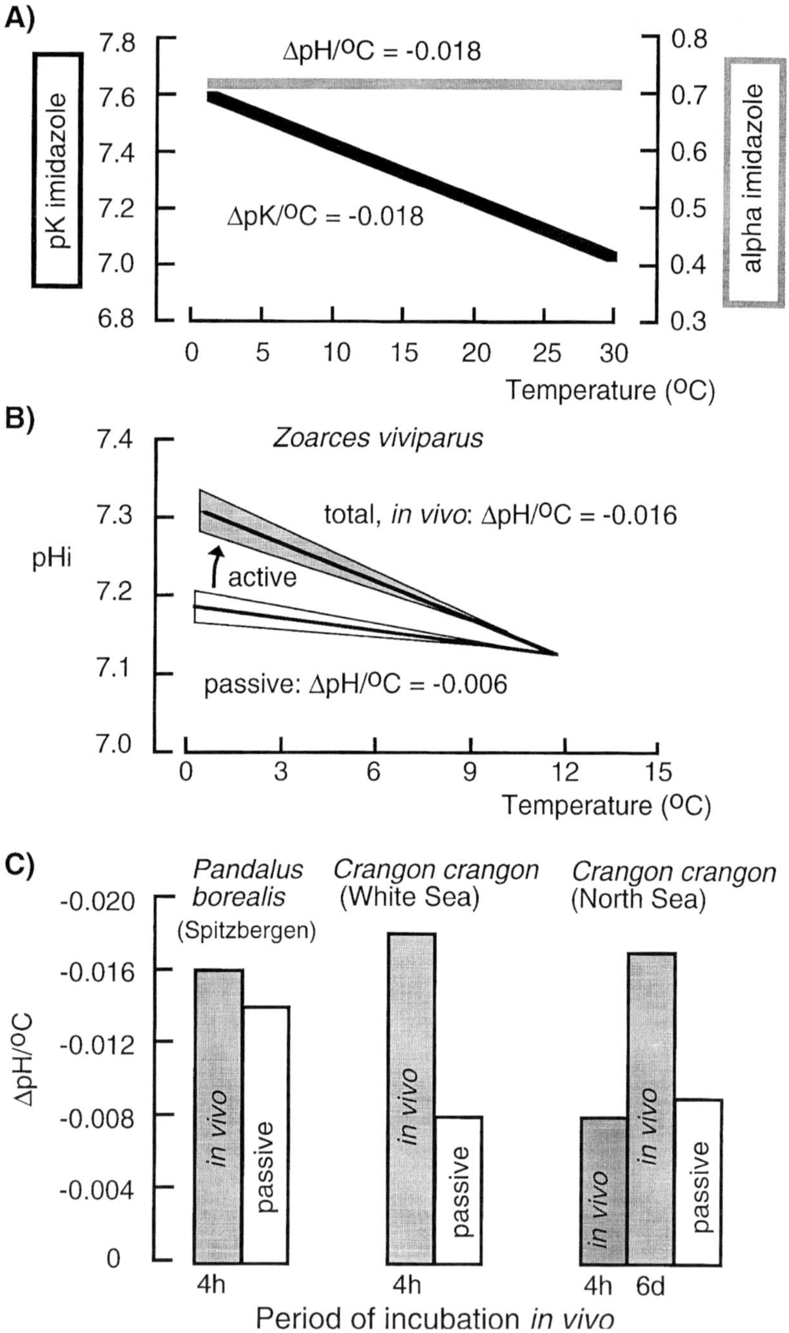
A)
pK imidazole
alpha imidazole
ΔpH/°C = -0.018
ΔpK/°C = -0.018
7.8
7.6
7.4
7.2
7.0
6.8
0.8
0.7
0.6
0.5
0.4
0.3
0 5 10 15 20 25 30
Temperature (°C)
B)
Zoarces viviparus
total, in vivo: ΔpH/°C = -0.016
active
passive: ΔpH/°C = -0.006
pHi
7.4
7.3
7.2
7.1
7.0
0 3 6 9 12 15
Temperature (°C)
C)
Pandalus borealis (Spitzbergen)
Crangon crangon (White Sea)
Crangon crangon (North Sea)
ΔpH/°C
-0.020
-0.016
-0.012
-0.008
-0.004
0
in vivo
passive
4h
4h
4h 6d
Period of incubation in vivo

1995*b*). In contrast, a reduction in metabolic activity could not be observed in *C. crangon* originating from the North Sea in winter (Sartoris & Pörtner, 1997*a*,*b*) and, possibly as a consequence, both winter and summer animals followed the α-stat pattern.

The question remains open whether lower pH for lower metabolism is true in all cases and whether a drop in pHi causes or is only a correlate of metabolic depression. To save energy, animals may tolerate passive changes of pHi during periods of relative inactivity, since the active transport of proton equivalents depends on energy supply. A down-regulation of ion exchange could lead to a reduction of pHi since pH would then approach thermodynamic equilibrium. In *S. nudus* extracellular pH is the key acid–base parameter eliciting metabolic depression during acidosis. Metabolic depression is not induced by moderate (0.3 pH units) decreases in pHi (Reipschläger & Pörtner, 1996). Since extracellular pH appears to play a predominant role in metabolic regulation, future work should investigate whether the observed changes in the relationship between intra- and extracellular pH with temperature (see above) affect metabolic rate.

The assumption that a deviation from α-stat at low temperatures is a correlate of metabolic depression deserves further consideration. In this context, it is interesting to determine the pH/temperature relationship in

Fig. 5. Changes in the pH of body fluids with temperature, the pattern of which is considered to be important for the maintenance of physiological functions. A) Effects of temperature on the dissociation constant for protons (pK) of the imidazole group in histidine, an important amino acid component of proteins. A parallel change in pK and cellular pH at -0.018 $°C^{-1}$ is postulated to occur in 'cold-blooded' animals and maintains the degree of dissociation, α-imidazole, which is thought to be essential for protein structure and function (alphastat hypothesis). B) Contributions of passive and active processes to adjustments in intracellular pH after temperature change in white muscle of *Zoarces viviparus*. pH_i decreases with rising temperature at a slope of -0.016 $°C^{-1}$. The framed areas represent the standard error of the slopes. **shaded frame**: pH/temperature relationship determined *in vivo* as the sum of active and passive mechanisms. **open frame**: pH/temperature relationship *in vitro*, representing the passive response of intracellular buffers to temperature change. C) Contributions of passive processes to the adjustment of intracellular pH *in vivo* (shaded) after temperature change (ΔpH $°C^{-1}$) in boreal (North Sea) and subarctic (White Sea) populations of *Crangon crangon* and *Pandalus borealis*. Intracellular pH was analysed after 4 hours (h) (*P. borealis* and *C. crangon* from the White Sea) and after 4 h and 6 days (d) (North Sea *C. crangon*) of incubation at various temperatures. Note that pH regulation was complete after 4 h in White Sea animals, whereas it was delayed and only found complete after 6 days in North Sea specimens (after Sartoris & Pörtner, 1997*a*; van Dijk *et al.*, 1997).

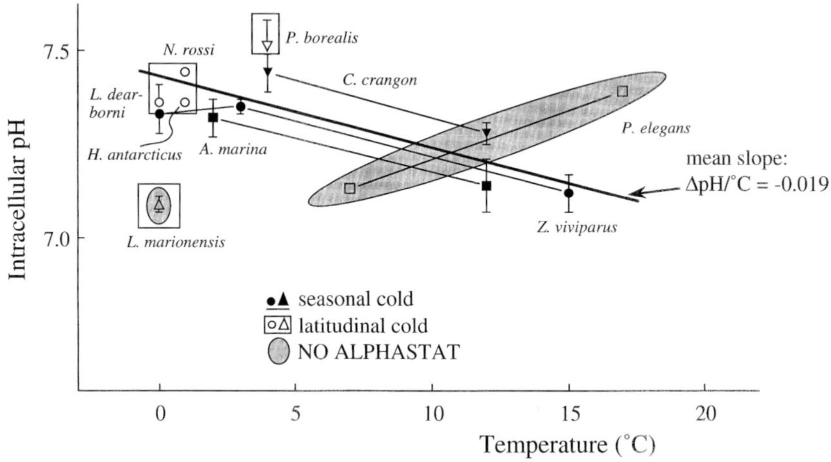

Fig. 6. Realisation of intracellular α-stat control during seasonal or latitudinal cold in marine ectotherms: Fish (Antarctica): *Harpagifer antarcticus* (white muscle, Egginton & Moerland, 1993, SD not shown), *Notothenia rossi* (red blood cells, Egginton *et al.*, 1991, SD not shown), *Lycodichthys dearborni* (white muscle, I. Hardewig, P. van Dijk & H.O. Pörtner, unpublished data), *Zoarces viviparus* (white muscle, North Sea; van Dijk *et al.*, 1997). Invertebrates: *Pandalus borealis, Crangon crangon* (tail muscle, Sartoris & Pörtner, 1997*a*), *Arenicola marina* (body wall muscle, Sommer *et al.*, 1996, 1997) *Palaemon elegans* (tail muscle, Thebault & Raffin, 1991), *Limopsis marionensis* (adductor muscle, H.O. Pörtner, S. Zielinski & L.S. Peck, unpublished data). Methods used by the various authors include the homogenate technique (Pörtner *et al.*, 1990) and ^{31}P-NMR.

species from a latitudinal temperature gradient as a test whether the activity level or a temperature-induced reduction of metabolism towards polar areas influences α-stat control. Actually, measurements in the deep water shrimp *Pandalus borealis* (Spitzbergen) and the Antarctic eelpout *Lycodichthys dearborni* indicate that cold-stenothermal animals follow α-stat regulation of intracellular pH (Fig. 6). The finding of α-stat-pHi by ^{31}P NMR in the Antarctic teleost *Harpagifer antarcticus* (Egginton & Moerland, 1993) supports this conclusion. However, a uniform picture is, again, not evident. For example, pH values in the Antarctic bivalve, *Limopsis marionensis,* do not reflect α-stat regulation (Fig. 6). It remains to be established whether this is a consequence of hypometabolism. Reduced activity at low temperatures could not explain the deviations from α-stat in the brown trout *Salmo trutta*, where a reduction in standard metabolic rate or swimming ability could not be observed in winter acclimated fish (Butler & Day, 1993). In this species pH was independent of temperature in the musculature, whereas α-stat regulation was observed in the blood plasma. In contrast, a reduction of swimming performance was found in rainbow trout at low temperature which was supposedly associated with relatively acidotic intra- and extracellular pH values deviating from the α-stat pattern only at low temperatures (Taylor, Taylor & Egginton, 1993, 1996).

α-stat pH control was observed in the body wall musculature of *A. marina*, whereas coelomic fluid pH remained independent of changing temperature (Sommer *et al.*, 1997). An important point to be considered in such analyses is that α-stat pH regulation only occurs in the range between low and high critical temperatures. pH becomes independent of temperature beyond these thresholds (Sommer *et al.*, 1997). Accordingly, the temperature window for α-stat regulation is expected to be much smaller in polar species than in temperate zone species as verified in the Antarctic bivalve *Limopsis marionensis* (H.O. Pörtner, S. Zielinski & L.S. Peck, unpublished data).

The mechanisms of α-stat control have not yet been sufficiently characterised. According to Reeves (1985), α-stat regulation consists of both passive and active components. The passive component depends upon the physicochemical composition of intra- and extracellular buffers and results from proton binding or release owing to the change in pK values and associated dissociation equilibria of the buffer components with temperature. The active component comprises adjustments in either ventilation or ion exchange or both. In air breathers, ventilation changes with temperature and causes a shift in $P\text{CO}_2$ associated with a pH shift. In water breathers, ion exchange predominantly determines the active component of the temperature-induced pH change. Distinguishing between respiratory and ion exchange mechanisms is very important in order to understand the velocity of α-stat regulation. Ventilatory adjustment is faster than the adjustment of a new steady state of acid–base homeostasis by means of ion transport across gills and cell membranes.

Previous model calculations of the relative contributions of various processes to pHi adjustment exclusively relied on the ΔpK/ΔT value of the imidazole group. However, ΔpK/ΔT depends upon local charge configurations in the environment of the imidazole group as well as on ionic strength and, therefore, varies between -0.016 and $-0.024\ ^{\circ}\text{C}^{-1}$ for histidine and free imidazole compounds (Heisler, 1986*a,b*), and ranges from -0.0010 to $-0.051\ ^{\circ}\text{C}^{-1}$ for histidine residues in proteins, leading to a large uncertainty about the accuracy of these model calculations.

These theoretical problems could be solved by experimental analyses of active and passive processes. The homogenate technique (Pörtner *et al.*, 1990) allows the rapid measurement of pHi in tissue samples and to distinguish active and passive elements in α-stat pH regulation. To quantify passive mechanisms, animals were exposed to control temperature and their tissues were analysed *in vitro* at different temperatures, thereby excluding the influence of biochemical reactions or ion exchange mechanisms. Therefore, measured pH changes result from passive, physicochemical buffering. In contrast, *in vivo* values determined in animals exposed to various temperatures prior to the collection of tissue samples can be interpreted to reflect the summed effects of active and passive processes (Fig. 5 B). The passive component comprises fast and temperature dependent proton binding or release by intracellular buffers, and the active component in water breathers represents ATP-dependent ion exchange which is considerably slower.

The results of these analyses suggest that the contribution and velocity of active processes to the pH shift differs between species and populations from different latitudes (Sartoris & Pörtner, 1997*a*; van Dijk, Hardewig & Pörtner, 1997; Fig. 5B and C). In the North Sea eelpout, *Zoarces viviparus*, 65% of the pH changes were elicited by active, energy requiring ion transport mechanisms, whereas only 35% were contributed by passive buffer processes. In the eurythermal shrimp *Crangon crangon*, from both the Russian White Sea and the North Sea, 50% of the pH change occurred by means of active mechanisms. Also, the passive ΔpHi/ΔT relationship was identical in summer and winter animals from the North Sea ($\Delta pHi/\Delta T = -0.009$ units $°C^{-1}$), suggesting that passive pHi adjustment does not depend upon the season and that the concentrations of relevant intracellular buffer substances remains unchanged (Sartoris & Pörtner, 1997*a*). In contrast, the active component of pH regulation amounted to only 10% in the Arctic deep sea shrimp *Pandalus borealis*, which is more sensitive to temperature fluctuations (stenothermal). In conclusion, the active component may be more prominent in eurythermal species. However, since these are the first results on the relative contribution of active and passive processes to α-stat regulation in polar animals, it may only be speculated that pH adjustment mostly occurs by passive mechanisms in cold stenothermal ectotherms, with a small contribution of active components. Future work is required to show whether the relative contribution of active and passive mechanisms to α-stat regulation is involved in determining the limits of temperature tolerance and is related to the geographical distribution of species. It may be important in this context that stenothermal animals are often stenohaline and, therefore, the capacity of ion regulatory mechanisms may be less developed than in euryhaline species which frequently are also eurythermal. These findings are also relevant with respect to the allocation of energy to acid-base regulation. Cold-stenothermal (and stenohaline) species may reduce the energy requirements of α-stat regulation by using predominantly passive mechanisms for pH adjustment, whereas eurythermal (and euryhaline) species emphasise ion regulatory mechanisms to allow for a flexible response to environmental change. A larger active than

Table 1. *Temperature-dependent elements influencing the Gibbs' free energy of ATP hydrolysis and thus cellular energy status* in vivo

1. pK values of H^+ and Mg^{2+} binding to
   ATP^{4-}, $HATP^{3-}$, ADP^{3-}, $HADP^{2-}$, AMP^{2-}, Pi^{2-}, PLA^-

2. Reaction equilibra (simplified) of

   ATPase
   $MgATP^{2-} + H_2O \rightarrow MgADP^- + HPO_4^{2-} + H^+$

   Arginine kinase
   $PLA^- + MgADP^- + H^+ \rightarrow MgATP^{2-} + L\text{-}Arg^+$

   Adenylate kinase
   $MgADP^- + MgADP^- \rightarrow MgATP^{2-} + MgAMP$

   ⇒ ***ATP free energy change***

$$dG/d\xi_{ATP} = \Delta G^\circ obs + RT \cdot \ln ([ADP]_{free,tot} \cdot [Pi]_{free,tot} / [ATP]_{free,tot})$$

$$dG/d\xi_{ATP} = \Delta G^\circ obs + RT \cdot \ln ([L\text{-}Arg]_{free,tot} \cdot [Pi]_{free,tot} / [PLA]_{free,tot} \cdot [H^+] \cdot Kapp_{AK})$$

Note:
Apparent equilibria determined experimentally will include all species complexed with H^+ and/or Mg^{2+}, but unbound to cellular protein $[X]_{free,\ tot}$.
Source: cf. Figure 7; for the calculation procedure see Pörtner *et al.* (1996).

passive component of α-stat regulation may therefore be a prerequisite to colonise shallow coastal waters.

An additional result of these comparisons was that, in White Sea *Crangon*, active pH regulation was faster and, therefore, reached the final pH earlier than in North Sea animals (Fig. 5 C). Obviously, the velocity of these active ion exchange processes was increased as a consequence of metabolic cold adaptation. Animals living at lower temperatures may, in general, be able to compensate acid–base disturbances faster. In the cold stenothermal *Pandalus borealis* the faster response can be attributed to the large passive fraction, while in the more eurythermal White Sea *Crangon crangon* cold adaptation is likely to increase the capacity of pH regulatory mechanisms.

Factors limiting tolerance to cold exposure: cellular energetics?

As outlined above, critical temperatures characterise the onset of functional hypoxia and an anaerobic metabolism and, thereby, a time-limited situation.

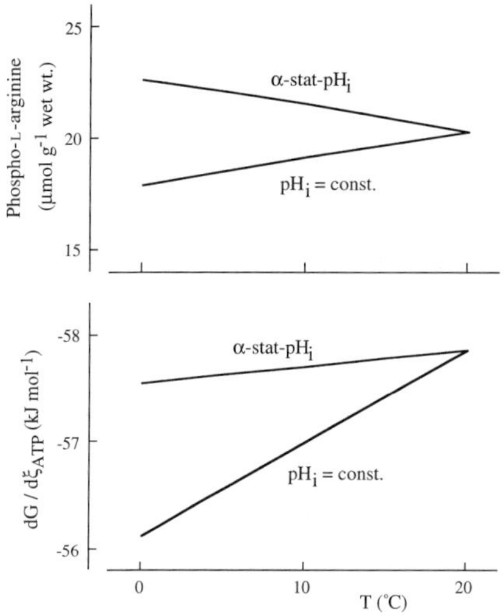

Fig. 7. Model calculations emphasise the importance of α-stat regulation for the maintenance of energy homeostasis during cooling. The calculation of the Gibbs' free energy change of ATP hydrolysis at different temperatures followed the rationale outlined by Pörtner *et al.* (1996). Numbers used are valid for the mantle muscle of the squid, *Lolliguncula brevis*. Levels of phospho-L-arginine were calculated from the relevant equilibrium of arginine kinase. Enzyme equilibria were calculated for each temperature by use of the van't Hoff equation. Fractional levels of reaction partners as they varied with fluctuating Mg^{2+} levels and pH were calculated from the respective dissociation equilibria (for details see Pörtner *et al.*, 1996). If it is assumed that the cells maintain constant levels of phospho-L-arginine plus L-arginine, as well as free ATP, ADP, Pi and Mg^{2+}, the concentrations of phospho-L-arginine and the level of Gibbs' free energy change of ATP hydrolysis will be maintained by α-stat pH regulation compared to constant pH. With α-stat control ATP free energy change will only fall by 0.3 kJ mol^{-1}, whereas a drop by about 1.7 kJ mol^{-1} is expected when pH remains constant.

Obviously, α-stat regulation of pH is also restricted to the specific window between critical temperatures of a species. The question arises which mechanisms limit tolerance to cold below the low Tc. As a precondition of low temperature tolerance, the maintenance of energy status as it depends upon temperature and the regulation of intracellular pH will be discussed. The onset of anaerobic metabolism is usually associated with a net decrease in cellular energy levels and may cause disturbances of ionic distribution and α-stat control below the Tc.

The actual energy level of a cell is quantified by the *in vivo* Gibbs' free energy change of ATP hydrolysis, a concept which has only recently been applied to studies of marine ectotherms (Pörtner, 1993; Pörtner *et al.*, 1993; Combs & Ellington, 1996; Pörtner, Finke & Lee, 1996, Pörtner *et al.*, 1997; Zielinski & Pörtner, 1996). The assessment of cellular energy levels depends upon the quantification of a series of factors compiled in Table 1 and the consideration of their temperature dependence in relation to the regulation of intracellular pH during temperature change (Fig. 7). The results of this analysis emphasise the importance of intracellular pH in energy homeostasis since α-stat pH regulation supports maintenance of the levels of the phosphagen, phospho-L-arginine and of the Gibbs' free energy change of ATP hydrolysis. This becomes evident from the depiction in Fig. 7 which shows that the maintenance of a constant pH with falling temperature will cause a decrease in phospho-L-arginine and ATP free energy change values.

The question arises of how the results of these model calculations compare with the *in vivo* situation. Figure 8 depicts the levels of ATP free energy change as evaluated for the body wall musculature of a marine sipunculid worm during cooling. Intracellular pH follows α-stat predictions until temperature falls below the critical threshold found between 4 and 0 °C for this species (see above). The ability of the animals to recover ceased after longer than 2 days of exposure to 0 °C. Intracellular pH returned to control values when recovery was still possible, but intracellular pH continued to fall at temperatures below the critical temperature, reaching even lower values in those animals which were brought back to 12 °C but were no longer able to recover. The correlated fall in ATP free energy change levels to a low, possibly critical value (Fig. 8 B) may characterise the limitation of survival (point of no return) and contribute to the development of lethal cold injuries (Zielinski & Pörtner, 1996). It remains to be investigated which cellular functions would be affected by the decrease in cellular energy levels. Similarly, low energy levels may indicate irreversible temperature stress in the Antarctic bivalve *Limopsis marionensis* where an upper critical threshold is surpassed when temperature rises from 0 to values between 2 and 4 °C (Pörtner, H.O., S. Zielinski & L.S. Peck, unpublished data).

Summary and conclusions

Critical temperature thresholds (Tc) can be defined for various invertebrate species which are characterised by the transition to an anaerobic mode of metabolism, once temperature reaches low or high extremes. Beyond critical temperatures, it is not the availability of ambient oxygen that is limiting. Both low and high Tcs are either set by the failure of oxygen uptake and transport in the blood, or by the insufficiency of ventilatory mechanisms in the cold. Metabolic cold adaptation can be understood as a downward shift of both low and high critical temperatures. Critical temperatures shift

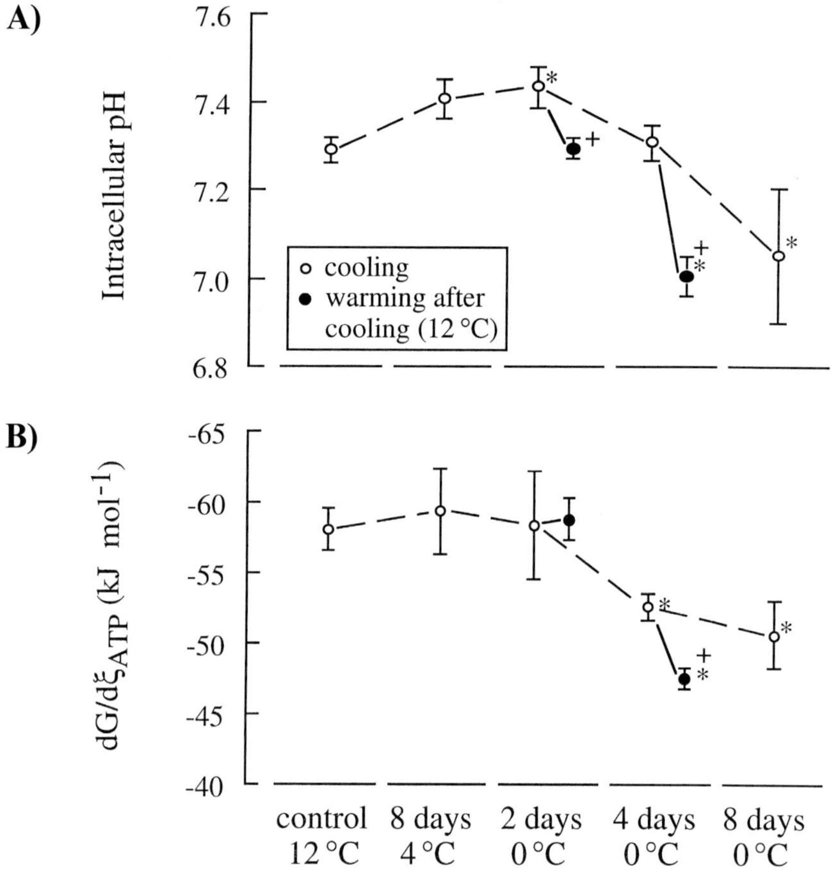

Fig. 8. A) Intracellular pH determined in the body wall musculature of the sipunculid worm, *Sipunculus nudus,* after different periods of exposure to control temperature (12 °C) and to decreasing temperatures (4 and 0 °C). The ability to recover from cold exposure was tested after 2 and 4 days at 0 °C, respectively. Animals were no longer able to recover after 4 days, when a severe metabolic acidosis developed during warming, which was partly caused by the irreversible accumulation of anaerobic end products (not shown). B) Energy content of ATP, quantified as the Gibbs' free energy change of ATP hydrolysis $dG/d\xi_{ATP}$, in the body wall musculature of *S. nudus* during long term exposure to cold temperatures and subsequent rewarming to 12 °C. Critical exposure is indicated by the fall in energy levels associated with the failure to recover from exposure to 0 °C for longer than 2 days (after Zielinski & Pörtner, 1996).

within, and differ between, populations depending on seasonal temperature adaptation and latitude. These differences may be related to genetic distances between populations. Low thresholds significantly above freezing observed in temperate zone species have been eliminated during evolutionary low temperature adaptation in polar ectotherms. In these organisms this process has led to a high sensitivity to elevated temperatures. The shift of the low Tc to below polar temperature is thought to require mitochondrial proliferation, but also a reduction in the basic cost of some ATP-dependent cellular functions. Both critical thresholds will be affected if the Tcs are set by mitochondrial density, which increases during cold adaptation but will then have to decrease during warm adaptation. The disadvantage of mitochondrial proliferation, in that a higher density of mitochondria causes a rise in energy turnover by itself, may be compensated to some extent. With a high mitochondrial density in polar species, the metabolic 'idling' of individual mitochondria may be reduced by membrane modifications which lead to a reduction in proton leakage and thus energy expenditure. Dissipative ion fluxes across cell membranes and epithelia are reduced at low temperatures allowing the energy turnover of Na^+/K^+-ATPase to decrease. Such a decrease appears to be the strategy found in animals exposed to permanent (polar) as opposed to seasonal cold and even more so in deep sea fish, where a mode of life more sluggish than found in Antarctic fish supports the development of such a strategy even further. The question of a potential reduction in the cost of acid–base regulation remains open.

Critical temperatures may also set the limits for an adjustment of pH regulation to changing temperatures. A rise in pH with falling temperature as required to maintain the degree of protonation (α) of imidazole moieties in proteins and, thus, protein function, is no longer observed beyond low or high critical temperatures. Alpha-stat regulation of pH supports the maintenance of cellular energy levels quantified as the Gibbs' free energy change of ATP hydrolysis. A decrease in these energy levels seen during lethal cold exposure below the Tc is interpreted to limit cold tolerance and impair ATP dependent cellular functions. Low levels of available energy may, for example, compromise acid–base and ionic regulation as well as muscular function relevant in the maintenance of ventilatory and circulatory function.

Future efforts must address the regulatory integration of the different biochemical and physiological processes depending on the temperature regime focusing on pH, both as a regulated parameter and a parameter effective in metabolic regulation. Those molecular mechanisms which are responsible for setting the critical temperatures need to be identified as well as those which are modified for a change in critical temperature values. Furthermore, it needs to be evaluated further to what extent the temperatures critical for physiological processes are related to distributional limits of a species in the natural environment. These efforts should also address to what extent the temperature-induced transition to anaerobic metabolism causes behavioural changes and migratory movements under environmental stress, as has recently been demonstrated for terrestrial ectotherms (Pörtner *et al.*, 1994).

Acknowledgements

Supported by grants of the Deutsche Forschungsgemeinschaft to H.O.P. Technical support for experimental studies reported in this review was provided by T. Hirse and G. Frank and is gratefully acknowledged. Alfred-Wegener-Institute publication no. 1133.

References

Archer, S.D. & Johnston, I.A. (1991). Density of cristae and distribution of mitochondria in the slow muscles of Antarctic fish. *Physiological Zoology,* **64**, 242–58.

Arntz, W.E., Brey, T. & Gallardo, V.A. (1994). Antarctic Zoobenthos. *Oceanography and Marine Biology: An Annual Review*, **32**, 241–304.

Beguin, P., Beggah, A.T., Chibalin, L.A., Vasilets, B.C., Jaisser, F. & Geering K. (1994). Phosphorylation of Na,K-ATPase by protein kinases: structure–functions relationship. In *The Sodium Pump*, ed. W. Schoner, pp. 682–5. Darmstadt: Steinkopf Verlag.

Blaustein, M.P. & Hamlyn, J.M. (1985). Role of an endogenous inhibitor of Na^+ pumps in the pathophysiology of essential hypertension. In *The Sodium Pump*, ed. L. Glynn & C. Ellory, pp. 629–39. Cambridge: The Company of Biologists Limited.

Bourne, P.K. & Cossins, A.R. (1981). The effects of thermal acclimation upon ion transport in erythrocytes. *Journal of Thermal Biology,* **6**, 179–81.

Brand, M.D. (1990). The contribution of the leak of protons across the mitochondrial inner membrane to standard metabolic rate. *Journal of Theoretical Biology*, **145**, 267–86.

Brand, M.D., Couture, P., Else, P.L., Withers, K.W. & Hulbert, A.J. (1991). Evolution of energy metabolism. Proton permeability of the inner membrane of liver mitochondria is greater in a mammal than in a reptile. *Biochemical Journal*, **275**, 81–86.

Brand, M.D., Steverding, D., Kadenbach, B., Stevenson, P.M. & Hafner, R.P. (1992). The mechanism of the increase in mitochondrial proton permeability induced by thyroid hormones. *Journal of Biochemistry*, **206**, 775–81.

Burton R.F. (1986). Ionic regulation in fish: the influence of acclimation temperature on plasma composition and apparent set points. *Comparative Biochemistry and Physiology*, **85A**, 23–8.

Busa, W.B. & Nuccitelli, R. (1984). Metabolic regulation via intracellular pH. *American Journal of Physio*logy, **246**, R409–38.

Butler, P. J. & Day, N. (1993). The relationship between intracellular pH and seasonal temperature in the brown trout *Salmo trutta*. *Journal of Experimental Biology*, **177**, 293–7.

Combs, C.A. & Ellington, W.R. (1996). Graded intracellular acidosis produces extensive and reversible reductions in the effective free energy change of ATP hydrolysis in a molluscan muscle. *Journal of Comparative Physiology*, **165B**, 203–12.

Cossins, A.R. & Bowler, K. (1987). *Temperature Biology of Animals.* London, New York: Chapman & Hall.

Cossins, A.R. & Kilbey, R.V. (1990). The temperature dependence of the adrenergic Na^+/H^+ exchanger of trout erythrocytes. *Journal of Experimental Biology*, **148**, 303–12.

Cossins, A.R., Bowler, K. & Prosser, C.L. (1981). Homeoviscous adaptation and its effects upon membrane-bound proteins. *Journal of Thermal Biology*, **6**, 183–7.

Cossins, A.R., Schwarzbaum, P.J. & Wieser, W. (1995). Effects of temperature on cellular ion regulation and membrane transport systems. In *Biochemistry and Molecular Biology of Fishes*, ed. P.W. Hochachka & T.P. Mommsen, vol. 5, pp. 101–26. Amsterdam: Elsevier Science BV.

DeWachter, B. & Pörtner, H.O. (1997). Lactate has a behavioural and metabolic signalling function in the shore crab, *Carcinus maenas. Journal of Experimental. Bio*logy, **200**, 1015–24.

Egginton, S. & Sidell, B.D. (1989). Thermal acclimation of subcellular structures of fish muscle. *American Journal of Physiology*, **256**, R1–10.

Egginton, S. & Moerland, T.S. (1993). Intracellular pH of fast muscle from an Antarctic teleost (*Harpagifer antarcticus*) estimated by *in vivo* ^{31}P NMR: *Journal of Physiology*, **473**, 234.

Egginton, S., Taylor, E.W., Wilson, R.W., Johnston, I.A. & Moon, T.W. (1991). Stress response in the Antarctic teleosts (*Notothenia neglecta* Nybelin and *N. rossii* Richardson). *Journal of Fish Biology*, **38**, 225–35.

Ellory, J.C. & Hall, A.C. (1987). Temperature effects on red cell membrane transport processes. In *Temperature and Animal Cells*, ed. K. Bowler & B.J. Fuller, vol. 21, pp. 53–66. Symposia of the Society for Experimental Biology.

Febry, R. & Lutz, P. (1987). Energy partitioning in fish: the activity-related cost of osmoregulation in a euryhaline cichlid. *Journal of Experimental Biology* **128**, 63–85.

Gibbs, A. (1995). Temperature, pressure and the sodium pump: The role of homeoviscous adaptation. In *Biochemistry and Molecular Biology of Fishes*, ed. P.W. Hochachka & T.P. Mommsen, vol. 5, pp. 197–212. Amsterdam: Elsevier Science BV.

Gibbs, A. & Somero, G.N. (1989). Pressure adaptation of Na^+-K^+-ATPase in gills of marine teleosts. *Journal of Experimental Biology*, **143**, 475–92.

Gonzalez-Cabrera, P.J., Dowd, F., Pedibhotla, V.K., Rosario, R., Stanley-Samuelson, D. & Petzel, D. (1995). Enhanced hypo-osmoregulation induced by warm-acclimation in Antarctic fish is mediated by increased gill and kidney Na^+/K^+-ATPase activities. *Journal of Experimental Biology*, **198**, 2279–91.

Guderley, H. & Johnston, I.A. (1996) Plasticity of fish muscle mitochondria with thermal acclimation. *Journal of Experimental Biology*, **199**, 1311–17.

Hall, A.C. & Willis, J.S. (1986). The temperature dependence of passive potassium permeability in mammalian erythrocytes. *Cryobiology*, **23**, 395–405.

Hand, S.C. & Hardewig, I. (1996). Downregulation of cellular metabolism during environmental stress: mechanisms and implications. *Annual Review of Physiology*, **58**, 539–63.

Hardewig, I., Addink, A.D.F., Grieshaber, M.K., Pörtner, H.O. & van den Thillart, G. (1991). Metabolic rates at different oxygen levels determined by direct and indirect calorimetry in the oxyconformer *Sipunculus nudus*. *Journal of Experimental Biology*, **157**, 143–60.

Heisler, N. (1986*a*). Buffering and transmembrane ion transfer processes. In *Acid–Base Regulation in Animals,* ed. N. Heisler, pp. 3–47, Amsterdam: Elsevier Science BV.

Heisler, N. (1986*b*). Comparative aspects of acid–base regulation. In *Acid–Base Regulation in Animals*, ed. N. Heisler, pp. 397–450. Amsterdam: Elsevier Science BV.

Heisler, N. & Neumann, P. (1980). The role of physico-chemical buffering and of bicarbonate transfer processes in intracellular pH regulation in response to changes of temperature in the larger spotted dogfish (*Scyliorhinus stellaris*). *Journal of Experimental Biology*, **85**, 99–110.

Hochachka, P.W. (1986). Defense strategies against hypoxia and hypothermia. *Science*, **231**, 234–41.

Hochachka, P.W. (1988). Channels and pumps – determinants of metabolic cold adaptation strategies. *Comparative Biochemistry and Physiology*, **90B**, 15–519.

Hochachka, P.W. & Somero, G.N. (1984). *Biochemical Adaptation*. 537 pp. Princeton University Press.

Johnston, I., Guderley, H., Franklin, C.E., Crockford, T. & Kamunde, C. (1994). Are mitochondria subject to evolutionary temperature adaptation? *Journal of Experimental Biology*, **195**, 293–306.

Kamiya, M. & Utida, S. (1969). Sodium–potassium-activated adenosintriphosphate activity in gills of fresh-water, marine and euryhaline teleosts. *Comparative Biochemistry and Physiology*, **31**, 671–74.

Latorre, R., Oberhauser, A., Labarca, P. & Alvarez, O. (1989). Varieties of calcium-activated potassium channels. *Annual Review of Physiology*, **51**, 385–99.

Levitan, I.B. (1994). Modulation of ion channels by protein phosphorylation and dephosphorylation. *Annual Review of Physiology*, **56**, 193–212.

McArdle, H.J. & Johnston, I.A. (1980). Evolutionary temperature adaptation of fish sarcoplasmatic reticulum. *Journal of Comparative Physiology*, **135B**, 157–64.

Malan, A. (1985). Intracellular pH in response to ambient changes: homeostatic or adaptive responses. In *Circulation, Respiration and Metabolism*, ed. R. Gilles, pp. 464–73. Berlin, Heidelberg: Springer-Verlag.

O'Dor, R.K., Pörtner, H.O. & Shadwick, R.E. (1990). Squid as elite athletes. In *Squid as Experimental Animals*, ed. D.L. Gilbert, W.J. Adelman & J.M. Arnold, pp. 481–503. New York: Plenum Press.

O'Grady, S.M. & DeVries, A.L. (1982). Osmotic and ionic regulation in polar fishes. *Journal of Experimental Marine Biology and Ecology*, **57**, 219–28.

Pörtner, H.O. (1993). Multicompartmental analyses of acid–base and metabolic homeostasis during anaerobiosis: invertebrate and lower vertebrate examples. In *Surviving Hypoxia: Mechanisms of Control and Adaptation* (eds P.W. Hochachka, P.L. Lutz, T. Sick, M. Rosenthal & G. van den Thillart, pp. 139–56.Boca Raton FL, USA: CRC Press Inc.

Pörtner, H.O., Boutilier, R.G., Tang, Y. & Toews, D.P. (1990). Determination of intracellular pH and Pco_2 after metabolic inhibition by fluoride and nitrolotriacetic acid. *Respiration Physiology*, **81**, 255–74.

Pörtner, H.O., Webber, D.M., O'Dor, R.K. & Boutilier, R.G. (1993). Metabolism and energetics in squid (*Illex illecebrosus, Loligo pealei*) during muscular fatigue and recovery. *American Journal of Physiology*, **265**, R157–65.

Pörtner, H.O., Branco, L.G.S., Malvin, G.M. & Wood, S.C. (1994). A new function for lactate in the toad *Bufo marinus*. *Journal of Applied Physiology*, **76**, 2405–10.

Pörtner, H.O., Finke, E. & Lee, P.G. (1996). Effective Gibbs' free energy change of ATP hydrolysis and metabolic correlates of intracellular pH in progressive fatigue of squid (*Lolliguncula brevis*) mantle muscle. *American Journal of Physiology*, **271**, R1403–14.

Pörtner, H.O., Reipschläger, A. & Heisler, N. (1997). Metabolism and acid–base regulation in *Sipunculus nudus* as a function of ambient carbon dioxide. *Journal of Experimental Biology*, in press.

Prosser, C.L., Mackay, W. & Kato, K. (1970). Osmotic and ionic concentrations in some Alaskan fish and goldfish from different temperatures. *Physiological Zoology*, **43**, 81–9.

Rady, A.A. (1993). Environmental temperature shift induced adaptive changes of carp (*Cyprinus carpio* L.) erythrocytes plasma membrane *in vivo*. *Comparative Biochemistry and Physiology*, **105A**, 513–18.

Rao, G.M.M. (1968). Oxygen consumption of rainbow trout (*Salmo gairdneri*) in relation to activity and salinity. *Canadian Journal of Zoology*, **46**, 781–6.

Raynard, R.S. & Cossins, A.R. (1991). Homeoviscous adaptation and thermal compensation of sodium pump of trout erythrocytes. *American Journal of Physiology*, **260**, R916–24.

Reeves, R.B. (1972). An imidazole alphastat hypothesis for vertebrate acid–base regulation: tissue carbon dioxide content and body temperature in bullfrogs. *Respiration Physiology*, **14**, 219–36.

Reeves, R.B. (1985). Alphastat regulation of intracellular acid-base state? In *Circulation, Respiration and Metabolism*, ed. by R. Gilles pp. 414–23. Berlin Heidelberg: Springer-Verlag.

Reipschläger, A. & Pörtner, H.O. (1996). Metabolic depression during environmental hypercapnia: the role of extra- versus intracellular pH in *Sipunculus nudus. Journal of Experimental Biology*, **199**, 1801–7.

Rubinsky, B., Arav, A. & Fletcher, G.L. (1991). Hypothermic protection – a fundamental property of 'antifreeze' proteins. *Biochemical and Biophysical Research Communications*, **180**, 566–71.

Sartoris, F.J. & Pörtner, H-O. (1997*a*). Temperature dependence of ionic and acid–base regulation in boreal and arctic *Crangon crangon* and *Pandalus borealis*. *Journal of Experimental Marine Biology and Ecology*, **211**, 69–83.

Sartoris, F.J. & Pörtner, H-O. (1997*b*). Elevated haemolymph magnesium protects intracellular pH and ATP levels during temperature change and anoxia in the common shrimp, *Crangon crangon*. *Journal of Experimental Biology*, **200**, 785–92.

Schwarzbaum, P.J., Wieser, W. & Niederstätter, H. (1991). Contrasting effects of temperature acclimation on mechanisms of ionic regulation in a eurythermic and a stenothermic species of freshwater fish (*Rutilus rutilus* and *Salvelinus alpinus*). *Comparative Biochemistry and Physiology*, **98A**, 483–9.

Schwarzbaum, P.J., Niederstätter, H. & Wieser, W. (1992*a*). Effects of temperature on the (Na^+ K^+)-ATPase and oxygen consumption in hepatocytes of two species of freshwater fish, roach (*Rutilus rutilus*) and brook trout (*Salvelinus fontinalis*). *Physiological Zoology*, **65**, 699–711.

Schwarzbaum, P.J., Wieser, W. & Cossins, A.R. (1992*b*). Species-specific responses of membranes and the Na^+ K^+ pump to temperature change in the kidney of two species of freshwater fish, roach (*Rutilus rutilus*) and Arctic charr (*Salvelinus alpinus*). *Physiological Zoology*, **65**,17–34.

Somero, G.N. (1991) Biochemical mechanisms of cold adaptation and stenothermality in Antarctic fish. In *Biology of Antarctic Fish*, ed. G. di Prisco, B. Maresca & B. Tota, pp. 232–47. Berlin: Springer-Verlag.

Somero, G.N., Giese, A.C. & Wohlschlag, D.E. (1968). Cold adaptation of the Antarctic fish, *Trematomus bernacchii*. *Comparative Biochemistry and Physiology*, **26**, 223–33.

Sommer, A., Hummel, H. & Pörtner, H.O. (1996). Adaptation of *Arenicola marina*, to changing temperatures: a comparison of boreal and subpolar populations. *Verhandlungen der Deutschen Zoologischen Gesellschaft* **89**.1, 186.

Sommer, A., Klein, B. & Pörtner, H.O. (1997). Temperature induced anaerobiosis in two populations of the polychaete worm *Arenicola marina*. *Journal of Comparative Physiology*, **167B**, 25–35.

Southward, A.J., Hawkins, S.J. & Burrows, M.T. (1995). Seventy years' observation of changes in distribution and abundance of zooplankton and intertidal organisms in the western English channel in relation to rising sea temperature. *Journal of Thermal Biology*, **20**, 127–55.

Staurnes, M., Rainuzzo, J.R., Sigholt, T. & Jorgensen, L. (1994). Acclimation of atlantic cod (*Gadus morhua*) to cold water: stress response, osmoregulation, gill lipid composition and gill Na^+-K^+-ATPase activity. *Comparative Biochemistry and Physiology*, **109A**, 413–21.

Stuenkel, E.L. & Hillyard, S.D. (1980). Effects of temperature and salinity on gill Na^+-K^+ ATPase activity in the pupfish, *Cyprinodon salinus. Comparative Biochemistry and Physiology*, **67A**, 179–82.

Taylor, S.E., Taylor, E.W. & Egginton, S. (1993). Seasonal changes in intracellular pH of rainbow trout. Abstracts A13.1, Society of Experimental Biology, Canterbury meeting.

Taylor, S.E., Taylor, E.W. & Egginton, S. (1996). Seasonal temperature acclimatisation of rainbow trout: cardiovascular and morphometric influences on maximal sustainable exercise level. *Journal of Experimental Biology*, **199**, 835–45.

Thebault, M.T. & Raffin, J.P. (1991). Seasonal variations in *Palaemon serratus* abdominal muscle metabolism and performance during exercise, as studied by ^{31}P NMR. *Marine Ecology Progress Series*, **74**, 175–83.

Thiel, H., Pörtner, H.O. & Arntz, W.E. (1996). Marine life at low temperatures – a comparison of polar and deep-sea characteristics. In *Deep-sea and Extreme Shallow-water Habitats: Affinities and Adaptations*, ed. Uiblein, F., J. Ott & M. Stachowitsch – *Biosystematics and Ecology Series* vol. 11, pp. 183–219. Vienna: Austrian Academy of Sciences.

Thomas, R.C. (1984). Experimental displacement of intracellular pH and the mechanism of its subsequent recovery. *Journal of Physiology*, **354**, 3P-22P.

Van Dijk, P., Hardewig, I. & Pörtner, H.O. (1997). The adjustment of intracellular pH after temperature change in fish: relative contributions of passive and active processes. *American Journal of Physiology*, **272**, R84–9.

Ventrella, V., Pagliarani, A., Prini, M., Trombetti, F. & Borgatti, A.R. (1993). Lipid composition and microsomal ATPase activities in gills and kidneys of warm- and cold-acclimated sea bass (*Dicentrarchus labrax* L.). *Fish Physiology and Biochemistry*, **12**, 293–304.

Werner, B. (1956). Über die Winterwanderung von *Arenicola marina* L. (Polychaeta sedentaria). *Helgoländer Wissenschaftliche Meeresuntersuchungen* **5**, 353–78.

Whiteley, N.M. & Taylor, E.W. (1993). The effects of seasonal variations in temperature on extracellular acid–base status in a wild population of the crayfish *Austropotamobius pallipes. Journal of Experimental Biology*, **181**, 295–311.

Whiteley, N.M., Naylor, J.K. & Taylor, E.W. (1995*a*). Extracellular and intracellular acid–base status in the fresh water crayfish *Austropotamobius pallipes* between 1 and 12 °C. *Journal of Experimental Biology*, **198**, 567–76.

Whiteley, N.M., Taylor, E.W. & El Haj, A.J. (1995*b*). Acid–base regulation in crustaceans living at low temperatures. *Physiological Zoology*, **68**, 190.

Zielinski, S. & Pörtner, H-O. (1996). Energy metabolism and ATP free-energy change of the intertidal worm *Sipunculus nudus* below a critical temperature. *Journal of Comparative Physiology*, **166B**, 492–500.

B.D. SIDELL and M.E. VAYDA

Physiological and evolutionary aspects of myoglobin expression in the haemoglobinless Antarctic icefishes

Fish fauna of the Southern Ocean present an unique combination of biological characteristics and evolutionary history compared with those from other marine systems. Two features figure prominently in setting these organisms apart from fishes of temperate zone and even polar boreal seas.

First, the level of endemism of Antarctic fishes is unparalleled in other ocean systems. Of the 250+ species of fish known to inhabit the Southern Ocean, the dominant group, in terms of both species numbers (>100) and abundance (50–90% of captures) are members of the perciform suborder Notothenioidei (Dewitt, 1971; Anderson, 1990; Eastman, 1993). With few exceptions, fishes of the six notothenioid families are indigenous to waters surrounding Antarctica where they have evolved during the last 25–40 My in isolation under conditions that are both thermally stable and severely cold. Within this monophyletic group are species displaying a wide diversity of ecologies and life histories, from sluggish demersal to active pelagic habits.

The second major feature that sets Antarctic notothenioid species apart from the ichthyofauna of other marine systems is their long geographical isolation in waters that are the most severely cold, thermally stable aquatic habitat on the planet. The best estimates are that thermal isolation of Antarctica began with the development of circumpolar currents in the late Oligocene and was followed shortly thereafter with the establishment of the Antarctic Convergence (about 20 million years ago) (Kennett, 1977, 1980). The demise of most non-notothenioid fishes and radiative expansion of this suborder in coastal Antarctica apparently began with the significant ocean cooling that predated these events (Anderson, 1990). At present, mean annual temperature in McMurdo Sound is −1.86 °C and varies only by about 0.1 °C seasonally (Littlepage, 1965). The Antarctic Peninsula shows only slightly greater variance with average summer and winter temperatures running between −1.1° and +0.3 °C (summer) and −1.1 °C in winter (Dewitt, 1971).

The very cold oxygen-rich waters of the Southern Ocean coincidentally provide both challenges and benefits with respect to respiratory requirements for oxygen. On the benefit side, exceptionally cold body temperature seems to have lowered the absolute oxygen demand for maintenance metabolism in Antarctic species relative to warm water counterparts (*e.g.* see Macdonald, Montgomery & Wells, 1987). Additionally, aqueous oxygen solubility increases with decreasing temperature and the water column south of the Antarctic Convergence is well mixed vertically; thus, all piscine habitats are well oxygenated. In many species of Antarctic fishes, haematocrit of the blood is either substantially reduced or red cells and haemoglobin (Hb) are absent (as in channichthyid icefishes). These conditions lower the otherwise high viscosity of blood at cold temperature (Hemmingsen & Douglas, 1977; Macdonald & Wells, 1991). One family of notothenioid fishes, the *Channichthyidae* or icefishes, are entirely devoid of haemoglobin. In these species, lack of haemoglobin is compensated by increased heart size, blood volume and cardiac output (Hemmingsen *et al.*, 1972; Hemmingsen, 1991). Cardiovascular adaptations considered to compensate for the absence of haemoglobin, however, would appear to contribute little to overcoming problems of intracellular oxygen movement in the highly aerobic tissues of this group. This latter problem will be focused on in the present chapter.

Do channichthyid icefishes express myoglobin?

Although the hearts of some northern temperate zone marine fishes appear to lack the protein (Driedzic & Stewart 1982), one of the diagnostic features of oxidative muscle tissues in the vast majority of vertebrate animals is high concentration of the intracellular oxygen-binding protein, myoglobin (Mb). Myoglobin both facilitates intracellular diffusion of oxygen and functions as an important intracellular reservoir for oxygen, which may be drawn upon during transient periods of hypoxic activity (Wittenberg & Wittenberg, 1989). Although the functions, genes and tissue-specific expression of haemoglobin and myoglobin are distinct, it was generally accepted until the mid-1980s that channichthyid icefish lack both these haemoproteins. Some confusion on this point, however, arose subsequent to the report of Douglas and co-workers that myoglobin is expressed in heart tissue from two icefish species, *Pseudochaenichthys georgianus* and *Chaenocephalus aceratus* (Douglas *et al.*, 1985). Our observations of the pale yellowish-green colour of cardiac tissue from the latter species (suggesting the absence of myoglobin) and the methodology upon which Douglas and coworkers based their conclusions prompted us to re-examine whether myoglobin was, indeed, expressed in this family of fishes.

Douglas and co-workers had concluded that myoglobin was present in the two icefish species that they examined, but their assessment lacked direct demonstration of the myoglobin polypeptide and was based solely upon the development of a spectrally detectable pyridine haemochromagen in high-speed supernatants from homogenised heart muscle tissue of the species mentioned above. Because of the high mitochondrial content (>40% of cell volume) of heart tissue from icefishes (Johnston & Harrison, 1987), there was a significant probability that homogenisation conditions used by Douglas and co-workers could have resulted in the release of haem-containing cytochromes into the high-speed supernatants, leading to an incorrect conclusion that myoglobin was present in these tissues. It has recently been possible to exploit availability of specific antibodies directed against myoglobin to ascertain definitively whether myoglobin protein is expressed in tissues from species of the *Channichthyidae* (Sidell *et al.*, 1997).

To date, 8 of the 15 known species of the *Channichthyidae* have been examined to determine the presence or absence of myoglobin protein in their oxidative muscle tissues. The approach for this phase of the work has been to perform immunoblots of soluble protein extracts using commercially available polyclonal and monoclonal antibodies raised against human myoglobin (Fig. 1). First it was established that each of the anti-Mb antibodies used cross-reacted strongly with myoglobin from tissues of closely related red-blooded notothenioids (same suborder as the channichthyid icefishes) that are known to express myoglobin in their heart ventricles. Results of these experiments reveal considerable variability in the expression of myoglobin among species of the icefish family.

Of the eight species of icefish examined, only three show absence of detectable myoglobin protein in extracts of heart ventricle (*Chaenocephalus aceratus, Champsocephalus gunnari* and *Pagetopsis macropterus* (Fig. 2)). Thus, one of the species characterised by Douglas and co-workers as Mb(+) apparently does not express the protein. Myoglobin protein, however, is present in heart ventricles of five others, *Pseudochaenichthys georgianus, Chionodraco rastrospinosus, Chionodraco hamatus, Chaenodraco wilsoni* and *Cryodraco antarcticus*. In addition to this unusual species-specific pattern of myoglobin expression in heart tissue among the channichthyid icefishes, it has also been possible to establish that myoglobin protein is not expressed in highly aerobic oxidative skeletal muscles from any member of this family. This latter curious tissue-specific pattern of myoglobin expression is a characteristic that is shared by all other notothenioid fishes (including 12 species of red-blooded notothenioids) that have been examined to date, suggesting that the event leading to loss of myoglobin expression in oxidative skeletal muscle occurred very early in the notothenioid lineage and prior to the divergence of the *Channichthyidae*. The peculiar variation among icefish species in cardiac expression of myoglobin has recently led to the investigation of the molecular genetic mechanisms that may underlie loss of myoglobin expression among the icefishes.

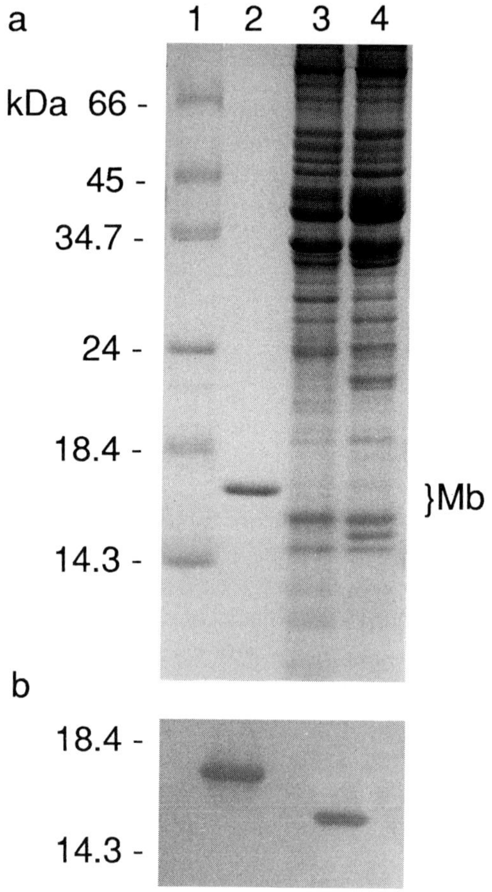

Fig. 1. Presence of myoglobin polypeptide in Antarctic icefishes. (Figure is from Sidell *et al.*, 1997) **a.** Soluble polypeptides were liberated from heart ventricles and resolved by SDS-PAGE: lane 1=protein molecular weight standards, lane 2=human heart myoglobin standard (1 μg), lane 3=*Chaenocephalus aceratus* (35 μg total protein), lane 4=*Chionodraco rastrospinosus* (35 μg total protein). Icefish myoglobin polypeptide has a molecular size of approximately 16 kDa (146 amino acids) while the more slowly migrating human myoglobin polypeptide has a molecular size of 17.8 kDa (153 amino acids). **b.** Duplicate gel to that shown in **a.** above was electroblotted to polyvinylidine diflouride membrane for Western blot analysis. The PVDF membrane was incubated with a mouse anti-human monoclonal antibody. Bound antibody was detected by a secondary antibody covalently linked to alkaline phosphatase and then incubated in stabilised Western-Blue substrate (Promega). Only the molecular weight range containing myoglobin is shown. Lane assignments are described in **a.** above.

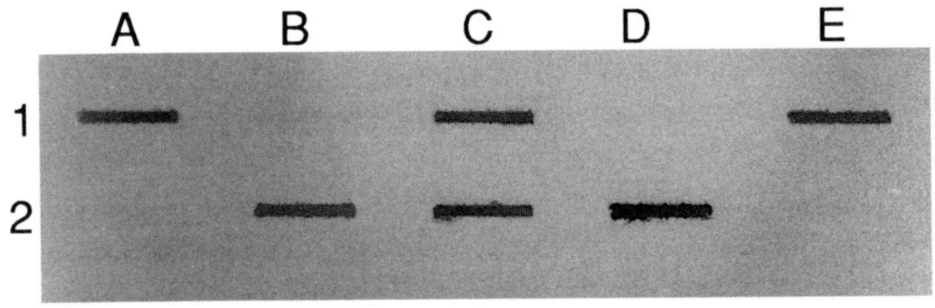

Fig. 2. Slot immunoblot of soluble polypeptides from heart ventricles of eight channichthyid icefishes (1A to 2C) and one red-blooded Antarctic nototheniid fish (2D). Equal amounts (5 μg) of total protein were loaded in each slot and immunoblotted as described in **Fig. 1** above. 1A=*Chionodraco rastrospinosus*, 1B=*Pagetopsis mactropterus*, 1C=*Pseudochaenichthys georgianus*, 1D=*Chaenocephalus aceratus*, 1E=*Chionodraco hamatus*, 2A=*Champsocephalus gunnari*, 2B=*Chaenodraco wilsoni*, 2C=*Cryodraco antarcticus*, 2D=*Notothenia coriiceps*. (Figure is from Sidell *et al.*, 1997.)

Mechanisms of loss of myoglobin protein expression

Several potential explanations for the absence of myoglobin protein in heart tissue of the Mb(−) icefish species may be identified *a priori*. First, it is possible that a catastrophic deletion of the structural gene for myoglobin from the genomic DNA accounts for the absence of the ultimate gene product. Secondly, failure to transcribe the myoglobin gene or process the myoglobin transcript (by a variety of potential mechanisms) may result in lack of competent mRNA for myoglobin. Finally, it is possible that mRNA encoding the myoglobin protein is produced, but that a subsequent block to translation of this message into functional protein accounts for the absence of myoglobin in the tissues. It is worthwhile pointing out that, in addition to actual blockage of transcriptional or translational processes, factors leading to instability and rapid degradation of either mRNA or the myoglobin protein may contribute to the observed pattern and would be very difficult to distinguish from block of either process without half-life studies in the presence of translational or transcriptional inhibitors.

To attempt to differentiate among the possible mechanisms described above, Mb-specific oligonucleotide primers were developed based on the polypeptide sequences of myoglobins from tuna (*Thunnus albacares*) and carp (*Cyprinus carpio*). Amplification was by PCR (Polymerase chain reaction) and a partial myoglobin cDNA was cloned from the closely related red-blooded Antarctic notothenioid, *Notothenia coriiceps* (see Vayda *et al.*, 1995, 1997 for specific methodological details). Sequencing confirmed that this cDNA encoded a myoglobin polypeptide with >79% identity to the

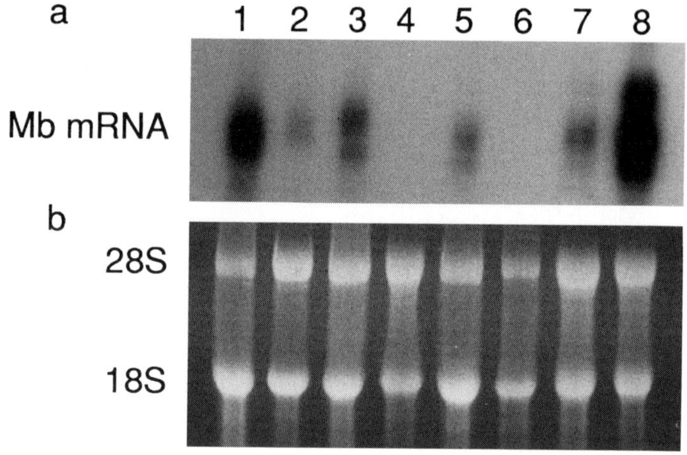

Fig. 3. Interspecific pattern of myoglobin mRNA expression in Antarctic fishes. (Figure is from Sidell *et al.*, 1997.) **a.** Northern blot analysis performed on total RNA extracted from heart ventricle. Myoglobin mRNA (0.9 kb) was detected by hybridisation with random-primed ^{32}P-labelled *Notothenia coriiceps* myoglobin probe. Lane 1.=*Chionodraco rastrospinosus*, lane 2=*Champsocephalus gunnari*, lane 3=*Pseudochaenichthys georgianus*, lane 4=*Chaenocephalus aceratus*, lane 5=*Cryodraco antarcticus*, lane 6=*Pagetopsis macropterus*, lane 7=*Chaenodraco wilsoni*, lane 8=*Notothenia coriiceps*. Analysis of *Chionodraco hamatus* total RNA also indicated presence of myoglobin mRNA in heart ventricle (data not shown). **b.** Ethidium bromide-staining of the 28S and 18S ribosomal RNA bands from total extracted RNA used in **a.** above.

polypeptide from tuna and carp. This partial nototheniid myoglobin cDNA was used to probe Northern blots of icefish heart and aerobic skeletal muscle mRNAs.

Given the absence of myoglobin protein in *C. aceratus, C. gunnari* and *P. macropterus*, initial expectation was that an equivalent absence of detectable myoglobin mRNA would be observed in tissues of these fishes. Indeed, no myoglobin mRNA was observed in aerobic skeletal muscle of any Antarctic notothenioid examined. Surprisingly, this was not the case uniformly for hearts from these three species. Although it was not possible to detect any evidence for mRNA encoding for myoglobin in heart tissues of *Chaenocephalus aceratus* or *Pagetopsis macropterus*, in preparations from all specimens of *Champsocephalus gunnari*, a mRNA has been observed consistently that hybridises to the myoglobin probe and exhibits the same size as message from Mb(+) icefish species (Fig. 3). Thus, for *C. gunnari*, the myoglobin structural gene is apparently successfully transcribed and mature

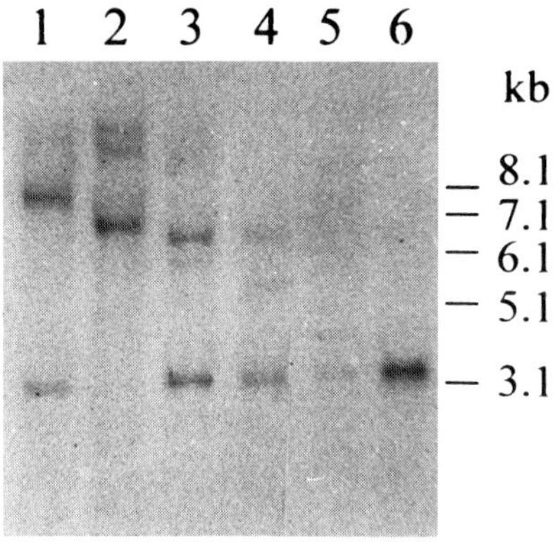

Fig. 4. Southern blot hybridisation of notothenioid genomic DNA with ^{32}P-labelled *Notothenia coriiceps* myoglobin cDNA. (Figure is from Sidell *et al.*, 1997). Genomic DNA isolated from *N. coriiceps* (lane 1), *Gobionotothen gibberifrons* (lane 2), *Chionodraco rastrospinosus* (lane 3), *Champsocephalus gunnari* (lane 4), *Pagetopsis macropterus* (lane 5) and *Chaenocephalus aceratus* (lane 6) were digested with *Hin*dIII, resolved by electrophoresis through a 0.7% agarose gel and hybridised with the random-primed myoglobin cDNA probe. Mobility of molecular size markers is shown at right.

myoglobin mRNA accumulates despite absence of detectable translated protein product in the heart. In the former two species, *C. aceratus* and *P. macropterus*, the possibility of loss of the structural gene for myoglobin from each species' genome could not be ruled out. To pursue this question further, genomic DNA has been prepared, isolated from *Chaenocephalus aceratus, Pagetopsis macropterus* and both red- and white-blooded species that express myoglobin. Southern blot analyses using the myoglobin cDNA probe developed from *N. coriiceps* reveals the presence of the myoglobin structural gene in all species examined to date (Fig. 4). Indeed, a genomic clone was isolated from *Chaenocephalus aceratus*, which includes the entire myoglobin gene. These results demonstrate that catastrophic deletion of the gene can be ruled out as a mechanism accounting for loss of myoglobin expression in species not expressing the pigment. Loss of myoglobin expression is distinct from the mechanism that accounts for the loss of haemoglobin expression among the channichthyid icefishes where the gene encoding for β-globin subunits of the molecule appears to be missing from genomic DNA (Cocca *et al.*, 1995).

The results described above resolve any confusion existing in the literature regarding myoglobin expression among the haemoglobinless icefishes. However, the pattern that has been observed is curious. Unlike previous conventional wisdom, it is clear that myoglobin is expressed by the majority of channichthyid species examined to date. Further, expression of the pigment is confined to heart ventricle only and myoglobin protein is absent from oxidative skeletal muscle of all notothenioid species, including the icefishes. Among the three icefish species that do not express myoglobin in their heart ventricles, at least two discrete classes of mechanism account for the loss of the protein. In *C. aceratus* and *P. macropterus*, mature messenger RNA encoding myoglobin is apparently not produced, despite the presence of the myoglobin gene in DNA of the species. Thus, no template exists for translation of myoglobin protein in these two species. In *Champsocephalus gunnari*, mature myoglobin mRNA is detectable, albeit at lower apparent levels than in Mb(+) species, but this message is not translated to produce myoglobin protein at detectable levels. The available phylogenetic relationships among species of this family can now be turned to gain understanding of when, during radiation of channichthyid species, mutations leading to the loss of myoglobin expression occurred.

Evolutionary mapping of the loss of myoglobin expression

Relationships among notothenioid families (Eastman, 1993) and within the *Channichthyidae* (Iwami, 1985) have been advanced, based on a cladistic analysis of morphological characteristics. The proposed relationships among the notothenioids have been confirmed largely by sequence analysis of the mitochondrial 16S rRNA gene (Bargelloni *et al.*, 1994). In this analysis, the small degree of sequence divergence observed is insufficient to resolve relationships among the *Channichthyidae* and it is Iwami's morphological tree that provides the only available phylogenetic framework upon which the question 'When was myoglobin expression lost in the channichthyid icefishes?' may be asked.

When the presence and absence of myoglobin protein and myoglobin mRNA that has been documented in our survey is mapped upon the Iwami phylogeny of the *Channichthyidae* (Fig. 5), several deductions can be drawn. First, because of the considerable phylogenetic distance between the three Mb(−) icefishes identified to date, it can be concluded that loss of myoglobin expression has occurred via a minimum of at least three independent events during the evolution of this family. Secondly, based upon the pattern of myoglobin protein and myoglobin mRNA presence and absence described earlier, it is clear that at least two entirely different types of genetic lesion have led to the Mb(−) condition among these species. The seemingly random pattern of loss in expression of this haemoprotein, which is normally important to the function of oxidative muscle tissues of vertebrate animals, raises fascinating questions regarding the physiological role of myoglobin in this exceptionally cold-bodied group of fishes.

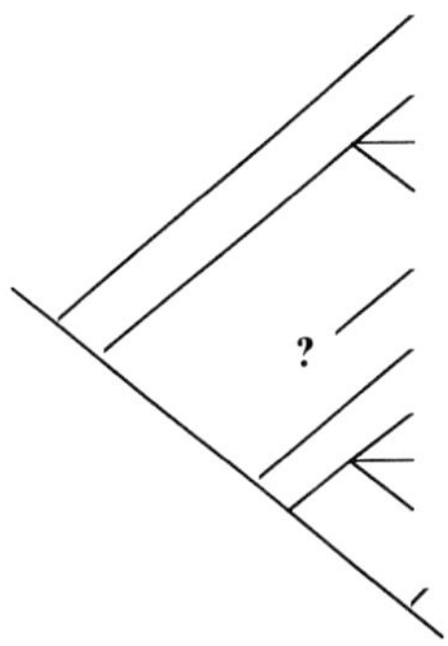

Genus	Total number of species	Species examined	Poly-peptide	RNA
Champsocephalus	2	*gunnari*	-	+
Pseudochaenichthyes	1	*georgianus*	+	+
Neopagetopsis	1			
Pagetopsis	2	*macropterus*	-	-
Dacodraco	1			
Channichthys	1			
Cryodraco	1	*antarcticus*	+	+
Chionobathyscus	1			
Chaenocephalus	1	*aceratus*	-	-
Chionodraco	3	*rastrospinosus*	+	+
		hamatus	+	+
Chaenodraco	1	*wilsoni*	+	+

Fig. 5. Expression of myoglobin polypeptide and mRNA in cardiac muscles with respect to phylogenetic relationships among channichthyid species. Phylogeny of family *Channichthyidae* is based upon morphological characters (Iwami, 1985). (Figure is from Sidell *et al.*, 1997.)

At least superficially, the pattern described above is consistent with the view that myoglobin may be a 'vestigial' protein and poorly functional (if at all) at very cold body temperatures. If such were the case, then any mutation resulting in loss of myoglobin expression would not have been subject to negative selective pressure during the evolution of the *Channichthyidae*. The validity of the cladistic tree presented in Fig. 5 is critical to this interpretation. Although morphological phylogenies are usually correct, some important morphological characters may be plastic or environmentally determined (e.g. see Bronmark & Miner, 1992). To provide an independent test of the morphologically based phylogeny of the channichthyid icefishes, collaboration between the authors and Irv Kornfield and co-workers at the University of Maine is developing an independent molecular phylogeny of the *Channichthyidae* using sequence analysis of the hypervariable mitochondrial displacement loop (D-loop) in these species. Preliminary results suggest substantial concordance with Iwami's morphologically based phylogeny of the family.

Several observations provide support *a priori* for the hypothesis that myoglobin function may be compromised at severely cold body temperature. First, at cold temperatures, other vertebrate myoglobins show an increase in their already very high affinity for oxygen and a dramatic lengthening of their kinetic off-constants for dissociation of oxygen (Stevens & Carey, 1981; Sato *et al.*, 1990). Secondly, the diffusion coefficient for the oxymyoglobin complex should be temperature dependent (Stevens & Carey, 1981; Dowd,

Murali & Seagrave, 1991), and presumably is influenced by the considerable increase in cytoplasmic viscosity encountered at cold body temperatures (Sidell & Hazel, 1987). Taken together, these factors suggest that conditions approaching cellular anoxia would be required to induce any dissociation of O_2 from the pigment, and that the facilitated flux of O_2 would be very restricted at best, unless specific modifications of the protein have occurred in Antarctic fishes to ensure improved myoglobin function at cold body temperature. Lack of a functional role for myoglobin also is consistent with the absence of the protein from highly oxidative skeletal muscle tissue of both icefishes and red-blooded notothenioids, a tissue with considerable aerobic potential as judged by both enzymatic and ultrastructural characteristics (Crockett & Sidell, 1990; Londraville & Sidell, 1990).

As a counterpoint to the view of non-functionality of myoglobin, considerable data on structure, metabolism and performance of muscular tissues from Antarctic fishes have been interpreted within the context of the presence/absence of functional myoglobin. In comparing enzymatic and ultrastructural characteristics of ventricular myocardia from haemoglobin- and myoglobin-containing *Notothenia neglecta* (now considered as the same species as *Notothenia coriiceps*) and haemoglobinless and myoglobinless *C. aceratus*, Johnston and Harrison (1987) noted a greater volume density of mitochondria in icefish heart than that of *N. neglecta*, yet roughly equivalent activities of cytochrome oxidase *per* gram of tissue. These workers concluded that the proliferation of mitochondria in icefish hearts has occurred to compensate for the absence of myoglobin by reduction in mean diffusional pathlength between the ventricular lumen and mitochondria. It is worth pointing out, however, that interpretation of the above interspecific comparison is complicated by simultaneous variation in more than just the presence and absence of myoglobin. In addition to possessing myoglobin in its heart, *N. neglecta* also is an haemoglobin-expressing red-blooded notothenioid, while *C. aceratus* lacks both of these haemoproteins. Having now established the existence of both Mb(+) and Mb(−) species among the haemoglobinless icefishes, the opportunity is there to isolate this variable in an attempt to resolve whether myoglobin is of functional significance to tissues of these cold-bodied vertebrates.

Is myoglobin functional in the channichthyidae?

Attempts have been made to resolve conflicting hypotheses regarding the physiological role of myoglobin in the icefishes by using a multi-tiered approach that combines studies of protein biochemistry, physiological performance and molecular genetics. First, in collaboration with Robert Cashon of Maine University, comparisons have been made of the oxygen binding and dissociation kinetics of Antarctic fish myoglobins with those of warmer-bodied vertebrates to ascertain whether functional characteristics of the protein in Antarctic fishes have been evolutionarily modified to improve its performance at the physiological temperatures characteristic of this group. Secondly, with Italian colleagues, Raffaele Acierno, Claudio Agnisola and Bruno Tota, advantage had been taken of the unique opportunity of Mb(+) and Mb(−) hearts within the icefish family to determine whether the presence of myoglobin confers greater mechanical performance capabilities on hearts of those species that possess the protein than is observed with hearts of those species in which it is absent. Finally, the above approaches have been complemented with molecular genetic analyses by cloning and sequencing both cDNAs for myoglobin derived from icefish species where the gene is active and genomic DNAs of both Mb(+) and Mb(−) species to determine the extent of sequence conservation in both coding and non-coding regions of the myoglobin gene. In the following sections, results will be described to date from these efforts that collectively support a functional role for myoglobin in tissues of Antarctic fish species.

Oxygen-binding kinetics of myoglobins from Antarctic fishes

Investigations with Robert Cashon have compared both apparent oxygen affinities and kinetic constants for association and dissociation of myoglobins from mammals (horse, sperm whale), a teleost capable of local muscular endothermy (yellowfin tuna, *Thunnus albacares*), a temperate zone marine teleost (mackerel, *Scomber scombrus*) and two Antarctic fish species, one red-blooded (*Notothenia coriiceps*) and one haemoglobinless icefish (*Chionodraco rastrospinosus*) (Cashon, Vayda & Sidell, 1997). At least at the level of apparent oxygen affinities, there appear few salient differences among myoglobins from these diverse vertebrate species. Tonometric estimates of P_{50} values (partial pressure of O_2 at which the protein is 50% saturated with oxygen), performed at 22 °C, range only between 0.6 and 1.0 mm Hg among these species, with the exception of *S. scombrus* whose myoglobin shows a P_{50} value of 3.7 mm Hg (Cashon *et al.*, 1997). Although these experiments were conducted well outside the physiological temperature range of Antarctic fishes, variance of <1 mm Hg in P_{50} values between myoglobins from mammalian and Antarctic fish species suggest little likelihood that this parameter of myoglobin function has been modified to yield differential performance of the protein between these groups. However, because of the dynamic nature of myoglobin's normal role in facilitating transcytoplasmic oxygen flux, affinity of the protein for O_2 is only part of the picture. It was

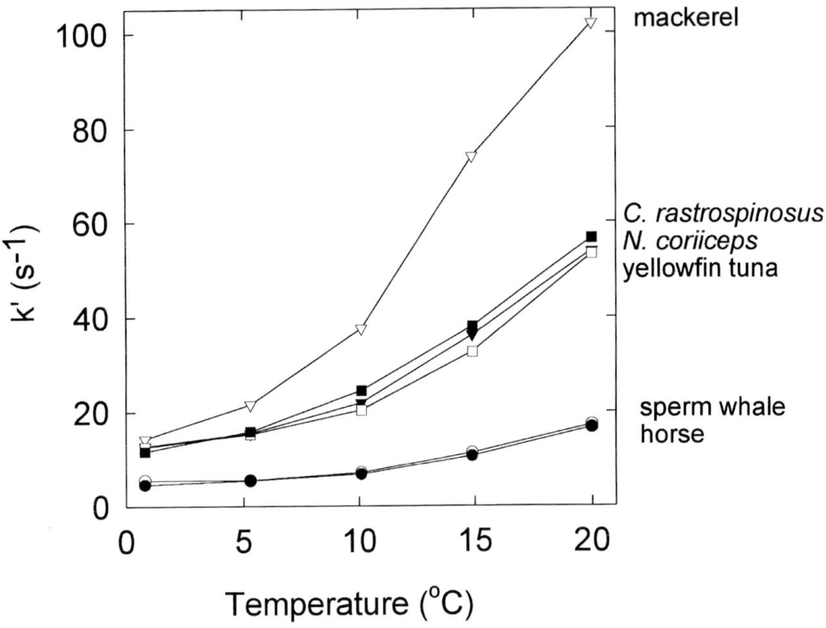

Fig. 6. Oxygen dissociation rate constants at different experimental temperatures for myoglobins from: horse, sperm whale, mackerel (*Scomber scombrus*), yellowfin tuna (*Thunnus albacares*), red-blooded Antarctic fish (*Notothenia coriiceps*) and the myoglobin-expressing icefish (*Chionodraco rastrospinosus*). Data were collected using an Applied Photophysics SF-17 microvolume stopped-flow spectrophotometer. (Figure is adapted from Cashon *et al.*, 1997.)

reasoned that the time constants for rates of both binding (association) and release (dissociation) of O_2 from myoglobin might be parameters of even more significance to the physiological role of myoglobin. A series of experiments have been performed to estimate these values via stopped-flow kinetics using instrumentation made accessible through the generosity of Dr A.I. Alayash (Center for Biologics Evaluation and Research, US FDA, Bethesda, MD).

Both rate constants for oxygen dissociation and for carbon monoxide association (considered to mimic oxygen association) have been estimated for myoglobins from each of the species mentioned above over a range of experimental temperatures from 0.8 to 20.0 °C (Figs 6 and 7). Several features of these data deserve consideration. First, it is clear that no consistent pattern of difference in characteristics between Antarctic fishes and warmer-bodied fishes is discernible from the results. Dissociation rate con-

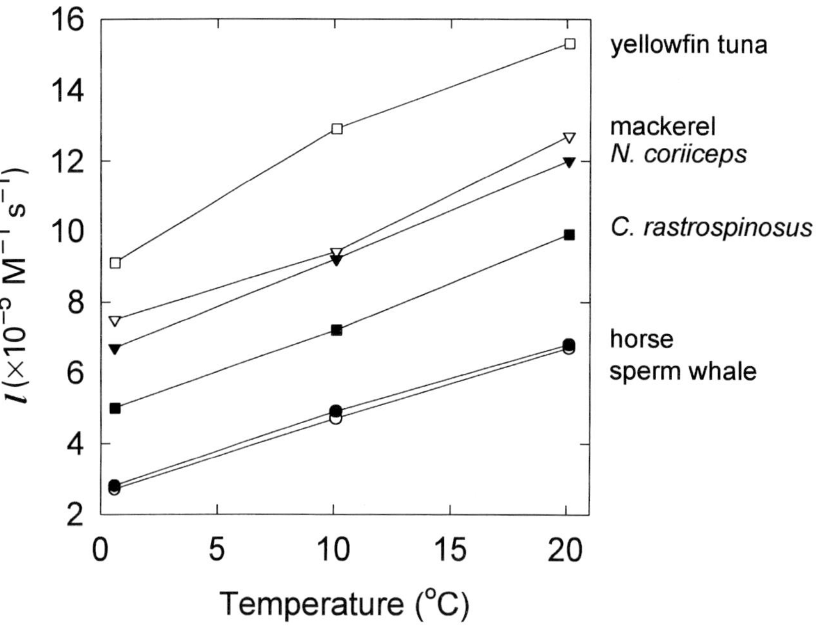

Fig. 7. The effect of experimental temperature on the carbon monoxide association constant of myoglobins from mammals, temperate zone fishes and Antarctic fishes (see Fig. 6 for species descriptions, instrumentation used and source of figure).

stants of yellowfin tuna and the two Antarctic species are very similar (Fig. 6). Carbon monoxide association rate constants show more variability, but mackerel and *C. rastrospinosus* results are very similar (Fig. 7). The second and important feature of the data is that, regardless of measurement temperature, rates of both association and dissociation consistently are much slower for myoglobins from mammalian species than those from the fishes as a group. Indeed, when data for the mammalian species are extrapolated to normal body temperature of 37 °C and compared with those measured for Antarctic fish myoglobins at the physiological temperature of 0.8 °C, the performance of myoglobins from these groups are not as vastly different as one would predict from the thermal characteristics of the mammalian myoglobins alone. In other words, myoglobins from fishes in general (including Antarctic fishes) are better suited for function at cold body temperature than those of mammals. These results suggest that myoglobins from Antarctic fishes may very well play a functional role at cold body temperature. The data are, however, rather distantly removed from the conditions of intact tissue.

Performance of isolated, perfused hearts from Mb(+) and Mb(−) icefishes

To gain a better insight into the possible role of myoglobin in helping to support the obligately aerobic work of cardiac muscle in the icefishes, the aid of colleagues Raffaele Acierno, Claudio Agnisola and Bruno Tota was enlisted. It was possible to execute a series of experiments to compare the mechanical performance characteristics of isolated, perfused hearts from channichthyid icefishes that either possess myoglobin (*Chionodraco rastrospinosus*) or do not express the protein (*Chaenocephalus aceratus*) (Acierno *et al.*, 1997). Both of these species are relatively sluggish and demersal in habit, eliminating disparities in lifestyle as a significant contributor to any differences that might be observed in performance of the hearts. In the preparation used, cardiac output (CO) was regulated by varying saline input pressure to the atrium and was stabilised at 100 ml $(\text{min kg})^{-1}$ at an afterload pressure of 2.5 kPa. These conditions reflect the resting values reported for *C. aceratus* (Hemmingsen *et al.*, 1972). The isolated hearts were challenged by gradually increasing static output pressure (afterload) in increments of 0.5 kPa. This design was chosen rather than volume-loading of the hearts because hearts of icefish have been shown to be susceptible to afterload challenge and to respond to volume-loading with variable flows that would result in differential delivery of oxygen to the tissue (Tota, Acierno & Agnisola, 1991). Initial results with saline-perfused hearts were compelling.

Myoglobin-containing hearts from *C. rastrospinosus* clearly were capable of maintaining cardiac output at higher afterload challenges than the myoglobin-lacking hearts of *C. aceratus* (Fig. 8). In other words, Mb(+) hearts were capable of greater pressure-work than Mb(−) hearts, strongly suggesting that presence of this oxygen-binding haemoprotein contributes to enhanced cardiac performance. Despite the provocative nature of these results, however, it was recognised that other non-Mb differences in the physiology or biochemistry of the hearts could be responsible for their disparate performance capabilities. Attempts were thus made to isolate more definitively myoglobin's role in accounting for the above differences in performance.

A series of experiments were conducted with isolated, perfused hearts that paralleled those described above. In these experiments, however, 5.0 mM $NaNO_2$ was incorporated into the saline perfusate. At this concentration, sodium nitrite selectively and reversibly poisons the ability of myoglobin to bind oxygen, while not interfering with the ability of mitochondria to respire (Bailey, Sephton & Driedzic, 1990). If myoglobin function was accountable for the differences in performance between hearts of *C. rastrospinosus* and *C. aceratus* observed during our saline perfusion experiments, it was reasoned

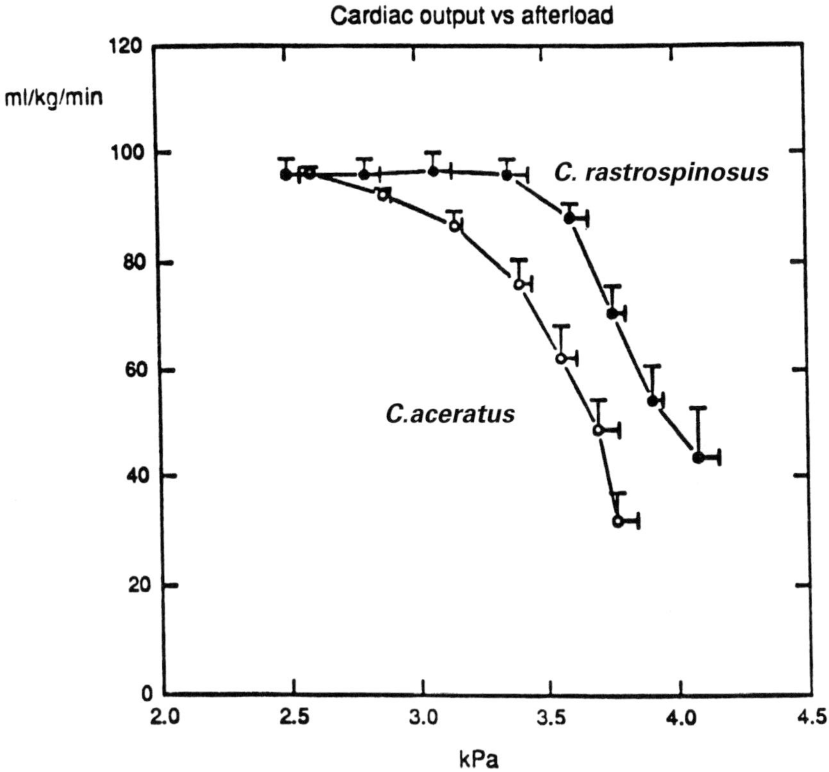

Fig. 8. Ability of Mb(+) (*Chionodraco rastrospinosus*) and Mb(−) (*Chaenocephalus aceratus*) isolated, perfused hearts of icefishes to perform in the face of increasing afterload challenge. Hearts were permitted to beat at their intrinsic myogenic rhythm which was not affected by afterload challenge. Initial cardiac output was set at 100 ml (min kg body wt)$^{-1}$, which is a physiological value for these species. Experimental temperature$=1.0\pm 0.5$ °C. (Figure is adapted from Acierno *et al.*, 1997.)

that inclusion of $NaNO_2$ in the perfusate should cause significant decrement in the mechanical performance of hearts from the former species while those from the latter should remain refractory to this treatment. Those expectations were met precisely by the results. The ability of Mb(−) hearts of *C. aceratus* to meet afterload challenges was unaffected by treatment with $NaNO_2$ (Fig. 9). In sharp contrast, $NaNO_2$-treated hearts from Mb(+) *C. rastrospinosus* were incapable of meeting the afterload challenge of untreated controls (Fig. 10). Indeed, performance of Mb-poisoned hearts of *C. rastrospinosus* was so greatly diminished that hearts of *C. aceratus*, which

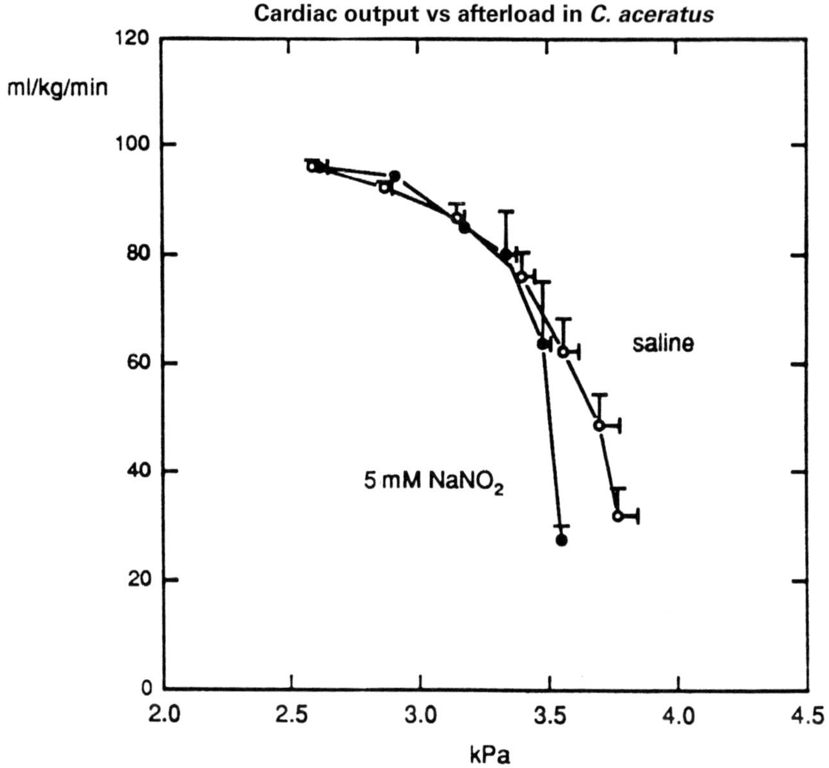

Fig. 9. Lack of effect 5 mM $NaNO_2$ on mechanical performance of hearts from Mb(−) *Chaenocephalus aceratus*. Sodium nitrite, at this concentration, selectively poisons function of myoglobin (see text). Because performance of hearts from *C. aceratus*, that lack myoglobin, is unaffected by this treatment, we can rule out non-myoglobin related effects of the compound. Experimental conditions are as described for Fig. 8 above. (Figure is adapted from Acierno *et al.*, 1997.)

inherently lack the protein, were capable of meeting a greater afterload challenge. These experiments provide very strong evidence that the presence of functional myoglobin contributes to cardiac performance at the cold physiological temperatures of channichthyid icefishes.

Nucleotide sequence conservation among myoglobins of the icefishes

The final component of the experiments was designed to probe whether myoglobin protein is functionally important to the channichthyid icefishes by

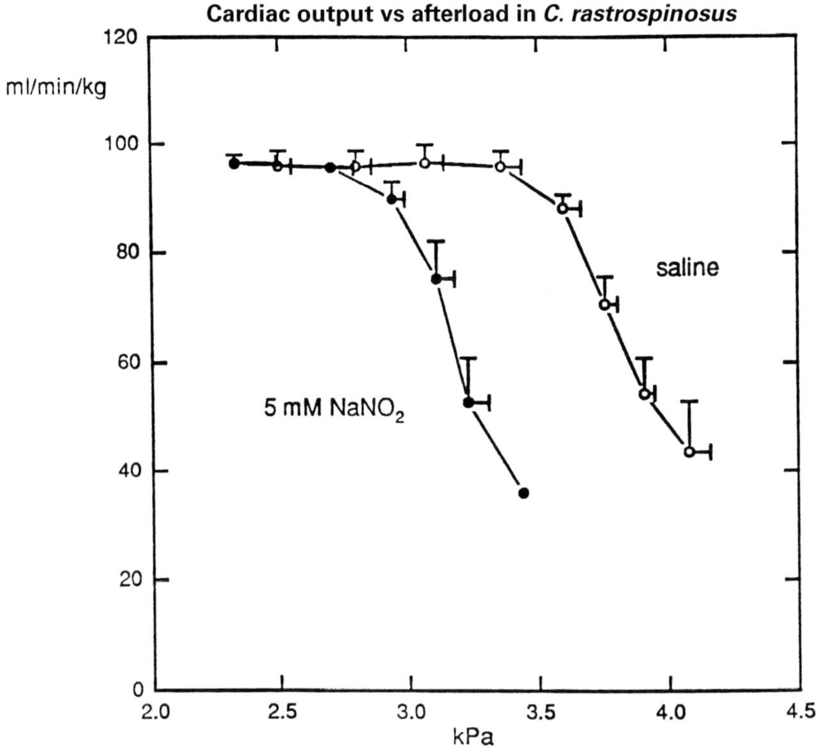

Fig. 10. Mechanical performance of hearts from Mb(+) *Chionodraco rastrospinosus* is dramatically impaired by selective poisoning of myoglobin with 5 mM $NaNO_2$. Experimental conditions are as described for Fig. 8 above. (Figure is adapted from Acierno *et al.*, 1997.)

assessing whether the gene sequence is under selective pressure. Specifically, both cDNA and genomic DNA have been cloned and sequenced from multiple species of channichthyid icefishes. Because cDNA is produced via reverse transcriptase reaction from the pool of cardiac messenger RNA, this sequence reflects the complete coding sequence for myoglobin plus upstream and downstream flanking sequences of the mature mRNA molecule. Sequence analysis of the myoglobin gene from genomic DNA libraries that have been prepared permits evaluation of nucleotide sequence divergence/conservation among species in both the regions of the gene that code for the protein (exons) and also in non-coding regions (introns) that are removed from the primary transcript during splicing of this RNA to form mature mRNA for myoglobin. Conservation of sequence in both cDNA and coding regions of genomic DNA would be consistent with a positive selection pressure being maintained on the gene for myoglobin. Results to date are very consistent with both oxygen-binding and perfused heart studies and indicate that selective pressure on myoglobin is being maintained within the *Channichthyidae*.

Thus far, cDNA sequences have been obtained for myoglobin from seven species of notothenioid fish, including five species of icefishes and two species of red-blooded nototheniid relatives, *Notothenia coriiceps* and *Gobionothen gibberifrons* (Vayda *et al.*, 1997). Nucleotide sequences of coding regions among these species are extremely well conserved and encode nearly identical myoglobin proteins. Among the channichthyid species, sequence variation of myoglobin cDNA was <2.5%. A similar value of <2.5% sequence divergence is obtained when cDNA sequences from channichthyids and red-blooded notothenioid species are compared.

One of the most interesting findings to develop from the molecular genetic studies yields insight into the absence of myoglobin in *Chaenocephalus aceratus*, a species that does not produce detectable protein or mRNA for myoglobin, but does contain the myoglobin gene in its genomic DNA. By aligning sequences from myoglobin cDNAs of icefishes with the complete myoglobin genomic DNA sequence, it is possible to identify the coding and non-coding regions of the mutant myoglobin gene. Comparison of intron-removed (*i.e.* coding) sequence of genomic DNA of *C. aceratus* with coding sequence of genes from other icefish species reveals that *C. aceratus* does not contain any frameshift mutations (nucleotide deletions or insertions) or point substitutions that would preclude expression of a functional myoglobin polypeptide. Indeed, only three nucleotide differences exist in coding sequences between Mb (+) *C. rastrospinosus* and Mb (−) *C. aceratus*; these substitutions would give rise to only two amino acid changes in the polypeptide. Further, there are no obvious aberrations in the gene structure of *C. aceratus* that would seem to preclude production of viable mRNA for myoglobin. Initiation and termination codons are the same as in Mb (+) *C. rastrospinosus* and, the 5′ untranslated region of the gene is identical in the two species. Intron sequences and polyadenylation signals likewise are similar in both species. In fact, the genes of these Mb (−) and Mb (+) species show 99% identity in their overall sequences.

The observations described above suggest three possible explanations for the lack of production of viable Mb mRNA or myoglobin protein in *Chaenocephalus aceratus*. First, there may be mutations further upstream in the regulatory region of the myoglobin gene in this species that have not yet been identified. Secondly, *C. aceratus* may be mutant in some transcription factor that is required for myoglobin gene expression. Thirdly, myoglobin may be expressed at some time earlier in the life cycle or during development of *C. aceratus*. The last possibility seems unlikely and still would not provide any type of explanation for variation in myoglobin expression between mature hearts of the two species. Likewise, if mutations exist in transcription factors, these undoubtedly would lead to pleiotropic effects on the expression of other heart-and/or muscle-specific genes; no such effects are observed. At present, it would seem that the most likely explanation for the lack of myoglobin gene expression in *C. aceratus* may be found in further upstream sections of the regulatory region of the gene.

Summary and conclusions

In this chapter, results of the line of investigations have been summarised aimed at shedding some light on physiological and evolutionary aspects of myoglobin expression within the channichthyid icefishes of the Southern Ocean. Data have been presented that definitively demonstrate that myoglobin polypeptide is expressed in many, but not all, icefish species. Presence of this protein among both icefishes and their red-blooded relatives, the *Nototheniidae*, however, is restricted to cardiac ventricle only. Northern blot analyses have revealed that mRNA encoding myoglobin is present in one of the three species that do not express the polypeptide. Positions of the Mb(−) icefish species within phylogenetic relationships of the *Channichthyidae* leads to the inescapable conclusion that mutations resulting in loss of myoglobin expression have occurred independently at least three different times and by two distinctly different mechanisms during evolution of this family.

Multiple losses of myoglobin expression are consistent with the protein no longer being functional at the very low physiological temperatures of Antarctic fishes. Results have been presented from studies of oxygen binding kinetics of myoglobin from Antarctic fishes, from experiments with isolated, perfused icefish hearts and from determinations of myoglobin cDNA and gene sequences; all three independent lines of evidence suggest that myoglobin does play a physiological role in the icefishes.

As is the case in most productive lines of scientific inquiry, the above results seem to be generating almost as many questions as answers. Given the apparent functional importance of myoglobin in tissues of those icefishes that express the protein, how do hearts of those species where myoglobin is absent maintain function? One might expect the answer to lie in the realm of either structural architecture of the heart at either tissue or subcellular level or in its metabolic characteristics. The presence of detectable mRNA encoding myoglobin in heart tissue of *Champsocephalus gunnari* where myoglobin polypeptide is not present also is enigmatic. One might speculate that some type of frame shift in the coding sequence of the message could result in either premature termination of translation or a defective translation product. Yet, preliminary data on sequence of myoglobin cDNA from *C.*

gunnari reveal a message that is nearly identical to that seen in channichthyids that do express the pigment. By continuing to combine studies at the levels of physiology, biochemistry and molecular biology and attempting to set each within an appropriate evolutionary context, it is hoped to unravel some of the remaining mysteries of this fascinating group of fishes.

Acknowledgements

The work reported here was supported by grants from the (US) National Science Foundation (OPP 92–20775, OPP 94–21657 to B.D.S.) and partially by the Italian National Antarctic Program to Bruno Tota. We also gratefully acknowledge the following colleagues for generously supplying tissue samples from icefish species that otherwise would have been unavailable to us: A.L. DeVries, G. diPrisco and R. Acierno. Finally, Antarctica presents many logistic challenges as a venue for pursuing scientific research. The contributions of masters and crew of *R/V Polar Duke* and support personnel of the United States' Palmer Station, Antarctica were as essential as those of the scientists who executed the experiments; we thank them for their excellent support.

References

Acierno, R., Agnisola, C., Tota, B. & Sidell, B.D. (1997). Myoglobin enhances cardiac performance in Antarctic icefish species that express the pigment. *American Journal of Physiology*, **273**, R100–6.

Anderson, M.E. (1990). The origin and evolution of the Antarctic ichthyofauna. In *Fishes of the Southern Ocean*, ed. O. Gon & P.C. Heemstra, pp. 26–33, Grahamstown, S. Africa: J.L.B. Smith Institute of Ichthyology.

Bailey, J.R., Sephton, D.H. & Driedzic, W.R. (1990). Oxygen uptake by isolated perfused fish hearts with differing myoglobin concentrations under hypoxic conditions. *Journal of Molecular and Cellular Cardiology*, **22**, 1125–34.

Bargelloni, L., Ritchie, P.A., Patarnello, T, Battaglia, B, Lambert, D.M. & Meyer, A. (1994). Molecular evolution at subzero temperatures: mitochondrial and nuclear phylogenies of fishes from Antarctica (Suborder Notothenioidei), and the evolution of antifreeze glycoproteins. *Molecular Biology and Evolution*, **11**, 854–63.

Bronmark, C. & Miner, J.G. (1992). Predator-induced phenotypical change in body morphology in crucian carp. *Science*, **258**, 1348–50.

Cashon, R.E., Vayda, M.E. & Sidell, B.D. (1997). Kinetic characterization of myoglobins from vertebrates with vastly different body temperatures. *Comparative Biochemistry and Physiology*, **117B**, 613–20.

Cocca, E, Ratnayake-Lecamwasam, M., Parker, S.K., Camardella, L., Ciaramella, M., diPrisco,G. & Detrich, H.W. III (1995). Genomic remnants of α-globin genes in the hemoglobinless Antarctic icefishes. *Proceedings of the National Academy of Sciences, USA*, **92**, 1817–21.

Crockett, E.L. & Sidell, B.D. (1990). Some pathways of energy metabolism are cold adapted in Antarctic fishes. *Physiological Zoology*, **63**, 472–88.

Dewitt, H.H. (1971). Coastal and deep-water benthic fishes of the Antarctic. In *Antarctic Map Folio Series. Folio 15*, ed. V.C. Bushnell, New York: American Geographical Society.

Douglas, E.L., Peterson, K.S., Gyso, J.R. & Chapman, D.J. (1985). Myoglobin in the heart tissue of fishes lacking hemoglobin. *Comparative Biochemistry and Physiology*, **81A**, 855–88.

Dowd, M.K., Murali, R. & Seagrave, R.C. (1991). Effect of temperature on the myoglobin-facilitated transport of oxygen in skeletal muscle. *Biophysical Journal*, **60**, 160–71.

Driedzic, W.R. & Stewart, J. (1982). Myoglobin content and the activities of enzymes of energy metabolism in red and white fish hearts. *Journal of Comparative Physiology*, **149B**, 67–73.

Eastman, J.T. (1993). *Antarctic Fish Biology: Evolution in a Unique Environment*. New York: Academic Press.

Hemmingsen, E.A. (1991). Respiratory and cardiovascular adaptation in hemoglobin-free fish: resolved and unresolved problems. In *Biology of Antarctic Fish*, ed. G. di Prisco, B. Maresca, & B. Tota, pp. 191–203. New York: Springer-Verlag.

Hemmingsen, E.A. & Douglas, E.L. (1977). Respiratory characteristics of the hemoglobin-free fish, *Chaenocephalus aceratus. Comparative Biochemistry and Physiology*, **33**, 733–44.

Hemmingsen, E.A., Douglas, E.L., Johansen, K. & Millard, R.W. (1972). Aortic blood flow and cardiac output in the hemoglobin-free fish *Chaenocephalus aceratus*. *Comparative Biochemistry and Physiology*, **43A**, 1045–51.

Iwami, T. (1985). Osteology and relationships of the family Channichthyidae. *Memoranda of the National Institute of Polar Research, Tokyo, Series E*, **36**, 1–69.

Johnston, I.A. & Harrison, P. (1987). Morphometrics and ultrastructure of myocardial tissue in notothenioid fishes. *Fish Physiology and Biochemistry*, **3**, 1–6.

Kennett, J.P. (1977). Cenozoic evolution of Antarctic glaciation, the circum-Antarctic Ocean and their impact on global paleoceanography. *Journal of Geophysical Research*, **82**, 3843–60.

Kennett, J.P. (1980). Paleoceanographic and biogeographic evolution of the Southern Ocean during the Cenozoic, and Cenozoic microfossil datums. *Palaeogeography, Palaeoclimatology and Palaeoecology*, **31**, 123–52.

Littlepage, J.L. (1965). Oceanographic investigation in McMurdo Sound, Antarctica. In *Biology of the Antarctic Seas*, Vol. II. pp. 1–37, ed. G.A. Llano, Antarctic Research Series. Washington, DC: American Geophysical Union.

Londraville, R.L. & Sidell, B.D. (1990). Ultrastructure of aerobic muscle in Antarctic fishes may contribute to maintenance of diffusive fluxes. *Journal of Experimental Biology*, **150**, 205–20.

Macdonald, J.A., Montgomery, J.C. & Wells, R.M.G. (1987). Comparative physiology of Antarctic fishes. In *Advances in Marine Biology*, ed. J.H.S. Blaxter, & A.J. Southward, Vol 24, pp. 321–88. London: Academic Press.

Macdonald, J.A. & Wells, R.M.G. (1991). Viscosity of body fluids from Antarctic notothenioid fish. In *Biology of Antarctic Fish*, ed. G. di Prisco, B. Maresca, & B. Tota, Berlin: Springer-Verlag.

Sato, F., Shiro, Y., Sakaguchi, Y., Iizuka, T. & Hayashi, H. (1990). Thermodynamic study of protein dynamic structure in the oxygen binding reaction of myoglobin. *Journal of Biological Chemistry*, **265**, 18823–8.

Sidell, B.D. & Hazel, J.R. (1987). Temperature affects the diffusion of small molecules through cytosol of fish muscle. *Journal of Experimental Biology*, **129**, 191–203.

Sidell, B.D., Vayda, M.E., Small, D.J., Moylan, T.J., Londraville, R.L., Yuan, M-L., Rodnick, K.J., Epplay, Z.A. & Costello, L. (1997). Variation in the expression of myoglobin among species of hemoglobinless Antarctic icefishes. *Proceedings of the National Academy of Sciences, USA*, **94**, 3420–4.

Stevens, E.D. & Carey, F.G. (1981). One why of the warmth of warm-bodied fish. *American Journal of Physiology*, **240**, R151–5.

Tota, B., Acierno, R. & Agnisola, C. (1991). Mechanical performance of the isolated and perfused heart of the haemoglobinless Antarctic icefish, *Chionodraco hamatus* (Lönnberg): effects of loading conditions and temperature. *Philosophical Transactions of the Royal Society, London B*, **332**, 191–8.

Vayda, M.E., Yuan, M-L., Small, D.J., Costello, L. & Sidell, B.D. (1995). Structure of the myoglobin gene of a hemoglobin-less Antarctic icefish. *Antarctic Journal of the United States*. in press.

Vayda, M.E., Yuan, M-L., Small, D.J., Costello, L. & Sidell, B.D. (1997). Extreme conservation of the myoglobin gene among Antarctic notothenioid fishes. *Molecular Marine Biology and Biotechnology*, in press.

Wittenberg, B.A. & Wittenberg, J.B. (1989). Transport of oxygen in muscle. *Annual Review of Physiology*, **51**, 857–78.

G. DI PRISCO, M. TAMBURRINI
and R. D'AVINO

Oxygen transport systems in extreme environments: multiplicity and structure–function relationship in haemoglobins of Antarctic fish

In the late Precambrian, 590 million years ago (mya), Antarctica was the central part of the supercontinent Gondwana, which remained intact for 400 million years (Ma). Fragmentation and the continental drift took Antarctica to its present position about 65 mya, at the beginning of the Cenozoic. The final separation from South America occurred at 22–25 mya, in the Oligocene–Miocene transition. The opening of the Drake passage produced the Circum-Antarctic Current and the development of the Antarctic Polar Front (or Antarctic Convergence). With reduction of heat exchange from northern latitudes, cooling of the environment proceeded to the present extreme conditions. The Antarctic ocean became gradually colder and seasonally ice covered. Although sea ice may have already been present at the end of the Eocene (40 mya), extensive ice sheets have probably formed only after the middle Miocene (14 mya) every 1–3 Ma. The latest ice-sheet expansion, with progressive cooling leading to the current climatic conditions, began 2.5 mya in the Pliocene. Antarctica is now a continent, almost fully coated with an ice sheet with average thickness of about 2000 m. It had enjoyed a much warmer climate than that of the dry, bitterly cold desert of the current times. Such variations produced many diversified forms of terrestrial and aquatic life (Eastman, 1991, 1993).

The Antarctic fish fauna

For the study of temperature adaptations, Antarctica is a unique natural laboratory. Although in this environment the absence of antifreeze protection (DeVries, 1983) would cause rapid freezing of fish from temperate waters, the oxygen-rich Antarctic waters support a wealth of marine life.

The first teleosts originated in the Jurassic, about 200 mya, and continued to evolve through the Cretaceous. Until the late Eocene they were quite cosmopolitan. In contrast, the modern Antarctic fish fauna is largely endemic and, unlike the populations of the other continental shelves, is dominated by a single group: the suborder Notothenioidei with 120 species in total. Of the 174 species living on the shelf or upper slope of the continent, 95 are notothenioids.

The lack of fossil record of Notothenioidei leaves a void of 38 Ma from the Eocene to present. There is lack of information on their site of origin, on the existence of a transition fauna and on the time of their radiation in the Antarctic. Indirect evidence suggests that notothenioids appeared in the early Tertiary, filling the ecological void on the shelf left by most of the other fish fauna (which experienced local extinction during maximal glaciation), and began to diversify in the middle Tertiary. The six notothenioid families fill a range of ecological niches normally occupied by taxonomically diverse fish communities in temperate waters. Only one of 11 species of Bovichtidae, the most primitive family, lives south of the Antarctic Polar Front; 15 of 49 species of Nototheniidae are non-Antarctic. Harpagiferidae, Artedidraconidae, Bathydraconidae and Channichthyidae are all Antarctic (Gon & Heemstra, 1990; Eastman, 1993). Notothenioids are red-blooded, with the exception of haemoglobin-less Channichthyidae.

In Antarctic waters, fish had to cope with temperatures below the freezing point of the body fluids and with high oxygen concentrations. They developed physiological and biochemical (often unique) specialisations. Cold adaptation was developed in the last 20–30 Ma during increasing isolation. By virtue of evolutionary responses to the environmental constraints, the Antarctic ocean is now the ideal habitat for the fish fauna. Fish are finely adjusted to the environment and intolerant of warmer temperatures; in fact, an increase in temperature of only a few centigrades has lethal effects (Somero & DeVries, 1967). They live in isolation south of the Polar Front, a barrier to migration in both directions and thus a key factor for fish evolution. Specialisations include neutral buoyancy, freezing avoidance, aglomerular kidneys, preference for metabolic fuels, cytoskeletal protein polymerisation, membrane structure, optimisation of haematological, enzymatic, muscular and nervous systems (Clarke, 1983; Hochachka & Somero, 1984; Macdonald, Montgomery & Wells, 1987; di Prisco, Maresca & Tota, 1988*a*, 1991*a*; di Prisco, 1991; Gon & Heemstra, 1990; Eastman, 1993). It is difficult to clearly assess whether a specialisation is adaptive or non-adaptive; this question has been extensively discussed by Eastman (1993).

In addition to low and constant temperature, seasonality may also influence biological processes such as feeding, growth and reproduction (see Clarke & North, 1991). It does not apply so much to temperature; but other variables, for instance, light and productivity, show marked seasonal fluctuations.

The oxygen-transport system of Antarctic fish

Antarctic fish differ from temperate and tropical species in having markedly reduced erythrocyte count and haemoglobin (Hb) content (and multiplicity) in the blood. This reduction counterbalances the increase in blood viscosity brought about, with potentially negative physiological effects (i.e. higher demand of energy needed for circulation), by the subzero seawater temperature (Wells, Macdonald & di Prisco, 1990). The critical analysis of Egginton & Davison (this volume) opens further discussion on the relationship between viscosity, haematology and vascular anatomy.

Oxygen carriers are one of the most interesting systems for studying the relationships between environmental conditions and molecular evolution. Hb, a direct link between the body and the exterior, has experienced a major evolutionary pressure to modify and adapt its functional features, in view of the variety of conditions to which all organisms that rely on Hb are exposed. In order to ensure adequate supply of oxygen, oxygen-carrying proteins have developed a common molecular mechanism based on ligand-linked conformational change in a multi-subunit structure. They generally exhibit a marked degree of co-operativity between oxygen-binding sites (homotropic interactions) which enables maximum oxygen unloading at high oxygen tension. In the simplest model, co-operativity is modulated by conformational transition between a low-affinity (T) and a high-affinity (R) state, which accounts for the sigmoidal shape of the binding curve. These two states are also involved in the modulation of the oxygen affinity brought about by several effectors (heterotropic interactions). Under physiological conditions, an effector may preferentially bind to the low-affinity or to the high-affinity state of the protein, lowering or enhancing the overall oxygen affinity of the molecule (for comprehensive reviews, see Dickerson & Geis, 1983; Bellelli & di Prisco, 1992).

Within the framework of this common mechanism, however, respiratory proteins have acquired special features to meet special needs. Thus, not only do they illustrate the variability possible within the same overall mechanism, but also represent a type-case of molecular adaptation to physiological requirements. For instance, a decrease in temperature induces an increase of oxygen affinity, since oxygen binding is generally exothermic. Thus, temperature and its interplay with heterotropic ligands in modifying the oxygenation–deoxygenation cycle in the respiring tissues have a role of the utmost importance, and thermodynamic analysis may indeed become the tool of choice to attain deeper insight into the physiology of a specific organism. At the molecular level, heat absorption and release can be considered a physiologically relevant modulating factor, similar to homo- and heterotropic ligands.

The overall thermodynamics of a biological macromolecule may vary in order to cope with special circumstances. In Hb this is achieved by linking the basic reaction with the binding of different ions and effectors whose thermodynamics contribute to the overall effect of temperature. In general, the thermal effects measured when oxygen binds to Hb are due to the contribution of intrinsic heat of oxygenation (binding of oxygen to the haem iron), heat of ionisation of oxygen-linked Bohr groups (endothermic), heat of oxygen solubilisation (exothermic), heat associated with the T → R allosteric transition and heat of binding of other ions (chloride, organophosphates).

The lack of Hb in Channichthyidae

Channichthyidae, the phylogenetically most highly developed notothenioids (Iwami, 1985; Eastman, 1993), are the only known vertebrates whose pale whitish blood is devoid of Hb (Ruud, 1954) and has a greatly reduced number of erythrocyte-like cells, two to three orders of magnitude less than temperate fish and one to two orders less than red-blooded notothenioids (Hureau *et al.*, 1977; Kunzmann, 1991). Hb has not been replaced by another oxygen carrier; oxygen is physically dissolved in the plasma, and the oxygen-carrying capacity of blood is only 10% of that of red-blooded fish. However, these fish are not at all disadvantaged by lack of Hb. Physiological adaptations such as low metabolic rate, large and well-perfused gills, cutaneous respiration, large blood volume, enlarged heart, increased stroke volume and capillary diameter, allow scaleless channichthyids to prosper without Hb (Hemmingsen & Douglas, 1970, 1977; Holeton, 1970; Macdonald *et al.*, 1987).

The co-existence of Hb-less and naturally cytopenic red-blooded species suggests that the need for an oxygen-carrier in a stable, cold environment is reduced also in red-blooded fish. Functional incapacitation of Hb and induced reduction of the haematocrit to 1–2% in the nototheniid *Pagothenia bernacchii* (Wells *et al.*, 1990; di Prisco, Macdonald & Brunori, 1992) caused no discernible harm to the fish, at least in the absence of metabolic challenges. Like channichthyids, red-blooded Antarctic fish can carry routinely needed oxygen dissolved in plasma. In scaled nototheniids there is also evidence of cutaneous respiration (Wells, 1987).

Channichthyids also lack muscle myoglobin (Hamoir, 1988). The evolutionary loss of the respiratory pigments is a highly specialised condition which raises several questions. If the Hb-less state is adaptive for one family, why not also for the others living in the same habitat? Wells (1990) has, in fact, suggested that the Hb-less state may be non-adaptive.

The fate of globin genes in channichthyids is an interesting question. Relieved of selective pressure for expression, they may have diverged from those of red-blooded notothenioids, or been lost altogether. The characterisation of globin DNA sequences from several species of red-blooded and Hb-less notothenioids has indicated that three channichthyids, spanning the clade from primitive to advanced genera, share retention of α-globin-related DNA sequences in their genomes and apparent loss, or rapid mutation, of β-globin genes (Cocca *et al.*, 1995). This common pattern suggests that loss of globin-gene expression is a primitive character, established in the common ancestral channichthyid about 25 mya (Eastman, 1993), prior to diversification within the clade. Deletion of the β-globin locus of the ancestor may have been the primary event leading to the Hb-less phenotype. The α-globin gene(s), no longer under selective pressure for expression, would then have accumulated mutations leading to loss of function without, as yet, complete loss of sequence information.

Red-blooded fish

Haematology of Antarctic fish has been extensively investigated in the past decades (Everson & Ralph, 1968; Hureau *et al.*, 1977; Wells *et al.*, 1980; Kunzmann, 1991).

Endemic Antarctic fish also display reduced Hb multiplicity. Thirty-four sedentary benthic species have a single major Hb (Hb 1) and often a second, minor component (Hb 2, about 5% of the total), having one of the chains (usually the β) in common (Table 1). Another component (Hb C) is present at less than 1% in all species. In contrast, the Hb system of temperate fish is often made of multiple, functionally distinct components, probably required to withstand environmental changes. Considering other non-endemic families, Antarctic Zoarcidae (suborder Zoarcoidei, with all-latitude distribution) possess four to five functionally distinct major Hbs (di Prisco & D'Avino, 1989; di Prisco *et al.*, 1990); also, two species of Macrouridae and Anotopteridae have three and four major Hbs, respectively (Kunzmann, 1991).

The degree of amino acid sequence identity (di Prisco *et al.*, 1991*b*) among the globins of Hb 1 of the Antarctic species is very high (80–90%; Tables 2 and 3). The chains of the minor components Hb 2 and Hb C, not in common with Hb 1, are in another group of highly similar sequences, which in turn have lower identity with major Hbs.

Fish Hbs often display regulation of oxygen binding by pH (Bohr effect; see Riggs, 1988), so that oxygen can be easily released under conditions of acidosis. Organophosphates can enhance this effect, leading to overstabilisation of the deoxy state and to loss of co-operative oxygen binding at lower pH, as shown by the decrease of the Hill coefficient nH to values close to one (Root effect; see Brittain, 1987). The presence of Bohr and Root effects has been investigated in 30 (plus an additional one) of the 34 notothenioids of Table 1 (Table 4). In Hb 1 and Hb 2, oxygen binding is generally strongly pH and organophosphate regulated (di Prisco, 1988; D'Avino & di Prisco, 1988, 1989; di Prisco *et al.*, 1988*b*, 1991*b*; D'Avino *et al.*, 1989, 1991; Kunzmann, 1991; Kunzmann, Caruso & di Prisco, 1991; Caruso *et al.*, 1991, 1992; Camardella *et al.*, 1992; di Prisco *et al.*, 1994). *Aethotaxis mitopteryx*

Table 1. *Hbs of Antarctic and non-Antarctic (*Notothenia angustata *and* Pseudaphritis urvillii*) Notothenioidei*[a]

Family	Species	Hb components[b]
Bovichtidae	*Pseudaphritis urvillii*	Hb 1 (95%), Hb 2 (5%)
Nototheniidae	*Notothenia coriiceps*	Hb 1 (95%), Hb 2 (5%)
	Notothenia rossii	Hb 1 (95%), Hb 2 (5%)
	Notothenia angustata	Hb 1 (95%), Hb 2 (5%)
	Nototheniops nudifrons	Hb 1 (95%), Hb 2 (5%)
	Nototheniops larseni	Hb 1 (95%), Hb 2 (5%)
	Gobionotothen gibberifrons	Hb 1 (90%), Hb 2 (10%)
	Pagothenia hansoni	Hb 1 (95%), Hb 2 (5%)
	Pagothenia bernacchii	Hb 1 (98%) (Hb 2?)
	Dissostichus mawsoni	Hb 1 (98%) (Hb 2?)
	Aethotaxis mitopteryx	one Hb (99%)
	Trematomus nicolai	Hb 1 (95%), Hb 2 (5%)
	Trematomus pennellii	Hb 1 (95%), Hb 2 (5%)
	Trematomus loennbergi	Hb 1 (95%), Hb 2 (5%)
	Trematomus eulepidotus	Hb 1 (95%), Hb 2 (5%)
	Trematomus lepidorhinus	Hb 1 (95%), Hb 2 (5%)
	Trematomus scotti	Hb 1 (95%), Hb 2 (5%)
Bathydraconidae	*Cygnodraco mawsoni*	Hb 1 (95%), Hb 2 (5%)
	Racovitzia glacialis	Hb 1 (90%), Hb 2 (10%)
	Parachaenichthys charcoti	one Hb (99%)
	Gymnodraco acuticeps	one Hb (99%)
	Bathydraco marri	one Hb (99%)
	Bathydraco macrolepis	one Hb (99%)
	Akarotaxis nudiceps	one Hb (99%)
	Gerlachea australis	one Hb (99%)
Artedidraconidae	*Artedidraco skottsbergi*	one Hb (99%)
	Artedidraco orianae	one Hb (99%)
	Artedidraco shackletoni	one Hb (99%)
	Histiodraco velifer	one Hb (99%)
	Pogonophryne scotti	one Hb (99%)
	Pogonophryne sp. 1	one Hb (99%)
	Pogonophryne sp. 2	one Hb (99%)
	Pogonophryne sp. 3	one Hb (99%)
Harpagiferidae	*Harpagifer antarcticus*	one Hb (99%)

Notes:
[a] The blood of all species contains traces (less than 1%) of Hb C.
[b] The number of components and their amounts were determined by cellulose acetate electrophoresis (D'Avino & di Prisco, 1989).

Table 2. α *chain sequence identity (%) between Antarctic and non-Antarctic Notothenioidei and other temperate fish haemoglobins*

Species	*Cyprinus carpio**[a]	*O. mykiss** Hb IV	*O. mykiss** Hb I	*N. angus** Hb 2	*N. cor.* Hb 2	*T. newn.* Hb 2	*P. antar.* Hb 3	*P. urvillii** Hb 1	*N. angus.** Hb 1	*P. antar.* Hb 1,2	*G. acut.*	*C. maws.* Hb 1,2	*A. mitopt.*	*P. bern.* Hb	*T. new.* Hb 1,C
N. coriiceps Hb 1[a]	59	57	55	63	63	61	69	80	99	95	82	83	83	89	87
T. newnesi Hb 1, C[a]	58	62	52	66	66	63	68	78	94	89	92	90	90	97	
P. bernacchii Hb 1[a]	64	62	57	70	70	65	69	80	96	92	91	91	90		
A. mitopteryx[a]	62	59	55	65	65	62	65	73	91	88	84	84			
C. mawsoni Hb 1, 2[a]	60	62	53	69	69	64	70	78	89	87	93				
G. acuticeps[a]	58	62	53	67	67	65	68	76	89	85					
P. antarcticum Hb 1, 2[b]	66	64	61	70	70	67	72	81	94						
*N. angustata** Hb 1[c]	65	62	58	69	69	65	70	78							
*P. urvillii** Hb 1[d]	60	59	60	65	65	62	65								
P. antarcticum Hb 3[b]	66	66	65	92	92	90									
T. newnesi Hb 2[a]	61	58	62	93	93										
N. coriiceps Hb 2[a]	63	63	62	100											
*N. angustata** Hb 2[c]	63	63	62												
*Oncorhynchus mykiss** Hb I[a]	66	60													
*O. mykiss** Hb IV[a]	63														

Notes:

The amino acid sequences are from: [a] di Prisco *et al.* (1991*b*); [b] Tamburrini *et al.* (1996); [c] Fago *et al.* (1992); [d] R. D'Avino & G. di Prisco (unpublished observations).

The asterisk denotes non-Antarctic species. The vertical line separates non-Notothenioidei from Notothenioidei.

Table 3. *β chain sequence identity (%) between Antarctic and non-Antarctic Notothenioidei and other temperate fish haemoglobins*

Species	*Cyprinus carpio**[a]	*O. mykiss** Hb IV	*O. mykiss** Hb I	*P. urvillii.** Hb 2	*C. maws.* Hb 2	*P. antar.* Hb2	*P. bern.* Hb C	*T. newn.* Hb C	*P. urvillii** Hb 1	*N. angus.* Hb 1	*P. antar.* Hb 1,3	*G. acut.*	*C. maws.* Hb 1	*A. mitopt.*	*P. bern.* Hb 1	*T. new..* Hb 1,2
N. coriiceps Hb 1,2[a]	57	63	53	70	65	72	70	70	77	93	86	80	88	82	90	86
T. newnesi Hb 1, 2[a]	57	62	53	67	64	69	68	68	77	85	84	80	84	83	93	
P. bernacchii Hb 1[a]	61	66	58	70	66	72	70	70	82	91	90	83	87	86		
A. mitopteryx[a]	58	60	54	66	62	69	66	67	77	84	84	77	80			
C. mawsoni Hb 1[a]	56	62	53	66	67	68	68	70	75	87	82	85				
G. acuticeps[a]	56	59	55	65	65	69	66	67	75	81	80					
P. antarcticum Hb 1, 3[b]	61	63	57	70	67	72	69	69	77	88						
*N. angustata** Hb 1[c]	61	64	57	71	66	73	71	71	81							
*P. urvillii** Hb 1[d]	63	66	61	69	66	70	69	69								
T. newnesi Hb C[a]	57	62	57	86	89	91	95									
P. bernacchii Hb C[a]	60	63	55	88	90	92										
P. antarcticum Hb 2[b]	63	66	58	84	86											
C. mawsoni Hb 2[a]	55	60	54	85												
*P. urvillii** Hb 2[d]	58	64	53													
*Oncorhynchus mykiss** Hb Ia	64	59														
*O. mykiss** Hb IV[a]	73															

Note:

The amino acid sequences are from: [a] di Prisco *et al.* (1991*b*); [b] Tamburrini *et al.* (1996); [c] Fago *et al.* (1992); [d] R. D'Avino & G. di Prisco (unpublished observations).

The asterisk denotes non-Antarctic species. The vertical line separates non-Notothenioidei from Notothenioidei.

Table 4. *Regulation of oxygen binding by pH and heterotropic physiological ligands in Hbs of Antarctic and non-Antarctic (*N. angustata *and* P. urvillii*) Notothenioidei*

Family	Species	Bohr and Root effects[a]; effect of organophosphates
Bovinchtidae	*Pseudaphritis urvillii*	strong in Hb 1, Hb 2
Nototheniidae	*Notothenia coriiceps*	strong in Hb 1, Hb 2
	Notothenia rossii	strong in Hb 1, Hb 2
	Notothenia angustata	strong in Hb 1, Hb 2
	Gobionotothen gibberifrons	strong in Hb 1, Hb 2
	Pagothenia hansoni	strong in Hb 1, Hb 2
	Pagothenia bernacchii	strong in Hb 1
	Dissostichus mawsoni	strong in Hb 1
	Aethotaxis mitopteryx	Root, absent; Bohr, weak
	Trematomus nicolai	strong in Hb 1, Hb 2
	Trematomus pennellii	strong in Hb 1, Hb 2
	Trematomus loennbergi	strong in Hb 1, Hb 2
	Trematomus eulepidotus	strong in Hb 1, Hb 2
	Trematomus lepidorhinus	strong in Hb 1, Hb 2
	Trematomus scotti[b]	strong in Hb 1, Hb 2
Bathydraconidae	*Cygnodraco mawsoni*	strong in Hb 1, Hb 2
	Racovitzia glacialis	strong (in haemolysate)
	Parachaenichthys charcoti	strong
	Gymnodraco acuticeps	absent
	Bathydraco marri	strong
	Bathydraco macrolepis[b]	strong, only with ATP
	Akarotaxis nudiceps[b]	strong
	Gerlachea australis[b]	strong
Artedidraconidae	*Artedidraco orianae*	weak (Root only with ATP)
	Artedidraco shackletoni[b]	weak, only with ATP
	Histiodraco velifer	weak
	Pogonophryne scotti	weak (Root only with ATP)
	Dolloidraco longedorsalis[b]	weak (in haemolysate)
	Pogonophryne sp. 1	weak (Root only with ATP)
	Pogonophryne sp. 2	weak (Root only with ATP)
	Pogonophryne sp. 3	weak (Root only with ATP)

Note:
[a] Bohr and Root effects were measured according to Giardina & Amiconi (1981) and D'Avino & di Prisco (1989). Bohr coefficients ($\Delta\log P_{50}/\Delta pH$) higher than -0.6 and lower than -0.3 denote 'strong' and 'weak' Bohr effects, respectively. 'Strong' and 'weak' Root effects correspond to Hb oxygenation at atmospheric pressure, pH 6.0, lower than 40% and higher than 80%, respectively.
[b] The Bohr effect was not measured.

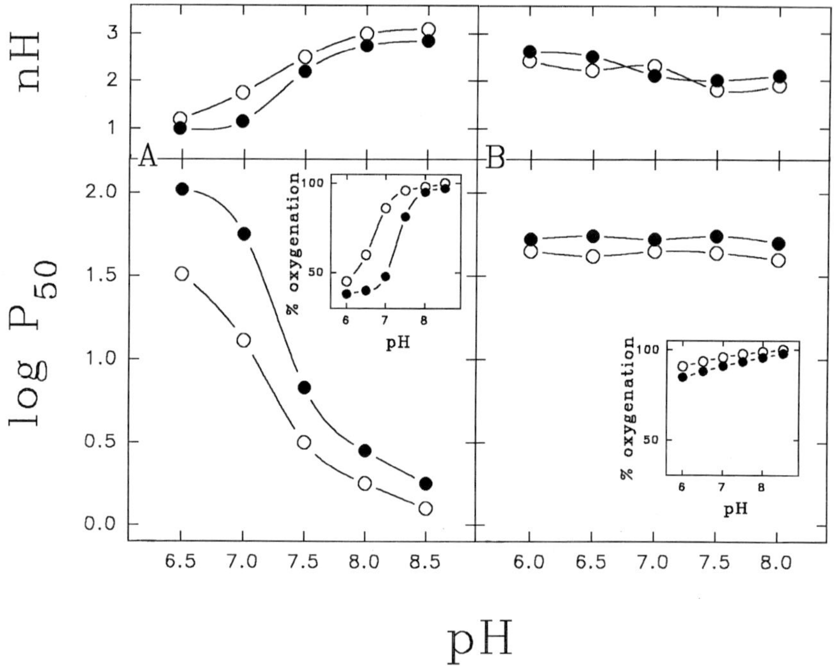

Fig. 1. Oxygen equilibrium isotherms as a function of pH (Bohr effect) of *P. bernacchii* Hb 1 (A; Camardella *et al.*, 1992) and of the single Hb of *G. acuticeps* (B; Tamburrini *et al.*, 1992), measured at 0 °C and 10 °C, respectively; the Hill coefficient (nH) denotes subunit cooperativity. Experiments were carried out in 100 mM Tris-HCl or Bistris-HCl buffers, in the absence (o) and presence (●) of 100 mM NaCl and 3 mM IHP (inositol hexakisphosphate). P_{50} (the partial pressure of oxygen required to achieve Hb half saturation) was measured in mmHg. The insets show the oxygen saturation curves in air as a function of pH (Root effect), in the absence (o) and presence (●) of 3 mM IHP.

(D'Avino *et al.*, 1992) and *Gymnodraco acuticeps* (Tamburrini *et al.*, 1992) have a single Hb with little or no Bohr effect and lacking the Root effect. Figure 1 illustrates the oxygen-binding features of a Bohr- and Root-effect Hb (Camardella *et al.*, 1992), and of a Hb lacking these effects (Tamburrini *et al.*, 1992).

Notothenioids with low Hb multiplicity are bottom dwellers. Two comments are pertinent here: (i) *A. mitopteryx* and *Dissostichus mawsoni* are actually bentho-pelagic, but the former is very sluggish and the latter is very moderately active; (ii) although a few of the other species are indicated as semipelagic in some reports (Montgomery & Wells, 1993; Eastman, 1993), this may reflect habits typical of early life stages, or occasional displacements

Table 5. *Notothenioidei of the family Nototheniidae with higher Hb multiplicity*

Species	Hb components[a]
T. newnesi (2 major Hbs)	
	Hb C (20%)
	Hb 1 (75%)
	Hb 2 (5%)
P. antarcticum (3 major Hbs)	
	Hb C (traces)
	Hb 1 (30%)
	Hb 2 (20%)
	Hb 3 (50%)
P. borchgrevinki (1 major Hb)	
	Hb C (traces)
	Hb 0 (10%)
	Hb 1 (70%)
	Hb 2 (10%)
	Hb 3 (10%)

Note:
[a] The number of components and their amounts were determined by cellulose acetate electrophoresis (D'Avino & di Prisco, 1989).

in the water column for feeding (DeWitt, Heemstra & Gon, 1990). In our experience, all species of Table 1 were caught by bottom trawling, or in gill nets, traps and hooks placed on the sea floor. For these reasons we consider these notothenioids as sedentary bottom dwellers.

As extensively discussed below, the Hb system has been studied in three pelagic species (D'Avino *et al.*, 1994; Tamburrini *et al.*,1996; M. Tamburrini & G. di Prisco, unpublished observations) and in two non-Antarctic notothenioids (Fago, D'Avino & di Prisco, 1992; R. D'Avino & G. di Prisco, unpublished observations).

Notothenioids with higher Hb multiplicity

Three notothenioids of the family Nototheniidae do not follow the pattern of low Hb multiplicity. *Pleuragramma antarcticum*, *Trematomus newnesi* and *Pagothenia borchgrevinki* have three to five functionally distinct Hbs (Table 5). Their lifestyle differs from that of the other sluggish benthic species.

Pleuragramma antarcticum

This species is a suitable target for studies on adaptation to extreme conditions; its life history and ecology is, in fact, exceptional (Hubold, 1985). Being the most abundant and the only fully pelagic species of high-Antarctic shelf systems, it combines the more general adaptations of all notothenioids with specialisations necessary for life in the water column.

The biomass of pelagic fish of the Weddell Sea is overwhelmingly dominated by *P. antarcticum* (Ekau, 1991; Hubold, 1991). This key species in the midwater ecosystem and food web of the Antarctic shelf has a circum-Antarctic distribution and migrates across different water masses (Andersen, 1984; Hubold, 1984, 1985; Kunzmann, 1990). While larval growth is fast, adults have an extremely slow growth (Ekau, 1988) and a very low-energy-consuming mode of life, combined with sluggish and pelagic or bentho-pelagic behaviour (Eastman & DeVries, 1982; Johnston, 1989; Kunzmann, 1990).

The Hb system of *P. antarcticum* (Tamburrini *et al.*, 1996) consists of three major components (Hb 1, Hb 2 and Hb 3); it displays the highest multiplicity among Notothenioidei which, with the exception of *T. newnesi* (D'Avino *et al.*, 1994), usually have a single major Hb (di Prisco *et al.*,1991*b*, Table 1). Hb 1 has the α chain in common with Hb 2, and the β in common with Hb 3. Hb 2 and Hb 3 have no chain in common. Thus, the Hb system is made of two α and two β chains. The complete amino acid sequence of the four chains has been established (Tamburrini *et al.*, 1996). Tables 2 and 3 summarise the degree of identity with Hb sequences of other fish. High identity is observed among the chains of Hb 1 and those of major Antarctic Hbs. The chains of Hb 2 and Hb 3 which are not in common with Hb 1 have high identity with those of other minor Antarctic Hbs. Similar to other Antarctic fish Hbs, all sequences have lower identity with non-Antarctic globins.

The Hbs display very strong effector-enhanced pH-dependence of oxygen affinity (Bohr effect) and of oxygenation at atmospheric pressure (Root effect), and high oxygen affinities.

Although these oxygen-binding features are similar, these Hbs show important differences in thermodynamic behaviour. In the pH and temperature ranges of 7.0–8.0 and 2–20 °C, in the absence and presence of effectors, they differ in the heat of oxygenation. Hb 1 and Hb 3 show a very high oxygenation enthalpy change at pH 8.0, further enhanced by chloride and organophosphates in the former, but drastically decreased in Hb 3. ΔH of Hb 2, in the presence and absence of effectors, is much lower (Table 6). A dramatic decrease is observed at lower pH in Hb 3, and also in Hb 2 (ΔH approaches zero in both); in contrast, Hb 1 retains high oxygenation enthalpy, especially when effectors are absent. These observations clearly

Table 6. *Heat of oxygenation of* P. antarcticum *Hbs (Tamburrini* et al., *1996)*

	100 mM NaCl, 3 mM ATP	ΔH (kcal/mol oxygen)[a]	
		pH 7.0	pH 8.0
Hb 1	absence	−12.8	−15.3
	presence	−8.6	−17.4
Hb 2	absence	−3.6	−6.4
	presence	−1.8	−8.1
Hb 3	absence	−0.1	−16.5
	presence	−4.1	−7.6

Note:
[a] The overall oxygenation enthalpy change ΔH (kcal/mol; 1 kcal=4.184 kJ), corrected for the heat of oxygen solubilisation (−3 kcal/mol), was calculated by the integrated van't Hoff equation:

$$\Delta H = -4.574\ [(T1 \cdot T2)\ /\ (T1 - T2)]\ \Delta \log P_{50}/1000$$

(P_{50} is the partial pressure of oxygen required to achieve Hb half saturation.)

indicate a stronger Bohr effect at physiological temperatures in Hb 1 (in the presence of effectors) and Hb 3 (in their absence).

The moderate effect of temperature on Hb 2 in the pH range 7.0–8.0 and on Hb 3 at pH 7.0 is in keeping with preventing lower temperatures encountered during migration to impair oxygen unloading. Reducing the thermal sensitivity of Hbs is one of the strategies adopted during evolution to increase the efficiency of carrying oxygen from the gills to respiring tissues; thermodynamic analysis has shown that the enthalpy change for oxygenation in species facing low temperature is often very low when compared to temperate organisms (di Prisco *et al.*, 1991c). In fact, also the Bohr effect of Hb 1 of *P. bernacchii*, a benthic fish, is almost temperature insensitive, and the overall ΔH of oxygen binding in the pH range 7.0–8.0 is even positive.

Thus, pelagic *P. antarcticum* is the only known species of the family Nototheniidae and of the suborder Notothenioidei having three major Hbs, characterised by strong Bohr and Root effects, and differently regulated by temperature. Sluggish *A. mitopteryx*, a very closely related benthopelagic species (Andersen, 1984), has a single Hb with a moderate Bohr effect and no Root effect (D'Avino *et al.*, 1992). These observations suggest that the two species must satisfy different oxygen demands, arising from special environmental conditions.

Trematomus newnesi

T. newnesi actively swims and feeds near the surface (Eastman, 1988). Its Hb system is made of Hb C, Hb 1 and Hb 2, and contains two α and two β chains, since Hb 1 has the α chain in common with Hb C and the β in common with Hb 2. Hb C and Hb 2 have no chain in common. The amino acid sequences of the four chains follow the usual identity pattern (Tables 2 and 3). High identity is observed among the chains of Hb 1 and those of major Antarctic Hbs. The chains of Hb C and Hb 2 not in common with Hb 1 have high identity with those of other minor Antarctic Hbs. All sequences have lower identity with non-Antarctic globins.

T. newnesi is the only species (D'Avino *et al.*, 1994) in which Hb C is not present in traces, but reaches 20–25% of the total. Unlike the other notothenioids, the oxygen binding of Hb 1 and Hb 2 is not regulated by pH and organophosphates; on the other hand Hb C displays effector-enhanced Bohr and Root effects. Thus *T. newnesi* is the only known notothenioid having two major Hbs, only one of which displays pH and organophosphate regulation.

Pagothenia borchgrevinki

Only preliminary evidence is available on the oxygen-transport system of this active cryopelagic species. Of the five Hbs (M. Tamburrini & G. di Prisco, unpublished observations), Hb 1 is the only major component, accounting for 70–80% of the total. Besides Hb C, present in traces, another component (Hb 0) displays strong, effector-enhanced Bohr and Root effects. Hb 1, Hb 2 and Hb 3 bind oxygen with weak pH-dependence and weak or no influence of organophosphates. Amino acid sequencing and thermodynamic characterisation of each Hb will hopefully shed light on another unusually complex and specialised oxygen-transport system.

Two non-Antarctic notothenioid species

Bovichtidae and Nototheniidae are the only notothenioid families inhabiting waters beyond the Antarctic and sub-Antarctic. The first one is considered the most primitive of the suborder (see Eastman, 1993), and is predominantly non-Antarctic; the second has a latitudinal range spanning from 35° S to 82° S (Eastman, 1993). In this order, according to the cladogram of Iwami (1985), these families were the first to diverge. Balushkin (1992) separates a new family, Pseudaphritidae, from Bovichtidae.

Similar to the Antarctic species, two non-Antarctic notothenioids have a major and a minor component, Hb 1 and Hb 2. These species differ markedly from each other (Fago *et al.*, 1992; R. D'Avino & G. di Prisco, unpublished observations).

Notothenia angustata

N. angustata, found in New Zealand waters, is a sedentary bottom dweller and experiences temperatures of about 5 °C at the lowest. It is not cold adapted, and does not synthesise antifreeze glycoproteins. Unlike Antarctic notothenioids, it is not neutrally buoyant and has pauciglomerular kidneys (Gon & Heemstra, 1990). The haematocrit, erythrocyte number, Hb concentration and mean corpuscular Hb content (MCHC) are higher than those of Antarctic notothenioids (Tetens, Wells & DeVries, 1984; Macdonald & Wells, 1991), as expected in a fish of lower latitudes; but Hb multiplicity and structural/functional features mimic those of the Antarctic species of the same family and suborder (Fago *et al.*, 1992).

N. angustata Hbs have high identity with the corresponding Antarctic sequences (Tables 2 and 3). Maximal identity, the highest ever found between notothenioid globin sequences, is observed with the nototheniid *Notothenia coriiceps*: the β chains in common are 93% identical, the α chains of Hb 1 differ in only two positions and the α chains of Hb 2 are identical.

Did separation of *N. angustata* from Antarctic notothenioids occur before or after development of cold adaptation? Andersen (1984) and Balushkin (1988) suggested that the two species diverged evolutionarily prior to the establishment of the Antarctic Polar Front. In this case, cooling would have exerted no pressure in determining the amino acid sequences, since Antarctic coastal waters cooled to 0 °C only 10 mya (Clarke,1983); the sequence similarity would then simply reflect an ancestral condition and the common phylogenetic origin of these species. However, recent studies indicate that the genome of *N. angustata* does have the antifreeze gene (Cheng, personal communication), suggesting that this species developed cold adaptation prior to migration from the Antarctic shelf. Thus the sequence similarity might indeed reflect cold adaptation of *N. angustata* prior to geographic radiation. In view of the concomitant adjustment of haematological parameters to favour oxygen transport in a milder environment, *N. angustata* appears as an ideal evolutionary link between cold-adapted and temperate species of the same family.

Pseudaphritis urvillii

P. urvillii is a catadromous species from coastal waters, estuaries and rivers of Australia. It is euryhaline and may migrate upstream as far as 120 km from the sea (Andrews, 1980). It has glomerular kidneys and no antifreeze glycoproteins; in freshwater it is almost neutrally buoyant (Eastman, 1993). It belongs to the family Bovichtidae (or Pseudaphritidae) and is considered a relict species. No information on the Hb system of this family has been available so far. The exceptionally high oxygen affinity displayed by Hb 1 (R. D'Avino & G. di Prisco, unpublished observations) appears noteworthy and is presumably linked to the peculiar life style of this fish.

Despite these differences with *N. angustata* (which indicate divergent evolutionary pathways), Tables 2 and 3 reveal high sequence identity (78% for the α chain and 81% for the β), higher than that observed among *P. urvillii* and other temperate fishes.

Although *P. urvillii* was never cold adapted, Hb 1 sequence identity with Antarctic notothenioids is higher (73%–80% for the α and 75%–82% for the β chains) than with temperate fish (59%–60% for the α and 61%–66% for the β chains, following the general trend shown by notothenioids). It should, however, be noted that the identity between *P. urvillii* and Antarctic Hb 1 is close to, or lower than, the low extreme of the ensemble of values determined among Antarctic notothenioids. These data argue in favour of a common origin, but also suggest that the major component has undergone modifications only to a limited extent. If sequence mutations in Antarctic fish are indeed related to the development of cold adaptation, this may imply that *P. urvillii* (unlike *N. angustata*) diverged and migrated during the first stages of the cooling process, probably before the event which gave origin to the biosynthesis of antifreeze glycoproteins.

Direct comparison with *Bovichtus elongatus* (the one Antarctic species of the same, or closely related, family Bovichtidae) must await availability of blood of the latter fish.

The minor Hb components of notothenioids

The high sequence identity among the minor component Hb 2 of *P. urvillii* and those (Hb 2 and Hb C) of the Antarctic species is very striking (Tables 2 and 3). The values range between 84% and 88% and are very similar to those observed among the minor components of Antarctic notothenioids. In view of this high identity and, conversely, of the low identity with the major Hbs, it is conceivable that the minor components did not undergo evolutionary pressure. They often differ from Hb 1 precisely in the residues found only in the Antarctic species (di Prisco *et al.*, 1991*b*). On the basis of these observations and taking into account the fact that in all notothenioids Hb 2 is functionally indistinguishable from the major component, we have hypothesised the minor components to be vestigial remnants. However in *T. newnesi*, in which Hb 1 (75%) and Hb 2 (5%) lack the Bohr and Root effects, Hb C (normally found in the other Antarctic species only in amounts of less than 1%) accounts for 20–25% (D'Avino *et al.*, 1994). It may be inferred that, even if Hb C has not been evolutionarily preferred, this component can none the less be 'activated' as a consequence of needs arising from the special lifestyle of this active fish, which presumably requires functionally distinct components.

It cannot be excluded that Hb 2 and Hb C are dominant components in the larval stage, still synthesised in limited amounts by adult fish. This would offer an alternative explanation for the high identity among minor components and, conversely, for the low similarity with major Hbs. These two Hb types would undergo very different evolutionary pressures, reflecting the uniform and basic needs that larvae have in order to survive, and the much more refined adjustments developed by adult fish. Amino acid sequences of larval Hbs from other temperate species are also required to test this hypothesis.

Concluding remarks

Adaptive evolution of Antarctic fish draws advantage from being staged within a simplified framework (reduced number of variables in a stable environment, dominated by a taxonomically uniform group). Although correlations of molecular data with ecology and lifestyle are difficult to establish, the primary importance of this objective is increasingly attracting the interest of scientists. The study of haematology and Hb has allowed us to reach two conclusions of general bearing: (*i*) the more the notothenioid families are phyletically derived, the lower is the erythrocyte number and Hb concentration and multiplicity; (*ii*) a correlation often appears between lifestyle and Hb multiplicity. Bottom- dweller notothenioids have a single major Hb. In contrast, three pelagic species (two of them active) have unique Hb systems of multiple, functionally distinct Hbs, which may be the result of adaptation to environmental changes experienced by these fish. In these cases a link with lifestyle becomes possible (di Prisco & Tamburrini, 1992), since the selective advantage of multiple Hb genes is clear, whereas it is often difficult to establish whether multiple genes are surviving selectively neutral gene duplication, with no obvious correlation with the environment.

The Hb system of active *T. newnesi* can ensure oxygen binding at the gills (via Hb 1) and controlled delivery to tissues (via Hb C) also when active behaviour produces acidosis. High levels of Bohr- and Root-effect Hb C, conceivably redundant in the other notothenioids, may compensate for the lack of proton/effector regulation of Hb 1.

The mode of life of *P. antarcticum*, a pelagic but sluggish fish, differs from that of *T. newnesi*. The main adaptive feature of the Hb system of this fish could conceivably be the response to the need to optimise oxygen loading/unloading during seasonal migrations through water masses which may have different and fluctuating temperatures, rather than to face acidosis. *P. antarcticum* relies on three Hbs, which differ functionally mainly in thermodynamic behaviour rather than in pH and organophosphate regulation. Among Notothenioidei this oxygen-transport system is remarkably unique, one of the most specialised ever found in fish.

On the basis of their relative amounts, none of the Hbs of *P. antarcticum*, unlike the minor components found in the benthic notothenioids, can be considered as an evolutionary (or larval) remnant devoid of physiological significance (di Prisco *et al.*, 1991*b*), even though the sequence data reveal high phylogenetic distance between Hb 1 and the globins of Hb 2 and Hb 3 that are not in common (Tables 2 and 3). The expression of multiple genes remains high also in the adult stage, in close similarity with juveniles (G. di Prisco, L. Camardella & M. Tamburrini, unpublished observations), suggesting refined mechanisms of regulation within the gene family. The three Hbs may serve different purposes during ontogeny, since young post-larvae, juveniles and adults have specific food sources with different environmental temperature preferences (Hubold, 1985). Moreover, ratios between multiple Hbs can vary seasonally in temperate and tropical fish, and synthesis on demand is possible (Love, 1980). Although this feature has never been investigated in *P. antarcticum*, it may well be typical of the oxygen-transport system of this fish.

Recent studies on non-Antarctic species are aimed at filling gaps in the available information on basic physiological parameters in Notothenioidei. The oxygen-transport system of two species which belong to this suborder but are non-Antarctic has been investigated at the molecular level for the first time. The comparison between cold-adapted and non-cold-adapted species may also help us to understand the evolutionary history of these fish, and the molecular approach to the study of the Hb systems of *N. angustata* and *P. urvillii* has provided useful information. Following the suggestion of Eastman (1993, p. 278), this is the first step of a logical extension of research to non-Antarctic notothenioids.

Acknowledgements

The work summarised in this chapter was sponsored by the Italian National Programme for Antarctic Research. The outstanding contribution of L. Camardella, V. Carratore, C. Caruso, E. Cocca, A. Fago, A. Riccio, M. Romano and the late B. Rutigliano is gratefully acknowledged. The authors thank Dr R. Williams for his help in capturing live specimens of *P. urvillii*, and Dr A.L. DeVries for the blood of *N. angustata*. GdP and MT are grateful to the Alfred Wegener Institute, Bremerhaven, Germany, for the invitation to participate in the expedition Ant X/3 (March–May 1992) in the northeastern Weddell Sea.

References

Andersen, N.C. (1984). Genera and subfamilies of the family Nototheniidae (Pisces, Perciformes) from the Antarctic and Subantarctic. *Steenstrupia*, **10**, 1–34.

Andrews, A.P. (1980). Family Bovichthyidae: Congolli. In *Freshwater Fishes of South-Eastern Australia*, ed. R.M. McDowall, pp. 167–8. Sydney: Reed.

Balushkin, A.V. (1988). Suborder Notothenioidei. In *A Working List of Fishes of the World*, ed. D.E. McAllister, pp. 1118–26. Ottawa: National Museum of Canada.

Balushkin, A.V. (1992). Classification, phylogenetic relationships and origins of the families of the suborder Notothenioidei (Perciformes). *Journal of Ichthyology*, **30**, 132–47.

Bellelli, A. & di Prisco, G. (1992). Thermodynamic and stereochemical modelling of vertebrate haemoglobin. In *Society for Experimental Biology Seminar Series 51: Oxygen Transport in Biological Systems*, ed. S. Egginton & H.F. Ross, pp. 103–34. Cambridge: Cambridge University Press.

Brittain, T. (1987). The Root effect. *Comparative Biochemistry and Physiology*, **86B**, 473–81.

Camardella, L., Caruso, C., D'Avino, R., di Prisco, G., Rutigliano, B., Tamburrini, M., Fermi, G. & Perutz, M.F. (1992). Haemoglobin of the Antarctic fish *Pagothenia bernacchii*. Amino acid sequence, oxygen equilibria and crystal structure of its carbonmonoxy derivative. *Journal of Molecular Biology*, **224**, 449–60.

Caruso, C., Rutigliano, B., Romano, M. & di Prisco, G. (1991). The hemoglobins of the cold-adapted Antarctic teleost *Cygnodraco mawsoni*. *Biochimica et Biophysica Acta*, **1078**, 273–82.

Caruso, C., Rutigliano, B., Riccio, A., Kunzmann, A. & di Prisco, G. (1992). The amino acid sequence of the single hemoglobin of the high-Antarctic fish *Bathydraco marri* Norman. *Comparative Biochemistry and Physiology*, **102B**, 941–6.

Clarke, A. (1983). Life in cold water: the physiological ecology of polar marine ectotherms. *Oceanography and Marine Biology: An Annual Review*, **21**, 341–453.

Clarke, A. & North, A.W. (1991). Is the growth of polar fish limited by temperature? In *Biology of Antarctic Fish*, ed. G. di Prisco, B. Maresca & B. Tota, pp. 54–69. Berlin, Heidelberg, New York: Springer-Verlag.

Cocca, E., Ratnayake-Lecamwasam, M., Parker, S.K., Camardella, L., Ciaramella, M., di Prisco, G. & Detrich, H.W. III (1995). Genomic remnants of α-globin genes in the hemoglobinless Antarctic icefishes. *Proceedings of the National Academy of Sciences of the USA*, **92**, 1817–21.

D'Avino, R. & di Prisco, G. (1988). Antarctic fish hemoglobin: an outline of the molecular structure and oxygen binding properties. – 1. Molecular structure. *Comparative Biochemistry and Physiology*, **90B**, 579–84.

D'Avino, R. & di Prisco, G. (1989). Hemoglobin from the Antarctic fish *Notothenia coriiceps neglecta.* 1. Purification and characterisation. *European Journal of Biochemistry*, **179**, 699–705.

D'Avino, R., Caruso, C., Romano, M., Camardella, L., Rutigliano, B. & di Prisco, G. (1989). Hemoglobin from the Antarctic fish *Notothenia coriiceps neglecta*. 2. Amino acid sequence of the α chain of Hb 1. *European Journal of Biochemistry*, **179**, 707–13.

D'Avino, R., Caruso, C., Camardella, L., Schinina, M.E., Rutigliano, B., Romano, M., Carratore, V., Barra, D. & di Prisco, G. (1991). An overview of the molecular structure and functional properties of the hemoglobins of a cold-adapted Antarctic teleost. In *Life Under Extreme Conditions. Biochemical Adaptations*, ed. G. di Prisco, pp. 15–33. Berlin, Heidelberg, New York: Springer-Verlag.

D'Avino, R., Fago, A., Kunzmann, A. & di Prisco, G. (1992). The primary structure and oxygen-binding properties of the high-Antarctic fish *Aethotaxis mitopteryx* DeWitt. *Polar Biology*, **12**, 135–40.

D'Avino, R., Caruso, C., Tamburrini, M., Romano, M., Rutigliano, B., Polverino de Laureto, P., Camardella, L., Carratore, V. & di Prisco, G. (1994). Molecular characterization of the functionally distinct hemoglobins of the Antarctic fish *Trematomus newnesi. Journal of Biological Chemistry*, **269**, 9675–81.

DeVries, A.L. (1983). Antifreeze peptides and glycoproteins in cold-water fishes. *Annual Review in Physiology*, **45**, 245–260.

DeWitt, H.H., Heemstra, P.C. & Gon, O. (1990). Nototheniidae. In *Fishes of the Southern Ocean*, ed. O. Gon & P.C. Heemstra, pp. 279–331. Grahamstown, South Africa: J.L.B. Smith Institute of Ichthyology.

Dickerson, R.E. & Geis, I. (1983). *Hemoglobin: Structure, Function, Evolution and Pathology*. Menlo Park, CA: Benjamin/Cummings.

di Prisco, G. (1988). A study of hemoglobin in Antarctic fishes: purification and characterisation of hemoglobins from four species. *Comparative Biochemistry and Physiology*, **90B**, 631–7.

di Prisco, G. (ed.) (1991). *Life under Extreme Conditions. Biochemical Adaptations*. Berlin, Heidelberg: Springer-Verlag.

di Prisco, G. & D'Avino, R. (1989). Molecular adaptation of the blood of Antarctic teleosts to environmental conditions. *Antarctic Science*, **1**, 119–24.

di Prisco, G. & Tamburrini, M. (1992). The hemoglobins of marine and freshwater fish: the search for correlations with physiological adaptation. *Comparative Biochemistry and Physiology*, **102B**, 661–71.

di Prisco, G., Maresca, B. & Tota, B. (eds.) (1988*a*). *Marine Biology of Antarctic Organisms. Comparative Biochemistry and Physiology*, **90B**, 459–637.

di Prisco, G., Giardina, B., D'Avino, R., Condo', S.G., Bellelli, A. & Brunori, M. (1988*b*). Antarctic fish hemoglobin: an outline of the molecular structure and oxygen binding properties – II. Oxygen binding properties. *Comparative Biochemistry and Physiology*, **90B**, 585–91.

di Prisco, G., D'Avino, R., Camardella, L., Caruso, C., Romano, M. & Rutigliano B. (1990). Structure and function of hemoglobin in Antarctic fishes and evolutionary implications. *Polar Biology*, **10**, 269–74.

di Prisco, G., Maresca, B. & Tota, B. (eds.) (1991*a*). *Biology of Antarctic Fish*. Berlin, Heidelberg, New York: Springer-Verlag.

di Prisco, G., D'Avino, R., Caruso, C., Tamburrini, M., Camardella, L., Rutigliano, B., Carratore, V. & Romano, M. (1991*b*). The biochemistry of oxygen transport in red-blooded Antarctic fish. In *Biology of Antarctic Fish*, ed. G. di Prisco, B. Maresca & B. Tota, pp. 263–81. Berlin, Heidelberg, New York: Springer-Verlag.

di Prisco, G., Condo', S.G., Tamburrini, M & Giardina, B. (1991*c*). Oxygen transport in extreme environments. *Trends in Biochemical Sciences*, **16**, 471–4.

di Prisco, G., Macdonald, J.A. & Brunori, M. (1992). Antarctic fishes survive exposure to carbon monoxide. *Experientia*, **48**, 473–5.

di Prisco, G., Camardella, L., Carratore, V., Caruso, C., Ciardiello, M.A., D'Avino, R,, Fago, A., Riccio, A., Romano, M., Rutigliano, B. & Tamburrini, M. (1994). Structure and function of hemoglobins, enzymes and other proteins from Antarctic marine and terrestrial organisms. In *Proceedings of the 2nd Meeting 'Biology in Antarctica'*, ed. B. Battaglia, P.M. Bisol & V. Varotto, pp. 157–77. Padova: Edizioni Universitarie Patavine.

Eastman J.T. (1988). Ocular morphology in Antarctic notothenioid fishes. *Journal of Morphology*, **196**, 283–306.

Eastman, J.T. (1991). The fossil and modern fish faunas of Antarctica: Evolution and diversity. In *Biology of Antarctic Fish*, ed. G. di Prisco, B. Maresca, & B. Tota, pp. 116–30. Berlin, Heidelberg, New York: Springer-Verlag.

Eastman, J.T. (1993). *Antarctic Fish Biology. Evolution in a Unique Environment*. San Diego, CA: Academic Press.

Eastman, J.T. & DeVries, A.L. (1982). Buoyancy studies of notothenioid fishes in McMurdo sound, Antarctica. *Copeia*, **1982(2)**, 385–93.

Ekau, W. (1988). Okomorphologie nototheniider Fische aus dem Weddellmeer, Antarktis. *Berichte zur Polarforschung*, **51**, 1–140.

Ekau W. (1991). Morphological adaptations and mode of life in high Antarctic fish. In *Biology of Antarctic Fish*, ed. G. di Prisco, B. Maresca & B. Tota, pp. 23–39. Springer-Verlag, Berlin, Heidelberg, New York.

Everson, I. & Ralph, R. (1968). Blood analyses of some Antarctic fish. *British Antarctic Survey Bulletin*, **15**, 59–62.

Fago, A., D'Avino, R. & di Prisco, G. (1992). The hemoglobins of *Notothenia angustata*, a temperate fish belonging to a family largely endemic to the Antarctic Ocean. *European Journal of Biochemistry*, **210**, 963–70.

Giardina, B. & Amiconi, G. (1981). Measurement of binding of gaseous and nongaseous ligands to hemoglobins by conventional spectrophotometric procedures. *Methods in Enzymology*, **76**,417–27.

Gon, O. & Heemstra, P.C. (eds) (1990). Fishes of the Southern Ocean. Grahamstown, South Africa: JLB Smith Institute of Ichthyology.

Hamoir, G. (1988). Biochemical adaptation of the muscles of the Channichthyidae to their lack of hemoglobin and myoglobin. *Comparative Biochemistry and Physiology*, **90B**, 557–9.

Hemmingsen, E.A. & Douglas, E.L. (1970). Respiratory characteristics of the hemoglobin-free fish *Chaenocephalus aceratus*. *Comparative Biochemistry and Physiology*, **33**, 733–44.

Hemmingsen, E.A. & Douglas, E.L. (1977). Respiratory and circulatory adaptations to the absence of hemoglobin in Chaenichthyid fishes. In *Adaptations within Antarctic Ecoystems*, ed. G.A. Llano, pp. 479–87. Washington: Smithsonian Institution.

Hochachka, P.W. & Somero, G.N. (1984). *Biochemical Adaptation*. Princeton, NJ: Princeton University Press.

Holeton, G.F. (1970). Oxygen uptake and circulation by a hemoglobinless Antarctic fish (*Chaenocephalus aceratus* Lonnberg) compared with three red-blooded Antarctic fish. *Comparative Biochemistry and Physiology*, **34**, 457–71.

Hubold, G. (1984). Spatial distribution of *Pleuragramma antarcticum* (Pisces: Nototheniidae) near the Filchner- and Larsen Ice Shelves (Weddell Sea/Antarctica). *Polar Biology*, **3**, 231–6.

Hubold, G. (1985). On the early life history of the high-Antarctic silverfish *Pleuragramma antarcticum*. In *Proceedings of the 4th SCAR Symposium on Antarctic Biology, Antarctic Nutrient Cycles and Food Webs*, ed. W.R. Siegfried, P.R. Condy & R.M. Laws, pp. 445–51. Berlin, Heidelberg, New York: Springer-Verlag.

Hubold, G. (1991). Ecology of notothenioid fish in the Weddell Sea. In *Biology of Antarctic Fish*, ed. G. di Prisco, B. Maresca & B. Tota, pp. 3–22. Berlin, Heidelberg, New York: Springer-Verlag.

Hureau, J-C., Petit, D., Fine. J.M. & Marneux, M. (1977). New cytological, biochemical and physiological data on the colorless blood of the Channichthyidae (Pisces, Teleosteans, Perciformes). In *Adaptations within Antarctic Ecoystems*, ed. G.A. Llano, pp. 459–77. Washington: Smithsonian Institution.

Iwami, T. (1985). Osteology and relationships of the family Channichthyidae. *Memories of the Institute of Polar Research, Tokyo, Series E*, **36**, 1–69.

Johnston, I.A. (1989). Antarctic fish muscles. Structure, function and physiology. *Antarctic Science*, **1(2)**, 97–108.

Kunzmann, A. (1990). Gill morphometrics of two Antarctic fish species: *Pleuragramma antarcticum* and *Notothenia gibberifrons*. *Polar Biology*, **11**, 9–18.

Kunzmann, A. (1991). Blood physiology and ecological consequences in Weddell Sea fishes (Antarctica). PhD Thesis, *Berichte zur Polarforschung*, **91**, 1–79.

Kunzmann, A., Caruso, C. & di Prisco, G. (1991). Haematological studies on a high-Antarctic fish: *Bathydraco marri* Norman. *Journal of Experimental Marine Biology and Ecology*, **152**, 243–55.

Love, R.M. (1980). *The Chemical Biology of Fishes. Advances 1968–77*, Vol. 2, p. 943. London: Academic Press.

Macdonald, J.A., Montgomery, J.C. & Wells, R.M.G. (1987). Comparative physiology of Antarctic fishes. *Advances in Marine Biology*, **24**, 321–88.

Macdonald, J.A. & Wells, R.M.G. (1991). Viscosity of body fluids from Antarctic notothenioid fish. In *Biology of Antarctic Fish*, ed. G. di Prisco, B. Maresca & B. Tota, pp. 163–78. Berlin, Heidelberg, New York: Springer-Verlag.

Montgomery, J.C. & Wells, R.M.G. (1993). Recent advances in the ecophysiology of Antarctic notothenioid fishes: metabolic capacity and sensory performance. In *Fish Ecophysiology*, ed. J.C. Rankin & F.B. Jensen, pp. 341–74. London: Chapman & Hall.

Riggs, A. (1988). The Bohr effect. *Annual Review of Physiology*, **50**, 181–204.

Ruud, J.T. (1954). Vertebrates without erythrocytes and blood pigment. *Nature*, **173**, 848–50.

Somero, G.N. & DeVries, A.L. (1967). Temperature tolerance of some Antarctic fishes. *Science*, **156**, 257–8.

Tamburrini, M., Brancaccio, A., Ippoliti, R. & di Prisco, G. (1992). The amino acid sequence and oxygen-binding properties of the single hemoglobin of the cold-adapted Antarctic teleost *Gymnodraco acuticeps. Archives of Biochemistry and Biophysics*, **292**, 295–302.

Tamburrini, M., D'Avino, R., Fago, A., Carratore, V., Kunzmann, A. & di Prisco, G. (1996). The unique hemoglobin system of *Pleuragramma antarcticum*, an Antarctic migratory teleost. *Journal of Biological Chemistry*, **271**, 23780–23785.

Tetens, V., Wells, R.M.G. & DeVries, A.L. (1984). Antarctic fish blood: respiratory properties and the effects of thermal acclimation. *Journal of Experimental Biology*, **109**, 265–79.

Wells, R.M.G. (1987). Respiration of Antarctic fish from McMurdo sound. *Comparative Biochemistry and Physiology*, **88A**, 417–24.

Wells, R.M.G. (1990). Hemoglobin physiology in vertebrate animals: a cautionary approach to adaptionist thinking. *Advances in Comparative and Environmental Physiology*, **6**, 143–61.

Wells, R.M.G., Ashby, M.D., Duncan, S.J. & Macdonald, J.A. (1980). Comparative studies of the erythrocytes and haemoglobins in nototheniid fishes from Antarctica. *Journal of Fish Biology*, **15**, 517–27.

Wells, R.M.G, Macdonald, J.A. & di Prisco, G. (1990). Thin-blooded Antarctic fishes: a rheological comparison of the haemoglobin- free icefishes, *Chionodraco kathleenae* and *Cryodraco antarcticus*, with a red-blooded nototheniid, *Pagothenia bernacchii. Journal of Fish Biology*, **36**, 595–609.

C. STORELLI, R. ACIERNO and M. MAFFIA

Membrane lipid and protein adaptations in Antarctic fish

The success of Notothenioids to survive in the Antarctic involves, among other things, adaptive modifications of proteins and membrane lipids, because only the conservation of their biochemical characteristics and physiological state allows physiological processes to proceed at the low temperature of the Antarctic sea.

Biological membranes are complex structures whose composition (mainly lipid and protein) and arrangement vary widely both between, and within, cells depending upon function (Gennis, 1989).

The phospholipid bilayer holds multiple essential properties for cell membrane function since it:

(i) acts as a physical barrier to solute diVusion;
(ii) regulates the utilization of energy in transmembrane ion gradients;
(iii) mediates the transmembrane movement of specific solutes;
(iv) provides an organizing matrix for the assembly of multicomponent metabolic and signal transduction pathways;
(v) supplies precursors for the generation of lipid-derived second messengers.

Biological membranes may contain different types of proteins related to different cellular functions such as transmembrane transport (channels and carriers), substrate hydrolysis (enzymes), hormone and neurotransmitter recognition (receptors) and protein contributions to the mechanical structure of the membrane. Membrane proteins may be associated with the lipid bilayer in different ways. They have been distinguished as peripheral or integral proteins according to whether they penetrate to a lesser or greater extent into the bilayer and can therefore be isolated by mild or more severe treatments.

The amount of lipid in biological membranes ranges between 20 and 80% of the dry weight and, because of the lipids, particular structural and physical properties of the membrane occur. The most common types of lipid in biological membranes are the glycerol-phosphatides; fatty acids, found in biological membranes, usually contain 16, 18, 20 or 22 carbon atoms, fully saturated or exhibiting between one and six *cis* double bonds, usually methylene interrupted. Bacteria may also contain a number of branched, cyclic or isoprenoid fatty acids in membrane lipids (Kaneda, 1977; Russel, 1984). Animal cell membranes usually contain cholesterol, while other cell types may contain sterols other than cholesterol (Nes & Nes, 1980; Bloch, 1983).

It is known that each class of lipid corresponds to a specific gel to liquid crystalline phase transition temperature and that, above that temperature, in the presence of water, phospholipids form stable bilayers (lamellar phase) which are the basic structure of biological membranes (Hazel, 1995). Therefore, a modification of lipid composition, and/or arrangement, is required to ensure the physiological efficiency of membranes at different environmental temperatures. This phenomenon has been described in temperature-acclimated poikilotherms and named 'homeoviscous adaptation' (Sinensky, 1974; Cossins & Sinensky, 1986).

As for the membrane lipid bilayer, low temperature may affect cell membrane function since it may directly affect the structure and function of proteins thus modifying their biological activity (both decreasing the rate of enzyme activity and/or denaturing them). In particular, temperature can affect protein activity in two different ways:

(i) by modifying protein structure, because all the energies that contribute to the conformational stability of proteins (hydrophobic forces; hydrogen-bonding and electrostatic contributions) vary in different ways with temperatures;
(ii) by influencing the rate of catalysed reactions, because the fraction of reactive substrate molecules, having an energy equal to or greater than activation energy (E_a), is decreased by reduction of temperature.

As a consequence, the rates of enzymatic reactions can slow down considerably at low temperatures and, therefore, in the absence of compensating mechanisms, physiological processes in cold-acclimated or adapted ectotherms ought to proceed more slowly than in ectotherms of temperate or tropical environments.

The major adaptive strategies which might counteract the effects of low temperature are:

(i) modification of the lipid microenvironment which controls the activity of the protein;
(ii) increase in the number of active enzyme molecules;
(iii) generation of new enzyme variants.

It is of extreme interest therefore, to understand how poikilotherms, such as Antarctic teleosts, have faced the challenge of maintaining physiological function at low temperatures.

Starting from the few data available in the literature obtained from different tissues of different animals (Clarke, 1987; Somero, 1991; Lund & Sidell, 1992; Nichols *et al.*, 1994), the aim of this chapter is to update information on lipid and protein cold adaptation mechanisms within the context of current knowledge, including the latest data obtained from the recent Italian expeditions to the Antarctic continent (Terranova Bay). In particular, the total lipid composition and some characteristics of the cell membrane functional proteins, alkaline phosphatase, leucine aminopeptidase, maltase, Na^+/ K^+-ATPase, Na^+-D-glucose cotransporter of the intestinal mucosa of the Antarctic teleost *Trematomus bernacchii*, are compared with those of the temperate fish *Anguilla anguilla*.

A further effort to understand cold adaptation mechanisms of the two main components of cell membranes and their possible relationship will be made by studying membrane lipid composition and properties of sugar and amino acid transporter proteins in isolated intestinal brush border membrane vesicles of Antarctic fish.

Owing to the reliance of actual data on few species, more comparative studies are needed in the future to extend the results to all Antarctic species.

Membrane lipid composition

Moving from the observation that *Escherichia coli* grown at 15 °C and 43 °C displayed very similar characteristics of membrane fluidity at their respective temperatures of growth, Sinensky (1974) introduced the concept of homeoviscous adaptation (HVA) to indicate the acclimatory or adaptive modification of poikilotherm cell membrane lipid composition that would compensate for a temperature effect on membrane fluidity. For many years, HVA, intended as a static description of lipid order, has been the main phenomenon invoked to explain the modifications that membranes undergo when exposed to temperature challenges (Cossins & Prosser, 1978; Cossins *et al.*, 1987; Farkas, Storebakken & Bhosie, 1988; Vrbjar *et al.*, 1992; Dahlhoff & Somero, 1993). Its main features have been identified as: (i) variation of unsaturated/saturated fatty acid ratio (or branched fatty acid amount in microorganisms); (ii) modification of the phosphatidylethanolamine/ phosphatidylcholine (PE/PC) ratio; and (iii) reduction of the plasmalogens/ diacylphospholipids ratio (Hazel, 1995).

More recently, moving from the observation that some temperature-induced lipid modifications do not follow the concept of HVA (McElhaney, 1984), a new paradigm, homeophasic adaptation (HPA), has been introduced to improve the assessment of the temperature-dependent lipid rearrangement. In particular, the increase of long-chain polyunsaturated fatty acids (PUFA) and the increase of the PE/PC ratio at low temperatures appear relatively difficult to explain in terms of HVA. Studies conducted on artificial monolayers disclosed that the presence of double bonds within the acyl chains of phospholipid fatty acids disturbs their packing order, consequently increasing the bilayer fluidity. A deeper investigation showed that the insertion of the first double bond in a saturated acyl chain produces the greatest enhancement of fluidity, while subsequent double bond addition causes proportionally smaller modifications (Demel *et al.*, 1972; Gosh, Williams & Tinoco, 1973). As a matter of steric hindrance, the first insertion of a double bond causes the highest increase of molecular area and, the closer the position of the double bond to the phospholipid polar head, the greater the effect in terms of molecular area extension. Thus, owing to the higher energetic cost of synthesising PUFAs, it is unclear why they should be preferred to monoenoic fatty acids to increase bilayer fluidity at low temperatures. Furthermore, even if possessing specific characteristics of a bilayer destabiliser phospholipid due to its conical shape (see Israelachvili, Marcelja & Horn, 1981 and Seddon, 1990 for more details), because of the capacity to form hydrogen bonds between its poorly hydrated headgroups (Silvius, 1986), PE usually tends to increase the gel/fluid phase transition temperature by increasing membrane order. McElhaney (1984) postulated that features of membrane organisation other than lipid order are subject to regulation when environmental conditions change. According to the model of the dynamically heterogeneous membrane state, admirably described by Bloom, Evans and Mouritsen (1991), the membrane bilayer would consist of a dynamic coexistence, over a finite range of temperature, of heterogeneous gel and fluid microdomains, in a dynamic balance that permits and regulates the functional efficiency of the membrane. In this sense, the high amount of PUFAs and PE in cold-acclimated poikilotherms may relate to the conservation of an optimal balance between membrane microdomains and, hence, the maintenance of a functional dynamic phase behaviour compatible with biological function (for an excellent review, see Hazel,1995).

Despite an abundance of data on membrane lipid modifications in temperature-acclimated poikilotherms (Kemp & Smith, 1970; Sinensky, 1974; Hazel, 1979, 1984; Hagar & Hazel, 1985; Schwalme, Mackay & Clandinin, 1993; Delgado *et al.*, 1994), little attention has been directed to date toward the adaptive (i.e. evolutionary adaptation) aspect of membrane lipid composition. Since the Early Miocene (around 20 to 25 my ago), cold water has been a characteristic of the Southern Ocean (Hudson, 1977; Savin, 1977) and, today, Antarctic fish commonly experience below zero environmental temperature throughout the entire year. Obviously, within a series of eco-physiological evolutionary aspects related to the thermodynamic disadvantages of life in the cold (biochemical efficiency of proteins, blood viscosity, neural transmission, etc.), particular adaptive compensatory mechanisms have been evolved to permit the biological function of cell membranes at below-zero temperatures.

The single, but indirect, report on membrane fluidity in Antarctic poikilotherms in the literature is the measurement of bilayer order by fluorescence anisotropy techniques (Behan-Martin *et al.*, 1993). From this work it appears that, despite large differences in environmental temperature, brain cell membrane order in the Antarctic teleost, *Notothenia neglecta,* is comparable to that of warm-adapted fish species and mammals at their respective body temperatures.

Data on membrane lipid composition in Antarctic poikilotherms are also limited. Neutral lipid composition of Antarctic fish adipose tissue, blood and oxidative muscle has shown a high percentage of UFA and monounsaturated fatty acid (MUFA) (Lund & Sidell, 1992). Based upon the *in vitro* measurement of the rates of oxidation of carbohydrates and fatty acids by oxidative muscle, Sidell and co-workers (1995) suggested that MUFA could be a preferred substrate for energy metabolism in Antarctic species. Nachman (1985) analysed the lipid composition of different Antarctic poikilotherms and reported an unusual predominance of even-carbon hydrocarbons of fatty acids. Nichols *et al.* (1994) reported the fatty acid composition in the muscle of Antarctic, sub-Antarctic and temperate fish. This comparison, carried out to evaluate the potential of Antarctic fish as a source of ω3 PUFA for the human diet, has shown non-significant differences between fish of different temperature habitats in MUFA and PUFA percentage composition. Interspecific differences, essentially linked to dietary preferences, have been suggested by the authors to explain their results. It is well known how dietary behaviour influences the lipid composition of some tissues. Muscular tissue, along with subcutaneous spaces, liver and mesentery, are usually the main locations of lipid storage in fish (Porteres, 1991; Argyropoulou, Kalogeropoulos & Alexis, 1992, Eastman, 1993; Berlin *et al.*, 1994). In Notothenioid Antarctic fish, which lack a swimbladder, the high lipid content of their tissues has been correlated to an increase of buoyancy (Eastman & DeVries, 1981; Clarke *et al.*, 1984; Nachman, 1985). To minimise misleading results due to catabolic implications (i.e. the proportion of MUFA in tissues with elevated energy requirements), dietary influences and buoyancy mechanisms, it was decided to investigate the lipid composition of the intestinal absorbing epithelium. Owing to its highly discriminative characteristics of ion, water and nutrient transport, the cell membrane of the intestinal mucosa must retain a well-defined and constant level of fluidity and, hence, of lipid composition.

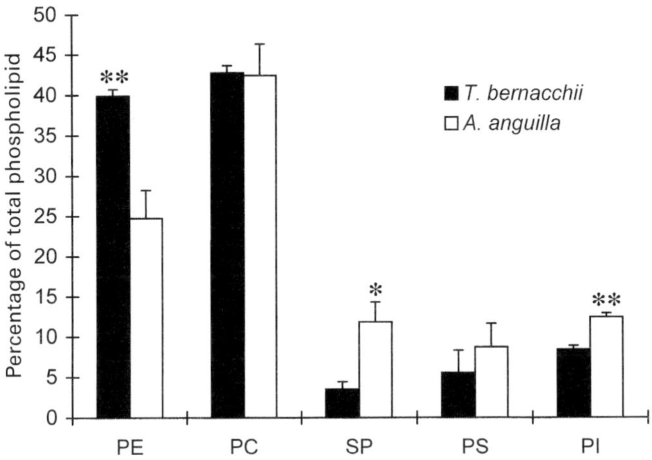

Fig. 1. Phospholipid composition of the intestinal mucosa of *T. bernacchii* and *A. anguilla*, expressed as percent of total phospholipid. PE=phosphatidylethanolamine, PC=phosphatidylcholine, SP=sphyngomyelin, PS=phosphatidylserine, PI=phosphatidylinositol. *=$p<0.05$; **=$p<0.01$ indicate significant difference between the two species (data from Acierno *et al.*, 1997).

Therefore, lipid composition of such a specialised tissue should, without doubt, remain unaffected, or scarcely affected, by the above reported variables.

The first aspect approached in the analysis of lipid composition has been the evaluation of phospholipid percent distribution in the enterocytes. The determination has been carried out on the Antarctic teleost *T. bernacchii* and the Mediterranean, temperate teleost *A. anguilla.* The most evident result arising from the comparison, as shown in Fig. 1, is that the Antarctic teleost possesses a higher percentage of the bilayer destabiliser PE and a lower percentage of sphingomyelin (SP), which is considered a bilayer stabiliser (Cullis *et al.*, 1985). In a recent review, Hazel (1995), with regard to the regulation of dynamic phase behaviour of membrane bilayers, emphasised that the presence of bilayer-destabilising and bilayer-stabilising phospholipid is required for optimal membrane function, indicating PE and PC as the main examples of the two classes, respectively. From Fig. 1 it is clear that *T. bernacchii* manifests a higher PE/PC ratio, compared to *A. anguilla,* a characteristic that occurs also in cold-acclimated animals as shown by Pruitt (1988). The high percentage of PE and the low percentage of SP found in the Antarctic teleost could then be related to the preservation of membrane function at low temperature and, in particular, to the ability of

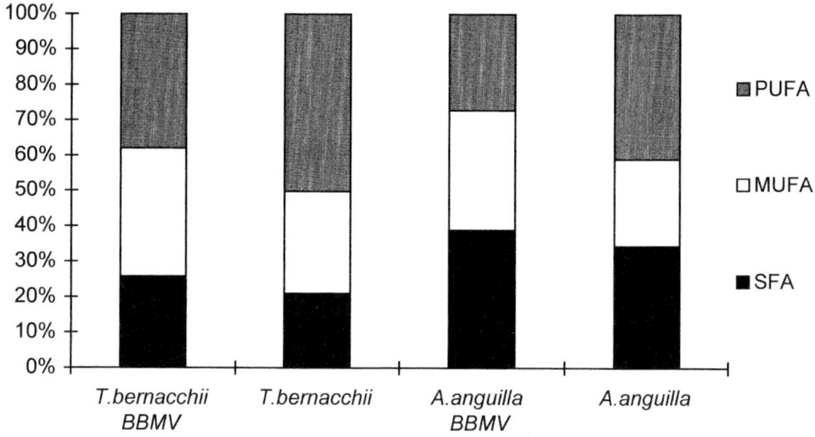

Fig. 2. Sum of saturated (SFA), monounsaturated (MUFA) and polyunsaturated (PUFA) fatty acids in homogenate and brush border membrane vesicles (BBMV) of the intestinal mucosa of *T. bernacchii* and *A. anguilla*, expressed as percent of total fatty acid. Difference between the two species is significant for SFA ($p<0.01$ both in homogenate and BBMV) and for PUFA ($p<0.05$ both in homogenate and BBMV).

regulating fusion events, critical for the control of intra- and trans-cellular membrane traffic.

Further characteristics of phospholipids in determining the dynamic behaviour of the membrane derive from their fatty acid composition. In biological membranes, the unsaturated/saturated fatty acid ratio has been found to be directly correlated to membrane viscosity (Cossins *et al.*, 1987; Hazel, 1989). The fatty acid composition of intestinal mucosa homogenate of *T. bernacchii* and *A. anguilla* has been analysed and compared; furthermore, to eliminate the contribution of non-membrane fatty acid in the results, the same comparison has been carried out on a highly purified cell membrane fraction: the intestinal brush border membrane vesicles (BBMV) of the same species. These comparisons reveal a quite constant pattern that agrees with the theoretical conclusion reported above (Table 1). As summarised in Fig. 2, the SFA/UFA ratio is always lower in *T. bernacchii* in comparison with *A. anguilla* and, moreover, the contribution of MUFA to the total UFA is significantly higher in the Antarctic teleost.

Beyond particular ecological and metabolic implications, all the results arising from our studies are consistent with a fluidising effect on membranes that would allow the cell to preserve a dynamic state essential for the maintenance of its biological functions.

Table 1. *Fatty acid composition of homogenate and brush border membrane vesicles (BBMV) of the intestinal mucosa of* T. bernacchii *and* A. anguilla, *expressed as percentage of total fatty acid*

	Homogenate			BBMV		
	Antarctic fish *T. bernacchii*	p	Temperate fish *A. anguilla*	Antarctic fish *T. bernacchii*	p	Temperate fish *A. anguilla*
14:0	0.90		0.31	4.32	*b*	0.61
15:0	0.25		0.04	0.31		0.43
16:0	11.38	*b*	22.42	14.25	*b*	22.68
16:1 n7	5.77	*a*	2.15	6.21	*a*	1.41
17:0	1.12	*b*	0.09	0.35	*a*	0.55
18:0	4.42	*b*	11.20	4.99	*a*	10.35
18:1 n9	8.75		12.87	16.87	*a*	7.14
18:1 n7	7.76	*a*	4.35	7.55	*b*	3.40
18:2 n6	0.58	*a*	1.47	1.09		0.99
18:3 n3	0.10		0.06	0.22	*b*	0.03
18:4 n3	0.40		0.31	0.40	*b*	0.03
19:0	0.06		0.11	0.49		0.70
20:1 n9	2.80		1.04	3.91	*b*	0.72
20:1 n7	2.58		2.95	0.80		0.84
20:2 n6	0.18		0.08	1.60		1.66
20:3 n3	0.07		0.04	0.10	*a*	0.02
20:4 n6	4.82		6.96	2.60	*b*	7.25
20:5 n3	16.22	*b*	5.04	10.54	*b*	2.68
21:5 n3	0.49		0.42	0.11	*b*	0.42
22:0	1.28		0.06	0.57	*a*	0.02
22:1 n9	0.07	*b*	1.16	0.94	*b*	20.48
22:5 n3	0.10	*b*	2.09	0.29	*b*	2.72
22:6 n3	27.36		24.50	20.99	*a*	11.41
24:0	1.57	*b*	0.09	0.48	*b*	3.44
24:1 n9	0.98		0.17	0.03		0.03

Note:
[a]=$p<0.05$; [b]=$p<0.01$ indicate significant difference between the two species.

Table 2. *Maximal enzymatic activity of alkaline phosphatase, leucine aminopeptidase, maltase, $Na^+/K^+-ATPase$ in* T. bernacchii *and* A. anguilla *intestinal mucosa*

	Antarctic fish *Trematomus bernacchii* (0 °C)	Temperate fish *Anguilla anguilla* (18 °C)
Alkaline phosphatase	267.38 ± 33.0^a	97.50 ± 2.97
Leucine aminopeptidase	14.78 ± 0.69	13.76 ± 0.89
Maltase	11.80 ± 3.40	13.90 ± 0.60
Na^+/K^+-ATPase	12.93 ± 5.18	9.73 ± 0.44

Notes:
Data reported in the table expressed as mU/mg protein for the enzymes are means ± standard errors (S.E.) of at least four different preparations. Alkaline phosphatase (EC 3.1.3.1), leucine aminopeptidase (EC 3.4.11.1) and maltase (EC 3.2.1.20) activities were measured by methods described by Berner, Kinne & Murer (1976), Haase *et al.* (1978), Storelli *et al.* (1986). Na^+/K^+-ATPase (EC 3.6.1.3) activity was determined by measuring the appearance of inorganic phosphate (method of Fiske and Subarrow as reported by Higgins, 1978) in the presence or in the absence of ouabain. ^{a}=$p<0.01$ indicates significant difference between the two species.

Membrane enzymatic proteins

Many enzymes of Antarctic fish have a higher catalytic rate than the homologous enzymes of warm-adapted species (Hazel & Prosser, 1974; Somero, 1991; Clarke, 1987; Feller *et al.*, 1991*a*; Feller, Thiry & Gerday, 1991*b*). Recent investigation confirms this statement. As shown in Table 2, the enzymatic activities of three hydrolases, alkaline phosphatase, leucine aminopeptidase and maltase, involved in the terminal steps of phosphate ester, peptide and carbohydrate digestion, respectively, of *T. bernacchii* intestine, are similar or higher than those found in the intestinal mucosa of *A. anguilla* when measured at the respective environmental temperature of the two fish species. Similar results were obtained by Genicot, Feller and Gerday (1988), when comparing trypsin activity from pyloric ceca of *Paranothotenia magellanica* and trout (*Oncorhynchus mykiss*). More recently, Na^+/K^+-ATPase of *T. bernacchii* intestine was measured at 0 °C and the results were not significantly different from those measured in *A. anguilla* intestine at 18 °C (Table 2). These results can be explained by:

(i) a quantitative increase of enzymatic molecules;
(ii) a generation of new enzyme variants well adapted to work at 0 °C.

An adaptive strategy involving the presence of a higher amount of enzymatic molecules in the cell will increase the rate of reaction without modifying the critical relationship between K_m and substrate concentration. Until recently, only a few studies measured the absolute amount of enzyme in cold-acclimated animals. For example, cytochrome oxidase content of skeletal muscle is significantly higher in goldfish acclimated to 5 °C than in animals acclimated to 25 °C (Wilson, 1973). Similar findings were observed for cytochrome *c* content of green sunfish skeletal muscle, demonstrating increases of about 50% following acclimation from 25 °C to 5 °C. In this case it has been demonstrated that the transfer to low temperature significantly reduced the rate of degradation of the enzyme, resulting in an increase in cellular concentration of cytochrome *c* (Sidell, 1977). In general, mitochondria concentrations increase after acclimation of eurythermal fish to low temperature. In muscle cells of *Carassius carassius* acclimated to 2 °C, the activity of mitochondrial enzymes is much higher than that found in the same fish acclimated to 28 °C. This finding is strictly connected with the observed increase in the number of mitochondria in the cells of low-temperature acclimated animals (Johnston & Maitland, 1980). Similar results were obtained in goldfish acclimated from 25 °C to 5 °C (Sidell, 1983; Tyler & Sidell, 1984).

The increase of enzyme concentration could be a useful strategy also in the evolutionary adaptation of fish to cold environments. The mitochondrial volume fraction in slow and cardiac muscle cells of the Antarctic hemoglobinless fishes, *Chaenocephalus aceratus* and *Chionodraco hamatus* is towards the upper range reported for teleosts (Johnston, 1981, 1983; Tota *et al.*, 1991). This feature suggests that the intracellular concentrations of enzymes associated with aerobic respiration can be very high in some Antarctic fish. The increase in enzyme concentration can compensate for the reduced rate of activity per site due to Q_{10} effects (Somero, 1991).

Besides the reduction of the rate of degradation, an adaptive strategy resulting in the increase of enzymatic molecules can be based upon a higher rate of protein synthesis in cold-adapted poikilotherms. In this regard, in several species of *Trematomus* from McMurdo Sound studied at −1.5 °C, the average '*in vivo*' liver and white muscle protein synthetic rates are, respectively, two- to three-fold higher than predicted values obtained from a temperature dependency plot with a Q_{10}=2.5, considering values derived from work on temperate species (Smith & Haschemeyer, 1980). However, part of this greater fraction of protein synthesis effort is devoted to the production of extracellular proteins (Haschemeyer & Mathews, 1980). Some of these

Table 3. *Protein/dry weight ratio in different tissues of temperate and Antarctic teleosts*

	Intestine	Kidney	Gills
Temperate fish *Anguilla anguilla*	918±175	549±83	438±34
Antarctic fish *Trematomus bernacchii*	649±189	1345±209 [b]	737±160
Antarctic ice-fish *Chionodraco hamatus*	548±108	597±88	865±139 [a]

Notes:
The data reported in the table are means ± S.E. of at least four different measurements. Data regarding the intestine refer to the mucosal component that was separated from the connectival and muscolaris components; the posterior part of the kidney of all teleost species was used for the analysis. Protein concentration was measured in homogenates by a Bio-Rad protein assay Kit 1 (using lyophilised gamma – globulin as a standard). [a]=$p<0.05$ between *C. hamatus* and *A. anguilla*; [b]=$p<0.01$ between *T. bernacchii* and the two other species.

export products may be related to survival under subzero conditions. For example, antifreeze glycopeptides account for 4 % of liver polypeptide synthesis and their half-life is only 4 weeks (Eastman, 1993).

In many cases, the rate of protein synthesis of Antarctic fish at subzero temperatures is lower than that performed in warm-water teleosts. A general increase in the number of enzymatic molecules used as an adaptive strategy of fish to cold-water does not seem therefore, to be always performed in these environmental conditions. Furthermore a general increase of all enzymatic species may face limitations such as a limited solvent capacity of the cell and/or a very high energetic cost of protein synthesis.

As a consequence of the increase in enzyme levels the relative percentage of proteins with respect to other biological components (lipids, nucleic acids, etc.) should be higher in Antarctic fish tissues with respect to those of temperate fish. As reported in Table 3 the amount of protein as a function of dry weight of three different tissues (intestine, gills and kidney) of two Antarctic species (*T. bernacchii* and *C. hamatus*) does not seem to be different from that measured in the tissues of the temperate species, *A. anguilla*. This observation seems to rule out the possibility that a specific increase of all proteins could be a general strategy utilized during cold-adaptation. However, for a definite answer to this question, studies must involve the quantification of the number of enzyme molecules in a given mass of tissue and the determination of their specific activities (k_{cat} values).

Evolution of new enzyme variants better able to operate at low temperature seems to be a more useful adaptive strategy involved in the maintenance of cellular function in cold-water fish. Work on *Nothothenia neglecta* provides evidence that, unlike the situation in temperate fishes, in some pathways only one set of isozymes, well-adapted to operate at low temperatures, is necessary for survival in thermally stable environments. Unlike the multiple isozymes of temperate teleosts, *N. neglecta* has only one isozymic variant of lactate dehydrogenase (LDH), which is not tissue specific, catalysing the reversible lactate to pyruvate step in the glycolytic pathway (Fitch, 1988, 1989).

The physiological importance of such enzyme variants in cold water can be demonstrated by studying both (i) molecular structure and (ii) the following kinetic properties of proteins:

(a) activation energies;
(b) affinity constants.

The following experimental observations strongly suggest that a different biochemical structure of an enzyme in cold-acclimated and adapted species is a useful response strategy. Acetylcholinesterase (AChE) isolated from the brain of rainbow trout, *O. mykiss*, acclimated to 2 °C and 17 °C, differed in electrophoretic mobility (Baldwin & Hochachka, 1970) and the temperature at which K_m reached its minimum value was very close to the respective acclimation temperature (Baldwin, 1971).

Structural differences can be shown by studying resistance of enzymes to high temperatures (thermostability). Studies on the thermostabilities of alkaline phosphatase and maltase show that, in Antarctic fish, enzyme thermostabilities differ from those found in eel (Maffia *et al.*, 1993). Different thermostability of trypsin activity between the Antarctic fish *Paranothothenia magellanica* and the trout *O. mykiss*, was also reported by Genicot *et al.* (1988). Brain microtubules of Notothenioids are cold stable since the tubulins remain polymerised at −2 °C. There have been adaptive structural changes in the tubulins of Notothenioids. For example, amino acid substitution may have increased the proportion and/or strength of hydrophobic interactions, thereby accounting for the ability of Notothenioid tubulins to remain polymerised at subzero temperatures (Detrich, 1991 *a*, *b*).

Also a lowering of activation energy (E_a), which is inversely correlated to catalytic efficiency of the enzyme, seems to be useful for Antarctic fish to offset potentially dangerous effects of low temperatures on enzymatic rates. Studies performed on the Mg^{2+}-Ca^{2+}-ATPase of white muscle of fish, showed a free energy of activation much lower in polar than in temperate or

Table 4. *Activation energy (E_a) values of different intestinal enzymes of* T. bernacchii *and* A. anguilla

	Antarctic fish *Trematomus bernacchii*	Temperate fish *Anguilla anguilla*
Leucine aminopeptidase[c]	7.5±1.5 (6) (2–37 °C)	14.4±1.1 (12)[b] (2–37 °C)
Alkaline phosphatase[c]	5.7±0.3 (8) (2–37 °C)	6.8±0.4 (11)[b] (2–37 °C)
Maltase[c]	7.3±2.0 (3) (2–37 °C)	16.8±0.4 (3)[b] (2–37 °C)
Na^+/K^+- ATPase	8.07±1.1 (4) (0–33 °C)	13.07±0.98 (4)[a] (−1–17 °C)

Notes:
Data reported in this Table, expressed as kcal/mol, are means ± S.E. of at least four different experiments and were calculated, according to the Arrhenius equation, as the slope of the straight lines best interpolating the experimental points. [a]=$p<0.05$; [b]=$p<0.01$ indicate significant difference between the two species ([c] data from Maffia *et al.*, 1993).

tropical fish (Clarke, 1987). Investigation on *T. bernacchii* intestine has shown that E_a values of alkaline phosphatase, maltase, leucine-aminopeptidase, measured by Arrhenius analysis, were lower than in *A. anguilla* intestine (Maffia *et al.*, 1993). Similar findings have been shown in previous comparative studies on the cytosolic enzymes lactate dehydrogenase and pyruvate kinase from *T. bernacchii* and *O. mykiss* (Somero, 1977). Results also suggest that, for Antarctic fish intestinal Na^+/K^+- ATPase, there is a decrease of E_a when compared to the eel intestinal enzyme. All activation energy values of the cited enzymes, measured in temperate and Antarctic fish, are reported in Table 4.

Another general adaptation strategy to low environmental temperature could be the maintenance of enzyme–substrate affinity in order to retain the maximal responsiveness to changes in substrate concentration and to the influences of allosteric modulators (Clarke, 1987). The affinity of the enzyme for its substrate, highly important in controlling metabolic flux, is often approximated by the apparent Michaelis–Menten constant $K_{m,app.}$ The value of this kinetic parameter of the reaction is conserved among species under physiological conditions of temperature, hydrostatic pressure, osmotic concentration, and pH that are highly perturbing of enzyme – ligand interactions (Hochachka & Somero, 1984). Investigation on the $K_{m,app}$ of the

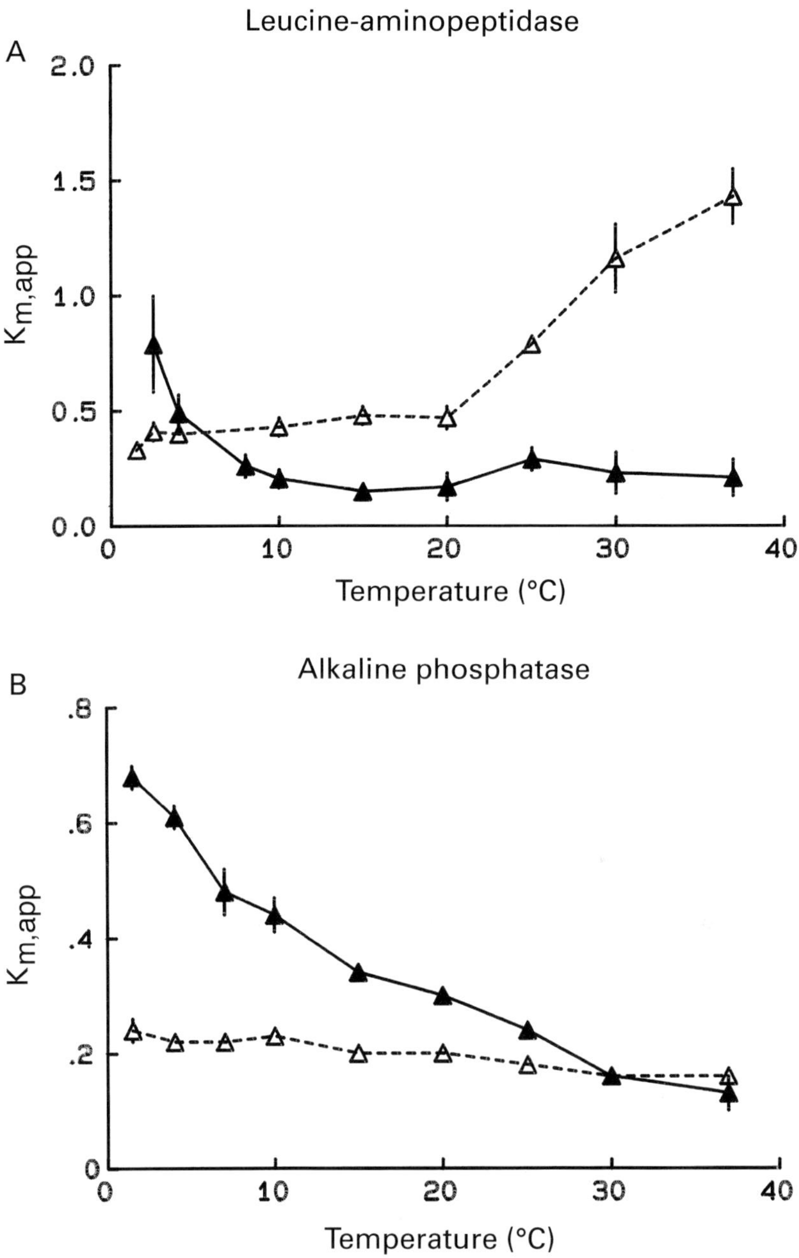

Fig. 3. Dependence of $K_{m,app}$ of leucine aminopeptidase (A) and alkaline phosphatase (B) of *T. bernacchii* (open symbols) and *A. anguilla* (closed symbols) upon temperature. Each data point is mean ± SE of three determinations. Leucine aminopeptidase: $p<0.01$ in the range 10 to 37 °C. Alkaline phosphatase: $p<0.01$ in the range 1.5 to 25 °C. (From Maffia *et al.*, 1993.)

intestinal enzyme alkaline phosphatase for *para*-nitro-phenyl-phosphate and of leucine-aminopeptidase for L-leucine-4–nitro-anilide confirms this adaptive strategy. Both enzymes show highest affinity for the substrate in the range of the respective environmental temperature of the two fishes (Fig. 3). Furthermore, the $K_{m,app}$ values of both the enzymes are differently influenced, in the two species, by the changes of the experimental temperature. This behaviour could be explained by differences in biochemical structure of the enzymes or other factors as, for example, temperature perturbation of lipid–protein interactions. Similar findings have been found for lactic dehydrogenase, pyruvate kinase, and acetylcholinesterase of *Pagothenia borchgrevinki,* which exibit minimum values of K_m for pyruvate, phosphenol-pyruvate and acetylcholine, respectively, in a temperature range near to that of its habitat (Somero, 1991). In Table 5, K_m values of different enzymes are reported. Interestingly, the enzymes of the Antarctic species showed an affinity for the substrate, at 0 °C, close to that of the temperate species measured at its environmental temperature.

Membrane transport proteins

Temperature plays a decisive role in the metabolic and ionic regulation of poikilotherms, one of the main reasons being that the temperature coefficients of passive flows and active transports may be quite different. In particular, while temperature has quite low effects on diffusion of metabolites or ionic species across biological membranes, the activities of individual transporters, much like enzymes, are likely to be strongly influenced by changes in temperature; thus, in order to maintain intracellular metabolite and ionic concentrations at low environmental temperatures of the Antarctic sea, the cold-adapted fish should compensate the imbalance between passive flows and active transports.

Until recently, no information regarding transport proteins in these animals was available. Now the existence of specific transporters for sugars and amino acids on the apical membranes of *T. bernacchii* intestine is reported, which operate at subzero temperatures, with the same characteristics of Na^+-dependence and electrogenicity previously reported in temperate fish (Ahearn & Storelli, 1994) or mammals (Stevens, Kaunitz & Wright, 1984; Wright, 1993). This study was performed by applying to Antarctic fish intestine a well-known technique of Mg^{2+}-fractional precipitation, previously applied to temperate species (Ahearn & Storelli, 1994), permitting the isolation of the brush border membrane as closed vesicles (Storelli, Vilella & Cassano, 1986). By using these membrane vesicles and radiolabelled D-glucose, the ligand binding properties, the specificity, and the effect of temperature on Na^+-dependent D-glucose transport of *T. bernacchii* intestine

Table 5. *K_m values of different intestinal hydrolases and of Na^+-D-glucose cotransport in BBMV of* T. bernacchii *and* A. anguilla

	Antarctic fish *Trematomus bernacchii* (0 °C)	Temperate fish *Anguilla anguilla* (20 °C)
Leucine aminopeptidase	0.33±0.06[a]	0.17±0.01[a]
Alkaline phosphatase	0.24±0.05[a]	0.30±0.02[a]
Maltase	1.85±0.44	1.12±0.19
Na^+-D-glucose transport	0.16±0.03[b]	0.10±0.02[c]

Notes:
Data reported, expressed as mmoles/l, are means ± S.E. of at least three different experiments, do not show any significant difference between species. (data from [a] Maffia *et al.*, 1993; [b] Maffia *et al.*, 1997; [c] Cassano *et al.*, 1988).

were studied. The results obtained suggest that $K_{m,app}$ is in the same range as that reported for temperate teleosts (Ahearn & Storelli, 1994) and for mammals (Maffia *et al.*, 1997) when measured at the respective environmental temperatures (see Table 5). Furthermore, the specificity for sugars seems to be similar to that shown in eel intestine. Finally, the activity of the carrier is maximal at −2 °C and decreases at higher temperatures as shown in Fig. 4. Interestingly, the inactivation temperature limit (around 6 °C) of Na^+-D-glucose cotransport of the Antarctic fish corresponds to that lethal for these animals. The reason for this temperature effect, completely different from that observed in *A. anguilla* (see Fig. 4), is at present unclear. It could be due to a protein unfolding or to a modification of lipid–protein interaction forces. A temperature-induced protein denaturation seems to be a less plausible explanation for the decrease of Na^+-dependent D-glucose transport, since the *in vitro* activity of enzymes such as leucine-aminopeptidase and maltase of the Antarctic fish intestine, that are not functionally influenced by the microenvironmental lipidic bilayer, increased exponentially with temperatures in a range between 2 and 37 °C (Maffia *et al.*, 1993). On the other hand, as previously reported in cold-acclimated poikilotherms, changes in membrane composition and physical characteristics (i.e. fluidity), which occur in response to short- and long-term temperature fluctuations, are known to influence the functional properties of several integral membrane

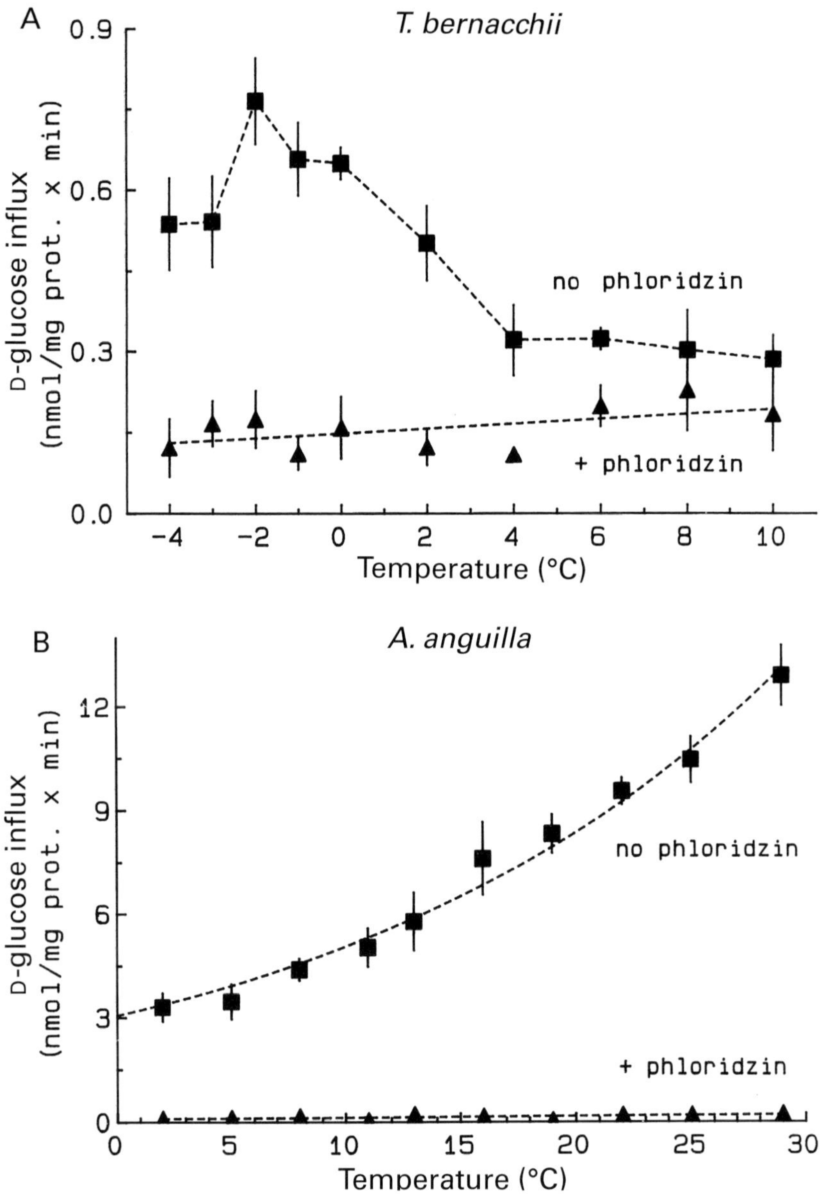

Fig. 4. Effect of temperature on D-glucose transport by (A) *T. bernacchii* and (B) *A. anguilla* intestinal BBMV. D-[^{14}C]-glucose uptake in BBMV was measured in the presence of a Na^+-gradient, both with or without the specific inhibitor of Na^+-D-glucose transport, phloridzin. Each data point is the mean ± SE of five determinations. (From Maffia *et al.*, 1997.)

proteins (Brasitus & Schachter, 1980; Hazel, 1979), including intestinal apical membrane transporters (Sadowski, Gibbs & Meddings, 1992; Houpe *et al.*, 1996). In this regard it could be important to emphasise that there is a different lipid composition of the apical plasma membranes of *T. bernacchii* with respect to that of *A. anguilla* (see Table 1). These considerations suggest that a loss of interaction between protein and lipid, following a perturbation of a lipidic microdomain, becomes a more likely explanation for the negative effect of temperatures on the Antarctic fish sugar transporter. An influence on protein activity due to a modification of lipid microenvironment was, in fact, reported by Hesketh *et al.* (1976).

However, at present, it has not been possible to distinguish the direct influence of temperature on nutrient transporter activity and the indirect influences caused by changes in membrane physical characteristics. None the less, the expressed capacity of Antarctic fish for homeophasic adaptation should surely induce an effect on transport processes. But a definitive answer to this question could be obtained by directly analysing the physical characteristics of the plasma membrane (fluidity) and the biochemical structure of the sugar transporter.

Conclusions

Data presented in this chapter indicate that adaptation at subzero (Antarctic) temperatures affects the composition (structure) and activity of both membrane, proteins and lipids. They suggest the involvement of lipid composition in the maintenance of an appropriate degree of membrane fluidity, ensuring physiological efficiency of the intestinal mucosa in the Antarctic teleost. The adaptive modification revealed by comparison with a temperate species parallels membrane lipid modification that occurs in cold-acclimated or cold-acclimatised poikilotherms.

As far as it concerns peripheral membrane proteins of the intestinal brush border such as maltase, leucine amino peptidase and alkaline phosphatase, results indicate that the higher activity expressed by the Antarctic fish enzymes at low temperatures may be related, rather than to an increase of the number of enzyme molecules, to the presence of different enzyme isoforms with physical (low E_a) and biochemical (conserved K_m) characteristics more efficient for work at low temperatures.

Finally, the comparison of sugar and amino acid (integral) transport systems, studied on intestinal brush border membrane vesicles isolated from cold and temperate adapted fishes, together with the lipid composition of the same membrane, strongly suggests a basic role played by the microenvironmental lipids in the cold-adapted transporting activity. Further experiments including transport protein isolation and reconstitution in liposomes or 'cloning' experiments are needed to completely elucidate these relationships.

References

Acierno, R., Maffia, M., Sicuro, P., Fiammata, L., Rollo, M., Ronzini, L. & Storelli, C. (1997). Lipid and fatty acid composition of intestinal mucosa of two Antarctic teleosts. *Journal of Comparative Physiology*, in press.

Ahearn, G.A. & Storelli, C. (1994). Use of membrane vesicle techniques to characterize nutrient transport processes of the teleost gastrointestinal tract. In *Biochemistry and Molecular Biology of Fishes*, ed. P.W. Hochachka & T.P. Mommsen, vol. 3, pp. 513–24. Elsevier Science.

Argyropoulou, V., Kalogeropoulos, N. & Alexis, M.N. (1992). Effect of dietary lipids on growth and tissue fatty acid composition of grey mullet *(Mugil cephalus). Comparative Biochemistry and Physiology*, **101A**, 129–35.

Baldwin, J. (1971). Adaptation of enzymes to temperature: acetylcholinesterases in the central nervous system of fishes. *Comparative Biochemistry and Physiology*, **40**, 181–7.

Baldwin, J. & Hochachka, P. 1970. Functional significance of isoenzymes in thermal acclimatization: acetylcholinesterases from trout brain. *Biochemical Journal*, **116**, 883–7.

Behan-Martin, M.K., Jones, G.R., Bowler, K. & Cossins, A.R. (1993). A near perfect temperature adaptation of bilayer order in vertebrate brain membranes. *Biochimica et Biophysica Acta*, **1151**, 216–22.

Berlin, E., McClure, D., Banks, M.A. & Peters, R.C. (1994). Heart and liver fatty acid composition and vitamin E content in miniature swine fed diets containing corn and menhaden oils. *Comparative Biochemistry and Physiology*, **109A**, 53–61.

Berner, W., Kinne, R. & Murer, H. (1976). Phosphate transport into brush-border membrane vesicles isolated from rat small intestine. *Biochemical Journal*, **160**, 467–74.

Bloch, K. (1983). Sterol structure and membrane function. *Critical Reviews in Biochemistry*, **14**, 47–92.

Bloom, M., Evans, E. & Mouritsen, O.G. (1991). Physical properties of the fluid lipid-bilayer component of cell membranes: a perspective. *Quarterly Review in Biophysics*, **24**, 293–397.

Brasitus, T.A., Schachter, D. (1980). Lipid dynamics and lipid–protein interaction in rat enterocyte basolateral and microvillus membranes. *Biochemistry*, **19**, 2763–9.

Cassano, G., Maffia, M., Vilella, S. & Storelli, C. (1988). Effects of membrane potential on Na cotransport in eel intestinal brush-border membrane vesicles: studies with a fluorescent dye. *Journal of Membrane Biology*, **101**, 225–36.

Clarke, A. (1987). The adaptation of aquatic animals to low temperatures. In *The Effect of Low Temperatures on Biological Systems*, ed. B.W.W. Grout & G.J. Morris, pp. 315–48. Australia: Edward Arnold.

Clarke, A., Doherty, N., DeVries, A.L. & Eastman, J.T. (1984). Lipid content and composition of three species of Antarctic fish in relation to buoyancy. *Polar Biology*, **3**, 77–83.

Cossins, A.R. & Prosser, C.L. (1978). Evolutionary adaptation of membrane to temperature. *Proceedings of the National Academy of Sciences, USA*, **75**, 2040–3.

Cossins, A.R. & Sinensky, M. (1986). Adaptations of membranes to temperature, pressure and exogenous lipids. In *Physiology of Membrane Fluidity*, ed. M. Shinitzsky, pp. 1–20. Boca Raton, FL: CRC Press.

Cossins, A.R., Behan, M.K., Jones, G. & Bowler, K. (1987). Lipid–protein interactions in the adaptive regulation of membrane function. *Biochemical Society Transactions*, **15**, 77–81.

Cullis, P.R., Hope, M.J., De Kruijf, B., Verkleij, A.J. & Tilcock, C.P.S. (1985). Structural properties and functional roles of phospholipids in biological membranes. In *Phospholipids and Cellular Regulation* ed. J.F. Kuo, vol. 1, pp. 1–59. Boca Raton, Florida: CRC Press.

Dahlhoff, E. & Somero, G.N. (1993). Effects of temperature on mitochondria from abalone (genus *Haliotis*): adaptive plasticity and its limits. *Journal of Experimental Biology*, **185**, 151–68.

Delgado, A., Estevez, A., Hortelano, P. & Alejandre, M.J. (1994). Analyses of fatty acids from different lipids in liver and muscle of sea bass (*Dicentrarchus labrax* L.). Influence of temperature and fasting. *Comparative Biochemistry Physiology*, **108B**, 673–80.

Demel, R.A., Geurts Van Kessel, W.S. & VanDeenen, L.L.M. (1972). The properties of polyunsaturated lecithins in monolayers and liposomes and the interactions of these lecithins with colesterol. *Biochimica et Biophysica Acta*, **266**, 26–40.

Detrich, H.W. III (1991*a*). Cold-stable microtubules from Antarctic fish. In *Life under Extreme Conditions: Biochemical Adaptation*, ed. G. Di Prisco, pp. 35–49. Berlin and Heidelberg: Springer-Verlag.

Detrich, H. W. III (1991*b*). Polymerization of microtubule proteins from Antarctic fish. In *Biology of Antarctic fish*, ed. G. Di Prisco, B. Maresca & B. Tota, pp. 248–62. Berlin, Heidelberg, New York: Springer.

Eastman, J.T. & DeVries, A.L. (1981). Buoyancy adaptations in a swim-bladderless Antarctic fish. *Journal of Morphology*, **167**, 91–102.

Eastman, J.T. (1993). Organ system adaptation in Notothenioids. In *Antarctic Fish Biology*, pp. 148–200. New York: Academic Press.

Farkas,T., Storebakken, T. & Bhosie, N.B. (1988). Composition and physical state of phospholipids in Calanoid copepods from India and Norway. *Lipids*, **23**, 619–22.

Feller, G., Thiry, M., Arpigny, G.L. & Gerday, C. (1991*a*). Cloning and expression in *Escherichia coli* of three lipase-encoding genes from the psychrotrophic Antarctic strain Moraxella TA144. *Gene*, **102**, 111–15.

Feller, G., Thiry, M. & Gerday, C. (1991*b*). Nucleotide sequence of the lipase gene lip2 from the Antarctic psychrotroph Moraxella TA144 and site-specific mutagenesis of the conserved serine and histidine residues. *DNA Cell Biol*ogy, **10**, 381–8.

Fitch, N.A. (1988). Lactate dehydrogenases in Antarctic and temperate fish species. *Comparative Biochemistry and Physiology*, **91B**, 671–6.

Fitch, N.A. (1989). Lactate dehydrogenase isozymes in the trunk and cardiac muscles of an Antarctic teleost fish, *Notothenia neglecta* Nybelin. *Fish Physiology and Biochemistry*, **6**, 187–95.

Genicot, S., Feller, G. & Gerday, C.H. (1988). Trypsin from Antarctic fish (*Paranothothenia magellanica* Forster) as compared with trout (*Salmo Gairdneri*) trypsin. *Comparative Biochemistry and Physiology*, **90B**, 601–9.

Gennis, R.B. (1989). Biomembranes. In *Molecular Structure and Function*, ed. C.R. Cantor, Berlin, Heidelberg, New York: Springer.

Gosh, D., Williams, M.A. & Tinoco, J. (1973). The influence of lecithin structure on their monolayer behaviour and interactions with cholesterol. *Biochimica et Biophysica Acta*, **291**, 351–2.

Haase, W., Schaffer, A., H. Murer, H. & Kinne, R. (1978). Studies on the orientation of the brush border membrane vesicles. *Biochemical Journal*, **172**, 57–62.

Hagar, A.F. & Hazel, J.R. (1985). The influence of thermal acclimation on the microsomal fatty acid composition and desaturase activity of rainbow trout liver. *Molecular Physiology*, **7**, 107–18.

Haschemeyer, A.E.V. & Mathews, R.W. (1980). Antifreeze glycoprotein synthesis in the Antarctic fish *Trematomus hansoni* by constant infusion *in vivo*. *Physiological Zoology*, **53**, 383–93.

Hazel, J.R. (1979). Influence of thermal acclimation on membrane lipid composition of rainbow trout liver. *American Journal of Physiology*, **236**, R91–101.

Hazel, J.R. (1984). Effects of temperature on the structure and metabolism of cell membranes in fish. *American Journal of Physiology*, **246**, R460–70.

Hazel, J.R. (1989). Cold adaptation in ectotherms: regulation of membrane function and cellular metabolism. In *Advances in Comparative and Environmental Physiology*, ed. L.C.H. Wang, vol. 4, pp.1–50. Berlin, Heidelberg, New York: Springer.

Hazel, J.R. (1995). Thermal adaptation in biological membranes: is homeoviscous adaptation the explanation? *Annual Review of Physiology*, **57**, 19–42.

Hazel, J.R. & Prosser, C.L. (1974). Molecular mechanisms of temperature compensation in poikiloterms. *Physiological Reviews*, **54**, 620–77.

Hesketh, K.R., Smith, G.A., Houslay, M.D., McGill, K.A., Birdsall, N.J.M., Metcalfe, J.C. & Warren, G.B. (1976). Annuluar lipids determine the ATPase activity of a calcium transport protein complexed with dipalmitoyllecithin. *Biochemistry*, **15**, 4145–51.

Higgins, A.J. (1978). Separation and analysis of membrane lipid component. In *Biological Membranes, A Practical Approach.* ed. J.B.C. Findlay & W. H. Evans, pp. 103–37. Oxford: IRL Press.

Hochachka, P.W. & Somero, G.N. (1984). *Biochemical Adaptation*. Princeton, NJ, USA: Princeton University Press.

Houpe, K.L., Malo, C., Oldham, P.B. & Buddington, R.K. (1996). Thermal modulation of channel catfish intestinal dimensions, BBM fluidity, and glucose transport. *American Journal of Physiology*, **270**, R1037–43.

Hudson, J.D. (1977). Oxygen isotope studies on Cenozoic temperatures, oceans and ice accumulation. *Scottish Journal of Geology*, **13**, 313–23.

Israelachvili, J., Marcelja, S. & Horn, R.G. (1981). Physical principles of membrane organization. *Quarterly Review in Biophysics*, **13**, 121–200.

Johnston, I.A. (1981). Structure and function of fish muscles. *Symposia of the Zoological Society of London*, **48**, 71–113.

Johnston, I.A. (1983). Cellular responses to altered body temperature: the role of alterations in the expression of protein isoforms. In *Cellular Acclimatization to Environmental Change*, ed. A.R. Cossins & P. Sheterline, pp.121–43. London: Cambridge University Press.

Johnston, I.A. & Maitland, B. (1980). Temperature acclimation in crucian carp: a morphometric analysis of fish muscle fibre ultrastructure. *Journal of Fish Biology*, **17**, 113–25.

Kaneda, T. (1977). Fatty acids of the genus Bacillus: an example of branched-chain preference. *Bacteriology Review*, **41**, 391–418.

Kemp, P. & Smith, M.W. (1970). Effect of temperature acclimatization on the fatty acid composition of goldfish intestinal lipids. *Biochemical Journal*, **117**, 9–15.

Lund, E.D. & Sidell, B.D. (1992). Neutral lipid composition of Antarctic fish tissues may reflect use of fatty acyl substrates by catabolic systems. *Marine Biology*, **112**, 377–82.

McElhaney, R.N. (1984). The structure and function of the *Acholeplasma laidlawii* plasma membrane. *Biochimica et Biophysica Acta*, **779**, 1–42.

Maffia, M., Acierno, R., Deceglie, G., Vilella, S. & Storelli, C. (1993). Adaptation of intestinal cell membrane enzymes to low temperatures in the Antarctic teleost *Pagothenia bernacchii*. *Journal of Comparative Physiology*, **163B**, 265–70.

Maffia, M., Acierno, R., Cillo, E. & Storelli, C. (1997). Na^+-D-glucose cotransport by intestinal brush border membrane vesicles of the Antarctic fish *Trematomus bernacchii*. *American Journal of Physiology*, in press.

Nachman, R.J. (1985). Unusual predominance of even-carbon hydrocarbons in an Antarctic food chain. *Lipids*, **20**, 629–33.

Nes, W.R. & Nes, W.D. (1980). *Lipids in Evolution*. New York: Plenum Press.

Nichols, D.S., Williams, D., Dunstan, G.A., Nichols, P.D. & Volkman, J.K. (1994). Fatty acid composition of Antarctic and temperate fish of commercial interest. *Comparative Biochemistry and Physiology*, **107B**, 357–63.

Porteres, G.A. (1991). Changes in fatty acid composition in *Ruditapes philippinarum* A & R (Veneridae) fed on industrial yeast. *Comparative Biochemistry and Physiology*, **100A**, 211–15.

Pruitt, N.L. (1988). Membrane lipid composition and overwintering strategy in thermally acclimated crayfish. *American Journal of Physiology*, **254**, R870–6.

Russel, N.J. (1984). Mechanisms for thermal adaptation in bacteria: blueprints for survival. *Trends in Biochemical Sciences*, **9**, 108–12.

Sadowski, D.C., Gibbs, D.J. & Meddings J.B. (1992). Proline transport across the intestinal microvillus membrane may be regulated by membrane physical properties. *Biochimica et Biophysica Acta*, **1105**, 75–83.

Savin, S.M. (1977). The history of the Earth's surface temperature during the past 100 million years. *Annual Review of Earth Planetary Science*, **5**, 319–55.

Schwalme, K., Mackay, W.C. & Clandinin, M.T. (1993). Seasonal dynamics of fatty acid composition in female northern pike *(Esox lucius)*. *Journal of Comparative Physiology*, **163B**, 277–87.

Seddon, J.M. (1990). Structure of the inverted hexagonal (HII) phase, and non-lamellar phase transitions of lipids. *Biochimica et Biophysica Acta*, **1031**, 1–69.

Sidell, B.D. (1977). Turnover of cytochrome *c* in skeletal muscle of green sunfish (*Lepomis cyanellus, R.*) during thermal acclimation. *Journal of Exerimental Zoology*, **199**, 233–50.

Sidell, B.D. (1983). Cellular acclimation to environmental change by quantitative alterations in enzymes and organelles. In *Cellular Acclimatization to Environmental Change*, ed. A.R. Cossius & P. Sheterline. SEBS 17, pp. 103–20. Cambridge: Cambridge University Press.

Sidell, B.D., Crockett, E.L. & Driedzic W.R. (1995). Antarctic fish tissues preferentially catabolize monoenoic fatty acids. *Journal of Experimental Zoology*, **271**, 73–81.

Silvius, J.R. (1986). Solid- and liquid-phase equilibria in phosphatidylcholine/phosphatidylethanolamine mixtures. A calorimetric study. *Biochimica et Biophysica Acta*, **857**, 217–28.

Sinensky, M. (1974). Homeoviscous adaptation – a homeostatic process that regulates viscosity of membrane lipids in *Escherichia coli. Proceedings of the National Academy of Sciences, USA*, **71**, 522–5.

Smith, M.A.K. & Haschemeyer, A.E.V. (1980). Protein metabolism and cold-adaptation in Antarctic fishes. *Physiological Zoology*, **53**, 373–82.

Somero, G.N. (1977). Temperature as a selective factor in protein evolution: the adaptational strategy of 'compromise'. *Journal of Experimental Zoology*, **194**, 175–88.

Somero, G.N. (1991). Biochemical mechanisms of cold adaptation and stenothermality in Antarctic fish. In *Biology of Antarctic Fish*, ed. G. Di Prisco, B. Maresca, & B. Tota, pp. 232–47. Berlin, Heidelberg, New York: Springer.

Stevens, B.R., Kaunitz, J.D. & Wright, E.M. (1984). Intestinal transport of amino acids and sugars: advances using brush border membrane vesicles. *Annual Review of Physiology*, **46**, 417–31.

Storelli, C., Vilella, S. & Cassano, G. (1986). Na^+-dependent D-glucose and L-alanine transport in eel intestinal brush border membrane vesicles. *American Journal of Physiology*, **251**, R463–9.

Tota, B., Agnisola, C., Schioppa, M., Acierno, R., Harrison, P. & Zummo, G. (1991). Structural and mechanical characteristics of the heart of the icefish *Chionodraco hamatus* (Lonnberg)

Tyler, S. & Sidell, B.D. (1984). Changes in mitochondrial distribution and diffusion distances in muscle of goldfish upon acclimation to warm and cold temperatures. *Journal of Experimental Biology*, **232**, 1–9.

Vrbjar, N., Kean, K.T., Szabo, A., Senak, L., Mendelsohn, R. & Keough, K.M.W. (1992). Sarcoplasmic reticulum from rabbit and winter flounder: temperature-dependence of protein conformation and lipid motion. *Biochimica et Biophysica Acta*, **1107**, 1–11.

Wilson, F.R. (1973). Enzyme changes in the gold-fish (*Carassius auratus L.*) in response to temperature-acclimation. I. *An immunochemical approach.* II. *Isozymes.* PhD thesis, University of Illinois.

Wright, E.M. (1993). The intestinal Na^+-D-glucose cotransporter. *Annual Review of Physiology*, **55**, 575–89.

R.A.H. VETTER and F. BUCHHOLZ

Kinetics of enzymes in cold-stenothermal invertebrates

This chapter will review the various mechanisms of temperature-dependent enzyme regulation, focusing on cold ocean invertebrates and crustaceans in particular. Temperature is one of the most important abiotic factors affecting physiological processes in marine poikilotherms (e.g. Bullock, 1955, Prosser, 1967; Hochachka & Somero, 1984). The distribution of species and hence the survival of organisms is often closely related to the ambient temperature regime (e.g. Dunbar, 1968; Somero, 1975). Therefore, complex metabolic processes must be adapted to the specific environment to ensure survival even under adverse cold conditions. Since the polar oceans comprise 20% of the world's ocean surface, and since the deeper ocean waters are usually colder than 5 °C (Clarke, 1983), cold water is the norm rather than the exception for marine life. Even in the deep-sea, physiological adaptations are mainly determined by low temperatures associated with declining depth, not by high pressure (e.g. Teal & Carey, 1967; Childress *et al.*, 1990). Low temperatures are disadvantageous to physiological processes due to the lower kinetic energy available, and the maintenance of metabolism at low temperatures requires effective biocatalysts, i.e. enzymes with high activities. Consequently, changes in enzyme characteristics or the production of new isoforms contribute to an appropriate temperature adaptation. In addition, enzymes must be regulated to avoid uncontrolled turnover of metabolites. This can be accomplished through feed-back control by metabolic endproducts (Fersht, 1985).

Mechanisms of biochemical adaptation

In general, metabolic temperature adaptation can be achieved in two different ways: The organism can either reduce metabolism to save energy in the cold (Clarke, 1983) or metabolic activity is increased at low temperatures, compensating for rate limiting effects and maintaining metabolism at an almost constant level (Hochachka & Somero, 1971, 1973). Irrespective of the adaptive mechanism, enzymes are always involved in biochemical adaptation. Numerous investigations have focused on enzymatic adaptations in marine ectotherms (e.g. Moon, 1972; Low & Somero, 1974, Hoffmann, 1976; Shaklee *et al.*, 1977; Guppy & Hochachka, 1979; Fideau-Alonso *et al.*, 1985; Hall, 1985; Johnston, Sidell & Driedzic, 1985; Clarke, 1987; Ferracin *et al.*, 1987; Dittrich, 1992*a,b*; Buchholz & Vetter, 1993; Klyachko, Polosukhina & Ozernyuk, 1993; Vetter, 1995*a,b*). In particular, enzymes of various fish were investigated with respect to the mechanisms of adaptation to low, particularly polar, temperatures (e.g. Baslow & Nigrelli, 1964; Somero, Giese & Wohlschlag, 1968, Somero & Hochachka, 1968; Clarke, 1987; Genicot, Feller & Gerday, 1988; Torres & Somero, 1988*a,b*, Somero, 1991; Karpov & Andreyeva, 1992; Maffia *et al.*, 1993; Goldspink, 1995). Physiological and biochemical studies have been performed to evaluate the effects of low temperature on metabolic rates and on the catalytic and regulatory properties of enzymes and on their structural changes (e.g. Jaenicke, 1990). Although recent studies reported a high abundance of polar invertebrates (Piepenburg, 1988; Voss, 1988) literature on physiological adaptation in invertebrates is still scarce. With the increasing political and commercial interest in polar, e.g. Antarctic ecosystems, studies on crustaceans, particularly the Antarctic krill *Euphausia superba*, were introduced to scientific research programmes (e.g. Karl *et al.*, 1986). Despite the enormous krill biomass in the Antarctic, the good prospects of a krill fishery faded, due to problems in processing and storage (Tokunaga, Iida & Nakamura, 1977; Choi & Kato, 1984; Adelung *et al.*, 1987). Autoproteolysis of frozen material occurred even at −20 °C (Suzuki & Kanna, 1977). Apparently, proteolytic enzymes of the Antarctic krill showed high activities (Hellgren, Mohr & Vincent, 1986; Campell *et al.*, 1987), even at very low temperatures (Dittrich, 1992*b*). Furthermore, the high metabolic activity of this polar crustacean is reflected in its respiration rate, which is comparable to those of temperate species (Nikolayeva & Ponomareva 1973; Kils, 1979, 1981; Opalinski, 1991; Torres *et al.*, 1994). Therefore, the efficiency of metabolic enzymes must be enhanced at low temperatures to ensure the high energy metabolism in this permanently swimming crustacean. In addition, it was found that a rise in temperature of only 4 °C led to an immense increase in moulting activity by decreasing the mean time of the moult cycle to about 50% (Poleck & Denys, 1982; Clarke & Morris, 1983; Buchholz, 1991). In this respect, this stenothermal species is able even to profit from the thermodynamic advantage of small increases in temperatures rather than compensating for the effect.

In the following years, fundamental research on enzymatic temperature adaptation focused on metabolic (e.g. glycolytic) and digestive (e.g. proteolytic and chitinolytic) enzymes. According to the different metabolic functions, the supply of energy and breakdown of nutrients, enzyme regulation and temperature adaptation are achieved by distinctive mechanisms in these two groups of enzymes. The mechanisms of biochemical adaptation will be reviewed with special regard to these differences between metabolic and digestive enzymes.

Enzymatic activity may be regulated by the amount of enzyme available, by various changes in kinetic characteristics, by the existence of enzymes with different properties (Hochachka & Somero, 1984), and by the influence of modulating metabolites (Fersht, 1985). The following chapter will discuss different ways of enzymatic temperature adaptation focusing on invertebrates, particularly on polar crustaceans.

Optimum temperature range

One proposed pattern of enzyme adaptations of ectotherms to cold environments comprises a shift of temperature optima to lower temperatures, i.e. into the physiological range (i.e. Russell, 1990). This was not found to be true in Antarctic invertebrates. As exemplified by the well-studied Antarctic krill, *E. superba*, not a single enzyme investigated exhibited a shift of maximal activity to low temperatures (Table 1). All temperature maxima were found between 25 °C (α-amylase) and 65 °C (glucanase), indicating a high thermostability of the enzyme proteins (Dittrich, 1992*a*,*b*; Buchholz & Vetter, 1993). This was unexpected, because high stability goes along with a reduced flexibility to conformational changes (Clarke, 1983). However, the functional capacity of any enzyme is ensured by a fine balance between stability, flexibility and the efficiency of catalysis (Hochachka & Somero, 1984). The remaining flexibility of the enzymes is obviously sufficient to maintain metabolism.

Activation energy

The estimation of temperature optima also yields data for assessments of the activation energy (E_a) of an enzyme. In connection with the values of entropy ($\Delta S^{\#}$) E_a-values can be used to calculate the Gibbs' free energy of activation of the reaction ($\Delta G^{\#}$) according to Johnston and Walesby (1977). Since the exact determination of $\Delta S^{\#}$ is rather difficult and time consuming, this value is usually assumed to be constant and only E_a-values are estimated as indicators of the catalytic efficiency of an enzyme (Low, Bada & Somero, 1973). Consequently, decreased E_a-values of enzymes from cold-adapted species can be interpreted with respect to temperature adaptation, because the energy barrier of catalysis is decreased. In this way, activation energy levels are adjusted to the reduced levels of kinetic energy available for catalysis. According to Somero and Siebenaller (1979) an increased E_a of the lactate dehydrogenase reaction from 48 kJ/mol to 55 kJ/mol resulted in a

Table 1. *Temperature and pH-optima of different enzymes from the Antarctic krill,* Euphausia superba

Enzyme	Temperature optimum	pH optimum	Source
α-Amylase	20–25 °C	7.2	Van Wormhoudt *et al.* (1983)
Cellulase	40 °C	6.5–7.0	Saborowski (*pers. comm.*)
endo-chitinase	50–55 °C	7.5	Saborowski *et al.* (1993)
exo-chitinase A	40 °C	4.0–4.5	Buchholz & Vetter (1993)
exo-chitinase B	36 °C	5.0–5.5	Buchholz & Vetter (1993)
exo-chitinase C	47 °C	4.5–5.0	Buchholz & Vetter (1993)
Citrate synthase	40 °C	7.7–8.3	Vetter (1995*a*)
Euphausiidin	40–45 °C	5.0–7.2	Paukova, Baidalinova & Mosolov (1992)
Laminarinase 1	60–65 °C	5.0	Turkiewicz (1995)
Laminarinase 2	40–45 °C	4.3–5.0	Suzuki *et al.* (1987)
Peptide hydrolase A	55 °C	8.0–8.5	Osnes & Mohr (1985*a,b*)
Peptide hydrolase B	60 °C	8.2	Osnes & Mohr (1985*a,b*)
Peptide hydrolase C	60 °C	8.2	Osnes & Mohr (1985*a,b*)
Protease A_1	48 °C	9.0	Chen, Yan & Chen (1980)
Protease B	50 °C	8.0	Chen *et al.* (1980)
Protease C	55–60 °C	7.5	Chen *et al.* (1980)
Protease D	55 °C	7.5–8.0	Chen *et al* (1980)
Serine proteinase	40 °C	6.0	Turkiewicz *et al.* (1986)
β-1,4-xylanase A	35–37 °C	5.8	Turkiewicz (1995)
β-1,4-xylanase B	35–37 °C	5.7–5.9	Turkiewicz (1995)

reduced relative velocity of only 54%. Therefore, an efficient adaptation to low temperatures might occur by reducing E_a to only a small extent.

Comparing E_a-values of metabolic, digestive and chitinolytic enzymes (Table 2), obvious differences were found. One of the controlling enzymes of energy metabolism, citrate synthase (CS), displayed a reduction of E_a only in the pelagic Antarctic krill, *E. superba* (10.9 kJ/mol), not in the Antarctic isopod *Serolis polita* (41.8 kJ/mol) or other euphausiids such as the Nordic krill *Meganyctiphanes norvegica* (45.1 kJ/mol) and the Mediterranean krill *M. norvegica (med.)* (43.3 kJ/mol). Given the high activation energy of *S. polita*, other adaptive mechanisms must be present in this benthic polar crustacean to facilitate a sufficient turnover rate at low temperatures. This will be discussed later. Of the digestive enzymes investigated, the activation energy was always markedly reduced in polar species, e.g. E_a of laminarinase and proteinase was less than 20% in *E. superba* (12.5 kJ/mol and 8.0 kJ/mol, respectively) compared to the temperate *Acartia hudsonica* (59.9 kJ/mol and

Table 2. *Activation energies, i.e. E_a-values ($kJ\,mol^{-1}$), of enzymes with different metabolic functions*

	Polar			Temperate				Tropical	
	Euphausia superba	*Serolis polita*	*Chorismus antarcticus*	*Acartia hudsonica*	*Cancer pagurus*	*Meganyctiphanes norvegica*	*Meganyctiphanes norvegica (med.)*	*Ocypode ryderi*	*Penaeus japonicus*
Citrate synthase	10.9[i]	41.8[i]				45.1[i]	43.3[i]		
Trypsin-like protease	27.9[b]		11.9[b]		27.1[b]	31.5[b]		28.5[b]	39.6[c]
α-Amylase	12.6[k]			30.3[e]					
Laminarinase	12.5[h]			59.9[f]					
Xylanase	26.2[h]								
Proteinase	8.0[h]			41.5[f]					
Chitinase	52.1[g]								43.4[d]
NAGase A	47.0[a]								
NAGase B	53.9[a]	58.0[a]				51.4[a]	59.1[a]	40.7[a]	
NAGase C	51.8[a]	52.3[a]				49.7[a]	50.5[a]		

Notes:
Top to bottom: energy metabolism, digestion, chitinolysis in crustaceans from different climatic zones (from left to right: polar, temperate, tropical).
Source: Data from: [a] Buchholz & Vetter, 1993; [b] Dittrich, 1992*b*; [c] Galgani *et al.*, 1983; [d] Kono, *et al.*, 1990; [e] Mayzaud, 1985*a*; [f] Mayzaud & Mayzaud, 1981; [g] Saborowski *et al.*, 1993; [h] Turkiewicz, 1995; [i] Vetter, 1995*a*; [k] Van Wormhoudt *et al.*, 1983.

41.5 kJ/mol, respectively). The reduction is less pronounced in a trypsin-like protease, but is still by as much as 50% in *Chorismus antarcticus* (11.9 kJ/mol) from the Antarctic (Table 2) compared to boreal *Cancer pagurus* (27.1 kJ/mol) and *M. norvegica* (31.5 kJ/mol) or tropical *Ocypode ryderi* (28.3 kJ/mol). Surprisingly, the value in a trypsin-like protease of *E. superba* (27.9 kJ/mol) was much higher than that of *C. antarcticus* and only slightly less than those of the temperate species. Nevertheless, the reduced activation energy appears essential to compensate for the rate limiting effects of polar temperatures on these digestive enzymes. In contrast, the E_a-values of chitinolytic enzymes were always in the range of 50 kJ/mol, irrespective of the climatic zone of the species investigated (Table 2). Alternative mechanisms of adaptation must be present in this group of enzymes, which will be discussed in the next chapter.

Enzyme substrate affinity

A possible mechanism of temperature adaptation consists of modulating enzyme substrate affinities seasonally or geographically, depending on environmental temperatures. An adaptive change of enzyme substrate affinities is reflected by the respective alterations of apparent Michaelis constants (K_m): the higher the affinity, the lower the K_m (London & Shaw, 1983). Low K_m values at ambient environmental temperatures were found not only in fish (e.g. Baldwin & Hochachka 1970) but also in various enzymes of invertebrates from different climatic zones, where the minima of temperature dependent K_m-values coincided with the ambient temperature range (e.g. Hoffmann 1976; Thebault & Bernicard, 1981; Van Praët, 1982; Van Wormhoudt, Mayzaud & Ceccaldi, 1983; Buchholz & Vetter, 1993; Dahlhoff & Somero, 1993*a*; Vetter, 1995*a*). In stenothermal polar crustaceans, *E. superba* and *S. polita*, K_m-values of chitinolytic enzymes such as *N*-acetyl-β-D-glucosaminidase (NAGase) were lowest at 0–5 °C (Fig. 1), whereas eurythermal temperate euphausiids, *M. norvegica*, showed a broad range of K_m-minima between 5 and 20 °C (Buchholz & Vetter, 1993). In Antarctic crustaceans, this compensation of K_m-values was found only in chitinolytic enzymes, but not in proteases (Dittrich, 1992*a*). However, the minima of K_m of NAGase encompass a broad range, i.e. over 5 to 20 °C the values do not vary (Fig. 1) and a reduction of K_m was not significant, when summer and winter specimens of *M. norvegica* were compared (Buchholz & Vetter, 1993). Accordingly, a pronounced regulatory significance with respect to temperature does not seem to exist. Equally, E_a-values are almost constant when compared in the species investigated (Table 2). As a consequence, temperature-dependent regulation is of minor importance. It is at least unclear in how far digestive enzymes or generally 'simple' hydrolases, i.e. proteases,

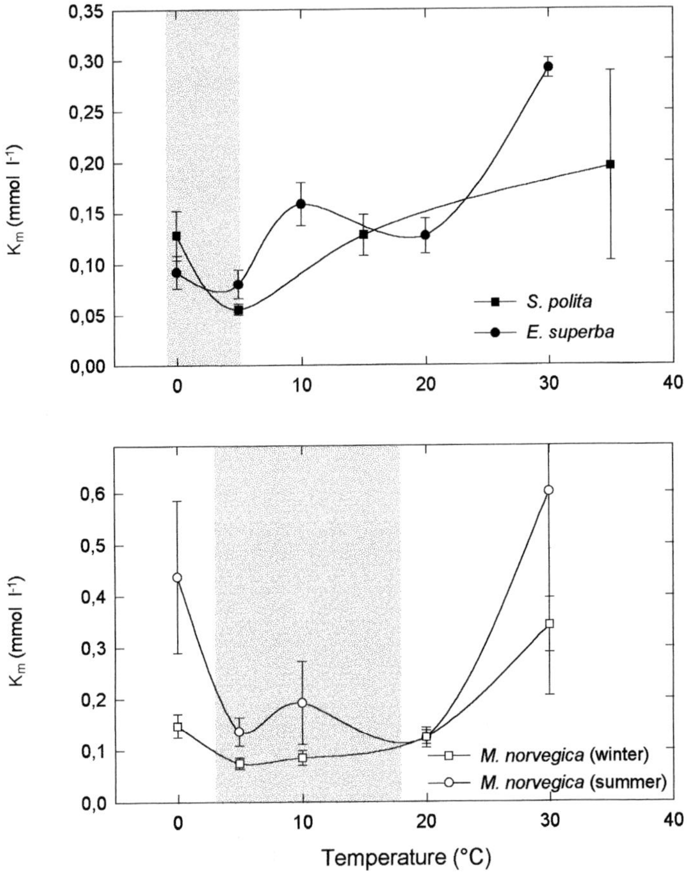

Fig. 1. Temperature dependence of K_m values of *N*-acetyl-β-D-glucosaminidase in *Euphausia superba* ● *Serolis polita* ■ (top) and *Meganyctiphanes norvegica* summer specimens ○ and winter specimens □ (bottom). Annual temperature ranges are shaded. (Figure modified after Buchholz & Vetter, 1993.)

chitinases, are regulated via modulation of K_m. In contrast, regulation of chitinases via hormonal control (Peters *et al.*, 1992) or nutrition (Mayzaud, 1985*b*; Buchholz & Saborowski, 1996) is much more pronounced and can easily supersede the former temperature-driven effects. The latter inductions may cause a ten-fold increase of total chitinolytic activity (Buchholz, 1989; Saborowski & Buchholz, 1991).

Besides the necessity of adapting genetically to the general climatic conditions depending on the latitude, the acclimatisation to seasonal or even daily

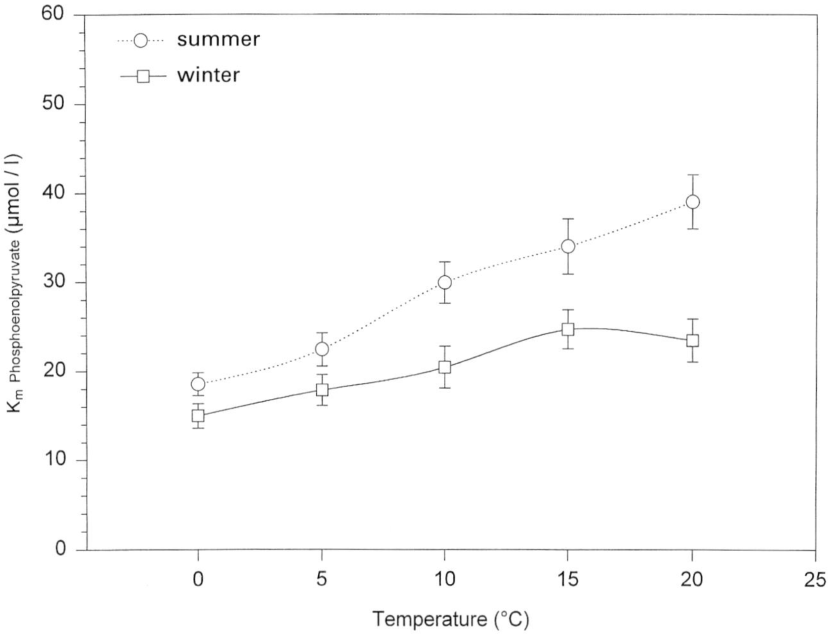

Fig. 2. Temperature dependence of K_m values of pyruvate kinase isoforms from abdominal muscle of *Meganyctiphanes norvegica* summer specimens ○ and winter specimens □. (Figure modified after Vetter & Buchholz, 1997.)

temperature changes is also essential, at least in eurythermal species. In the case of pyruvate kinase (PK) of *M. norvegica* (Fig. 2), seasonal temperature acclimatisation occurred by another process. PK was present in two organ-specific isoforms. In both isoforms, K_m values were less dependent on the assay temperature compared to PK from other sources (e.g. Hoffmann, 1976), resulting in flat horizontal curves or only weakly linear slopes instead of U-shaped curves (Vetter & Buchholz, 1997). Nevertheless, seasonal acclimatisation by varying K_m appeared to be feasible in the isoform from the abdominal muscle. Winter values were reduced to 50% over the whole temperature range compared to summer specimens resulting in an increased enzyme–substrate affinity in the cold season. The other isoform from the hepatopancreas did not contribute to seasonal temperature adaptation by varying K_m values (Vetter & Buchholz, 1997). The molecular mechanisms of adapting K_m values to seasonally different conditions are not well investigated yet. Different isoforms could be involved (Hochachka & Somero, 1984), but also differences in hydrogen bonds (Klyachko *et al.*, 1993) or the phosphorylation status (Plaxton & Storey, 1984) may contribute to varying enzyme–substrate affinities.

In addition, the eurythermal species, *M. norvegica*, showed a considerable shift of K_m minima depending on the acclimation temperature as exemplified by CS. This was not found in the stenothermal krill *E. superba*. Maintenance at various temperatures for 11 days revealed substantial differences. In *E. superba* K_m-values were similar in specimens taken from the field and in aquaria- maintained individuals, irrespective of the maintenance temperature (Vetter, 1995*a*). In contrast, K_m values of eurythermal *M. norvegica* were strongly temperature dependent. The resulting minima corresponded to the respective maintenance temperature in each case. Consequently, enzyme–substrate affinity was highest at ambient temperatures. Variation of K_m can be seen as an effective mechanism of acclimation of CS in eurythermal but not in stenothermal krill. Substantial seasonal and diurnal temperature differences for *M. norvegica* necessitate efficient enzyme regulation to compensate for environmental temperature changes. In contrast, seasonal temperature changes are only small in the natural habitat of *E. superba* (−2 °C to +2 °C) and daily changes are negligible (Dunbar, 1968). As a consequence, short-term acclimation is not required so that the changes in acclimation temperature are not compensated (Vetter, 1995*a*).

The effect of temperature adaptation by increasing enzyme substrate affinities at ambient temperatures can be summarised as follows: different groups of enzymes exhibited this adaptive mechanism, e.g. chitinolytic enzymes (e.g. Buchholz & Vetter, 1993), digestive enzymes such as β-1,3-glucanase (e.g. Turkiewicz, 1995) and metabolic enzymes such as PK and malate dehydrogenase (MDH) (e.g. Hoffmann 1976; Dahlhoff & Somero, 1993*a*), but the effect is found neither in every enzyme of a single organism, nor in the same enzyme of various species (e.g. Dittrich, 1992*a*,*b*; Buchholz & Vetter, 1993; Vetter & Buchholz, 1997; Vetter *et al.*, 1997). Therefore, there is no general principle when this mechanism applies, because it is not restricted to a selected group of enzymes.

In addition, the enhancement of enzyme substrate affinity may be accompanied by changes in the membrane fluidity. In general, membrane fluidity is increased at high temperatures (Hazel & Williams, 1990). However, a compensatory effect occurs: exemplified on different *Haliotis* species membrane fluidity was compensated on the same level in the ambient temperature range (Dahlhoff & Somero, 1993*b*). The same compensatory effect was also found in K_m-values of MDH in the same species (Dahlhoff & Somero, 1993*a*). These findings suggest a connection between membrane fluidity and enzyme substrate affinity of enzymes bound to the membrane or in mitochondrial enzymes. Although there is no proof for a direct influence of this so-called homeoviscous response, a variation of enzyme substrate affinity may be facilitated by a variation of membrane fluidity. This was also found in the adaptation of deep-sea species, where enzymes exhibited reduced perturbation of function by pressure owing to the adaptation of cell membrane fluidity to deep-sea pressures and temperatures (Somero, 1992).

Total amount of enzyme molecules

Further acclimatisation to seasonal or daily temperature changes might also occur by a quantitative adaptation, i.e. at low temperatures, specific enzyme activity is enhanced by additional enzyme synthesis to increase the amount of catalytic molecules (e.g. Rao, 1962; Jones & Sidell, 1982; Clarke, 1983; Hochachka & Somero, 1984). Although there is little direct evidence to confirm changes in the absolute molar concentration of enzymes (Mayzaud van Wormhoult & Roche-Mayzaud, 1987), clear indications exist for an increased rate of protein synthesis during cold acclimation, e.g. in mussels (Rao, 1967) and in the killifish *Fundulus heteroclitus* (Crawford & Powers, 1989). In *F. heteroclitus*, a correlation between mRNA levels and enzyme activity was found in the B_4-isoform of lactate-dehydrogenase from two populations at different latitudes. The concentrations of this enzyme isoform and its mRNA were approximately twice as high in the northern (colder) population compared to the southern (warmer) population (Crawford & Powers 1989). These data strongly suggest that increased activity in the cold is induced by *de novo* synthesis of enzyme protein in these fish.

Nevertheless, it may be asked whether stenothermal invertebrates, which encounter smaller temperature amplitudes, are also capable of acclimating to different environmental temperatures in this way. In a comparative study of Nordic krill *M. norvegica* (eurythermic) and Antarctic krill *E. superba* (stenothermic) only *M. norvegica* showed differences in the specific activity of CS (Vetter, 1995*a*). After 11 days, cold acclimated specimens (0 °C) had three times higher values compared to warm acclimated ones (16 °C). Low environmental temperatures caused either an increased rate of CS synthesis or the production of CS with a higher specific activity. Comparing specific enzyme activities at environmental temperatures, i.e. when enzyme analyses were performed at acclimation temperatures, CS activity remained constant between 0 °C and 10 °C and decreased slightly with increasing temperature to 15 °C (Vetter, 1995*a*). In contrast, this compensation was not found in *E. superba*, where maintenance experiments had no effect on the specific activity of CS (Vetter, 1995*a*). These findings indicate a direct relationship between temperature changes in the natural habitat and regulation by specific activities. In eurythermal *M. norvegica* specific activity and enzyme substrate affinity of CS (described previously) were increased to compensate for the loss of activity with decreasing temperatures. These compensatory mechanisms were not found in the stenothermal Antarctic krill *E. superba*. Rising

Table 3. *Inhibitor constants (K_i values of competitive inhibition of ATP) of citrate synthase from different species*

Species	K_i (mmol l^{-1})
Serolis polita	2.867±0.594
Idotea baltica	0.594±0.114
Euphausia superba	0.779±0.097
Meganyctiphanes norvegica	0.927±0.234
Meganyctiphanes norvegica (med.)	1.036±0.399

Source: Data from Vetter, 1995*b*.

temperatures lead to increasing specific activity and enzyme substrate affinities remained unaffected (Vetter, 1995*a*). Recent direct determination in our lab using immunotechniques confirm these findings.

Affecting metabolites

Enzyme activity can also be regulated by feed-back mechanisms of effectors, e.g. by metabolic intermediates. Such mechanisms, either activating or inhibitory, often affect the enzyme-substrate affinity of regulatory enzymes, e.g. phosphofructokinase, PK, CS, etc. (e.g. Freed, 1971; Hardewig *et al.*, 1991; Simpfendörfer *et al.*, 1994, Vetter, 1995*b*). In addition, the influence of these metabolites can vary with temperature thus resulting in efficient temperature adaptation. This adaptive mechanism will be reviewed in detail for CS of crustaceans from different climatic zones. Since ATP is the most important effector of CS (Hathaway & Atkinson 1965; Jangaard, Unkeless & Atkinson, 1968), temperature adaptation is made possible by adjusting the degree of inhibition by ATP (Vetter, 1995*b*). Although K_i values of competitive inhibition appeared to be independent of assay temperatures, variation of inhibitor constants was found between different species. In benthic isopods the polar *S. polita* displayed a fivefold higher value compared to boreal *Idotea baltica* from the Baltic (Table 3). Consequently, the inhibition of CS by ATP is small and unimportant in the polar species due to its high K_i value, because even high ATP concentrations will only slightly decrease CS activity (Vetter, 1995*b*). In contrast, the higher rate of inhibition, i.e. the low K_i value found in *I. baltica*, avoids high turnover rates of the basal metabolism in this temperate species. Pelagic euphausiids from different climatic zones did not exhibit differences in K_i values. In all species investigated the inhibitor constants were at low levels (Table 3). Low K_i values allow a higher degree of regulation

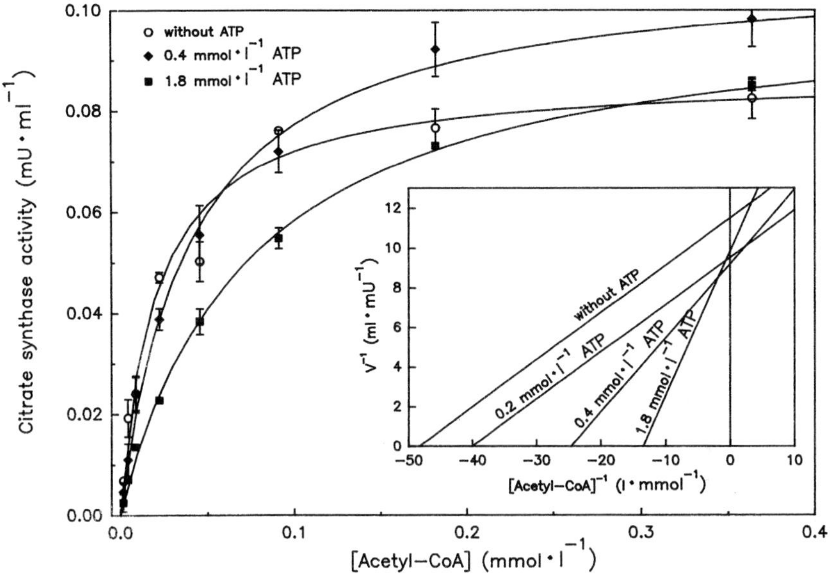

Fig. 3. Michaelis–Menten and Lineweaver–Burk plot (inset) of citrate synthase of euphausiids at varying ATP concentrations determined at 0 °C. In the direct plot, values of 0.2 mmol l^{-1} ATP were omitted for clarity. (Figure modified after Vetter, 1995*b*.)

depending on the concentration of the inhibiting metabolite (Fersht, 1985). Obviously, regulation of CS activity by ATP is essential in the permanently swimming euphausiids. Additional control of CS is improved by an activating effect of low ATP concentrations (Vetter, 1995*b*). Small quantities of ATP ($\leqslant$ 0.2 mmol l^{-1}) lead to increased maximal velocities (V_{max}) of CS (Fig. 3). Due to a simultaneous increase of K_m, the initial slope of the Michaelis–Menten plot remains constant. This effect resembles the mechanism of an uncompetitive inhibition (Kontro & Oja, 1981), but in the reversed sense, i.e. the enzyme is activated. Therefore, it is called an uncompetitive activation. The existence of this mechanism can be evaluated from the Lineweaver–Burk plot by a parallel shift of the line to the right (Fig. 3). At the enhanced level of V_{max} competitive inhibition occurred with increasing ATP concentrations indicated by typical revolving lines around the intercept of the ordinate in the Lineweaver–Burk plot (Fig. 3). This linkage between activating and inhibitory effects of different ATP concentrations has important consequences for the physiology of the pelagic krill. With an increasing demand for energy, ATP needs to be produced. Small amounts of this nucleotide increase V_{max} of CS. This increase has no effect at low acetyl-CoA concentrations, but at higher acetyl-CoA levels enzyme activity is elevated. Accordingly, only sufficient acetyl-CoA concentrations will increase the turnover rate of CS. As a result, high energy demands lead to a faster channelling of acetyl-CoA into the citric acid cycle and energy supply is rapidly increased. As soon as the ATP concentration reaches a threshold value (at least more than 0.2 mmol l^{-1}) competitive inhibition is initiated. At high substrate concentrations the inhibition is negligible because acetyl-CoA and ATP compete for the same binding site on CS and the affinity of the active centre is higher for acetyl-CoA compared to ATP. With decreasing acetyl-CoA the catalytic rate is rapidly reduced and energy production diminishes (Vetter, 1995*b*). With this dual effect of ATP the efficiency of CS is increased and the ability to regulate CS is preserved. This mechanism can be understood as an adaptation to the active, pelagic way of life of the euphausiids by optimising CS activity in relation to the high and fluctuating energy demand of these species (Clarke & Morris, 1983; Kils, 1983).

In fact, ATP concentrations of crustacean muscles may vary due to the physiological conditions (Beis & Newsholme, 1975). In non-exhausted, i.e. ATP-'loaded', abdominal muscle maximum ATP-concentrations were found between 7 μmol/g wet weight (approx. 8.7 mmol/l) in lobster (Beis & Newsholme, 1975) and 14.5 μmol/g wet weight (approx. 18.1 mmol/l) in Antarctic krill *E. superba* (Buchholz, 1988). Although it is difficult to evaluate the ATP-concentration in the mitochondrial matrix, it might be assumed that existing ATP levels suffice to inhibit CS efficiently under resting conditions. During flight swimming by tail flipping the energy demand increases, resulting in rapidly decreasing ATP concentrations followed by a drastically reduced inhibition to ensure high CS activity.

In conclusion, various mechanisms are involved in effector-mediated adaptation of CS, which depends on the lifestyle and the ambient temperature of different species. In the permanently swimming euphausiids, the influence of ATP on CS activity is independent of temperature. In contrast, in the slow-moving isopods inhibitor constants and the ambient temperature are interrelated in temperature adaptation: the colder the climate, the lower the inhibition by ATP. In euphausiids, adaptation to polar temperatures is achieved by a reduction of the activation energy instead of minimal ATP inhibition to retain the capability of ATP regulation. However, the disadvantage of higher inhibition is partially compensated by the activating effect of ATP, which was not found in the isopods.

Conclusions

For a better understanding of the eco-physiological effects of temperature on cold-stenothermal invertebrates, the investigation of enzyme character-

Table 4. *Mechanisms of adaptation to low temperatures found in different types of invertebrate enzymes*

Mechanism of temperature adaptation	Type of enzyme		
	Chitinolytic	Proteolytic	Metabolic
Shift of maximum temperature	—	—	—
Decrease of activation energy	—	●	●
Increase of enzyme substrate affinity	●	—	●
Increase of specific activity	—	—	●
Decrease of inhibitory effects	—	—	●

Note:
●=mechanism present; – =mechanism not described previously.

istics is an essential tool to elucidate mechanisms of temperature adaptation. For a comparative approach to assess adaptive capacities and evaluate the effects of long-term adaptation to cold, a comparison of enzymes in closely related species from different climatic zones is useful. In single enzymes of energy metabolism, which are necessarily regulated to a high extent, i.e. CS or PK, various mechanisms are involved in parallel depending on the life-style, the climatic zone and the ambient temperature regime. Long-term adaptation to cold temperature can be optimised by varying kinetic parameters (Table 4), such as a decrease in activation energy and in Michaelis constants or increasing total enzyme activity. This temperature dependent increase may be due to *de novo* synthesis of enzyme molecules. In addition, effects of activating or inhibiting metabolites are also involved in temperature adaptation of key metabolic enzymes. Short-term acclimation to different temperatures does not significantly alter enzyme characteristics of stenothermal invertebrates, whereas eurythermal species show adaptive capacities even within brief acclimation periods.

In comparison to enzymes of metabolic pathways, digestive enzymes, such as hydrolases and proteases, are often regulated to a smaller extent by temperature in the same animal species. Even if a single kinetic parameter is varied, activity is usually more strongly modulated by factors other than temperature, e.g. by hormones or food availability. Consequently, temperature adaptation seems to be less important in these types of catabolic enzymes. As a corollary, adaptation to different temperatures must be achieved for key enzymes of metabolic pathways, characterised by a complex combination of adaptive mechanisms to optimise the metabolic machinery under changing temperature regimes.

In conclusion, metabolic enzymes, in particular, are regulated by a combination of different mechanisms of adaptation, i.e. the reduction of activation energy, temperature-dependent enzyme substrate affinities, and a variation in inhibiting and activating effects of metabolites to solve the problems caused by the ambient temperature and the lifestyle of different animals.

References

Adelung, D., Buchholz, F., Culik, B. & Keck, A. (1987). Fluoride in tissues of krill *Euphausia superba* Dana and *Meganyctiphanes norvegica* M. Sars in relation to the moult cycle. *Polar Biology*, **7**, 43–50.

Baldwin, J. & Hochachka, P.W. (1970). Functional significance of isozymes in thermal acclimation: acetylcholinesterase from trout brain. *Biochemical Journal.* **116**, 883–7.

Baslow, M.H. & Nigrelli, R.F. (1964). The effect of thermal acclimation on brain cholinesterase activity of the killifish, *Fundulus heteroclitus*. *Zoologica NY*, **49**, 41–51.

Beis, I. & Newsholme, E.A. (1975). The contents of adenine nucleotides, phosphagens and some glycolytic intermediates in resting muscles from vertebrates and invertebrates. *Biochemical Journal*, **152**, 23–32.

Buchholz, F. (1988). Zur Lebensweise des antarktischen und des nordischen Krills *Euphausia superba* und *Meganyctiphanes norvegica*. Vergleichende Untersuchungen der Häutungsphysiologie und des Wachstums im Freiland und Labor. *Berichte des Instituts für Meereskunde der Christian-Albrechts-Universität Kiel* (185), 245 pp.

Buchholz, F. (1989). Moult cycle and seasonal activities of chitinolytic enzymes in the integument and digestive tract of the Antarctic krill, *Euphausia superba*. *Polar Biology*, **9**, 311–17.

Buchholz, F. (1991). Moult cycle and growth of Antarctic krill *Euphausia superba* in the laboratory. *Marine Ecology Progress Series*, **69**, 217–29.

Buchholz, F. & Vetter, R.A.H. (1993). Enzyme kinetics in cold water: characteristics of *N*-acetyl-β-D-glucosaminidase activity in the Antarctic krill, *Euphausia superba*, compared with other crustacean species. *Journal of Comparative Physiology*, **163B**, 28–37.

Buchholz, F. & Saborowski, R. (1996). A field study on the physiology of digestion in the Antarctic krill, *Euphausia superba*, with special regard to chitinolytic enzymes. *Journal of Plankton Research*, **18**, 895–906.

Bullock, T.H. (1955). Compensation for temperature in the metabolism and activity of poikilotherms. *Biological Reviews*, **30**, 311–42.

Campbell, D., Hellgren, L., Karlstam, B. & Vincent, J. (1987). Debriding ability of a novel multi-enzyme preparation isolated from Antarctic krill (*Euphausia superba*). *Experientia*, **43**, 578–9.

Chen, C.S., Yan, T.R. & Chen, H.Y. (1980). Purification and properties of trypsin-like enzymes and a carboxypeptidase A from *Euphausia superba*. *CAPD Fisheries Series*, **2**, 29–45.

Childress, J.J., Cowles, D.L., Favuzzi, J.A. & Mickel, T.J. (1990). Metabolic rates of benthic deep-sea decapod crustaceans decline with increasing depth primarily due to the decline in temperature. *Deep Sea Research*, **37**, 929–49.

Choi, S.H. & Kato, H. (1984). Changes in cooked odor of Antarctic krill during frozen storage. *Agricultural Biology and Chemistry*, **48**, 545–7.

Clarke, A. (1983). Life in cold water: the physiological ecology of polar marine ectotherms. *Oceanography and Marine Biology: An Annual Review*, **21**, 341–453.

Clarke, A. (1987). The adaptation of aquatic animals to low temperatures. In *The Effects of Low Temperatures on Biological Systems*, ed. B.W.W. Grout & G.J. Morris, pp. 315–48. Baltimore, MD: Edward Arnold.

Clarke, A. & Morris, D.J. (1983). Towards an energy budget for krill: the physiology and biochemistry of *Euphausia superba* Dana. *Polar Biology*, **2**, 69–86.

Crawford, D.L. & Powers, D.A. (1989). Molecular basis of evolutionary adaptation at the lactate dehydrogenase-B locus in the fish *Fundulus heteroclitus*. *Proceedings of the National Academy of Sciences, USA*, **86**, 9365–9.

Dahlhoff, E. & Somero, G.N. (1993*a*). Kinetic and structural adaptations of cytoplasmic malate dehydrogenases of eastern pacific abalone (genus *Haliotis*) from different thermal habitats: biochemical correlates of biogeographical patterning. *Journal of Experimental Biology*, **185**, 137–50.

Dahlhoff, E. & Somero, G.N. (1993*b*). Effects of temperature on mitochondria from abalone (genus *Haliotis*): adaptive plasticity and its limits. *Journal of Experimental Biology*, **185**, 151–68.

Dittrich, B. (1992*a*). Thermal acclimation and kinetics of a trypsin-like protease in eucarid crustaceans. *Journal of Comparative Physiology*, **162B**, 38–46.

Dittrich, B.U. (1992*b*). Life under extreme conditions: aspects of evolutionary adaptation to temperature in crustacean proteases. *Polar Biology*, **12**, 269–74.

Dunbar, M.J. (1968). *Ecological Development in Polar Regions – A Study in Evolution*. Englewood Cliffs NJ: Prentice-Hall.

Ferracin, A., Vitelli, A., Guida, S. & Prosperi, T. (1987). Thermal behaviour of a heterothermic enzyme: A4 lactate dehydrogenase from *Triturus cristatus*. *Biochemical Systems in Ecology*, **15**, 695–8.

Fersht, A. (1985). *Enzyme Structure and Mechanism*. New York: Freeman.

Fideau-Alonso, M.D., Perez, M.L.P., Amil, M.R. & Santos, M.J.H. (1985). Regulation of liver sea bass pyruvate kinase by temperature, substrates and some metabolic effectors. *Comparative Biochemistry and Physiology*, **82B**, 841–8.

Freed, J.M. (1971). Temperature effects on muscle phosphofructokinase of the Alaskan king crab *Paralithodes camtschatika. Comparative Biochemistry and Physiology*, **39B**, 765–74.

Galgani, F., Benyamin, Y., Van Wormhoudt, A. & Ceccaldi, J. (1983). Variations des activités digestives en fonction des facteurs du milieu chèz les crustacés. *Actes de Colloques IFREMER.* **1**, 227–92.

Genicot, S., Feller, G. & Gerday, C. (1988). Trypsin from Antarctic fish (*Paranotothenia magellanica* Forster) as compared with trout (*Salmo gairdneri*) trypsin. *Comparative Biochemistry and Physiology*, **90B**, 601–9.

Goldspink, G. (1995). Adaptation of fish to different environmental temperature by qualitative and quantitative changes in gene expression. *Journal of Thermal Biology*, **20**, 167–74.

Guppy, M. & Hochachka, P.W. (1979). Pyruvate kinase functions in hot and cold organs of tuna. *Journal of Comparative Physiology*, **129**, 185–91.

Hall, J.G. (1985). The adaptation of enzymes to temperature: catalytic characterization of glucosephosphate isomerase homologues isolated from *Mytilus edulis* and *Isognomon alatus,* bivalve molluscs inhabiting different thermal environments. *Molecular Biology and Evolution*, **2**, 251–69.

Hardewig, I., Kreutzer, U., Pörtner, H.O. & Grieshaber, M.K. (1991). The role of phosphofructokinase in glycolytic control in the facultative anaerobe *Sipunculus nudus. Journal of Comparative Physiology*, **161B**, 581–9.

Hathaway, J.A. & Atkinson, D.E. (1965). Kinetics of regulatory enzymes: Effect of adenosine triphosphate on yeast citrate synthase. *Biochemical and Biophysical Research Communications*, **20**, 661–5.

Hazel, J.R. & Williams, E.E. (1990). The role of alterations on membrane lipid composition in enabling physiological adaptation of organisms to their physical environment. *Progress in Lipid Research*, **29**, 167–227.

Hellgren, L., Mohr, V. & Vincent, J. (1986). Proteases of Antarctic krill – a new system for effective enzymatic debridement of necrotic ulcerations. *Experientia*, **42**, 403–4.

Hochachka, P.W. & Somero, G.N. (1971). Biochemical adaptation to the environment. In *Fish Physiology* ed. W.S. Hoar & D.J. Randall, vol. VI, pp. 99–156. New York: Academic Press.

Hochachka, P.W. & Somero, G.N. (1973). *Strategies of Biochemical Adaptation*. Philadelphia: Saunders.

Hochachka, P.W. & Somero, G.N. (1984). *Biochemical Adaptation.* Princeton: Princeton University Press.

Hoffmann, K.H. (1976). Catalytic efficiency and structural properties of invertebrate muscle pyruvate kinases: correlation with body temperature and oxygen consumption rates. *Journal of Comparative Physiology*, **110**, 185–95.

Jaenicke, R. (1990). Protein structure and function at low temperature. *Philosophical Transactions of the Royal Society of London B*, **326**, 535–51.

Jangaard, N.O., Unkeless, J. & Atkinson, D.E. (1968). The inhibition of citrate synthase by adenosine triphosphate. *Biochimica et Biophysica Acta*, **151**, 225–35.

Johnston, I.A. & Walesby, N.J. (1977). Molecular mechanisms of temperature adaptation in fish myofibrillar adenosine triphosphatases. *Journal of Comparative Physiology*, **119B**, 195–206.

Johnston, I.A., Sidell, B.D. & Driedzic, W.R. (1985). Force-velocity characteristics and metabolism of carp muscle fibres following temperature acclimation. *Journal of Experimental Biology*, **119**, 239–49.

Jones, P.L. & Sidell, B.D. (1982). Metabolic responses of striped bass (*Morone saxatilis*) to temperature acclimation. II) Alterations in metabolic carbon sources and distributions of fiber types in locomotory muscle. *Journal of Experimental Zoology*, **219**, 163–71.

Karl, H., Schreiber, W., Oehlenschläger, J., Fichtl, P., Manthey, M. & Rehbein, H. (1986). Continuous processing of raw Antarctic krill into food products for human consumption on a pilot-plant scale. *Archiv für Fischereiwissenschaft*, **37**/1, 187–98.

Karpov, A.K. & Andreyeva, A.P. (1992). Peculiarities of the kinetics of the lactate dehydrogenase reaction in some fish species of the family Gadidae. *Journal of Ichthyology*, **32**, 144–9.

Kils, U. (1979). Performance of Antarctic krill *Euphausia superba*, at different levels of oxygen saturation. *Meeresforschung*, **27**, 35–48.

Kils, U. (1981). The swimming behaviour, swimming performance and energy balance of Antarctic krill, *Euphausia superba*. *BIOMASS Science Series*, **3**, 122 pp.

Kils, U. (1983). Swimming and feeding of Antarctic krill, *Euphausia superba* – some outstanding energetics and dynamics – some unique morphological details. *Berichte zur Polarforschung Sonderheft*, **4**, 130–55.

Klyachko, O.S., Polosukhina, E.S. & Ozernyuk, N.D. (1993). Temperature causes structural and functional changes in lactic dehydrogenase of fish skeletal muscles. *Biophysics*, **38**, 613–18.

Kono, M., Matsui, T., Shimizu, C. & Koga, D. (1990). Purifications and some properties of chitinase from the liver of a prawn, *Penaeus japonicus*. *Agricultural Biology and Chemistry*, **54**, 2145–7.

Kontro, P. & Oja, S.S. (1981). Enzyme kinetics. In *Methods in Neurobiology*, ed. R. Lahue, pp. 265–337. New York: Plenum Press.

London, J.W. & Shaw, L.M. (1983). Reaction kinetics. In *Methods of Enzymatic Analysis*, ed. H.U. Bergmeyer, 3rd edn, vol. I, pp. 68–82. Weinheim: Verlag Chemie.

Low, P.S. & Somero, G.N. (1974). Temperature adaptation of enzymes: A proposed molecular basis for the different catalytic efficiencies of enzymes from ectotherms and endotherms. *Comparative Biochemistry and Physiology*, **49B**, 307–12.

Low, P.S., Bada, J.L. & Somero, G.N. (1973). Temperature adaptation of enzymes: role of the free energy, the enthalpy, and the entropy of activation. *Proceedings of the National Academy of Sciences, USA*, **70**, 430–2.

Maffia, M., Ancierno, R., Deceglie, G., Vilella, S. & Storelli, C. (1993). Adaptation of intestinal cell membrane enzymes to low temperatures in the Antarctic teleost *Pagothenia bernacchii*. *Journal of Comparative Physiology*, **163B**, 265–70.

Mayzaud, O. (1985*a*). Purification and kinetic properties of the α-amylase from the copepod *Acartia clausi* (Giesbrecht, 1889). *Comparative Biochemistry and Physiology*, **82B**, 725–30.

Mayzaud, P. (1985*b*). Nutrition metabolism of the Antarctic krill. *Oceanis: Série de Documents Oceanographiques*, **11**, 101–15.

Mayzaud, P. & Mayzaud, O. (1981). Kinetic properties of digestive carbohydrases and proteases of zooplankton. *Canadian Journal of Fisheries and Aquatic Sciences*, **38**, 535–43.

Mayzaud, P., Van Wormhoudt, A. & Roche-Mayzaud, O. (1987). Spatial changes in the concentrations and activities of amylase and trypsin in *Euphausia superba*. A comparison between activity measurement and immunoquantitation. *Polar Biology* **8**, 73–80.

Moon, T.W. (1972). The functional adaptations of enzymes to temperature: comparison of NADP linked isocitrate dehydrogenase of trout liver and pig heart. *Comparative Biochemistry and Physiology*, **43B**, 525–38.

Nikolayeva, G.G. & Ponomareva, L.A. (1973). Oxygen consumption by tropical euphausiids. *Oceanology*, **13**, 408–11.

Opalinski, K.W. (1991). Respiratory metabolism and metabolic adaptations of Antarctic krill *Euphausia superba*. *Polski Archiwum Hydrobiologii*, **38**, 183–263.

Osnes, K.K. & Mohr, V. (1985*a*). Peptide hydrolases of Antarctic krill, *Euphausia superba*. *Comparative Biochemistry and Physiology*, **82B**, 599–606.

Osnes, K.K. & Mohr, V. (1985*b*). On the purification and characterization of three anionic, serine-type peptide hydrolases from Antarctic krill, *Euphausia superba*. *Comparative Biochemistry and Physiology*, **82B**, 607–19.

Paukova, L.M., Baidalinova, L.S. & Mosolov, V.V. (1992). Some properties of enzyme preparation Euphausiidin. *Applied Biochemistry and Microbiology*, **28**, 480–3.

Peters, G., Vetter, R.A.H., Saborowski, R., Mentlein, R. & Buchholz, F. (1992). On the quantification of rates of synthesis of chitinolytic enzymes in the Antarctic krill, *Euphausia superba*, using polyclonal antibodies. *Verhandlungen der Deutschen Zoologischen Gesellschaft*, **85**.1, 159.

Piepenburg, J. (1988). Zur Zusammensetzung der Bodenfauna in der westlichen Framstraße. *Berichte zur Polarforschung*, **52**, 1–118.

Plaxton, W.C. & Storey, K.B. (1984). Purification and properties of aerobic and anoxic forms of pyruvate kinase from red muscle tissue of the channelled whelk, *Busycotypus canaliculatum*. *European Journal of Biochemistry*, **143**, 257–65.

Poleck, T.P. & Denys, C.J. (1982). Effect of temperature on the moulting, growth and maturation of the Antarctic krill *Euphausia superba* (Crustacea: Euphausiacea) under laboratory conditions. *Marine Biology*, **70**, 255–65.

Prosser, C.L. (1967). Molecular mechanisms of temperature adaptation in relation to speciation. In *Molecular Mechanisms of Temperature Adaptation*, ed. C.L. Prosser, Vol. 84, pp. 351–76. Washington DC: American Association for the Advancement of Science (AAAS).

Rao, K.P. (1962). Physiology of acclimation to low temperature in poikilotherms. *Science NY*, **137**, 682–3.

Rao, K.P. (1967). Biochemical correlates of temperature acclimation. In *Molecular Mechanisms of Temperature Adaptation*, ed. C.L. Prosser, Vol. 84, pp. 227–44. Washington DC: American Association for the Advancement of Science (AAAS).

Russell, N.J. (1990). Cold adaptation of microorganisms. *Philosophical Transactions of the Royal Society of London B*, **326**, 595–608.

Saborowski, R. & Buchholz, F. (1991). Induktion von Enzymen im Verdauungstrakt des antarktischen Krills *Euphausia superba*. *Verhandlungen der Deutschen Zoologischen Gesellschaft*, **84**, 422

Saborowski, R., Buchholz, F., Vetter, R.A.H., Wirth, S.J. & Wolf, G.A. (1993). Soluble, dye-labelled chitin derivative adapted for the assay of krill chitinase. *Comparative Biochemistry and Physiology*, **105B**, 673–8.

Shaklee, J.B., Christiansen, J.A., Sidell, B.D., Prosser, C.L. & Whitt, G.S. (1977). Molecular aspects of temperature acclimation in fish: contributions of changes in enzyme activities and isozyme patterns to metabolic reorganization in the green sunfish. *Journal of Experimental Zoology*, **201**, 1–20.

Simpfendörfer, R., Vial, M.V., Nash, D., Gonzalez, M.L. & Lopez, D. (1994). Pyruvate kinase from the adductor muscle of *Mytilus chilensis*: specimens obtained from different intertidal levels. *Revista de Biologia Marina*, **29**, 113–25.

Somero, G.N. (1975). Temperature as a selective factor in protein evolution: the adaptational strategy of 'compromise'. *Journal of Experimental Zoology*, **194**, 175–88.

Somero, G.N. (1991). Biochemical mechanisms of cold adaptation and stenothermality in Antarctic fish. In *Biology of Antarctic Fish*, ed. B. DiPrisco, B. Maresca, & B. Tota, pp. 232–47. Berlin: Springer Verlag.

Somero, G.N. (1992). Biochemical ecology of deep-sea animals. *Experientia*, **48**, 537–43.

Somero, G.N. & Hochachka, P.W. (1968). The effect of temperature on the catalytic and regulatory functions of pyruvate kinase of the rainbow trout and the Antarctic fish *Trematomus bernacchii*. *Biochemical Journal* **110**, 395–400.

Somero, G.N. & Siebenaller, J.F. (1979). Inefficient lactate dehydrogenases of deep-sea fishes. *Nature*, **282**, 100–2.

Somero, G.N., Giese, A.C. & Wohlschlag, D.E. (1968). Cold adaptation of the Antarctic fish *Trematomus bernacchii*. *Comparative Biochemistry and Physiology*, **26**, 223–33.

Suzuki, T. & Kanna, K. (1977). Denaturation of muscle proteins in krill and shrimp during frozen storage. *Bulletin of the Tokai Regional Fisheries Research Laboratory*, **91**, 67–72.

Suzuki, M., Horii, T., Kikuchi, R. & Ohnishi, T. (1987). Purification of laminarinase from Antarctic krill *Euphausia superba*. *Bulletin of the Japanese Society of Scientific Fisheries*, **53**, 311–17.

Teal, J.M. & Carey, F.G. (1967). Effects of pressure and temperature on the respiration of euphausiids. *Deep Sea Research*, **14**, 725–33.

Thebault, M.-T. & Bernicard, A. (1981). Bases Moleculaires de l'Adaptation Thermique Chez les Invertebres Marins. *Oceanis*, **7**, 613–21.

Tokunaga, T., Iida, H. & Nakamura, K. (1977). Formation of dimethyl sulfide in Antarctic krill, *Euphausia superba*. *Bulletin of the Japanese Society of Scientific Fisheries*, **43**, 1209–17.

Torres, J.J. & Somero, G.N. (1988*a*). Vertical distribution and metabolism in Antarctic mesopelagic fishes. *Comparative Biochemistry and Physiology*, **90B**, 521–8.

Torres, J.J. & Somero, G.N. (1988*b*). Metabolism, enzymic activities and cold adaption in Antarctic mesopelagic fishes. *Marine Biology*, **98**, 169–80.

Torres, J.J., Aarset, A.V., Donnelly, J., Hopkins, T.L., Lancraft, T.M. & Ainley, D.G. (1994). Metabolism of Antarctic micronektonic Crustacea as a function of depth of occurrence and season. *Marine Ecology Progress Series*, **113**, 207–19.

Turkiewicz, M. (1995). Characterization of some digestive enzymes from *Euphausia superba* Dana. In *Microbiology of Antarctic Marine Environments and Krill Intestine, Its Decomposition and Digestive Enzymes,* ed. S. Rakusa-Suszczewski, & S.P. Donachie, pp. 197–254. Dziekanów Lesny: Institute of Ecology Polish Academy of Sciences.

Turkiewicz, M., Galas, E., Kalinowska, H., Romanowska, I. & Zielinska, M. (1986). Purification and characterization of a proteinase from *Euphausia superba* Dana (Antarctic krill). *Acta Biochimica Polonica*, **33**, 87–100.

Van Praët, M. (1982) Amylase and trypsin- and chymotrypsin-like proteases from A*ctinia equina* L.; their role in the nutrition of this sea anemone. *Comparative Biochemistry and Physiology*, **72B**, 523–8.

Van Wormhoudt, A., Mayzaud, P. & Ceccaldi, H.J. (1983). Purification et proprietés de l' α-amylase *Euphausia superba*. *Oceanis*, **9**, 287–8.

Vetter, R.A.H. (1995*a*) Ecophysiological studies on citrate-synthase: (I) enzyme regulation of selected crustaceans with regard to temperature adaptation. *Journal of Comparative Physiology*, **165B**, 46– 55.

Vetter, R.A.H. (1995*b*). Ecophysiological studies on citrate-synthase: (II) enzyme regulation of selected crustaceans with regard to life-style and the climatic zone. *Journal of Comparative Physiology*, **165B**, 56–61.

Vetter, R.A.H. & Buchholz, F. (1997). Catalytic properties of two pyruvate kinase isoforms in Nordic krill, *Meganyctiphanes norvegica,* with respect to seasonal temperature adaptation. *Comparative Biochemistry and Physiology*, **116A**, 1–10.

Vetter, R.A.H., Saborowski, R., Peters, G. & Buchholz, F. (1997). Temperature adaptation and regulation of citrate-synthase in the Antarctic krill compared with other crustaceans from different climatic zones. In *Antarctic Communities: Species, Structure and Survival*, ed. D.W.H. Walton, pp. 295–9. Cambridge: Cambridge University Press.

Voss, J. (1988). Zoogeographie und Gemeinschaftsanalyse des Makrobenthos des Weddell-Meeres (Antarktis). *Berichte zur Polarforschung*, **45**, 1–145.

A. VIARENGO, D. ABELE-OESCHGER
and B. BURLANDO

Effects of low temperature on prooxidant processes and antioxidant defence systems in marine organisms

Oxidative stress and antioxidant defence

It has only recently become apparent that the highly energy-efficient aerobic metabolism carries the risk of oxidative stress, i.e. uncontrolled liberation of deleterious reactive oxygen species (ROS) into tissues and body fluids, owing to an imbalance between ROS producing prooxidant processes and ROS scavenging antioxidant defence. By superimposing the controlled basic ROS liberation, which regularly accompanies uptake, transport and reduction of oxygen, the phenomenon of oxidative stress literally represents the drawback of oxygen consuming heterotrophic life. As one of the basic effects, oxidative stress leads to an impairment of the well-adjusted cellular redox balance. The latter is based on the ratio of reduced to oxidised glutathione, which is enzymatically coupled to the NAD : NADH system and acts as a redox buffer, protecting functional thiol groups from oxidation to disulphide (Sies, 1985). This redox buffer is essential in counterbalancing oxygen radicals and will be overpowered if the oxidative stress prevails for too long. Oxidative stress accompanies various generally stressful conditions like hypoxia, ischaemia, and intoxication and can interfere with a number of metabolic reactions via oxidation of proteins, membrane lipids, and nucleic acids.

Apart from more general physiological consequences like ageing and the promotion of carcinogenesis, oxidative stress may affect elementary and life-supporting cellular structures as oxygen binding blood pigments and ion and proton transporters in plasma membranes.

Cellular antioxidant defence is generally accomplished by a complex anti-oxidant system, the capacity of which can be modulated and adjusted during periods of increased oxidative stress (by augmenting antioxidant enzyme activities). On the other hand, the capacity to respond to certain levels of oxidative stress is also limited by physiological preconditions like age and food availability of an individual organism.

Although the database describing oxidative stress in cold-adapted marine organisms is still small, in this article we attempt to outline what is known on prooxidant and antioxidant processes in animals from seasonally or permanently cold marine environments. Specific questions are: does the cold ocean environment imply oxidative challenges which differ from temperate and warm water environments? And, did the evolutionary adaptation to permanently cold environments influence and change the capacity of ectotherms to balance oxidative stress?

Internal production of reactive oxygen species (ROS)

In aerobic cells, a number of metabolic pathways perform one-electron oxygen reductions, thereby yielding different reactive oxygen species (Halliwell & Gutteridge, 1984), including, from lower to higher reactivity, hydrogen peroxide (H_2O_2), the superoxide anion radical ($O_2^{-\bullet}$), the hydroxyl radical ($^{\bullet}OH$), and singlet oxygen $^1O_2$. Although potentially very toxic, these compounds are normally neutralised by a complex of antioxidant systems (Diguiseppi & Fridovich, 1984). Yet, ROS production can under unfavourable conditions overwhelm antioxidant defence, leading to multiple changes of cellular homeostasis and physiological functions, generally referred to as 'oxidative stress'. These alterations include cell redox potential, membrane lipid peroxidation, calcium metabolism, acid–base status, enzyme inactivation, and DNA damage (Slater, 1984; Halliwell & Aruoma, 1992).

Participation of oxidative damage in many pathologies has long been recognised (Gilbert, 1981), while more recently enhanced ROS production due to environmental contaminants has also been indicated to endanger aquatic organisms (Livingstone *et al.*, 1990; Winston & Di Giulio, 1991). ROS production occurs via different enzymatic and non-enzymatic reactions, involving metabolic intermediates as well as xenobiotics. One of the major sources of internal H_2O_2 and $O_2^{-\bullet}$ formation is the mitochondrial electron transport system, where active oxygen derives from quinone and cytochrome oxidation (Loschen *et al.*, 1974; Cadenas *et al.*, 1977). Another important site of ROS production is the cytochrome P_{450}-dependent electron transport chain in the endoplasmic reticulum. Cytochrome P_{450}-mediated ROS production involves different processes, due to the uncoupling of P_{450} monooxygenase reactions (Karuzina & Archakov, 1994, see below in the Section on cyt P_{450}) or due to the redox cycling of xenobiotic compounds (e.g. quinones, nitroaromatics) catalysed by flavoprotein reductases such as cytochrome P_{450} reductase (Washburn & Di Giulio, 1989; Lemaire *et al.*, 1994). In a redox cycle, the xenobiotic undergoes one-electron reduction by the reductase and the intermediate product can reduce O_2 to $O_2^{-\bullet}$ (Kappus & Sies, 1981; Livingstone *et al.*, 1990). The net result is oxyradical production at the expense of cellular reducing equivalents, such as NADPH.

Another important source of oxidative damage to cells is the redox cycling of transition metals, mainly Cu and Fe, which in the non-protein-bound form react with H_2O_2 to produce highly reactive $^{\bullet}OH$ radicals via Fenton-type reactions (Winterbourn, 1995):

$$Me^{n+} + H_2O_2 \rightarrow Me^{(n+1)+} + {}^{\bullet}OH + OH^{-}.$$

The latter process can become particularly harmful in the case of tissue overload of Cu and Fe, essential metals whose ionic concentrations in the cytosol depend on different heavy metal sequestering and detoxifying activities of cells (Simkiss, Taylor & Mason, 1982, Viarengo & Nott 1993).

Accumulation of hydrogen peroxide in seawater and its effects on animal physiology

UV radiation (UVR) induces formation of ROS in surface waters through absorbance of UV-A and UV-B photons by photo-sensitive dissolved organic matter (DOM), DOM photochemical degradation, and liberation of excited electrons, which initiate reduction of molecular oxygen:

$$DOM + UV\text{-}photon \rightarrow DOM^{*} + e^{-}_{(aq.)}$$

$$e^{-}_{(aq.)} + O_2 \rightarrow O_2^{-\bullet}.$$

Superoxide anion radicals abstract a second electron from DOM and protonate to yield H_2O_2:

$$O_2^{-\bullet} + DOM^{*} + 2H^{+} \rightarrow H_2O_2 + \text{refractory DOM}.$$

Hydrogen peroxide is not a radical, but it shows a considerable toxicity, as the weak O-O bond increases its reactivity and activation can lead to the liberation of $^{\bullet}OH$:

$$H_2O_2 + \text{energy} \rightarrow 2\,{}^{\bullet}OH$$

UV-induced H_2O_2 formation in surface waters has been shown to be a temperature-independent process (Abele-Oeschger, Tüg & Röttgers, 1997*a*). Availability of DOM, together with the co-occurring H_2O_2-decomposing processes, are presumably key factors limiting photochemical H_2O_2 accumulation in oligotrophic cold ocean surface waters. Various estimations of the persistence of H_2O_2 in seawater have been made. A half-life of 60 hours was determined in filtered seawater by Petasne and Zika (1987), but as H_2O_2 is apparently degraded via biological decomposition, this time-span decreases to between 7 and 10 hours in unamended seawater (Cooper & Lean, 1989). Temporal variabilities of H_2O_2 in ocean surface waters are linked to seasonal and daily fluctuations of solar radiation (Cooper & Lean, 1989; Fujiwara *et al.*, 1993). When photo-production of H_2O_2 exceeds biological decomposition, net accumulation amounts to levels between 0.1 and 0.3 μmol l^{-1} in temperate surface waters (Szymczak & Waite, 1988; Price, Worsfold & Mantoura, 1992), whereas at lower latitudes maximum surface concentrations of 0.03 μmol l^{-1} are reported (Resing *et al.*, 1993; Weller & Schrems, 1993).

In polar marine environments, H_2O_2 concentrations in shelf and sea ice samples are higher than in open ocean surface waters, because they receive H_2O_2 from atmospheric wet deposition of snow. The concentrations actually measured in ice samples vary between a mean of 1.2 μmol l^{-1} in Antarctic ice (Sigg, 1990; Eiken *et al.*, 1994) and mean values between 4 and 5 μmol l^{-1} in Greenland ice (Sigg & Neftel, 1991).

Concentrations of H_2O_2 in rainwater and snow are highly variable and can well reach micromolar levels. Measurements in rainwater yielded concentrations between 280 μmol l^{-1} (Cooper, Saltzman & Zika, 1987) and 34 μmol l^{-1} (Cooper & Lean, 1989). Abele-Oeschger *et al.*, 1997*c* document an episodic increase of H_2O_2 levels up to 2 μmol l^{-1} in intertidal rock pool water at King George Island, South Shetlands, Antarctica in November 1995, which they could link to freshly fallen snow, wherein H_2O_2 levels amounted to 13 μmol l^{-1}. Although H_2O_2 data from another Antarctic site (Palmer Station, Antarctic Peninsula, Tien & Karl, 1993) question the importance of atmospheric wet deposition as a source for H_2O_2, a time-series of concentrations from intertidal waters at King George Island (Abele-Oeschger *et al.*, 1997*c*) clearly shows that a period of snowfall entails a significant increase of the H_2O_2 concentrations.

When H_2O_2 reaches micromolar concentrations in surface or intertidal pool water it can substantially interfere with vital physiological functions of marine invertebrates (Abele-Oeschger & Buchner 1995; Abele-Oeschger, Oeschger & Theede, 1994; Abele-Oeschger, Sartoris & Pörtner, 1997*b*). Being an uncharged molecule, H_2O_2 can enter cells by diffusion and cause oxidative damage by liberating noxious $^{\bullet}OH$. Hydrogen peroxide also enhances damage induced by high energy radiation in a synergistic way (Halliwell & Gutteridge, 1989). It has been shown, for some intertidal invertebrates, that when ambient peroxide accumulation exceeds tolerable levels, usually between 2 and 20 μmol l^{-1}, activity levels of H_2O_2 decomposing enzymes, such as catalase, can be increased to a certain extent, while metabolic rates decline (Abele-Oeschger *et al.*, 1994; Buchner, Abele-Oeschger & Theede, 1996). Also, Abele-Oeschger and Buchner (1995) found a significant reduction of the filtration rates by 40% in the surface-dwelling bivalve *Cerastoderma edule* from the German Wadden Sea, when exposing the animals to 5 μmol l^{-1} H_2O_2. Reduced oxygen consumption was observed with the polychaete *Nereis diversicolor* at H_2O_2 concentrations between 0.5 and 5 μmol l^{-1} (Abele-Oeschger *et al.*, 1994), as well as the mudshrimp *Crangon crangon* at 20 μmol l^{-1} (Abele-Oeschger *et al.*, 1997b).

In a first approach to study the effect of H_2O_2 on aspects of invertebrate cellular homeostasis, Abele-Oeschger *et al.* (1997*b*) found a significant decrease in intracellular pH (pH_i) in *Crangon crangon* abdominal muscle tissue, irrespective of whether whole animals or isolated muscles were incubated. As neither the accumulation of anaerobic metabolites, nor enhanced ATP hydrolysis could account for the acidification, H_2O_2-mediated direct or indirect impairment of the transmembrane proton transport via Na^+/H^+ exchanger could be responsible for the drop in pH_i.

Effects of UV-radiation and atmospheric ozone depletion on hydrogen peroxide accumulation

Energy flux of UV radiation (UVR) per unit of earth surface is determined by the solar zenith angle (SZA), the density of the prevailing ozone layer and the sky cloud cover (Frederick & Snell, 1988; Vincent & Roy, 1993). Ozone depletion in the Southern hemisphere was first reported in 1985, but can be traced back to the late 1970s. Maximal ozone depletion is observed in Antarctica during the 'ozone hole' period at the beginning of October with minimal ozone density ranging as low as 145 Dobson units (Roy *et al.*, 1994), where 100 Dobson Units are defined as a 1 mm ozone layer at 1 bar air pressure and 22 °C air temperature. Atmospheric ozone does not attenuate UV-A radiation (320–400 nm), whereas it has a pronounced absorptive effect on high energy UV-B radiation (290–320 nm). Thus a 10% decrease in ozone density will increase surface irradiance at 300 nm by 100% (Frederick & Lubin, 1994).

For the Antarctic ecosystem, ozone depletion during early spring means higher UV-B radiation during a time of lower SZA. When column ozone returns to its normal level (350–400 Dobson Units, Karentz, 1994) during late spring, i.e. November or December, SZA is approaching summer solstice, which means longer duration of daytime light and therewith again higher doses of daily irradiation. Thus the Antarctic ozone hole means a prolonged period of elevated surface UV-B irradiation, a situation which will worsen with the duration of austral spring ozone deficiency.

With respect to the efficiency of different light ranges for photochemical H_2O_2 formation under conditions *in situ* Abele-Oeschger *et al.* (1997*a*) found a medium efficiency of $7.85 \cdot 10^{-4}$ molecules H_2O_2 per photon in the UV-B range (295–320 nm), while an average of $6.84 \cdot 10^{-5}$ molecules H_2O_2 were formed per UV-A photon (335–370 nm). This implies an 11 times higher efficiency of UV-B light for the photochemical production of H_2O_2. According to the calculations of these authors, a 10% ozone depletion could result in an approximately 30–40 % increase of potential H_2O_2 production in Antarctic intertidal waters. At the same time, UV-A radiation, to which can be ascribed repairing effects with respect to, e. g. UV-B-induced thymine-dimers, will stay unaltered.

According to a prognosis of Thompson, Owens and Stewart (1989), continuing carbon monoxide, nitric oxide and methane emissions will lead to a 100 % increase of the H_2O_2 concentrations in the urban boundary layer between 1980 and 2030. In more remote regions and lower latitudes, the increase in H_2O_2 over the same time period is estimated to range between 20 and 40%. From the aforedescribed interaction of UVR and oxygen, it can be inferred that the combined effect of the stratospheric ozone depletion and elevated tropospheric water vapour due to global warming will prospectively increase H_2O_2-induced oxidative stress in Antarctic surface waters and intertidal environments.

Physiological detoxification of ROS

Cellular ROS production is balanced by a complex system of enzymatic and low molecular weight antioxidants. Low molecular weight scavengers include hydrophilic (e.g. reduced glutathione=GSH, ascorbic acid, urate) and lipophilic compounds (e.g. α-tocopherol, carotenoids), generally accomplishing a non-selective scavenging of radical species.

Another potential ROS scavenger is nitric oxide (NO) (Moncada, Palmer & Higgs, 1991), which can inactivate $O_2^{-\bullet}$ by yielding peroxynitrite ($ONOO^-$) (Saran, Michel & Bors, 1990):

$$NO + O_2^{-\bullet} \rightarrow ONOO^-.$$

Nitric oxide has been reported to protect cells from oxidative damage (Wink *et al.*, 1995), but $ONOO^-$ was also found to be cleaved into NO_2 and $^{\bullet}OH$, and this was supposed to play a role in microvascular injury during ischaemia-reperfusion (Beckman *et al.* 1990). Therefore, the real contribution of the ROS scavenging activity of NO to the cellular redox balance remains to be verified.

Marine organisms possess mycosporine-like amino acids (MAAs, Karentz, 1994) which are known as UV-screeners and, recently, direct antioxidant potential has been ascribed to specific MAAs (Dunlap & Yamamoto, 1995).

In contrast to the low molecular weight scavengers, antioxidant enzymes can specifically remove oxyradicals and hydroperoxides which initiate lipid peroxidation chain reactions (Ahmad 1995). The most important of these enzymes are superoxide dismutase (SOD; EC 1.15.1.1), which catalyses the dismutation of $O_2^{-\bullet}$:

$$O_2^{-\bullet} + O_2^{-\bullet} + 2H^+ \rightarrow H_2O_2 + O_2$$

catalase (CAT; EC 1.11.1.6) which reduces H_2O_2:

$$H_2O_2 + H_2O_2 \rightarrow 2H_2O + O_2$$

selenium-dependent glutathione peroxidase (Se-GPX; EC 1.11.1.9) and other glutathione peroxidase activities, which reduce peroxides by oxidizing reduced glutathione (GSH):

$$ROOH + 2GSH \rightarrow ROH + GS\text{-}SG + H_2O;$$

where ROOH is an organic hydroperoxide, ROH the hydroxylated product, and GS-SG oxidised glutathione.

Besides primary antioxidant enzymes, other enzymes also prevent ROS formation (Ahmad, 1995). Among these latter, DT-diaphorase (DTD; EC 1.6.99.2) catalyses the direct two-electron reduction of quinones (Q) to hydroquinones (QH_2):

$$Q + NAD(P)H + H^+ \rightarrow QH_2 + NAD^+(P)$$

thereby preventing their redox cycling (see above in the Section on internal production of ROS). However, it should be noted that depending on the nature of the substrate, DTD can either act as an antioxidant or promote ROS production (Cadenas, 1995). For instance, DTD from fish liver was found to catalyse NADH-dependent ROS production for different quinones, thus indicating that this enzyme can be involved in quinone toxicity (Lemaire *et al.*, 1996).

Antioxidant enzymes can exhibit characteristic distribution among tissues and subcellular compartments. For instance, it has been shown, in polychaete worms, that lipid-rich and well-perfused tissues exhibit higher enzyme activities (Abele-Oeschger & Oeschger, 1995). Intracellular distribution shows elevated enzyme activities in the peroxisomes (catalase) as well as in the cytosol (SOD, glutathione peroxidase and glutathione reductase), while the mitochondria exhibit only low antioxidant potential (Livingstone *et al.*, 1992; Buchner *et al.*, 1996). Moreover, evidence concerning an involvement of different metal binding proteins (e.g. transferrin, metallothioneins, ceruloplasmin) in antioxidant defence is accumulating (Felton, 1995).

Effects of seasonal cold on antioxidant systems

It is now well known that the antioxidant systems of higher organisms also occur in the cells and tissues of fish and invertebrates (e.g. Viarengo *et al.*, 1989; Livingstone *et al.*, 1992; Gamble *et al.*, 1995; Förlin, Lemaire & Livingstone, 1995). Studies on marine invertebrates have shown seasonal variations in the levels of antioxidant enzymes and ROS scavengers. In the digestive gland of *Mytilus* sp., the activities of SOD, catalase and Se-GPX showed similar annual trends, with minimum values in February (Fig. 1). These enzyme data correlated well with the seasonal variabilty of GSH and vitamin E (Fig. 2), whereas carotenoids displayed a more complex trend,

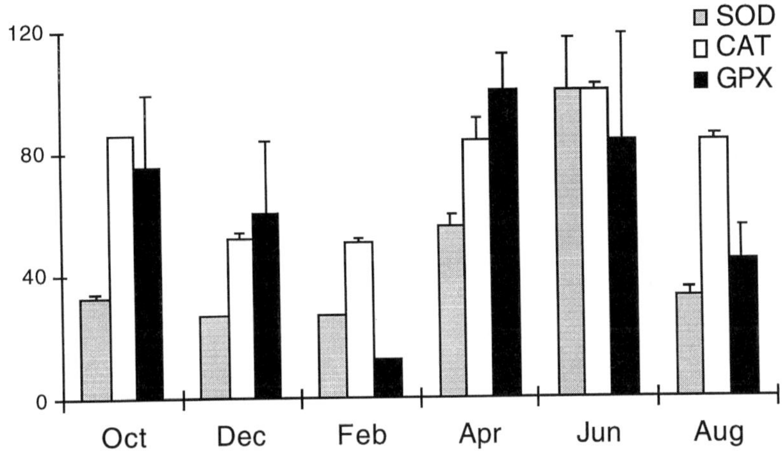

Fig. 1. Seasonal variations of antioxidant enzyme activities in the mussel *Mytilus edulis*. For each enzyme, data are expressed as a percentage of the maximum activity recorded: superoxide dismutase, SOD=3500 U / g wet weight, catalase, CAT=3000 moles H_2O_2 /min · g ww, glutathione peroxidase, GPX=95 nmoles NADPH / min · g ww (Viarengo *et al.*, 1991).

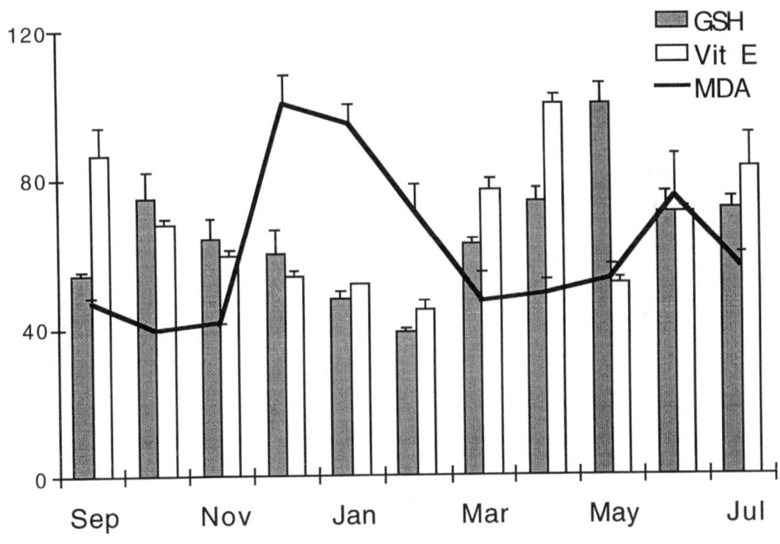

Fig. 2. Seasonal variations of ROS scavengers and malondialdehyde in the Mediterranean mussel *Mytilus galloprovincialis*. For each compound data are expressed as a percentage of the maximum content recorded: glutathione, GSH=550 μg / g ww, vitamin E, Vit E=13.5 μg / g ww, malondialdehyde , MDA=260 nmoles / g ww (Viarengo *et al.*, 1991).

Table 1. *Summer and winter activities of catalase and superoxide dismutase (SOD) in some intertidal invertebrates from the German Wadden coast*

	Catalase		SOD	
	winter	summer	winter	summer
Cerastoderma edule	57.7±16.2	137.5±58.7	nd	2.8±0.6[a]
Nereis diversicolor	16.7±9.5	57.3±14	1.9±0.9	3.4±1.7[b]
Arenicola marina	480±160	1165.±207	16.1±4.7	23.3±3.3[c]
Heteromastus filiformis	20.7±6.1	47 7	17.0±14.5	62.6±14.5[d]

Note:
Data are in U per mg , nd: not determined. *C. edule*: gill, *N. diversicolor* and *H. filiformis*: body wall tissue, *A. marina*: chloragog. Data sources: [a]Abele-Oeschger & Buchner (1995), [b]Abele-Oeschger *et al.* (1994), [c]Buchner & Abele-Oeschger (1995), [d]Großpietsch, H. (1996). One unit of catalase decomposes 1 mol l^{-1} H_2O_2 min^{-1} at pH 7.0 and at 20 °C. One unit of SOD activity is defined as the amount of enzyme which at 25 °C and pH 8.2 inhibits pyrogallol autoxidation by 50%.

probably due to the seasonality of algal blooms (Viarengo *et al.*, 1991). This study also confirmed that variations within the antioxidant status can involve alterations of an organism's sensitivity to radical damage. Malondialdehyde (MDA), which is a product of lipid peroxidation, showed higher levels during the cold season (Fig. 2), indicating enhanced lipid peroxidation in coincidence with a seasonal decrease of the antioxidant defence (Viarengo *et al.*, 1991). The pattern of seasonal variations found in mussels was confirmed by the finding of low winter activities of SOD and catalase in different invertebrates from the German Wadden Sea (Table 1), thus indicating that a decrease of enzymatic antioxidant activities during the cold season could represent a common trend in the physiology of temperate marine invertebrates.

The reduction of antioxidant defence in winter could be particularly disadvantageous to organisms, since lower winter temperatures are accompanied by higher O_2 levels in seawater and consequently also in the tissue of marine organisms (e.g. at a salinity of 34 g l^{-1} 100% air saturation is reached at 8.1 ml O_2 l^{-1} at 0 °C, 6.36 ml l^{-1} at 10 °C, and 5.2 ml l^{-1} at 20 °C, Grasshoff, Ehrhardt & Kremling, 1983). However, a study of the *in vitro* effects of O_2 and temperature on lipid peroxidation in digestive gland extracts of the scallop *Pecten jacobaeus* demonstrated that at 0 °C, i.e. in the presence of higher dissolved O_2 levels, peroxidation occurred at a lower rate than at 25 °C

(Viarengo *et al.*, 1995). These results indicate that temperature variations are more important than O_2 variations in determining the net degree of lipid peroxidation.

As a corollary, it can be inferred that, in temperate marine ectotherms, low winter enzymatic antioxidant defence could be partly related to a temperature-dependent reduction of metabolic rates (Clarke, 1983), accompanied by reduced internal ROS production. However, more research would be needed to clarify the combined effects of higher O_2 solubility and decreased O_2 consumption on pro- and antioxidant processes. In any case, available data suggest that, during winter, the cells of marine invertebrates are more sensitive to anthropogenic or natural agents able to stimulate ROS production (Viarengo *et al.*, 1991).

Besides a direct influence of temperature on metabolic rates, other factors are likely to be involved in annual fluctuations of pro- and antioxidant processes, such as modifications of the physiological status during the reproductive cycle of marine fish and invertebrates. Among exogenous factors, the seasonality of food supply presumably plays a pivotal role, particularly in suspension feeders, through a modification of the energy balance and tissue composition of the animals (Widdows, 1978; Barnes & Clarke, 1995). This may also apply to the tissue concentrations of antioxidants compounds, as it is well known that the tissue contents of carotenoids and vitamin E in different organisms are closely dependent on the level of food intake (Ong & Packer, 1992).

Effects of permanent cold on antioxidant and prooxidant mechanisms

Antioxidant mechanisms in the Antarctic fauna

It has been proposed by Wohlschlag (1960) that cold-adapted marine ectotherms show temperature compensation allowing high metabolic performance at low environmental temperatures. More recently, this view has been challenged (Holeton, 1974; Clarke, 1983). However, a certain degree of adaptation to low temperatures is known for different metabolic pathways, involving the occurrence of specific isozymes/allozymes or adjustments of enzyme quantities (Hochachka & Somero, 1973, 1984).

As for the antioxidant defence system in cold-adapted animals, the database is still very limited. Yet, different studies, comparing Antarctic and temperate marine ectotherms, provide insight into the possible occurrence of metabolic cold adaptations for antioxidant enzymes. For example, a comparison between antioxidant enzymes of Antarctic and sub-Antarctic populations of the ciliated protozoa *Euplotes rariseta* revealed higher levels of SOD activity in the Antarctic population (Albergoni *et al.*, 1994).

Viarengo *et al.* (1995) compared antioxidant defence levels in the digestive gland and gills of the Antarctic scallop (*Adamussium colbecki*) and of a Mediterranean species (*Pecten jacobaeus*). In this study, both Antarctic and temperate animals were collected during summer periods, in order to reduce seasonality effects. Catalase and Se-GPX activities were higher in the Antarctic mollusc as compared to the Mediterranean species, whereas SOD levels were lower. The latter difference might indicate higher rates of direct H_2O_2 production by enzymes like amino oxidase, compared to SOD-catalysed $O_2^{-\bullet}$ dismutation in the Antarctic *A. colbecki*, although temperature effects on $O_2^{-\bullet}$ dismutation or on enzyme properties cannot be excluded.

As stated above, except for episodic accumulation due to atmospheric wet deposition, H_2O_2 levels in Antarctica appear considerably lower than what is measured in North Sea Wadden areas (Abele-Oeschger *et al.*, 1997*a*). However, in the gills of the Antarctic intertidal limpet *Nacella concinna* catalase activity, on a tissue protein basis, was found to be equally as high as in Wadden Sea intertidal molluscs during summer (*Mya arenaria, Cerastoderma edule, Scrobicularia plana*). In addition, SOD activity was higher in gills of *N. concinna* compared to the boreal intertidal molluscs (Abele-Oeschger *et al.*, 1997*c*). Neither the polar (*N. concinna*) nor the Wadden Sea species studied (*M. arenaria, C. edule, S. plana*) could be called (strictly) stenothermal, i.e. the animals encounter variable temperatures in their habitats, but the temperature ranges in which these variations occur are distinct. Hence, these comparisons of enzyme activities in similar tissues of Antarctic and boreal temperate molluscan species, may help to describe the effects of different temperature regimes for oxidative stress phenomena.

The fact that *N. concinna* gills contain similar activities of both antioxidant enzymes as molluscs adapted to higher temperatures, can be seen as a mere feature of metabolic cold adaptation. On the other hand, it shows that Antarctic limpets are prepared to face ROS accumulation in seawater as well as a possible increase of metabolic ROS production during periods of water-warming. On sunny days, water temperatures in the Antarctic intertidal environment may temporarily rise to values above 10 °C. Peck (1989) showed that *N. concinna* exhibits partial acclimation to elevated temperatures, i.e. a temperature increase is followed by an overshoot in oxygen consumption, after which the metabolic rate declines again to lower levels, but still above the rates seen before the temperature increase. The elevated SOD activity in *N. concinna* may be a reaction to temperature-induced oxidative stress, i.e. ROS liberation during the initial overshoot in oxygen consumption or the subsequent rise in steady-state metabolism. In summary, cold-adapted intertidal *N. concinna* seem to show metabolic cold adaptation with respect to H_2O_2 and $O_2^{-\bullet}$ metabolizing enzymes, which enables them to react to sudden changes of H_2O_2 water concentrations, as well as of environmental temperatures.

Altogether, the available knowledge of antioxidant systems in Antarctic marine ectotherms suggests the occurrence of cold compensation in the antioxidant pathways of these organisms. However, due to the limited database and also to the fact that not all results are consistent with this idea, e.g. the previously reported comparison between SOD activities in *A. colbecki* and *P. jacobaeus* (Viarengo *et al.*, 1995), more research is required to justify a generalisation. However, in Antarctic organisms the occurrence of well-developed antioxidant defences does not seem confined to marine species, as a recent study concerning the larvae of an Antarctic fly (*Belgica antarctica*) reports higher levels of SOD and catalase activities compared to the larvae of temperate insects (*Tipula paludosa* and *Tenebrio molitor*) (Gruborlajsic *et al.*, 1995).

It should be noted that antioxidant enzyme activities of marine ectotherms were generally assayed at room temperature and not at the *in situ* temperatures of the investigated organisms. In studies of antioxidant enzyme seasonality in temperate organisms (see above), this might have led to an overestimation of the real winter antioxidant status, thus underestimating the actual reduction in winter antioxidant capacity. Also for polar organisms, a quantitative appraisal of antioxidant cold adaptation would require enzyme assays conducted at temperatures *in situ*. However, support for the idea of antioxidant cold compensation in Antarctic ectotherms originates from the finding that the temperate *P. jacobaeus* and the Antarctic *A. colbecki* showed similar tissue contents of non-enzymatic antioxidant compounds (Viarengo *et al.*, 1995), whose determination is not affected by assay temperatures.

A comprehensive view of antioxidant patterns in cold ocean environments must also consider the possibility of seasonal cycles. As reported above, the occurrence of seasonal cycles of antioxidant enzymes and compounds in temperate marine organisms has been established. By contrast, we still lack these kinds of data with respect to cold-adapted animals, leaving us prone to speculation. It is important to observe that the available data about antioxidants enzymes and compounds in Antarctic marine ectotherms only concern spring or summer periods, when organisms show elevated metabolic and growth rates (body size increase), and high feeding activity. Data on winter antioxidant status in Antarctic species are still missing. Seawater temperatures in the Antarctic ocean do not show major seasonal differences. Yet, for many Antarctic organisms, especially invertebrates, food intake is extremely reduced in winter (Clarke, 1988), which should lead to a significant decrease of metabolic rates and consequently of the antioxidant status (e.g. carotenoids, vitamin E). Considering annual variations of other abiotic factors such as UV radiation, seasonal variability of the antioxidant defence system in Antarctic invertebrates appears as a convincing possibility.

Effects of cold adaptation on lipid peroxidation

Lipid peroxidation is one of the most studied features of ROS damage to cells (Slater, 1984). The term 'lipid peroxidation' refers to a complex series of enzymatic or non-enzymatic steps, essentially involving radical chain reactions based on hydrogen abstraction from the unsaturated carbon atoms of phospholipid fatty acid chains. The process is largely stimulated within cells during uncontrolled free radical production, and leads to the formation of a mixture of toxic carbonyl compounds, *viz.* MDA and other aldehydes and chetones (Dianzani & Ugazio, 1978).

As the occurrence of unsaturated fatty acids is an essential prerequisite for lipid peroxidative processes, it follows that the lipid composition of cellular membranes plays a critical role in determining membrane sensitivity to radical injury. In Antarctic fish, membrane fluidity seems to be ensured by higher contents of low molecular-weight lipids combined with higher levels of lipid unsaturation, as compared to temperate fish (Maffia *et al.* 1994, see also Storelli, this volume). Such a pattern is similar to cold tolerance mechanisms in temperate marine fish (Tooke, Holland & Gabbott, 1985), but it does not seem to be the only cold adaptive strategy in the cell membranes of marine animals.

In a study of seasonal variations of cell membrane lipids in the mussel *Mytilus galloprovincialis*, it has been shown that adaptation to lower winter temperatures involves higher levels of unsaturated fatty acids and cholesterol, but also of branched-chain fatty acids (Accomando *et al.*, 1996). In addition, in the Antarctic scallop *A. colbecki* the occurrence of branched-chain saturated fatty acids (e.g. tridecanoic-4, 8, 12-trimethyl acid) represents a main strategy for cell membrane adaptation to permanent cold (Viarengo *et al.*, 1994). Considering the critical role played by unsaturated fatty acids in the process of lipid peroxidation, the adaptive strategy followed by bivalve molluscs, and in particular by the Antarctic *A. colbecki*, may also relate to the need of protecting cell membranes from ROS damage. This view is sustained by yet another study by Viarengo *et al.* (1995), in which comparable increases of MDA levels were found in *A. colbecki* and *P. jacobaeus*, upon experimental stimulation of lipid peroxidation, hence suggesting that cold adaptation of the cell membranes in the Antarctic *A. colbecki* does not involve higher susceptibility to oxidative injury.

Specific adaptations of Antarctic fish

Blood pigment contents

A well-known evolutionary trend of Antarctic fish involves the reduction of erythrocyte numbers and haemoglobin content (Everson & Ralph, 1968, see also di Prisco, this volume). Blood pigment reduction is compensated by higher O_2 solubility in the water and in animal tissues (Wells *et al.*, 1980), as well as an increased blood volume and a high cardiac output at low pressure (Clarke, 1983). Extreme adaptation has occurred in the haemoglobin- and myoglobin-free Channichthyidae, the family of Antarctic icefish (Ruud, 1954).

Apart from considerations involving O_2 transport, respiration and blood fluidity, this evolutionary trend is also likely to have an effect with respect to prooxidant processes. The iron centered haemoglobin molecule can autoxidize to methaemoglobin, a process which is an important source of $O_2^{-\bullet}$ and H_2O_2 in vertebrate erythrocytes (Misra & Fridovich, 1972; Winterbourn, 1985). Haemoglobin autoxidation is stimulated by peroxides (Winterbourn, McGrath & Carrell, 1976) and can only be tolerated owing to the presence of high levels of ROS-scavenging enzymes in erythrocytes (Elstner, 1990). Therefore, the reduction or absence of blood pigments in Antarctic fish could help to reduce oxidative damage resulting from haemoglobin autoxidation, thus compensating for possible negative effects of elevated oxygen concentrations in seawater at low temperatures. The presence of significantly higher levels of SOD and catalase activities in the tissue of a red-blood Antarctic fish (*Pagothenia bernacchii*), compared to a haemoglobin-free species (*Chionodraco hamatus*) (Albergoni *et al.*, 1992, 1994) supports this view. In addition, lower levels of haemoglobin could also result in a reduction of toxic effects, such as ROS production, due to Fe^{2+} accumulation in liver and spleen during blood pigment turnover.

Cytochrome P450 and MFO patterns

Cytochrome P_{450} is a highly polymorphic haem protein which forms an integral part of a microsomal mono-oxygenase system, also known as the mixed function oxygenase system (MFO) (Kappas & Alvarez, 1975; Estabrook, 1978; Stegeman & Hahn, 1994). This system plays an important role in physiological processes, such as steroid hormone metabolism, and is moreover involved in the detoxification of organic xenobiotic compounds, including polynuclear aromatic hydrocarbons and organochlorine compounds (Stegeman, 1989; Coon *et al.*, 1992; Walker & Livingstone, 1992).

The reaction catalysed by cytochrome P_{450} is an oxidation process, such as:

$$RH + O_2 + 2e^- + 2\,H^+ \rightarrow ROH + H_2O.$$

This reaction occurs through a series of steps, including formation of a complex between the ferricytochrome and the substrate:

$$RH + P_{450}\text{-}Fe^{3+} \rightarrow P_{450}\text{-}(RH)\text{-}Fe^{3+}$$

reduction of the latter compound by NADPH-cytochrome P_{450} reductase:

$$P_{450}\text{-}(RH)\text{-}Fe^{3+} + NADPH \rightarrow P_{450}\text{-}(RH)\text{-}Fe^{2+} + NADP + H^+$$

and binding of molecular oxygen:

$$P_{450}\text{-(RH)-Fe}^{2+} + O_2 \rightarrow P_{450}\text{-(RH)-Fe}^{2+}\text{-}O_2.$$

The compound formed in the latter reaction is in equilibrium with a ferri-superoxide complex P_{450}-(RH)-Fe^{3+}-$O_2^{-\bullet}$ which undergoes a second NAD(P)H-dependent reduction to form a ferro-superoxide compound, and eventually a ferri-monooxygen compound:

$$P_{450}\text{-(RH)-Fe}^{3+}\text{-}O_2^{-\bullet} + \text{NAD(P)H} \rightarrow P_{450}\text{-(RH)-Fe}^{2+}\text{-}O_2^{-\bullet} + \text{NAD(P)} + H^+;$$

$$P_{450}\text{-(RH)-Fe}^{2+}\text{-}O_2^{-\bullet} + 2\,H^+ \rightarrow P_{450}\text{-(RH)-Fe}^{3+}\text{-O} + H_2O;$$

the latter complex being the active form in operating the substrate hydroxylation.

Uncoupling of this monooxygenase process leads to the release of partially reduced reactive oxygen species, such as H_2O_2 and $O_2^{-\bullet}$:

$$P_{450}\text{-(RH)-Fe}^{2+}\text{-}O_2^{-\bullet} \leftrightarrow P_{450}\text{-(RH)-Fe}^{3+}\text{-}O_2^{2-} \xrightarrow{H^+ \;\; H_2O_2} P_{450}\text{-(RH)-Fe}^{3+}$$

$$P_{450}\text{-(RH)-Fe}^{3+}\text{-}O_2^{-\bullet} \rightarrow P_{450}\text{-(RH)-Fe}^{3+} + O_2^{-\bullet};$$

whereby the substrate promotes cytochrome P_{450} autoxidation and ROS production, instead of undergoing monoxygenation.

Certain organic xenobiotic compounds are potent inducers of MFO activities (Payne 1984; Livingstone, 1988; Stegeman, 1989), thereby increasing the potential for P_{450}-dependent ROS production (Livingstone *et al.*, 1989). Due to their substrate inducibility and redox cycling properties, MFO enzyme systems have been intensively studied in ecophysiological and ecotoxicological research, which also led to their widespread use as biomarkers in biomonitoring programmes (Payne, 1984; Viarengo & Canesi, 1991; Livingstone, 1993).

The worldwide dispersion of halogenated organic pollutants has resulted in the occurrence of measurable quantities of these compounds in polar environments (Courtney & Langston, 1981; Tanabe, Hidaka & Tatsukawa, 1983), as well as a detectable bioaccumulation in the tissue of cold ocean organisms (Tanabe, Mori & Tatsukawa, 1984; Subramanian *et al.*, 1986; Ernst & Klages 1991; Focardi, Lari & Marsili, 1992*a*). MFO systems have also been investigated in Antarctic fish, revealing very low basal activities compared to Mediterranean fish (Focardi *et al.*, 1989). Moreover, an experimental study on Antarctic fish showed almost no MFO inducibility after injection of drugs and environmental pollutants in the species *Chionodraco* *hamatus* and *Pagothenia bernacchii*. Different experimental groups, each treated with a different xenobiotic compound, i.e. phenobarbital (85 mg/kg), 3-methylcholanthrene (15 mg/kg), or Arochlor 1260 (130 mg/kg) by caudal vein injection for 24 hours, showed no induction or even a decrease of P_{450}-dependent activities. Only in *C. hamatus* was a significant response obtained after exposure to 3-methylcholanthrene, one of the most effective MFO inducers in temperate fish (Focardi *et al.* 1992*b*).

The observed low MFO levels in Antarctic fish were explained by the very limited environmental exposure to organic xenobiotic contaminants (Focardi *et al.*, 1989), although MFO activities at subsaturating substrate concentrations could be proportionally higher in Antarctic species due to apparent K_m compensation. Total cytochrome P_{450} content could be a more helpful tool in determining the actual potential of Antarctic fish MFO activities. However, on the basis of this study, it is tempting to speculate that if low MFO activities were a common feature of Antarctic fish, this could have relevant consequences. First, these organisms could be unfit to metabolise all the possible contaminants, which possibly entails cytotoxic pathways different from those of temperate fish, where the toxicity of organic xenobiotic compounds is generally related to their P_{450}-dependent oxidation. Secondly, the low activity of MFO in Antarctic fish could also involve a reduction of prooxidant processes related to P_{450}-promoted ROS production.

Summary and conclusions

It is now well established that marine organisms are generally endowed with a battery of enzymatic and chemical antioxidants, which closely resembles the antioxidant defence system of higher animals. Such a background of knowledge has been useful in pointing out possible adaptive features of antioxidant defence systems in cold-adapted ectotherms. Studies on temperate sea invertebrates have documented seasonal variations of antioxidants, with lower levels during the cold season, possibly reflecting a temperature-dependent metabolic slow-down. In invertebrates from intertidal and subtidal Antarctic environments, the summer levels of antioxidant enzymes (catalase, superoxide dismutase, glutathione peroxidase) and ROS scavengers (glutathione, vitamin E, carotenoids) are similar to the summer levels of comparable temperate species, indicating enzymes of the antioxidant pathways of Antarctic invertebrates to be fully cold compensated. In different cold-adapted animals the maintenance of membrane fluidity involves a rise in unsaturated fatty acids, possibly enhancing the sensitivity to lipid peroxidation. However, adaptation to seasonal cold in a Mediterranean mussel, and to permanent cold in an Antarctic scallop, involves increased contents of branched-chain fatty acids, which reduces potential oxidative damage. Biochemical features of Antarctic fish, such as a lack of blood pigment and low cytochrome P_{450} activities may presumably cause a reduction of pro-oxidant processes, as compared to temperate organisms.

The data reported in this review have to be considered essentially as a starting point towards a better understanding of cold adaptation with respect to prooxidant and antioxidant mechanisms. In particular, besides the need of a better characterisation of antioxidant pathways, it would be useful also to explore the relationships between the antioxidant status and other factors such as seasonality of food availability, fluctuations in metabolic rate and reproductive cycles, in both temperate and polar sea organisms. Also, from an environmental point of view, an assessment of the sensitivity to oxidative stress in cold-adapted marine organisms should be considered an important contribution to the evaluation of the effects of global contamination and increasing human presence in polar environments.

Acknowledgements

Thanks go to Tanja Buchner and Heike Großpietsch for collaboration with D. A. O., as well as to Professor Hans-Otto Pörtner for extensive discussions. The work was supported by the Deutsche Forschungsgemeinschaft contract Ab 64/1–3 (D. A. O.) and by the Italian National Programme for Antarctic Research (A. V. and B. B.).

References

Abele-Oeschger, D. & Buchner, T. (1995). Effect of environmental hydrogen peroxide accumulation on filtration rates of the intertidal bivalve clam *Cerastoderma edule*. *Verhandlüngen der Deutschen Zoologischen Gesellschaft*, **88,** 93.

Abele-Oeschger, D. & Oeschger, R. (1995). Antioxidant protection in 3 different developmental stages of polychaete spawn. *Ophelia*, **43**, 101–10.

Abele-Oeschger, D., Oeschger, R. & Theede, H. (1994). Biochemical adaptations of *Nereis diversicolor* (Polychaeta) to temporarily increased hydrogen peroxide levels in intertidal sandflats. *Marine Ecology Progress Series*, **106,** 101–10.

Abele-Oeschger, D., Tüg, H. & Röttgers, R. (1997*a*). Dynamics of UV-driven hydrogen peroxide formation on an intertidal sandflat. *Limnology and Oceanography* (in press).

Abele-Oeschger, D., Sartoris F.J. & Pörtner, H.O. (1997*b*). Hydrogen peroxide causes a decrease in aerobic metabolic rate and in intracellular pH in the shrimp *Crangon crangon*. *Comparative Biochemistry and Physiology*, **117C**, 123–9.

Abele-Oeschger, D., Pörtner, H-O., Ferreyra, G., Duttweiler, F. & Peck, L. (1997*c*). Pro- and antioxidant processes in the Antarctic limpet *Nacella concinna* (in review).

Accomando, R., Viarengo, A., Trielli, F., Benatti, U., Damonte, G. & Gotelli, P. (1996). Seasonal variations in lipid composition of cell membranes from the digestive gland of *Mytilus galloprovincialis* Lam. *European Society for Comparative Physiology and Biochemistry 17th Annual Conference*. University of Antwerp (RUCA), Antwerp, Belgium, p. 29.

Ahmad, S. (1995). Antioxidant mechanisms of enzymes and proteins. In: *Oxidative Stress and Antioxidant Defences in Biology*, ed. S. Ahmad, pp. 238–72. New York: Chapman & Hall.

Albergoni, V., Piccinni, E., Cassini, A., Tallandini, L., Turchetto, M., Coppellotti, O., Favero, N., Guidolin, L. & Irato, P. (1992). Physiological and toxicological aspects and physiological and biochemical responses to heavy metal contamination in Antarctic organisms. Physiological and biochemical features of adaptation to the increase of dissolved oxygen. In *Atti del Primo Convegno di Biologia Antartica*, ed. B. Battaglia, P.M. Bisol & V. Varetto, pp. 275–97. Padova: Edizioni Universitarie Patavine.

Albergoni, V., Cassini, A., Coppellotti, O., Favero, N., Piccinni, E., Tallandini, L. & Turchetto, M. (1994). Heavy metals and antioxidant enzymes in Antarctic protozoa and fish. In *Proceedings of the 2nd Meeting on Antarctic Biology*, ed. B. Battaglia, P.M. Bisol & V. Varotto, pp. 267–90. Padova: Edizioni Universitarie Patavine.

Barnes, D. K. A. & Clarke, A. (1995). Seasonality of feeding activity in Antarctic suspension feeders. *Polar Biology*, **15**, 335–40.

Beckman, J. S., Beckman, T. W., Chen, J., Marshall, P. A. & Freeman, B. A. (1990). Apparent hydroxyl radical production by peroxynitrite: implications for endothelial injury from nitric oxide and superoxide. *Proceedings of the National Academy of Sciences, USA*, **87**, 1620–4.

Buchner, T. & Abele-Oeschger, D. (1995). Seasonal differences of the enzymatic antioxidant defence in the lugworm *Arenicola marina* (Polychaeta). *Verhandlungen der Deutschen Zoologischen Gesellschaft*, **88,** 96.

Buchner, T., Abele-Oeschger, D. & Theede, H. (1996). Aspects of antioxidant status in the polychaete *Arenicola marina*: tissue and subcellular distribution, and reaction to environmental hydrogen peroxide and elevated temperatures. *Marine Ecology Progress Series*, **143**, 141–50.

Cadenas, E. (1995). Antioxidant and prooxidant functions of DT-diaphorase in quinone metabolism. *Biochemical Pharmacology*, **49**, 127–40.

Cadenas, E., Boveris, A., Ragan, C. I. & Stoppani, A. O. M. (1977). Production of superoxide radicals and hydrogen peroxide by NADH-ubiquinone reductase and ubiquinol-cytochrome C reductase from beef heart mitochondria. *Archives in Biochemistry and Biophysics*, **180**, 248–57.

Clarke, A. (1983). Life in cold water: the physiological ecology of polar marine ectotherms. *Oceanography and Marine Biology: An Annual Review*, **21**, 341–453.

Clarke, A. (1988). Seasonality in the Antarctic marine environment. *Comparative Biochemistry and Physiology*, **90B**, 461–73.

Coon, M. J., Ding, X., Pernecky, S. J. & Vaz, A. D. N. (1992). Cytochrome P_{450}: progress and prediction. *FASEB Journal*, **6**, 669–73.

Cooper, W. J. & Lean, D. R. S. (1989). Hydrogen peroxide concentration in a Northern lake: photochemical formation and diel variability. *Environmental Science and Technology*, **23,** 1425–8.

Cooper, W. J., Saltzman, E. S. &, Zika, R. G. (1987). The contribution of rainwater to variability in surface ocean hydrogen peroxide. *Journal of Geophysical Research*, **92,** 2970–80.

Courtney, W. A. M. & Langston, W. J. (1981). Organochlorines in Antarctic marine systems. *British Antarctic Survey Bulletin*, **53,** 255–7.

Dianzani, M. U. & Ugazio, G. (1978). Lipid peroxidation. In *Biochemical Mechanisms of Liver Injury*, ed. T.F. Slater, pp. 139–89. New York: Academic Press.

Diguiseppi, J. & Fridovich, I. (1984). The toxicology of molecular oxygen. *CRC Critical Reviews in Toxicology*, **12**, 315–42.

Dunlap, W. C. & Yamamoto, Y. (1995). Small-molecule antioxidant in marine organisms: antioxidant activity of mycosporine glycine. *Comparative Biochemistry and Physiology*, **112B**, 105–14.

Eiken, H., Lange, M. A., Hubberten, H-W. & Wadhams, P. (1994). Characteristics and distribution patterns of snow and meteoric ice in the Wedell Sea and their contribution to the mass balance of sea ice. *Annales Geophysicae*, **12,** 80–93.

Elstner, F. E. (1990). *Der Sauerstoff*. Mannheim, Wien, Zürich: Wissenschaftsverlag.

Ernst, W. & Klages, M. (1991). Bioconcentration and biotransformation of ^{14}C-γ-hexachlorocyclohexane and ^{14}C-γ-hexachlorobenzene in the Antarctic amphipod *Orchomene plebs* (Hurley, 1965). *Polar Biology*, **11**, 249–52.

Estabrook, R. W. (1978). Microsomal electron-transport reactions: an overview. In *Methods in Enzymology*, Biomembranes, ed. S. Fleisher & L. Packer, Vol. 52, part C, pp. 43–7. New York: Academic Press.

Everson, I. & Ralph, R. (1968). Blood analyses of some Antarctic fish. *Bulletin of the British Antarctic Survey*, **15**, 59–62.

Felton, G. W. (1995). Antioxidant defences of vertebrates and invertebrates. In *Oxidative Stress and Antioxidant Defences in Biology*, ed. S. Ahmad, pp. 356–434. New York: Chapman & Hall.

Focardi, S., Fossi, C., Leonzio, C. & Di Simplicio, P. (1989). Mixed function oxidase activity and conjugating enzymes in two species of Antarctic fish. *Marine Environmental Research*, **28**, 31–3.

Focardi, S., Lari, L. & Marsili, L. (1992*a*). PCB congeners, DDTs and hexachlorobenzene in Antarctic fish from Terra Nova Bay (Ross Sea). *Antarctic Science*, **4**, 151–4.

Focardi, S., Fossi, C., Lari, L., Marsili, L., Leonzio, C. & Casini, S. (1992*b*). Induction of Mixed Function Oxygenase (MFO) system in two species of Antarctic fish from Terra Nova Bay (Ross Sea). *Polar Biology*, **12**, 721–5.

Förlin, L., Lemaire, P. & Livingstone, D. R. (1995). Comparative studies of hepatic xenobiotic metabolizing and antioxidant enzymes in different fish species. *Marine Environmental Research*, **39**, 201–4.

Frederick, J. E. & Snell, H. E. (1988). Ultraviolett radiation levels during the Antarctic spring. *Science*, **241**, 438–40.

Frederick, J. E. & Lubin, D. (1994). Solar ultraviolet irradiance at Palmer Station, Antarctica. In *Ultraviolet Radiation in Antarctica: Measurements and Biological Effects. Antarctic Research Series*, **62**, pp. 43–52.

Fujiwara, K., Ushiroda, T., Takeda, K., Kumamoto, Y. & Tsubota, H. (1993). Diurnal and seasonal distribution of hydrogen peroxide in seawater of the Seto Inland Sea. *Geochemical Journal*, **27,** 103–15.

Gamble, S. C., Goldfarb, P. S., Porte, C. & Livingstone, D. R. (1995). Glutathione peroxidase and other antioxidant enzyme function in marine invertebrates (*Mytilus edulis, Pecten maximus, Carcinus maenas* and *Asterias rubens*). *Marine Environmental Research*, **39**, 191–5.

Gilbert, D. L. (1981). *Oxygen and Living Processes: An Interdisciplinary Approach.* New York: Springer Verlag.

Grasshoff, K., Ehrhardt, M. & Kremling, K. (1983). *Methods of Seawater Analysis*. Weinheim Chemie.

Großpietsch, H. (1996). Modulation von Oxidationsschutzmechanismen bei *Heteromastus filiformis* (Capitellidae, Polychaeta) in Abhängigkeit von verschiedenen Umgebungsparametern. Diploma-Thesis, Bremen University, Germany.

Gruborlajsic, G., Block, W., Jovanovic, A. & Worland, R. (1995). Antioxidant enzymes in larvae of the Antarctic fly, *Belgica antarctica. Cryo Letters*, **17**, 39–42.

Halliwell, B. & Aruoma, O. I. (1992). DNA damage by oxygen-derived species: its mechanism and measurement using chromatographic methods. In *Molecular Biology of Free Radicals Scavenger Systems*, ed. J.G. Scandalios, Vol. 9, pp. 47–67. Plainview, NY. Cold Spring Harbor Laboratory Press.

Halliwell, B. & Gutteridge, J. M. C. (1984). Oxygen toxicity, oxygen radicals, transition metals and disease. *Biochemical Journal*, **219**, 1–14.

Halliwell, B. & Gutteridge, J. M. C. (1989). *Free Radicals in Biology and Medicine.* 2nd edn, Oxford: Clarendon Press.

Hochachka, P. W. & Somero, G. N. (1973). *Strategies of Biochemical Adaptation*. Philadelphia: W.B. Saunders.

Hochachka, P. W. & Somero, G. N. (1984). *Biochemical Adaptation*. Princeton: Princeton University Press.

Holeton, G. F. (1974). Metabolic cold adaptation of polar fish: fact or artefact? *Physiological Zoology*, **47**, 137–52.

Kappas, A. & Alvares, A. P. (1975). How the liver metabolizes foreign substances. *Scientific American*, **232**, 22–31.

Kappus, H. & Sies, H. (1981). Toxic drug effects associated with oxygen metabolism: redox cycling and lipid peroxidation. *Experientia*, **37**, 1233–58.

Karentz, D. (1994). Ultraviolet tolerance mechanisms in Antarctic marine organisms. *Antarctic Research Series* **62**, 93–110.

Karuzina, I. I. & Archakov, A. I. (1994). The oxidative inactivation of cytochrome P450 in monooxygenase reactions. *Free Radical Biology Medicine*, **16**, 73–97.

Lemaire, P., Matthews, A., Förlin, L. & Livingstone, D. R. (1994). Stimulation of oxyradical production of hepatic microsomes of Flounder (*Platichthys flesus*) and Perch (*Perca fluviatilis*) by model and pollutant xenobiotics. *Archives in Environmental Contamination Toxicology*, **26**, 191–200.

Lemaire, P., Sturve, J., Förlin, L. & Livigstone, D. R. (1996). Studies on aromatic hydrocarbon quinone metabolism and DT-diaphorase function in liver of fish species. *Marine Environmental Research*, **42**, 317–21.

Livingstone, D. R. (1988). Responses of microsomal NADPH-cytochrome P-450 reductase activity and cytochrome P-450 in digestive gland of *Mytilus edulis* and *Littorina littorea* to environmental and experimental exposure to pollutants. *Marine Ecology Progress Series*, **46**, 37–43.

Livingstone, D. R. (1993). Biotechnology and pollution monitoring: use of molecular biomarkers in the aquatic environment. *Journal of Chemical Technology and Biotechnology*, **57**, 195–211.

Livingstone, D. R., Kirchin, M. A. & Wiseman, A. (1989). Cytochrome P_{450} and oxidative metabolism in molluscs. *Xenobiotica*, **19**, 1041–62.

Livingstone, D. R., Garcia, Martinez, P., Michel, X., Narbonne, J. F., O'Hara, S., Ribera, D. & Winston, G. (1990). Oxyradical production as a pollution-mediated mechanism of toxicity in the common mussel *Mytilus edulis* L. *Aquatic Toxicology*, **15**, 231–6.

Livingstone, D. R., Lips, F., Garcia, Martinez, P. & Pipe, R. K. (1992). Antioxidant enzymes in the digestive gland of the common mussel *Mytilus edulis*. *Marine Biology*, **112**, 265–76.

Loschen, G., Azzi, A., Richter, C. & Flohe, L. (1974). Superoxide radicals as precursors of mitochondrial hydrogen peroxide. *FEBS Letters*, **42**, 68–72.

Maffia, M., Acierno, R., De Vicienti, A. & Storelli, C. 1994. Lipid composition of intestinal cell membranes of the Antarctic fish *Pagothenia bernacchii*: comparative studies with the temperate fish *Anguilla anguilla*. In *Proceedings of the 2nd Meeting on Antarctic Biology*, ed. B. Battaglia, P. M. Bisol & V. Varotto, pp. 187–92. Padova: Edizioni Universitarie Patavine.

Misra, H. P. & Fridovich, I. (1972). The generation of superoxide radicals during the autoxidation of hemoglobin. *Journal of Biological Chemistry*, **247**, 6960–2.

Moncada, S., Palmer, R. M. J. & Higgs, E. A. (1991). Nitric oxide: physiology, pathophysiology and pharmacology. *Pharmacology Reviews*, **43**, 109–42.

Ong, A. S. H. & Packer, L. (1992). Lipid-soluble antioxidants: biochemistry and clinical applications. *Molecular Cell Biology Updates*. Basel: Birkhauser.

Payne, J. F. (1984). Mixed function oxygenases in biological monitoring programs: review of potential usage in different phyla of aquatic animals. In *Ecotoxicological Testing for the Marine Environment*, vol. **1**, pp. 625–55.

Peck, L. S. (1989). Temperature and basal metabolism in two Antarctic marine herbivores. *Journal of Experimental Marine Biology and Ecology*, **127,** 1–12.

Petasne, R. G. & Zika, R. G. (1987). Fate of superoxide in coastal seawater. *Nature*, **325,** 516–18.

Price, D., Worsfold, P. J. & Mantoura, R. F. C. (1992). Hydrogen peroxide in the marine environment: cycling and methods of analysis. *Trends in Analytical Chemistry*, **11**, 379–84.

Resing, J., Tien, G., Letelier, R. & Karl, D. M. (1993). Hydrogen peroxide in the Palmer LTER region: II. Water column distribution. *Antarctic Journal of the US*, **28,** 227–9.

Roy, C. R., Gies, H. P., Tomlinson, D. W. & Lugg, D. L. (1994). Effects of ozone depletion on the ultraviolet radiation environment at the Australian stations in Antarctica. *Antarctic Research Series*, **62,** 1–15.

Ruud, J. T. (1954). Vertebrates without erythrocytes and blood pigment. *Nature*, **173**, 848–50.

Saran, M., Michel, C. & Bors, W. (1990). Reaction of NO with $O_2^{-\bullet}$. Implications for the action of endothelium-derived relaxing factor (EDRF). *Free Radical Research Communications*, **10**, 221–6.

Sies, H. (1985). *Oxidative Stress*. London, New York: Academic Press.

Sigg, A. (1990). *Wasserstoffperoxid-Messungen an Eisbohrkernen aus Grönland und der Antarktis und ihre atmosphärenchemische Bedeutung.* Ph.D. Thesis, Zürich University, Switzerland, 1–140.

Sigg, A. & Neftel, A. (1991). Evidence of a 50% increase in H_2O_2 over the past 200 years from a Greenland ice core. *Nature*, **351,** 557–9.

Simkiss, K., Taylor, M. & Mason, A. Z. (1982). Metal detoxification and bioaccumulation in molluscs. *Marine Biology Letters*, **3**, 187–201.

Slater, T. F. (1984). Free-radical mechanisms in tissue injury. *Biochemical Journal*, **222**, 1–15.

Stegeman, J. J. (1989). Cytochrome P-450 forms in fish: catalytic, immunological and sequence similarities. *Xenobiotica*, **19**, 1093–110.

Stegeman, J. J. & Hahn, M. E. (1994). Biochemistry and molecular biology of monooxygenases: current perspectives on forms, functions, and regulation of cytochrome P450 in aquatic species. In *Aquatic Toxicology. Molecular, Biochemical, and Cellular Perspectives*, ed. D.C. Malins & G.K. Ostrander, pp. 87–206. Boca Raton, Florida: Lewis Publishers.

Subramanian, A. N., Tanabe, S., Hidaka, H. & Tatsukawa, R. (1986). Bioaccumulation of organochlorines (PCBs and pp'DDE) in Antarctic Adelie penguins (*Pygoscelis adeliae*) collected during a breeding season. *Environmental Pollution*, **40**, 173–89.

Szymczak, R. & Waite, T. D. (1988). Generation and decay of hydrogen peroxide in estuarine waters. *Australian Journal of Freshwater Research*, **39**, 289–99.

Tanabe, S., Hidaka, H. & Tatsukawa, R. (1983). PCBs and chlorinated hydrocarbon pesticides in Antarctic atmosphere and hydrosphere. *Chemosphere*, **12**, 277–88.

Tanabe, S., Mori, T. & Tatsukawa, R. (1984). Bioaccumulation of DDTs and PCBs in the Southern minke whale (*Balaenoptera acutorostrata*). *Memoirs of National Institute of Polar Research* (Special Issue), **32**, 140–50.

Thompson, A. M., Owens, M. A. & Stewart, R. W. (1989). Sensitivity of tropospheric hydrogen peroxide to global chemical and climate change. *Geophysical Research Letters*, **16**, 53–6.

Tien, G. & Karl, D. M. (1993). Hydrogen peroxide in the Palmer LTER region: III. Local sources and sinks. *Antarctic Journal of the US*, **28**, 229–30.

Tooke, N. E., Holland, D. L. & Gabbott, P. A. (1985). Phospholipid fatty acid composition and cold tolerance in two species of barnacle, *Balanus balanoides* (L.) and *Elminius modestus* Darwin. II. Isolation and phospholipid fatty acid composition of subcellular membrane fractions. *Journal of Experimental Marine Biology and Ecology*, **87**, 255–69.

Viarengo, A., Canesi, L., Garcia, Martinez, P., Peters, L. D. & Livingstone, D. R. (1995). Pro-oxidant processes and antioxidant defence systems in the tissues of the Antarctic scallop (*Adamussium colbecki*) compared with the Mediterranean scallop (*Pecten jacobaeus*) *Comparative Biochemistry and Physiology*, **111B**, 119–26.

Viarengo, A., Accomando, R., Roma, G., Benatti, U., Damonte, G. & Orunesu, M. (1994). Differences in lipid composition of cell membranes from Antarctic and Mediterranean scallops. *Comparative Biochemistry and Physiology*, **109B**, 579–84.

Viarengo, A. & Canesi, L. (1991). Mussels as biological indicators of pollution. *Aquaculture*, **94**, 225–43.

Viarengo, A., Canesi, L., Pertica, M. & Livingstone, D. R. (1991). Seasonal variations of the antioxidant defence systems and lipid peroxidation of the digestive gland of mussels. *Comparative Biochemistry and Physiology*, **100C**, 187–90.

Viarengo, A., Pertica, M., Canesi, L., Accomando, R., Mancinelli, G. & Orunesu, M. (1989). Lipid peroxidation and level of antioxidant compounds (GSH, vitamine E) in the digestive gland of mussels of three different age groups exposed to anaerobic and aerobic conditions. *Marine Environmental Research*, **28**, 291–5.

Viarengo, A. & Nott, A. (1993). Mechanisms of heavy metal cation homeostasis in marine invertebrates. *Comparative Biochemistry and Physiology*, **104C**, 355–72.

Vincent, F. W. & Roy, S. (1993). Solar ultraviolet-B radiation and aquatic primary production: damage, protection, and recovery. *Environmental Review* **1**, 1–12.

Walker, C. H. & Livingstone, D. R. (eds.) (1992). *Persistent Pollutants in Marine Ecosystems*. SETAC Special Publications, Oxford: Pergamon Press.

Washburn, P. C. & Di Giulio, R. T. (1989). The stimulation of superoxide production by nitrofurantoin, *p*-nitrobenzoic acid, and *m*-dinitrobenzene in hepatic microsomes of three species of freshwater fish. *Environmental Toxicology and Chemistry*, **8,** 171–80.

Weller, R. & Schrems, O. (1993). H_2O_2 in the marine troposphere and seawater of the Atlantic Ocean. *Geophysical Research Letters*, **20,** 125–8.

Wells, R. M. G., Ashby, M. D., Duncan, S. J. & Macdonald, J. A. (1980). Comparative studies of the erythrocytes and hemoglobins in notothenid fish from Antarctica. *Journal of Fish Biological*, **17**, 517–27.

Widdows, J. (1978). Combined effects of body size, food concentration and season on the physiology of *Mytilus edulis*. *Journal of the Marine Biology Association UK*, **58**, 109–24.

Wink, D. A., Cook, J. A., Pacelli, R., Liebmann, J., Krishna, M. C. & Mitchell, J. B. (1995). Nitric oxide (NO) protects against cellular damage by reactive oxygen species. *Toxicology Letters*, **82-3**, 221–6.

Winston, G. W. & Di Giulio, R. T. (1991). Prooxidant and antioxidant mechanisms in aquatic organisms. *Aquatic Toxicology*, **19**, 137–61.

Winterbourn, C. C. (1995). Toxicity of iron and hydrogen peroxide: the Fenton reaction. *Toxicology Letters*, **82-3**, 969–74.

Winterbourn, C. C. (1985). Reactions of superoxide with hemoglobin. In *CRC Handbook of Methods for Oxygen Radical Research*, ed. R.A. Greenwald, pp. 137–41. Florida: CRC Press Inc.

Winterbourn, C. C., McGrath, B. M. & Carrell, R. W. (1976). Reactions involving superoxide and normal and unstable hemoglobins. *Biochemical Journal*, **155**, 493–502.

Wohlschlag, D. E. (1960). Metabolism of an Arctic fish and the phenomenon of cold adaptation. *Ecology*, **41**, 287–92.

PART III Exploitative adaptations

G.L. FLETCHER, S.V. GODDARD,
P.L. DAVIES, Z. GONG, K.V EWART,
and C.L. HEW

New insights into fish antifreeze proteins: physiological significance and molecular regulation

Discovery of antifreeze proteins

It has been 40 years since Scholander and colleagues first travelled to Labrador to resolve what was thought to be a simple question: how do marine teleosts avoid freezing in icy seawater at temperatures (−1.8 °C) approximately 1 °C below the freezing point of their body fluids (−0.7 °C) (Scholander *et al.*, 1957)? They made two important observations as a result of their expeditions. The first was that some fish exist in a supercooled state; that is, at temperatures below their freezing points. However, if these fish are brought into contact with ice, they immediately freeze and die. The combination of supercooling and ice contact is lethal. The second observation was that other fish are able to survive at very low temperatures because their blood has the same freezing point as seawater. The blood solute responsible for lowering the freezing point was named 'antifreeze' by these pioneering researchers, but they were unable to determine its nature (Gordon, Andur & Scholander, 1962).

It took almost 10 years before the nature of these freeze-protecting plasma solutes was discovered by DeVries and Wohlschlag (1969) and characterised by DeVries, Komatsu and Feeney (1970). It is now known that they are polypeptides or glycopeptides, which are primarily synthesised in the liver and secreted into the blood. Still collectively termed 'antifreeze', these proteins are 200–300 times better at lowering the freezing point than would be expected on the basis of the number of antifreeze molecules present in solution (colligative properties) alone. Thus, fish have evolved a mechanism to reduce the freezing point of their body fluids without changing the osmotic properties of these fluids appreciably.

Antifreeze types

Four distinct classes of antifreeze have been characterised to date (Table 1). These blood serum antifreeze types consist of sets of closely related

Table 1. *Structural characteristics of the antifreeze proteins (AFP) and glycoproteins (AFGP)*

	AFGP	AFP Type I	AFP Type II	AFP Type III
Molecular mass (Daltons)	2600 – 33000	3300 – 4500	11 000 – 24 000	6500
Primary structure	(alanine-alanine-threonine)$_n$ disaccharide	alanine-rich multiple of eleven aa repeats	cystine-rich, disulphide linked	general
Carbohydrate	yes	no	no (exception: smelt has <3% carbohydrate)	no
Secondary structure	expanded	α helical amphiphilic	β sheet	β sandwich
Tertiary structure	not determined	100% helix	not determined	not determined
Biosynthesis	multiprotein	prepro AFP	prepro AFP (?)	pro AFP
Protein components	8	7	2 – 6	12
Gene copies	not determined	80–100	15	30–150
Fish species in which antifreeze type is found	Antarctic notothenioids; Northern cods (Atlantic cod)	right-eyed flounders (winter flounder); sculpins (shorthorn)	sea raven; smelt; herring	ocean pout; wolffish

Note:
Fish species listed=Atlantic cod: *Gadus morhua*; Winter flounder: *Pleuronectes americanus*; Shorthorn sculpin: *Myoxocephalus scorpius*; Sea raven: *Hemitripterus americanus*; Smelt: *Osmerus mordax*; Herring: *Clupea harengus harengus*; Ocean pout: *Macrozoarces americanus*; Wolffish: *Anarhichas lupus*.

proteins resolvable by high performance liquid chromatography (HPLC) into as few as 2 and as many as 12 independently active components (Hew & Fletcher, 1985; Davies & Hew, 1990). Type I AFP are the alanine rich (60 mol%) antifreeze peptides typified by those found in the winter flounder (*Pleuronectes americanus*). Antifreeze of this type has been crystallised and found to be completely α-helical (Yang *et al.*, 1988; Sicheri & Yang, 1995). The Type II antifreeze of sea raven (*Hemitripterus americanus*), smelt (*Osmerus mordax*) and Atlantic herring (*Clupea harengus harengus*) is characterised by disulphide bridges and an extensive β-structure. These Type II AFP share structural homology with the carbohydrate recognition domain of C-type lectins (Ng, Trinh & Hew, 1986; Ewart & Fletcher, 1990; Ewart, Rubinsky & Fletcher, 1992). In addition, the AFP from herring and smelt are dependent on Ca^{2+} for activity (Ewart *et al.*, 1996). The third type of peptide antifreeze (AFP III) is found in the ocean pout (*Macrozoarces americanus*) and other zoarcids (Hew *et al.*, 1984). It has no bias in its amino acid composition and contains a β-sandwich secondary structure (Sonnichsen *et al.*, 1993). The antifreeze glycoproteins (AFGP) are found in the Antarctic notothenioids (see Wöhrmann, this volume) and in the northern cods: – the Gadidae. They consist primarily of a repeating tripeptide unit (alanine–alanine–threonine)$_n$ with a carbohydrate unit attached to each threonine (DeVries *et al.*, 1970; Fletcher, Hew & Joshi, 1982; Wöhrmann, this volume). Evolution of the antifreeze proteins and glycoproteins (abbreviated collectively to AF(G)P) and their genes has been reviewed by Davies, Ewart and Fletcher (1993).

Mechanisms of action

Antifreeze and ice

Despite their structural diversity, all of the antifreeze proteins act in a similar fashion. They have an affinity for ice crystals and exert their effect by a mechanism known as adsorption-inhibition. The antifreeze molecules become attached to the growing prism faces of developing ice crystals and inhibit addition of water molecules to the ice lattice. This action results in a depression of freezing point proportional to the concentration of antifreeze present, and consequently, the environmental temperature must be lowered further before ice crystal growth can resume. Due to the way antifreeze molecules bind to ice crystal faces, the crystal habit is modified and long, spicular crystals are produced rather than the expected hexagonal ice crystals. This inhibition of ice crystal growth and resulting depression of freezing point is believed to be the major means by which antifreeze proteins protect the body fluids of fish from freezing (DeVries, 1988; Hew & Yang, 1992).

As a result of their particular affinity for, and interaction with, ice crystals, the freeze depressing action of the antifreeze proteins is non-colligative, i.e. not based simply on the number of antifreeze molecules present. However, their effect on the melting point is purely colligative. Thus, there can be a considerable difference between the freezing and melting points of solutions containing antifreeze proteins, and this temperature differential is referred to as the solution thermal hysteresis. The ability of antifreeze proteins to lower the freezing point and increase the freezing resistance of fish is obviously of critical importance to the survival of teleost fish in ice-laden oceans. However, studies carried out in recent years on a variety of cell types are providing evidence to suggest that the antifreeze proteins and glycoproteins may also play a significant role in survival at low (hypothermic) temperatures by interacting with cell membranes rather than ice crystals.

Demonstrations of hypothermic protection at the cellular level

In a series of experiments, Rubinsky and colleagues reported that all of the known antifreeze types could improve the cold tolerance of cold-sensitive mammalian cells. For example, in one study, bovine oocytes were incubated at 4 °C for 24 h in the presence or absence of antifreeze proteins, following which they were rewarmed. The ability of the oocytes to go through the final stages of maturation followed by fertilisation was then observed and compared with fertilisation rates of freshly collected, mature oocytes. It was found that for oocytes incubated with antifreeze proteins, the percentage capable of fertilisation closely resembled that of fresh oocytes. Those incubated at 4 °C without antifreeze lost their membrane integrity, and died (Rubinsky, Arav & Fletcher, 1991). The mechanism whereby the antifreeze proteins protected the cell membranes from cold damage is unknown.

One possible mechanism could involve the blockage of ion channels. Hochachka (1986, 1988*a*, *b*) proposed that a major cause of cell damage at low temperatures is the decoupling between metabolic activities and membrane function. This results in the dissipation of ion gradients across the cell membrane, increased intracellular Ca^{2+} concentration, activation of membrane phospholipid hydrolysis, and cell death. Could antifreeze proteins be protecting cells from damage at hypothermic temperatures by reducing passive ion flow across the cell membrane? Experiments by Negulescu *et al.* (1992) demonstrating that Type III AFP from ocean pout blocked passive Ca^{2+} entry into rabbit parietal cells at room temperature support this hypothesis, and suggest that the protective effect of AF(G)P may be related to their ability to block the entry of Ca^{2+} into the cell.

A further protective property of antifreeze proteins, apparently unrelated to their possible ion channel blocking action, has been described by Hays *et*

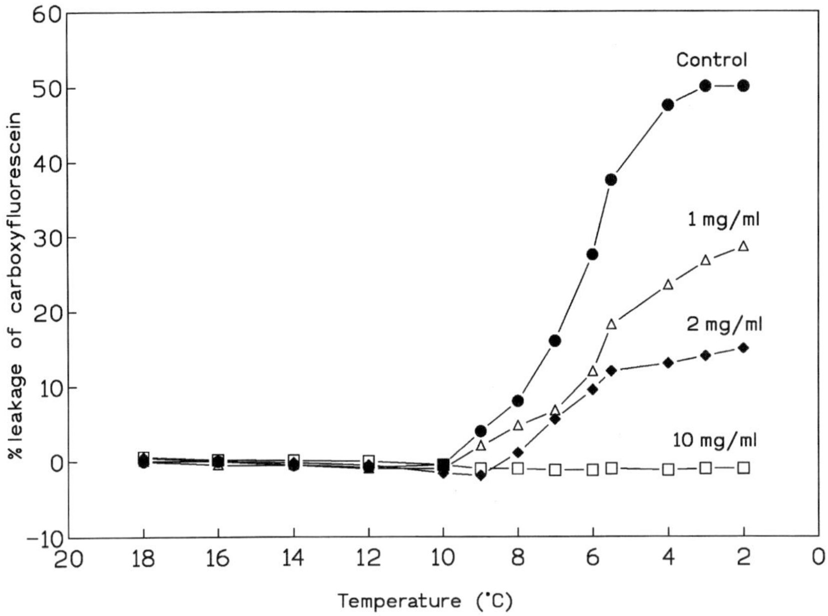

Fig. 1. Percentage leakage of fluorescent contents (carboxyfluorescine) from DEPC (dielaidoylphosphatidylcholine) liposomes in the presence (1, 2, 10 mg/ml) or absence (control) of AFGP fraction 8 purified from *Trematomus borchgrevinki*. The three concentrations of AFGP 8 were added to the experimental runs at 18 °C, after which the liposomes were cooled through the phase transition temperature and leakage of contents recorded. Modified from Hays *et al.*, 1996.

al. (1996). Cellular damage at low temperatures can be incurred when cells are cooled down through the lipid membrane phase transition temperature. During this liquid-crystalline to gel phase transition, membranes become leaky, resulting in the loss of intracellular contents. Hays *et al.* (1996) demonstrated that antifreeze glycoproteins can protect the integrity of lipid bilayers and inhibit leakage of liposome contents during the thermotropic phase transition. The results of one such experiment by these authors, illustrating the concentration dependent effect of AFGP on liposome leakage as they are cooled through the phase transition temperature (12 °C), are presented in Fig. 1.

Further evidence that the antifreeze proteins can act to preserve cells from cold damage by interacting with membrane phospholipids during phase transition was presented by Tablin *et al.* (1996) using human blood platelets. Platelets are highly thermosensitive, and respond to lowered temperatures by

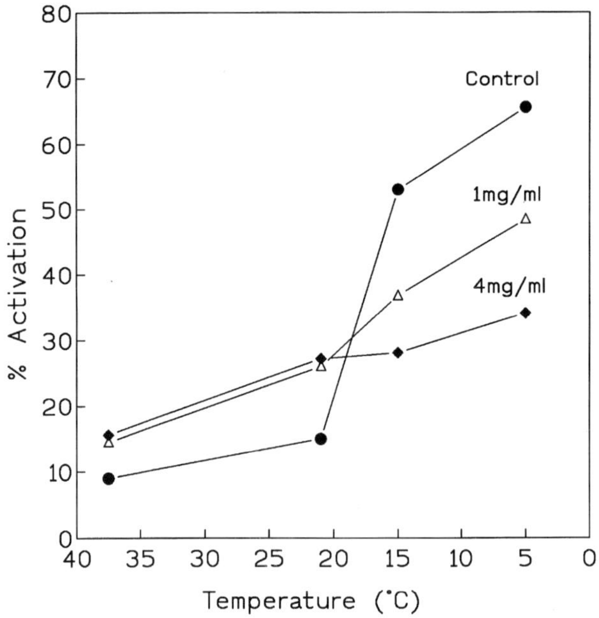

Fig. 2. Percentage cold-activation of platelets when exposed to low temperatures in the presence (1.4 mg/ml) or absence (control) of antifreeze glycoproteins (from Tablin *et al.*, 1996). Washed platelets were incubated at either 37 °C, 22 °C, 15 °C or 4 °C for one hour, following which percentage activation was assessed.

undergoing shape change with accompanying membrane changes, resulting in secretion of the intracellular contents, a process referred to as cold-activation. Tablin *et al.* (1996) found that platelet cold-activation occurred at a temperature very close to the lipid phase transition temperature (15–18 °C). However, when the platelets were cooled in the presence of AFGP, their cold-activation was markedly decreased in a dose-dependent manner (Fig. 2). Leakage of platelet contents could be the result of packing imperfections in the membranes of the dense tubular system and, subsequently, at the plasma membrane during phase transition. While the exact mechanism by which AFGP suppress platelet cold-activation requires further clarification, the results of these experiments suggest interaction between AFGP and platelet membranes.

Antifreeze and cold acclimation in fish

The ability of AF(G)P to protect whole fish from freezing has been shown experimentally (Fletcher, Kao & Fourney, 1986). However, the functional significance of the more recently described properties of the antifreeze proteins has yet to be documented.

Experimental results demonstrating that AF(G)P can, in some way, assist in the stabilisation of cold-sensitive mammalian cell membranes suggest that antifreeze might play a physiological role in the cold acclimation process. However, data in support of this hypothesis are scant. One line of supporting evidence relates to observations that a number of fish species inhabiting the NW Atlantic produce insufficient AF(G)P to protect them from freezing at the temperatures (down to −1.8 °C) they encounter in the ocean. For example, antifreeze levels in yellowtail flounder (*Pleuronectes ferruginea)*, American plaice (*Hippoglossoides platessoides*) and adult cod (*Gadus morhua*) are only sufficient to reduce the freezing temperature of their body fluids to approximately −1.1 °C (Scott *et al.*, 1988; Goddard *et al.*, 1994). However, although these fish are undercooled by 0.5 °C during winter, they are in little danger of freezing because they reside at depths where nucleating ice crystals are unlikely to be present (> 40 m in the NW Atlantic, although in other areas, e.g. some regions of the Weddell Sea, ice crystals can be present at depths >200 m). The above observations suggest that, in these species, AF(G)P are either a vestigial adaptation to a geological time where freeze resistance was essential, or they play some other physiological role, possibly in cold adaptation.

Two additional lines of evidence come from laboratory studies. Goddard Morgan and Fletcher (1997) carried out experiments to determine whether the presence of endogenously produced AFGP influenced the behaviour of cod in relation to cold water. Juveniles with, and without, plasma antifreeze were acclimated to 2 °C and then placed in a low temperature thermal gradient of −0.5 to 1.0 °C. As would be expected, both groups of cod preferred the warmer end of the gradient. However, juveniles possessing AFGP explored the cold areas of the gradient more frequently than did those without AFGP. This could indicate that forays into cold water were physiologically less stressful to juvenile cod possessing AFGP. Since juvenile cod overwinter inshore in subzero waters rather than migrate to deeper warmer waters, this result suggests that antifreeze may play a role in the cold-adaptive process.

Despite the suggestive nature of the preceding observations they do not provide evidence of a direct causal relationship between antifreeze and cold adaptation. To address this question directly, Hobbs and Fletcher (1996) examined the influence of intraperitoneally administered AFP (Type I) on the low temperature tolerance of goldfish, a species that does not produce antifreeze. The basic experimental design was to administer AFP to goldfish acclimated to 28 °C and then place them directly in cold water. The results were two-fold; goldfish injected with AFP avoided cold coma for longer periods of time, and they recovered from this coma in greater numbers than controls. It can be concluded from these experiments that AFP improved the ability of goldfish to survive cold shock. However, further studies into the mechanisms of action of AF(G)P are required for us to develop a complete understanding of their physiological role in the fish that produce them in nature.

The winter flounder: a model for understanding antifreeze protein production and gene regulation

Although there are many questions still to be answered with regard to the mechanisms of action of AF(G)P, many aspects of antifreeze production have been examined in a considerable number of teleost species. In particular, the winter flounder (*Pleuronectes americanus*), a species of subpolar/temperate habit, has received intensive investigation. AFP from the winter flounder has been studied in terms of its protein structure and function, gene organisation, gene expression and regulation, tissue distribution, hormonal control of production, seasonal cycle of production, and population differences. For this reason, the flounder serves as an excellent model to better understand the possibly diverse roles played by antifreeze proteins in the survival of teleosts at low temperatures.

The winter flounder is primarily a shallow water (1–30 m) resident of the coastal areas of Atlantic Canada and northeastern United States (Scott & Scott, 1988). In Newfoundland, these fish are subjected to temperatures ranging from 16 °C during late summer to −1.6 °C in February and March. During the winter, the inshore areas bear ice of freshwater and Arctic origins. In addition, ice crystals can be driven deep into the water column during storm activity. Thus, the winter flounder inhabits an environment that would be lethal to most fishes.

The annual cycle of antifreeze production has been studied extensively in the winter flounder. All of the fish in a given population have been found to be highly synchronous in their annual cycle and there is very little year-to-year variation (Fletcher, 1977). Thus, the flounder is, in many ways, an ideal fish for such a study. In Newfoundland, the annual cycle of plasma antifreeze levels correlates closely with the annual cycle of seawater temperatures (Fig. 3). Antifreeze proteins appear in the plasma during November when the water temperature is approximately 4–6 °C, reach peak levels of 10–15 mg/ml during winter, and disappear in May when the temperature usually rises above 0 °C. The peak antifreeze levels during winter reduce the plasma freezing temperature to −1.7 °C. Since blood plasma freezing temperatures are a good indicator of the whole fish freezing temperature, it is evident that AFP increase the flounder's freezing resistance almost to the

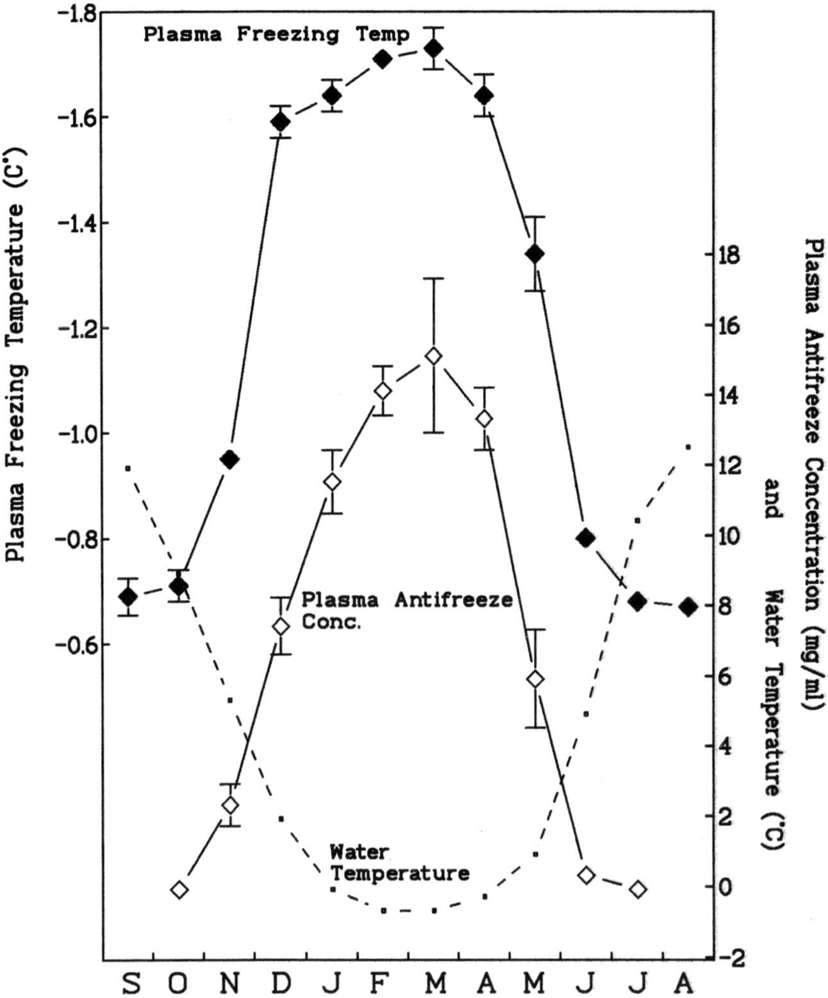

Fig. 3. Annual cycle of blood plasma freezing temperatures and antifreeze protein concentration in the plasma of the Newfoundland winter flounder. Blood samples were obtained from winter flounder collected from the field at monthly intervals. Data points are means of 5–10 samples ± 1 SE.

freezing point of seawater (Fletcher *et al.*, 1986). It is readily apparent from studies of these annual cycles that the winter flounder prepares for winter by producing antifreeze proteins well before they are needed, and that AFP are retained in the fish until the danger of freezing has passed.

The environmental factors regulating this annual cycle have been reviewed in detail by Chan, Fletcher and Hew (1993). Water temperature does not play a major role in initiating, or preventing the initiation of AFP mRNA or AFP synthesis in the fall, nor does it appear to play a role in terminating AFP production in the spring. However, temperature must be sufficiently low (6–8 °C) for AFP mRNA to accumulate and direct the synthesis of normal winter levels of AFP (Fletcher, 1981; Price *et al.*, 1986; Vaisius, Martin-Kearley & Fletcher, 1989).

Photoperiod is the main environmental factor regulating antifreeze production in the winter flounder due to its control, via the central nervous system, of the timing of growth hormone production in the pituitary gland. Growth hormone appears to inhibit antifreeze synthesis in the flounder, and the disappearance of growth hormone from the circulation under the influence of shortening daylength is of critical importance in the timing of initiation of AFP synthesis in the fall (for review see Fletcher *et al.*, 1989) (Fig. 4). Hypophysectomy (removal of the source of growth hormone) at any time of the year results in a dramatic increase in liver AFP mRNA levels followed by the appearance of AFP in the plasma. This process can be reversed in the hypophysectomised flounder by subsequent intramuscular injections of pituitary extracts or purified growth hormone, both of which block transcription of the AFP genes, causing a decline in the levels of liver AFP mRNA, and a concomitant decline in the plasma AFP levels (Fletcher *et al.*, 1989).

The environmental and hormonal control of antifreeze synthesis and accumulation result in an annual cycle of production that correlates very well with the need for antifreeze protection in the flounder. This is also true for other species with different life history patterns. For example, northern Atlantic cod (*Gadus morhua*) living off the coasts of Newfoundland and Labrador have a sporadic need for antifreeze during the winter, depending on choice of overwintering location. Adult cod are highly migratory, and may overwinter offshore (>2 °C) or inshore (subzero). In this species, antifreeze production is controlled primarily by temperature rather than photoperiod (hormonal control is not yet understood). Thus, unlike the flounder which is an obligate producer of antifreeze in winter, adult cod only produce antifreeze when stimulated to do so by exposure to water temperatures declining through 0 °C (Fletcher, King & Kao, 1987).

Antifreeze genes: regulation of antifreeze production at the molecular level

Antifreeze gene dosage: correlation with environment

Research to date suggests a strong positive correlation between antifreeze gene dosage, antifreeze protein levels, and the environmental freezing conditions to which the fish are exposed.

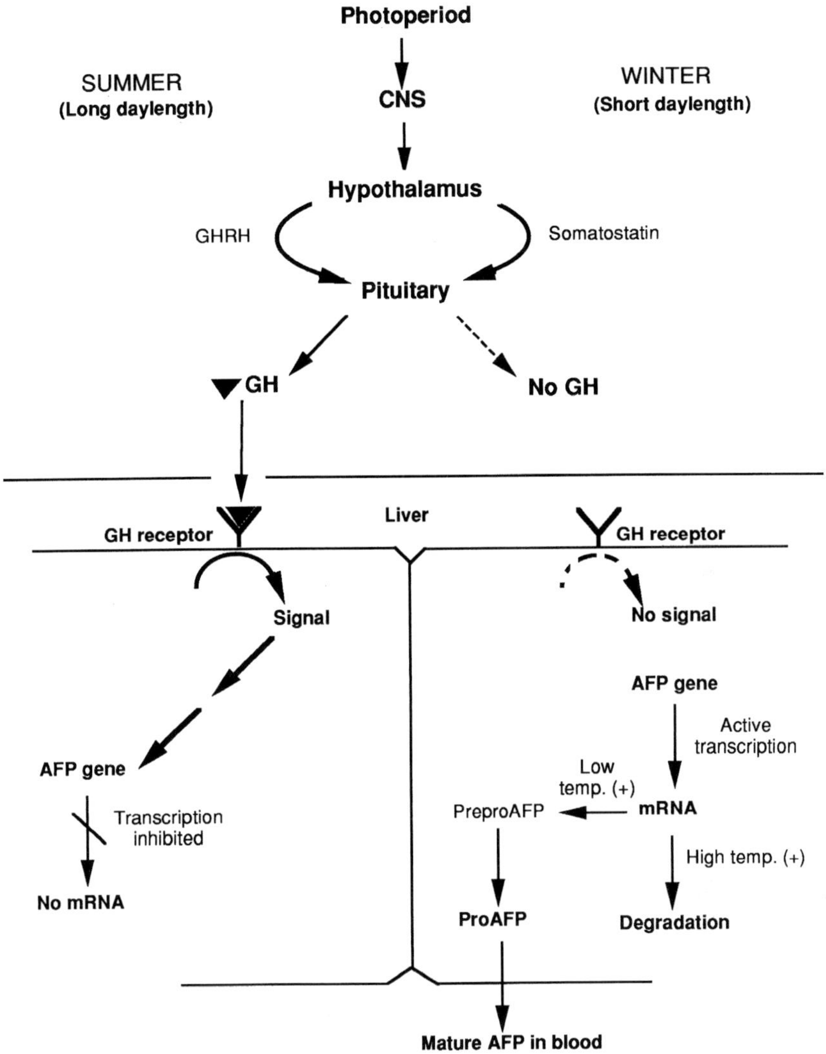

Fig. 4. Model of current thinking on the factors regulating antifreeze production in the liver of the winter flounder. Production and release of growth hormone (GH) by the pituitary is under the positive control of growth hormone releasing hormone (GHRH) and the negative control of somatostatin, both of which are released from the hypothalamus due to the influence of photoperiod on the central nervous system (CNS). GH in turn appears to control the production of antifreeze proteins (AFP) by inhibiting transcription of the AFP genes in the liver. Thus in summer, under the influence of photoperiod, GH is produced and antifreeze is not present in the blood. During the short days of winter, GH production ceases and antifreeze gene transcription and secondary processing can proceed in the liver, resulting in the presence of mature antifreeze in the blood.

Newfoundland winter flounder residing in shallow (<30 m) coastal waters produce high concentrations of plasma antifreeze (10–15 mg/ml) during winter, and have approximately 100 copies of the AFP genes, many of which are arranged as tandem repeats, with the rest linked, but irregularly spaced. The tandem repeat genes code for the most abundant AFP found in the blood plasma (Scott , Hew & Davies, 1985; Chan *et al.*, 1993).

The yellowtail flounder (*Pleuronectes ferrugenia*), a close relative of the winter flounder, is also exposed to subzero temperatures during winter. However, by virtue of its deeper water habitat (50–100 m) it faces little or no danger of freezing by contact with nucleating ice crystals. This species has one-third of the AFP gene copy number observed in the winter flounder (Fig. 5), and produces considerably less plasma AFP (2–4 mg/ml) (Fig. 6) (Scott *et al.*, 1988).

A further, more striking example of the relationship between gene dosage and AFP levels can be found in the ocean pout (*Macrozoarces americanus*). Plasma AFP levels in Newfoundland populations of ocean pout are considerably higher than those found in ocean pout from a more southerly population residing in New Brunswick waters, where sub-zero water temperatures and ice occur considerably less frequently than in Newfoundland (Fig. 7) (Fletcher *et al.*, 1985). Genomic Southern blots reveal that the basis for these population differences in plasma AFP levels is probably AFP gene dosage; Newfoundland ocean pout have approximately 150 gene copies, whereas pout from New Brunswick have 30–40 (Fig.8) (Hew *et al.*, 1988). Again, a good correlation exists between the fishes' antifreeze production system and the need for antifreeze as dictated by environmental conditions.

Antifreeze gene expression in epithelial (barrier) tissues

Several studies indicate that biological membranes can be very effective at blocking ice propagation to undercooled fluids. One such investigation was carried out by Valerio, Kao and Fletcher (1992) wherein the temperatures at which ice would propagate across isolated winter flounder skin were determined using a modified Ussing Chamber. It was demonstrated that not only is fish skin an effective barrier to ice propagation, but that the effectiveness of this barrier can be improved by the addition of antifreeze proteins. Thus, *in vivo*, it is evident that antifreeze proteins can act within epithelial tissues and assist in blocking ice propagation from a subzero, ice-laden environment into the fish. Further evidence that skin may be an important site of antifreeze action comes from reports demonstrating the presence of AFP in the skin, but not in the blood plasma, of European shorthorn sculpin (*Myoxocephalus scorpius*) and cunner (*Tautogolabrus*

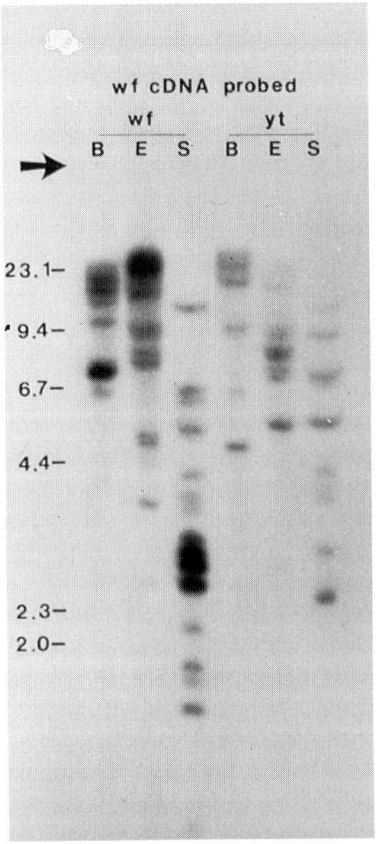

Fig. 5. Genomic Southern blots of winter flounder (wf) and yellowtail flounder (yt) DNA showing AFP gene dosage of the two species. Genomic DNA was digested with restriction enzymes Bam HI (B), Eco RI (E), or Sst I (S) and probed with a winter flounder AFP cDNA clone. The arrow indicates the origin of the gel, and numerals on the left indicate DNA size fragments (kilobases).

adspersus) exposed to cold (Schneppenheim & Theede, 1982; Valerio, Kao & Fletcher, 1990).

One surprising outcome of the study by Valerio *et al.* (1992) was that the skin ice propagation temperature could be as much as 1 °C lower than the plasma and skin freezing points. Indeed, during winter, ice propagation temperatures are lower than the freezing point of sea water (Fig. 9). Since the temperature at which winter flounder freeze in the presence of ice is essentially the same as that of their blood plasma, it is evident that the nucleating ice crystals must enter the fish at a site other than the skin, the

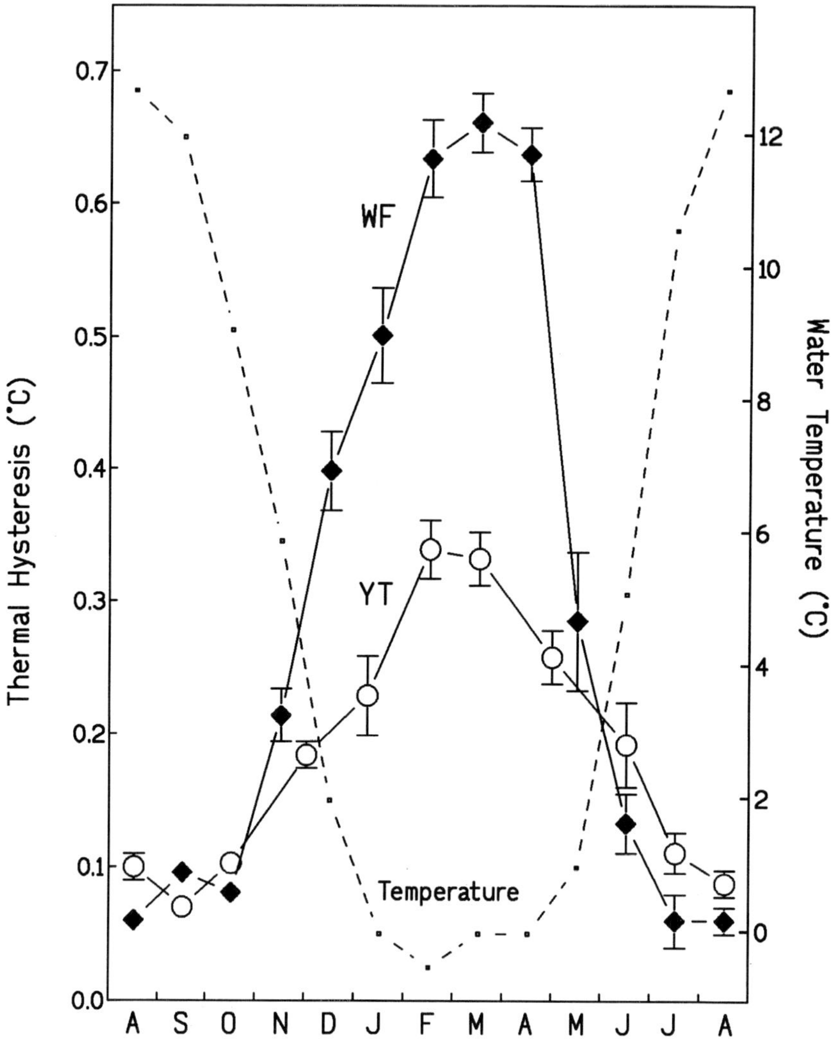

Fig. 6. Annual cycle of antifreeze production in the winter flounder (WF) and the yellowtail flounder (YT). Values of thermal hysteresis (°C) obtained as in Fletcher *et al.*, 1985, are presented as means ± 1 SE.

most likely route being via the gills. One hypothesis put forward to explain the difference between the skin freezing point and its ice propagation temperature was that the solute concentration in the interstitial space of the epithelial cell layer was greater than that of the skin extracellular space overall. The possibility that one such solute could be AFP prompted us to

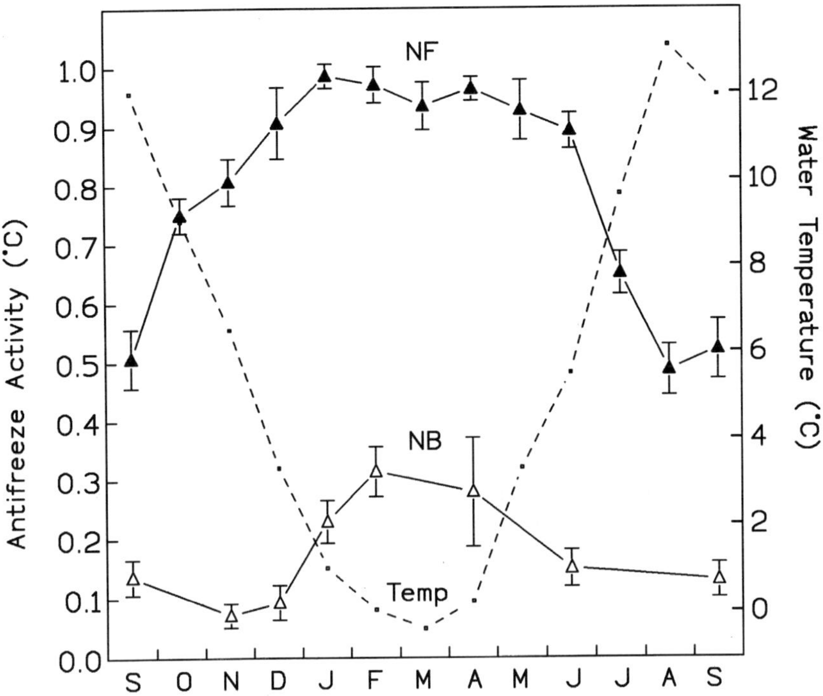

Fig. 7. Antifreeze production in ocean pout collected from Newfoundland (NF) and New Brunswick (NB) coastal waters and held under identical conditions of temperature and photoperiod for an annual cycle. Values of antifreeze activity (°C) obtained as in Fletcher *et al.*, 1985, are presented as means ± 1 SE.

examine epithelial tissues for antifreeze more closely than we had in the past.

There is abundant evidence to indicate that the liver is the major site of plasma AFP synthesis in the winter flounder (Hew *et al.*, 1986). However, observations indicating that AFP are present in flounder skin during July when they are absent from the plasma (Valerio *et al.*, 1992) prompted us to investigate the possible expression and differential regulation of AFP genes in tissues other than the liver.

The expression of AFP mRNA transcripts was examined in a variety of winter flounder tissues by Northern blot analysis using a liver AFP cDNA probe (Gong, Fletcher & Hew, 1992). AFP mRNA was detected in most tissues examined, indicating that AFP synthesis was widespread throughout the fish (Fig. 10). However, the abundance of AFP mRNA in the liver was at least 10–20 times greater than that found in any of the

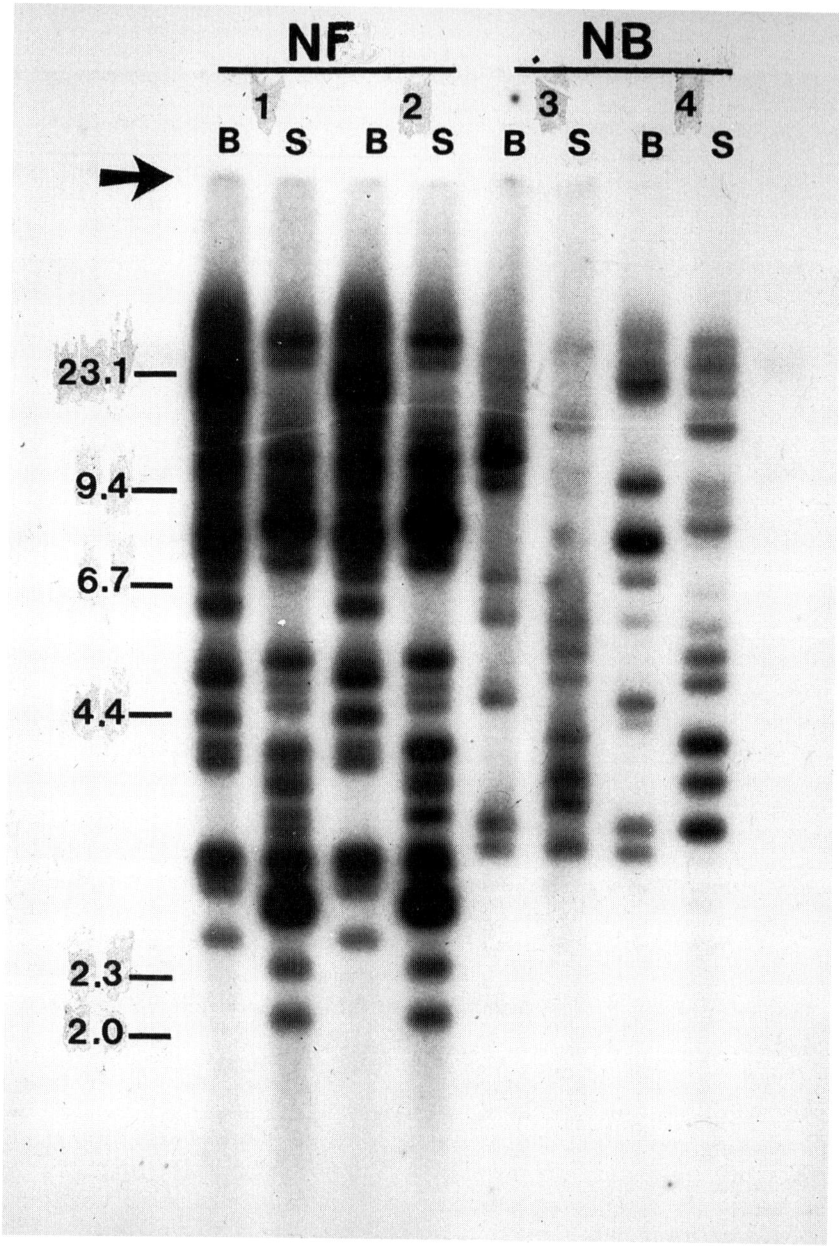

Fig. 8. Ocean pout antifreeze genes: population differences in the antifreeze multigene family. Southern blot of testes DNA from two Newfoundland (NF) and two Passamaquoddy Bay, New Brunswick (NB) ocean pout digested with restriction enzymes Bam HI (B) or Sst (S) and investigated with a cloned ocean pout antifreeze protein cDNA probe (Li *et al.*, 1985). Arrow indicates origin. Numbers on the left indicate DNA size fragments (kilobases).

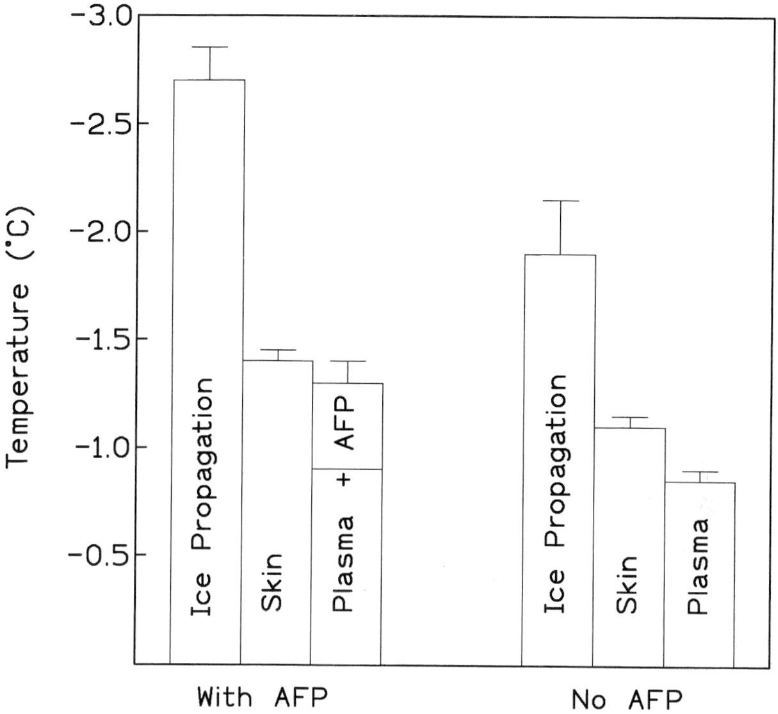

Fig. 9. Temperatures at which ice propagates through fish (winter flounder, *Pleuronectes americanus*) skin, and the freezing points of skin tissues and plasma. Values (presented as means ± 1 SE) were measured in two groups of fish (N=10 in each group), one with antifreeze protein (AFP) present in the plasma, and one in which it was absent. Winter flounder were sampled during the spring and were separated into two groups based on analysis of antifreeze activity in the plasma (see Valerio, Kao & Fletcher, 1992). In the group with antifreeze, the contribution of antifreeze to overall plasma freezing point [plasma+AFP] is represented by the upper portion of the histogram.

other tissues, indicating that the liver is likely to be the major source of plasma AFP.

Among the remaining tissues, the highest AFP mRNA levels were associated with external epithelia: gill filaments, fins, scales, and skin. Moreover, the levels of AFP mRNA observed suggested them to be amongst the more abundant mRNA species in these tissues, and thus of considerable physiological significance.

An examination of the AFP mRNA levels in gill filaments over an annual cycle revealed a seasonal cycle similar to that observed in the liver, with peak

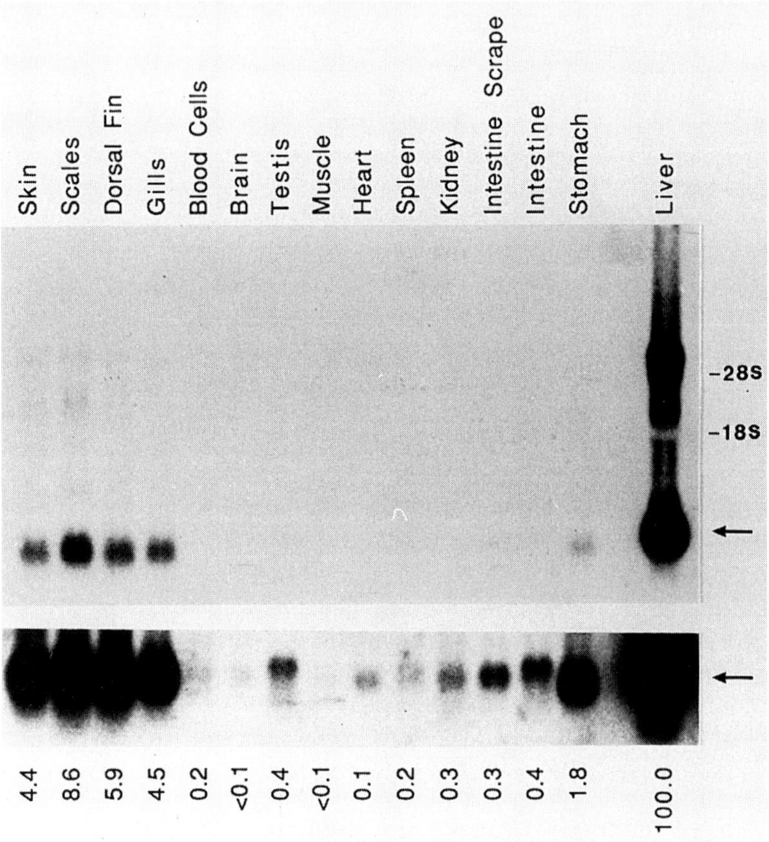

Fig. 10. Expression of AFP mRNA in 15 different winter flounder tissues. Total RNAs (10 μg RNA per lane) were assayed by Northern blot analysis using a radioactive liver AFP cDNA probe. For quantification, the 0.5 kb RNA bands, as indicated by the arrow, were excised and radioactivity was counted in a scintillation counter. The relative concentration of AFP mRNA for each tissue, calculated as a percentage of AFP mRNA in the liver, is indicated under each lane. The upper panel represents a 1 h autoradiogram of the membrane, while the lower panel is a 19 h autoradiogram of the same membrane showing only the 0.5 kb transcript (arrow again indicates 0.5 kb band). The 28S and 18S ribosomal RNA bands are indicated on the right.

levels occurring in December and minimum values in July. However, the magnitude of seasonal change differed considerably between the tissues. Liver AFP mRNA levels in December were at least 300 times greater than in July, whereas gill AFP mRNA levels showed only a ten–fold change over the same period. This difference suggested that the liver and gill AFP mRNA are not regulated in the same way.

Table 2. *Factors by which AFP mRNA levels in various tissues are increased after hypophysectomy*

Tissue	X-Fold Increase
Liver	42.2
Gill	2.5
Heart	1.9
Spleen	0.9
Intestine	2.6
Kidney	1.9
Skin	0.6
Scales	2.2

A further look into the regulation of AFP mRNA levels was carried out using hypophysectomised winter flounder. Consistent with studies discussed earlier, hypophysectomy (and the resultant drop in growth hormone levels) produced a considerable increase in liver AFP mRNA levels (40-fold). However, there was little or no change in the AFP mRNA levels in any of the other tissues examined (Table 2). This result supports the hypothesis that the mechanisms controlling expression of AFP mRNA levels in the epithelial tissues are different from the control mechanisms (photoperiod, growth hormone levels) operating in the liver (Gong *et al.*, 1995).

Skin-type antifreeze proteins and genes

The expression of AFP mRNA in skin epithelia of the winter flounder, *Pleuronectes americanus* (the species studied in all the following investigations), and the evidence indicating that its regulation differs from that of liver AFP mRNA, suggested the possibility that skin AFP are distinct polypeptides encoded by a different set of genes. This prompted us to examine the skin AFP and their mRNA in detail, in order to compare them with liver AFP mRNA.

Earlier studies of liver AFP mRNA in the winter flounder can be summarised as follows: the major plasma AFP are synthesised in the liver as preproAFP, precursor proteins consisting of 82 amino acid residues. The pre-sequence (a 23 amino acid secretory signal peptide) is removed cotranslationally, and the proAFP is secreted into the blood where the 22-amino acid prosequence including the C-terminal glycine is removed within 24 hours to yield a 37 amino acid mature AFP (Fig. 11) (Hew *et al.*, 1986).

L AFP HPLC-8 *m a l s l f t v g g l i f l f w t m r i t e a* s p d p a a k a a p a a a a a p a

L AFP HPLC-8 a a a p D T A S D A A A A A A L T A A N A A A A A K L T

S AFP1 M D A P A R A A A A T A A A A K A A A E A T

L AFP HPLC-8 A D N A A A A A A A T A R g

S AFP1 K A A A A K A A A A T K A A A H

Fig. 11. Amino acid sequence comparison of liver type (underlined) and skin type (not underlined) AFP from the winter flounder. Liver type AFP (HPLC-8), shows the pre- (lower case italics) and pro- (lower case) sequences followed by the mature protein (upper case). Skin type antifreeze (sAFP1 – Gong *et al.*, 1996) lacks a pre- and pro-sequence, the mature protein being produced directly within the cell. Areas of sequence homology are boxed for ease of identification.

Analysis of a skin cDNA library revealed the presence of AFP mRNA that differed considerably from any described for the liver. Nine distinct cDNA clones were sequenced, most of which encoded polypeptides of 37–40 amino acids that were homologous to mature AFP found in plasma. However, unlike the liver AFP mRNA, the skin AFP mRNA lacked pre- and pro-sequences (see Fig. 11). Amino acid sequencing of AFP purified from skin extracts demonstrated that several were identical to those encoded by skin AFP cDNA clones. Thus, the AFP mRNA in skin are translated to produce AFP products (Gong *et al.*, 1996). Moreover, the results of the seasonal study of AFP mRNA levels suggest that AFP are synthesised in gill epithelia throughout the year (Gong *et al.*, 1995).

With the availability of skin-specific, as well as liver-specific AFP cDNA probes, the tissue distribution of these two types of AFP mRNA could be examined in detail. The results demonstrated that liver type AFP mRNA expression was largely confined to liver tissue. In contrast, skin type AFP mRNA was present in all tissues examined, including liver. However, by far the most abundant expression was evident in the exterior epithelia (Gong *et al.*, 1996).

The difference in primary structure between the skin and liver type AFP, together with their differing tissue specific expression indicates the presence of two sets of AFP genes present in the winter flounder genome: a liver type, and a skin type. Furthermore, the discovery of nine different skin type AFP cDNA clones in the skin cDNA library suggests that the skin type AFP belong to a multigene family similar to that of the liver type AFP genes.

In order to confirm this conclusion, winter flounder genomic DNA was examined using Southern blot analysis. Two identical blots were prepared and probed with either a liver type or a skin type AFP probe. Both probes hybridised to multiple bands, indicating the presence of multiple genes. The skin probe hybridised to most, if not all of the liver AFP DNA fragments, however, there were also many skin AFP specific fragments (Fig. 12). Thus, it is evident that the skin type AFP are encoded by a multiple gene family that is similar in number (30–40) to that of the liver type AFP gene family.

Skin-type antifreeze: physiological significance

Skin AFP were isolated from winter flounder scale tissue, a tissue containing substantial amounts of extracellular space. It is, therefore, impossible to know the physiological concentration of AFP within the tissue component parts since we do not know exactly where it is located. However, the abundance of AFP mRNA present in skin tissue suggests that substantial amounts of AFP are produced.

Since the skin AFP are active in terms of binding to ice crystals and depressing the freezing temperature non-colligatively, it seems likely that these AFP are concentrated in the exterior epithelial cell layer as a first line of defence against the propagation of ice from the environment into the fish. The lack of a sequence for a secretory signal peptide (pre-sequence) associated with the skin AFP genes suggests that the skin type AFP remain and function intracellularly (Gong *et al.*, 1996). If this is the case, then what role are these AFP playing in protecting the fish at low temperatures? It seems unlikely that they are needed to prevent ice propagation from the environment into the epithelial cells themselves since cells are well protected, due to the nature of the cell membrane and the spatial requirements for ice crystal growth (see Valerio *et al.*, 1992). They could be functioning to stabilise the membranes of intracellular organelles, or to reduce the possibility of cell damage due to spontaneous ice nucleation within undercooled cells. At low temperatures, all cells in the body would be undercooled to the same extent, and thus, one would expect to find intracellular AFP in every cell if its primary function were intracellular. Skin-type AFP mRNA has been detected in liver, stomach, intestine, spleen, kidney, skin, scales, fins, and gills. This ubiquitous distribution suggests that skin type AFP play a role in the overall cold adaptation of the fish. However, the high concentrations of skin type AFP mRNA found in the external epithelia point to a more specific function in tissues constantly exposed to intimate contact with external ice crystals.

Since ice propagation across epithelial cell layers is probably limited

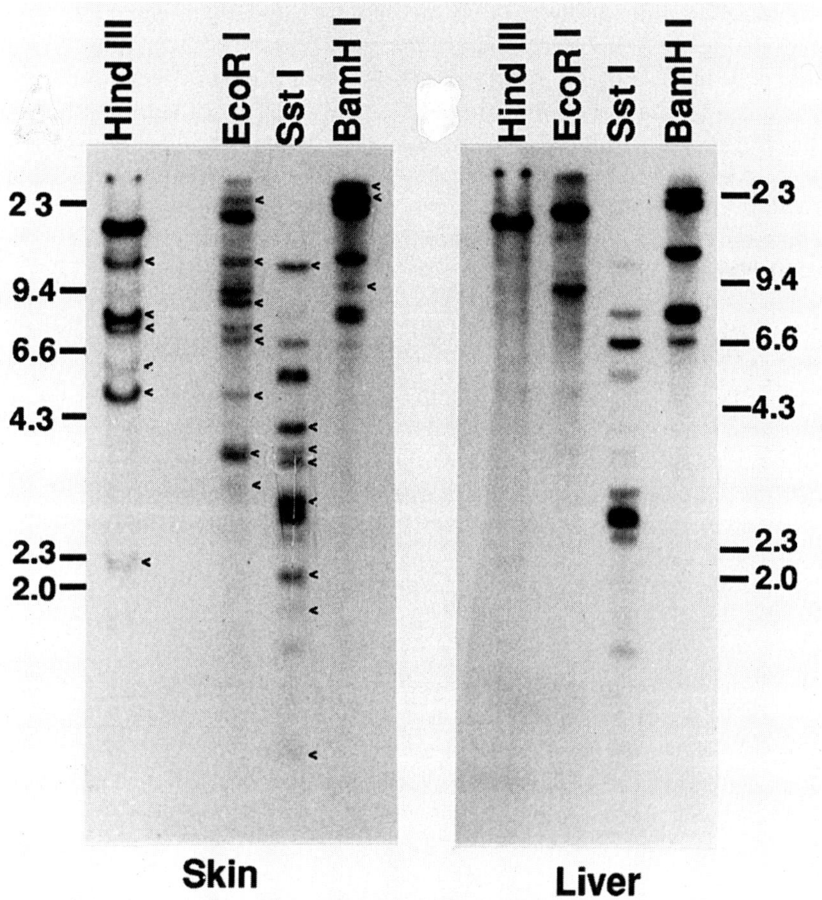

Fig. 12. Genomic Southern blot analysis of the winter flounder AFP gene family. Genomic DNA (testis) was cut by four different restriction enzymes: *Hin*dIII, *Eco*RI, *Sst*I, and *Bam*HI. Two identical blots were made and probed with skin- type and liver-type AFP probes, respectively. Both probes hybridised to multiple bands. While the liver AFP probe only hybridised to a limited number of DNA fragments, the skin probe recognised most, if not all liver AFP fragments in addition to some fragments specific to the skin (indicated by *carets*). Skin-specific fragments are defined as those absent from, or appearing only weakly in the liver blot. In total there appear to be approximately 80–100 copies of the AFP genes in the flounder genome (Gong *et al.*, 1995). Numbers on left and right indicate DNA size fragments (kilobases).

to the intercellular spaces, it seems logical to suppose that high concentrations of AFP located in these spaces would be of maximum benefit to fish in icy, low temperature environments. Thus, an additional function for skin AFP can be postulated if one assumes that a mechanism exists whereby the AFP makes its way out of the cell and into these intercellular spaces. The externalisation of a variety of proteins lacking signal peptides has been described by Mignatti, Morimoto and Rifkin (1992). Thus, although the skin AFP lack a secretory signal peptide, it is possible that these small polypeptides use an alternative pathway for secretion out of the cell, and function to lower the freezing temperature in the interstitial space between the skin and gill epithelial cells. This could account, at least in part, for the effectiveness of the skin as a barrier to ice propagation (see Fig. 9) (Valerio *et al.*, 1992). Quite apart from assisting the epithelium in blocking ice propagation into the fish, the chief function of the skin type AFP may be to protect the epithelial cells themselves from any damaging effects of ice formation in the intercellular space. Damage to gill epithelial cells could have particularly far reaching consequences for the fish.

Conclusions

Antifreeze proteins or glycoproteins are produced in the liver and secreted into the blood, from whence they are distributed throughout the extracellular space, thus surrounding the body cells with a fluid designed to inhibit the propagation and growth of ice crystals. Until recently, it was generally believed that synthesis of the antifreezes was confined almost exclusively to the liver, and that the sole physiological function of the antifreeze proteins and glycoproteins was to protect marine teleosts from freezing by reducing the freezing temperature of their extracellular fluids to levels mirroring those of their environment.

However, the recent discovery in the winter flounder of a second set of antifreeze genes that are principally expressed in external epithelia suggests that a central supply of AFP from the liver might not afford sufficient protection to cell layers that come into direct contact with ice. Exactly how these skin type AFP protect epithelia remains to be determined. However, the lack of a signal peptide in the primary sequence suggests that they remain and function intracellularly.

In addition to conferring freeze protection to the whole fish, and protecting epithelial cell layers, the evidence indicating that AF(G)P can interact with cell membranes and protect them from cold (rather than freeze) damage suggests that these proteins may also play a role in the cold acclimation process.

These recent discoveries have given new insights into the physiological functions of the antifreeze proteins and glycoproteins. However, they have also revealed that we are further than we thought from attaining a complete understanding of their mechanisms of action.

Acknowledgements

The authors would like to thank Drs F. Tablin, J. H. Crowe, and L.M. Hays for providing us with copies of their manuscripts on the effects of AF(G)P on liposomes and platelets prior to publication. Thanks, also, to Linda Mark for preparation of this manuscript. Research supported by NSERC and MRC Canada.

References

Chan, S. L., Fletcher, G. L. & Hew, C. L. (1993). Control of antifreeze protein gene expression in winter flounder. *In Biochemistry and Molecular Biology of Fishes,* ed. P.W. Hochachka & T.P. Mommsen, vol. 2, pp. 293–305 Elsevier Science Publishers.

Davies, P. L. & Hew, C. L. (1990). Biochemistry of fish antifreeze proteins. *FASEB Journal*, **4,** 2460–8.

Davies, P. L., Ewart, K. V. & Fletcher, G. L. (1993). The diversity and distribution of fish antifreeze proteins: new insights into their origins. *In Biochemistry and Molecular Biology of Fishes*, ed. P.W .Hochachka & T.P. Mommsen, vol. 2, pp. 279–91. Elsevier Science Publishers.

DeVries, A. L. (1988). The role of antifreeze glycopeptides and peptides in the freezing avoidance of Antarctic fishes. *Comparative Biochemistry and Physiology*, **90B,** 611–21.

DeVries, A. L. & Wohlschlag, D. E. (1969). Freezing resistance in some Antarctic fishes. *Science*, **163**, 1074–5.

DeVries, A. L., Komatsu, S. K. & Feeney, R. E. (1970). Chemical and physical properties of freezing point-depression glycoproteins from Antarctic fishes. *Journal of Biological Chemistry*, **245**, 2901–13.

Ewart, K. V. & Fletcher, G. L. (1990). Isolation and characterization of antifreeze proteins from smelt (*Osmerus mordax*) and Atlantic herring (*Clupea harengus harengus*). *Canadian Journal of Zoology*, **68**, 1652–8.

Ewart, K. V., Rubinsky, B. & Fletcher, G. L. (1992). Structural and functional similarity between fish antifreeze proteins and calcium-dependent lectins. *Biochemical and Biophysical Research Communications*, **185**, 335–40.

Ewart, K. V., Yang, D. S. C., Ananthanarayanan, V. S., Fletcher, G. L. & Hew C. L. (1996). Ca^{2+}-dependent antifreeze proteins. *Journal of Biological Chemistry*, **271,** 16,627–32.

Fletcher, G. L. (1977). Circannual cycles of blood plasma freezing point and Na^+ and Cl^- concentrations in Newfoundland winter flounder (*Pseudopleuronectes americanus*): correlation with water temperature and photoperiod. *Canadian Journal of Zoology*, **55,** 789–95.

Fletcher, G. L. (1981). Effects of temperature and photoperiod on the plasma freezing point depression, Cl^- concentration, and protein 'antifreeze' in winter flounder. *Canadian Journal of Zoology*, **59,** 193–201.

Fletcher G. L., Hew, C. L. & Joshi, S. B. (1982). Isolation and characterization of antifreeze glycopeptides from the frostfish, *Microgadus tomcod. Canadian Journal of Zoology*, **60,** 348–55.

Fletcher, G. L., Hew, C. L., Li, X. M., Haya, K. & Kao, M. H. (1985). Year-round presence of high levels of plasma antifreeze peptides in a temperate fish, ocean pout (*Macrozoarces americanus*). *Canadian Journal of Zoology*, **63**, 488–93.

Fletcher, G. L., Idler, D. R., Vaisius, A. & Hew, C. L. (1989). Hormonal regulation of antifreeze protein gene expression in winter flounder. *Fish Physiology and Biochemistry*, **7**, 387–93.

Fletcher, G. L., Kao, M. H. & Fourney, R. M. (1986). Antifreeze peptides confer freezing resistance to fish. *Canadian Journal of Zoology*, **64**, 1897–901.

Fletcher, G. L., King, M. J. & Kao, M. H. (1987). Low temperature regulation of antifreeze glycopeptide levels in Atlantic cod (*Gadus morhua*). *Canadian Journal of Zoology*, **65,** 227–33.

Goddard, S. V., Wroblewski, J. S., Taggart, C. T., Howse, K. A., Bailey, W. L., Kao, M. H. & Fletcher, G. L. (1994). Overwintering of adult Northern Atlantic cod (*Gadus morhua*) in cold inshore waters as evidenced by plasma antifreeze glycoprotein levels. *Canadian Journal of Fisheries and Aquatic Sciences*, **51**, 2834–42.

Goddard, S. V., Morgan, M. J. & Fletcher, G. L. (1997). The influence of plasma antifreeze glycoproteins on temperature selection by Atlantic cod (*Gadus morhua*) in a thermal gradient. *Canadian Journal of Fisheries and Aquatic Sciences*, **54**, Suppl. No. 1, 88–93.

Gong, Z., Fletcher, G. L. & Hew, C. L. (1992). Tissue distribution of fish antifreeze protein mRNAs. *Canadian Journal of Zoology*, **70**; 810–14.

Gong, Z., King, M. J., Fletcher, G. L. & Hew, C. L. (1995). The antifreeze protein genes of the winter flounder, *Pleuronectes americanus*, are differentially regulated in liver and non-liver tissues. *Biochemical and Biophysical Research Communications*, **206**, 387–92.

Gong, Z., Ewart, K. V., Hu, Z., Fletcher, G. L. & Hew, C. L. (1996). Skin antifreeze protein genes of the winter flounder, *Pleuronectes americanus*, encode distinct and active polypeptides without the secretory signal and prosequences. *Journal of Biological Chemistry*, **271,** 4106–12.

Gordon, M. S., Andur, B. N. & Scholander, P. F. (1962). Freezing resistance in some northern fishes. *Biological Bulletin*, **122**, 52–62.

Hays, L. M., Feeney, R. E., Crowe, L. M., Crowe, J. H. & Oliver, A. E. (1996). Antifreeze glycoproteins inhibit leakage from liposomes during thermotropic phase transitions. *Proceedings of the National Academy of Sciences*, USA, **93,** 6835–40 .

Hew, C. L. & Fletcher, G. L. (1985). Biochemical adaptation to the freezing environment – structure, biosynthesis, and regulation of fish antifreeze polypeptides. In *Circulation, respiration and metabolism*, ed. R. Gilles, pp. 553–63. Berlin and Heidelberg: Springer-Verlag.

Hew, C. L. & Yang, D. S. C. (1992). Protein interaction with ice. *Journal of Biochemistry*, **203,** 33–42.

Hew, C. L., Slaughter, D., Joshi, S. B., Fletcher, G. L. & Ananthanarayanan, V. S. (1984). Antifreeze polypeptides from the Newfoundland ocean pout, *Macrozoarces americanus:* presence of multiple and compositionally diverse components. *Journal of Comparative Physiology*, **155B**, 81–8.

Hew, C. L., Wang, N. C., Joshi, S., Fletcher, G. L., Scott, G. K., Hayes, P. L., Buettner, B. & Davies, P. L. (1988). Multiple genes provide the basis for antifreeze protein diversity and dosage in the ocean pout *Macrozoarces americanus*. *Journal of Biological Chemistry*, **263**, 12049–55.

Hew, C. L., Wang, N. C., Yan, S., Cai, H., Sclater, A., & Fletcher, G. L. (1986). Biosynthesis of antifreeze polypeptides in the winter flounder – characterization and seasonal occurrence of precursor polypeptides. *European Journal of Biochemistry*, **160**, 267–72.

Hobbs, K. D. & Fletcher, G. L. (1996). Antifreeze proteins increase the cold tolerance of goldfish. *Bulletin of the Canadian Society of Zoologists*, **27**, 64.

Hochachka, P. W. (1986). Defence strategies against hypoxia and hypothermia. *Science*, **231**, 234–41.

Hochachka, P. W. (1988*a*). Metabolic-, channel-, and pump-coupled functions: constraints and compromises of coadaptation. *Canadian Journal of Zoology*, **66**, 1015–27.

Hochachka, P. W. (1988*b*). Channels and pumps – determinants of metabolic cold adaptation strategies. *Comparative Biochemistry and Physiology*, **90B**, 515–19.

Mignatti, P., Morimoto, T. & Rifkin, D. B. (1992). Basic fibroblast growth factor, a protein devoid of secretory signal sequence, is released from cells via a pathway independent of the endoplasmic reticulum–Golgi complex. *Journal of Cellular Physiology*, **151,** 81–93.

Negulescu, P. A., Rubinsky, B., Fletcher, G. L. & Machen, T. E. (1992). Fish antifreeze proteins block Ca^{++} entry into rabbit parietal cells. *American Journal of Physiology*, **263**, C1310– 13.

Ng, N. F., Trinh, K. Y. & Hew, C. L. (1986). Structure of an antifreeze polypeptide from the sea raven (*Hemitripterus americanus*). *Journal of Biological Chemistry* **261**, 15690– 5.

Price, J. L., Gourlie, B. B., Lin, Y. & Huang, R. C. C. (1986). Induction of winter flounder antifreeze protein messenger RNA at 4 °C *in vivo* and *in vitro*. *Physiological Zoology*, **59**, 679–95.

Rubinsky, B., Arav, A. & Fletcher, G. L. (1991). Hypothermic protection – a fundamental property of 'antifreeze' proteins. *Biochemical and Biophysical Research Communications*, **180**, 566–71.

Schneppenheim, R. & Theede, H. (1982). Freezing-point depressing peptides and glycoproteins from Arctic-boreal and Antarctic fishes. *Polar Biology*, **1**, 115–23.

Scholander, P. F., VanDam, L., Kanwisher, J. W., Hammell, H. T. & Gordon, M. S. (1957). Supercooling and osmoregulation in Arctic fish. *Journal of Cellular and Comparative Physiol*ogy **49**, 5–24.

Scott, W. B. & Scott, M. G. (1988). Atlantic fishes of Canada. *Canadian Bulletin of Fisheries and Aquatic Sciences*, **219**, 731.

Scott, G. K., Davies, P. L., Kao, M. H. &. Fletcher, G. L. (1988). Differential amplification of antifreeze protein genes in the Pleuronectinae. *Journal of Molecular Evolution*, **27**, 29–35.

Scott, G. K., Hew, C. L. & Davies, P. L. (1985). Antifreeze protein genes are tandemly linked and clustered in the genome of the winter flounder. *Proceedings of the National Academy of Sciences*, USA, **82**, 2613–17.

Sicheri, F. & Yang, D. S. C. (1995). Ice-binding structure and mechanism of an antifreeze protein from winter flounder. *Nature*, **375,** 427–31.

Sonnichsen, F. D., Sykes, B. D., Chao, H. & Davies, P. L. (1993). The non-helical structure of antifreeze protein type III. *Science* **259,** 1154–7.

Tablin, F., Oliver, A. E., Walker, N. J., Crowe, L. M. & Crowe, J. H. (1996). Membrane phase-transition of intact human platelets – correlation with cold-induced activation. *Journal of Cellular Physiology*, **168,** 305–13.

Valerio, P. F., Kao, M. H. & Fletcher, G. L. (1990). Thermal hysteresis activity in the skin of cunner, *Tautogolabrus adspersus*. *Canadian Journal of Zoology*, **68**, 1065–7.

Valerio, P. F., Kao, M. H. & Fletcher, G. L. (1992). Fish skin: an effective barrier to ice crystal propagation. *Journal of Experimental Biology*, **164,** 135–51.

Vaisius, A., Martin-Kearley, J. & Fletcher, G. L. (1989). Antifreeze protein gene transcription in winter flounder is not responsive to temperature. *Cellular and Molecular Biology*, **35**, 547–54.

Yang, D. S. C., Sax, M., Chakrabartty, A. & Hew, C. L. (1988). Crystal structure of an antifreeze polypeptide and its mechanistic implications. *Nature*, **333**, 232–7.

A.P.A. WÖHRMANN

Antifreeze glycopeptides and peptides in the freezing avoidance of Weddell Sea fishes: its relation to mode of life, depth distribution and evolution

Low temperatures have pervaded the Southern Ocean for some 40 million years (Kennett, 1977). Recent data suggest relatively constant low temperatures (+3 °C to −2 °C, Hellmer & Bersch, 1985) for the last 13 million years (Eastman & Grande, 1989). This phenomenon may have led to the high degree of stenothermy and endemism of the fish fauna (DeVries & Eastman, 1981). The temperature of large portions of Southern Ocean is less than 0 °C with only small differences between water temperatures in winter and summer. South of 60° S there is usually less than 2 °C difference in the mean surface water temperature between the warmest and coldest months (Deacon, 1984), and the water masses throughout most of the Southern Ocean, from the surface to great depths, differ by only 4–5 °C (Knox, 1970). Close to the continental shelf waters are always at or below the freezing point of seawater (−1.86 °C) and are thermally stable throughout the year.

A variety of physical characteristics of cold sea-water has certainly influenced the evolutionary adaptation of the fish fauna in the Southern Ocean. However, it must be considered that over 90% of the global seawater is below 5 °C. Therefore, for marine organisms, cold water is not an unusual environment and low temperature as such is unlikely to be a limiting factor for biosynthetic processes (for review, see Clarke, 1983, 1987). In addition to low temperature, the continual presence of ice in subzero water presents fishes with another physiological challenge. Coping with the presence of ice has been a significant factor in the evolutionary adaptation of fishes. The presence of antifreeze peptides (AFP) and glycopeptides (AFGP) in fish is one important adaptation for survival in freezing seawater. Because of this, Antarctic fishes tend to occupy most ecological niches of the Antarctic Ocean including the surface and midwaters which are often rich in food and also ice-laden year round (for review see DeVries, 1988; Cheng & DeVries, 1991; Eastman, 1993).

It has been shown in earlier investigations that the content of AFGPs depends upon the habitats of the fishes (for review see DeVries, 1988; Cheng & DeVries, 1991). Species like the nototheniid *Trematomus pennellii* as well as the channichthyid *Pagetopsis macropterus* or the bathydraconid *Gymnodraco acuticeps* possess large amounts of AFGP. There is a correlation between the freezing point of the blood, the concentration and composition of antifreezes, the freezing point of fish in the presence of ice and the habitat zonation of the fish (e.g. water depth). Fishes in shallow waters, which may come into contact with anchor ice, have high concentrations of both high and low molecular weight AFGP. Fishes freeze at lower temperatures than those living in ice-free deep water such as the notothenioids *Neopagetopsis ionah*, *Pogonophryne* spp., and the non-notothenioid species (e.g. *Macrourus holotrachys* or *Lycenchelys hureaui*) all of which will freeze in surface water at −1.9 °C. In their natural deep water habitat, these species are in little danger of freezing because of the higher temperature (e.g. +0.5 °C on the continental slope in the Lazarev Sea) and because hydrostatic pressure lowers their freezing point to −2.3 °C at 500 m depth, a temperature well below that of seawater. Nevertheless, antifreeze molecules could be found in all species of fish investigated (Wöhrmann, 1993, 1996).

The aim of this chapter is to review the characteristics of antifreezes as one of the components found in the Antarctic marine fauna, the Weddell Sea fishes. The discussion is supported by results from published literature as well as the author's own experimental work. Special attention is paid to the vertical distribution, the mode of life, metabolism and the evolution of these teleosts in relation to their habitat, in an attempt to substantiate correlations between these parameters and processes and antifreeze properties.

The structure of antifreeze

Glycopeptide and peptide antifreeze compounds have evolved in a number of unrelated lineages of cold water teleosts including notothenioids, zoarcids, cottids, gadids, pleuronectids, clupeids and osmerids (Davies, Hew & Fletcher, 1988; DeVries, 1988; Cheng & DeVries; 1991; Kao *et al.* 1986; Ewart & Fletcher, 1990; Wöhrmann, 1996).

Over the past 25 years, DeVries and colleagues have isolated antifreezes in 19 species of notothenioids (Eastman, 1993) and in three non-notothenioid Antarctic species. The antifreeze glycopeptides consist of a series of eight distinct glycopeptides ranging in molecular weight from 2600 to 33,700 daltons (Fig. 1). For convenience, they are numbered and classified as large (1–5; 33.7 – 10.5 kDa) and small (6–8; 7.9 – 2.6 kDa) molecules. The primary structure of these antifreeze glycopeptides, abbreviated as AFGPs, of all investigated species is made up of a repeat of the tripeptide unit alanyl–alanyl–threonine, in which the disaccharide β-D-galactopyranosyl-(1→3)-2-

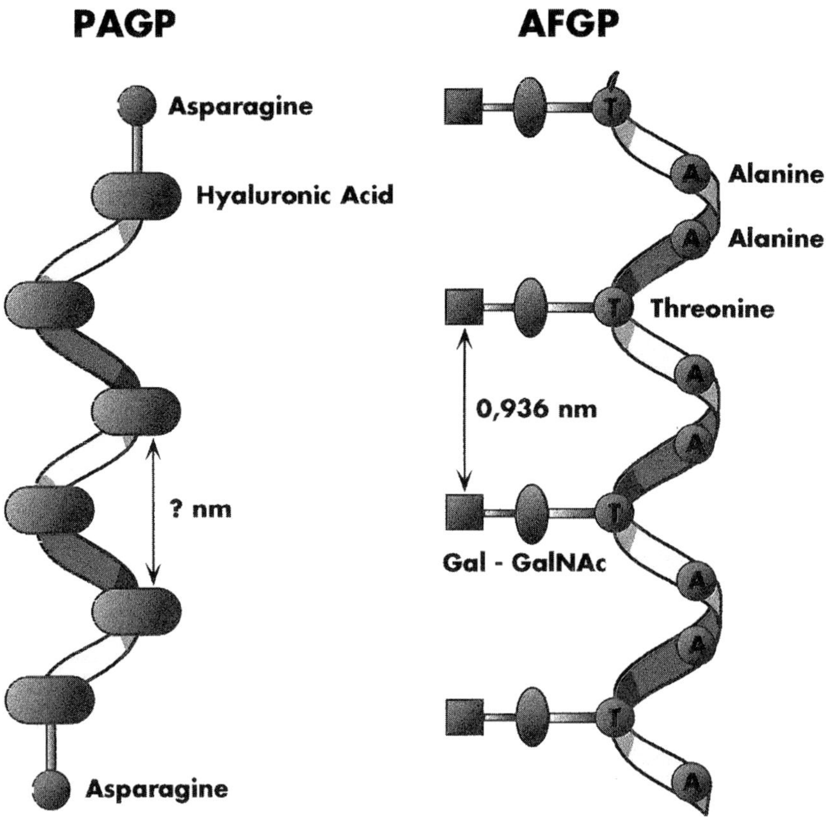

Fig. 1. Model of the secondary structure of the antifreeze glycopeptides (AFGPs) of notothenioids and of the *Pleuragramma*-antifreeze glycopeptide (PAGP). The AFGP consists of amino acids in the sequence [alanyl-alanyl-threonine]$_n$. Each threonine is bound to a disaccharide through glycosidic linkage. In low molecular weight AFGPs 6–8 (MW 7900 – 2668 daltons), proline is periodically substituted for alanine at position 1 of the tripeptide. In PAGP the disaccharide unit hyaluronic acid consists of D-glucoronic acid and *N*-acetyl-D-glucosamine with β(1→3) glycosidic linkage. The peptides are presumably linked to the glycan by asparagine through *N*-glycosidic linkage.

acetamido-2-deoxy-α-D-galactopyranose is attached to the hydroxyl oxygen of the threonine residue through an O-glycosidic linkage (Komatsu, DeVries & Feeney, 1970; Shier, Lin & DeVries, 1975; Feeney & Yeh, 1978). In smaller-sized AFGPs proline replaces some of the alanine residues. The AFGP structure shows similarities with other blood proteins and the Thomsen–Friedenreich Antigen glycoprotein (Cao *et al.*, 1995; Karsten *et al.*, 1995).

Unlike the low molecular weight cryoprotectants (e.g. glycerol) of cold hardy insects (Zachariassen, 1985), AFGPs lower the freezing point in a non-colligative manner. Since their effectiveness is not strictly dependent on the number of particles in solution, they do not upset osmotic gradients within the fish. The antifreeze activities of the AFGPs ranges from about 0.52 °C–1.20 °C at a concentration of 20 mg/ml; on a molar basis this represents about 100–300 times the (thermodynamic) freezing point depression expected on the basis of particle number. Thus high molecular weight AFGPs are most effective in lowering the freezing point, probably because their size ensures a more efficient adsorption to ice (Raymond & DeVries, 1977). The small molecular weight forms of AFGP (7–8; MW 2668 and 3274) comprise most of the circulating antifreeze (Burcham *et al.*, 1984) but show only two-thirds of the antifreeze activity of the larger AFGPs (Ahlgren & DeVries, 1984). The chemical structure of these AFGPs is highly conserved among all high-Antarctic notothenioids studied to date (Ahlgren & DeVries, 1984; Wöhrmann, 1996).

A novel antifreeze glycopeptide from *Pleuragramma antarcticum* could be isolated (Wöhrmann & Haselbeck, 1992; Wöhrmann *et al.*, 1997*a*). The *Pleuragramma*-antifreeze glycopeptide (PAGP) was shown in the biochemical analysis to be a proteoglycan bearing carbohydrate chains of the hyaluronic acid type (Fig. 1). Hyaluronic acid usually consists of 250 to 25000 kDa disaccharide units, held together by glycosidic $\beta(1\rightarrow4)$ links. The basic disaccharide unit consists of D-glucoronic acid and *N*-acetyl-D-glucosamine with a $\beta(1\rightarrow3)$ glycosidic linkage. Hyaluronic acid as a polyanion binds to cations. This explains the strong binding affinity of PAGP to DEAE anion exchangers (Wöhrmann, 1995; Wöhrmann & Haselbeck, 1992). X-ray structural analysis of fibres has shown that Ca^{2+} hyaluronate forms expanded left-handed single strain helices, containing three disaccharide units per turn. This secondary structure could explain the extraordinary antifreeze activity (Fig. 2 and Fig. 4). On the other hand, similarities to AFGPs become obvious (left-handed helices composed of glycopeptides). The peptides, presumably linked to the glycan by Asn-GlcNAc, confer a certain flexibility to the, otherwise rigid, hyaluronic acid molecule.

The non-glycosylated antifreeze peptide AFP type III found in Antarctic and Arctic eelpouts (Zoarcidae) (Hew *et al.*, 1984; Schrag *et al.*, 1987; Cheng & DeVries, 1989; Fletcher *et al.*, this volume) lacks distinctive features in its composition and sequence, and an AFP in the Antarctic liparidid *Paraliparis devriesi* (Jung *et al.*, 1995) remained not further characterised. All non-notothenioid species investigated so far (five species of five families, excluding gadiforms) possess non-glycosylated antifreezes at very low concentrations (Wöhrmann, 1993).

Studies on the purified AFGPs revealed that chemical modifications (by acetylation, periodate oxidation and complexation with borate), or the removal (by alkaline β-elimination and by digestion with endo-α-*N*-acetylgalactosaminidase) of the sugar residues result in a loss of antifreeze activity. Similarly, inactivation of PAGP was observed following splitting of the carbohydrate chain hyaluronic acid and the peptide group by the enzyme hyaluronidase. Furthermore, the peptide is split off the glycan structure by means of hydrazinolysis (Wöhrmann *et al.*, 1997*a*). For lowering the freezing point both the glycan and the peptide structure are necessary.

The function of antifreeze

The various forms of antifreeze peptides and glycopeptides vary considerably from each other in numerous molecular and biochemical aspects other than size so that some of these differences probably contribute to the broad range in antifreeze activity. A comparison of thermal hysteresis curves observed for solutions containing different antifreeze compounds is presented in Fig. 2. These curves were highly reproducible even when different preparations of purified antifreezes were used.

On a mass basis, AFGP from *Pleuragramma* exhibited the highest while AFGP from *Neopagetopsis* showed the lowest activity. The activity of the antifreeze glycopeptide of *Muraenolepis marmoratus* (probably AFGP) and of the proteoglycan PAGP were considerably lower than all of the AFGP considered. The antifreeze activity of *M. marmoratus* glycopeptides is essentially the same as that reported for the low molecular weight AFGP isolated from notothenioids (Knight, DeVries & Oolman, 1984; Wöhrmann, 1993). The antifreeze activity curve observed for *Pleuragramma* AFGP is comparable to that reported for high molecular weight AFGPs of other notothenioids such as *Dissostichus mawsoni* (Schrag *et al.*, 1982; Knight *et al.*, 1984) and to the curve reported for the AFP of the Arctic ocean pout (Kao *et al.*, 1986).

Although the maximum antifreeze activity of two deep-living species, *Pogonophryne macropogon* and *Bathydraco marri*, were found to be identical (0.85°C) and the antifreeze activity curve comparable (Fig. 2), the antifreeze composition differed considerably (Fig. 3). The elution profile of antifreezes on anion exchangers shows AFGPs in the buffer front. In the case of *B. marri* an additional antifreeze peptide could be isolated with 160 mM NaCl. The relative composition of AFGPs with different size is shown by the SDS-polyacrylamide electrophoresis blotted on nitrocellulose and detected with the digoxigenin-labelled lectin peanut agglutinin (PNA) which recognises the disaccharide Galβ1→3GalNAc in glycopeptides. The AFGP fraction of *P. macropogon* is dominated by the high molecular weight AFGPs. However, in addition to the AFP, AFGPs 5–6 (MW 10 500 and 7900 Da) show the highest

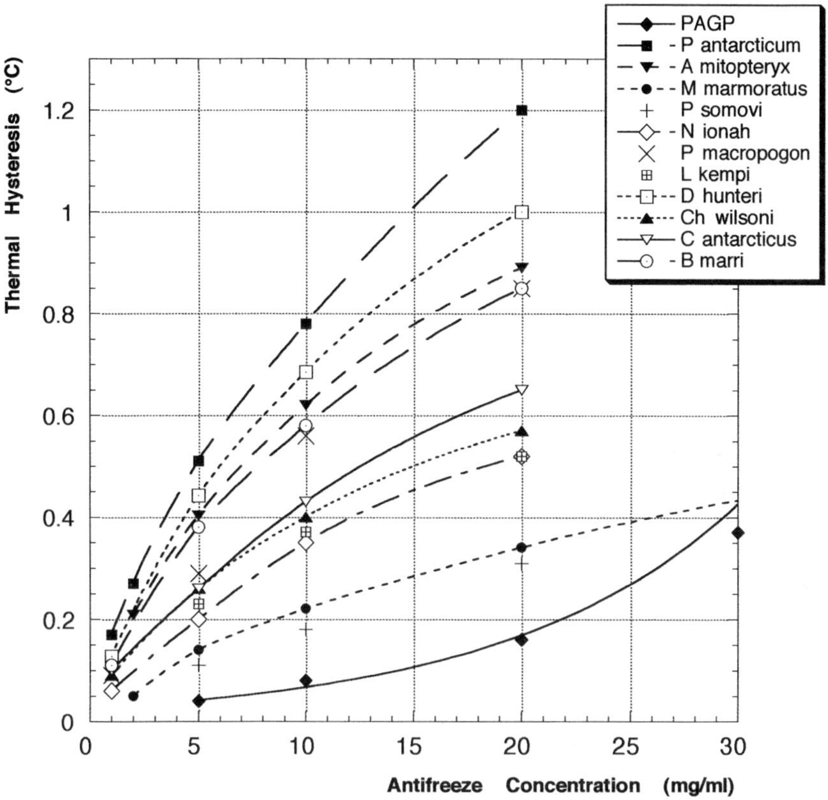

Fig. 2. Comparison of thermal hysteresis curves on a weight/volume basis of antifreeze peptides and glycopeptides obtained from *Pleuragramma antarcticum* (PAGP and AFGP 1–5), *Aethotaxis mitopteryx* (AFGP), *Muraenolepis marmoratus* (probably AFGP), *Paraliparis somovi* (AFP, not further characterised), *Neopagetopsis ionah* (AFGP), *Pogonophryne macropogon* (AFGP), *Lepidonotothen kempi* (AFGP), *Dacodraco hunteri* (AFGP), *Chaenodraco wilsoni* (AFGP), *Cryodraco antarcticus* (AFGP), and *Bathydraco marri* (AFGP).

concentrations in *B. marri*. Probably, higher AFGP levels in *B. marri* are explained by the difference in ambient water temperatures at which *P. macropogon* and *B. marri* were caught (i.e. at +0.75 °C and −1.0 °C, respectively).

When the thermal hysteresis curves were compared on a molar basis, a direct correlation was evident between antifreeze activity and molecular size (Fig. 4). A positive correlation between molecular weight and thermal hysteresis could also be found by examining the thermal hysteresis curves of AFP I from the Arctic shorthorn sculpin (Kao *et al.*, 1986). When compared

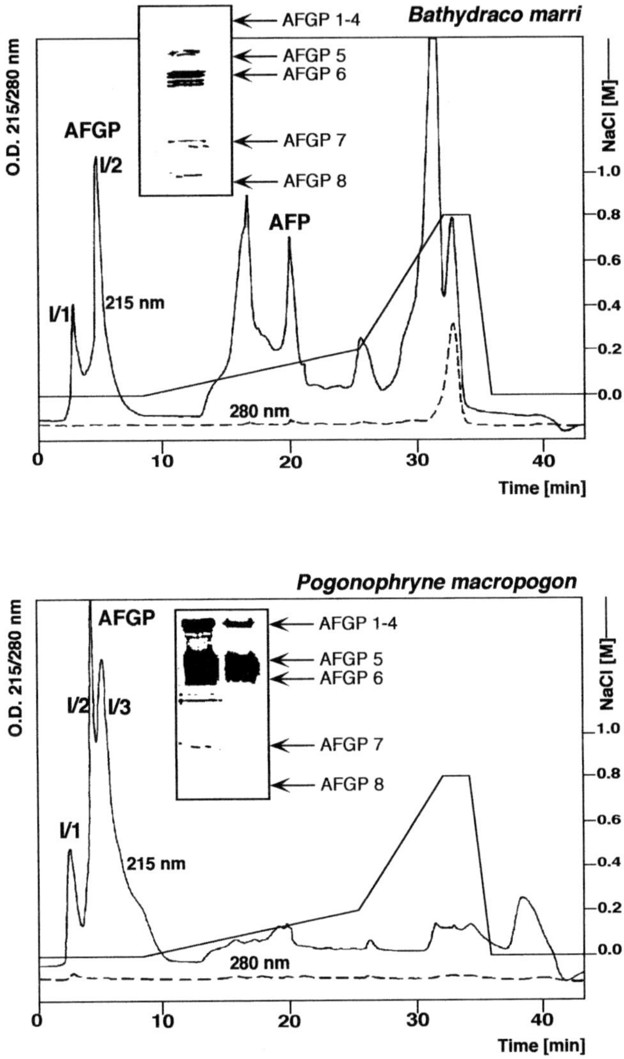

Fig. 3. HPLC elution profile via DEAE anion-exchange of antifreezes from the deep living *Bathydraco marri* and *Pogonophryne macropogon*. One ml of protein solution (1 mg ml^{-1}) was injected. Starting buffer: 20 mM Tris-HCl, pH 9.5 with a salt concentration gradient (0–0.8 M NaCl); absorbance was read at 215 and 280 nm. Pooled fractions (peak I/1–I/3) run on SDS-polyacrylamide gel electrophoresis. Antifreeze glycopeptides were blotted on a nitrocellulose membrane and detected by the lectin peanut agglutinine (PNA) which recognises the disaccharide moiety of AFGP (Wöhrmann, 1995). The antifreeze glycopeptides consist of a series of eight distinct glycopeptides ranging in molecular weight from 33 700 to 2668 daltons (AFGP 1–8).

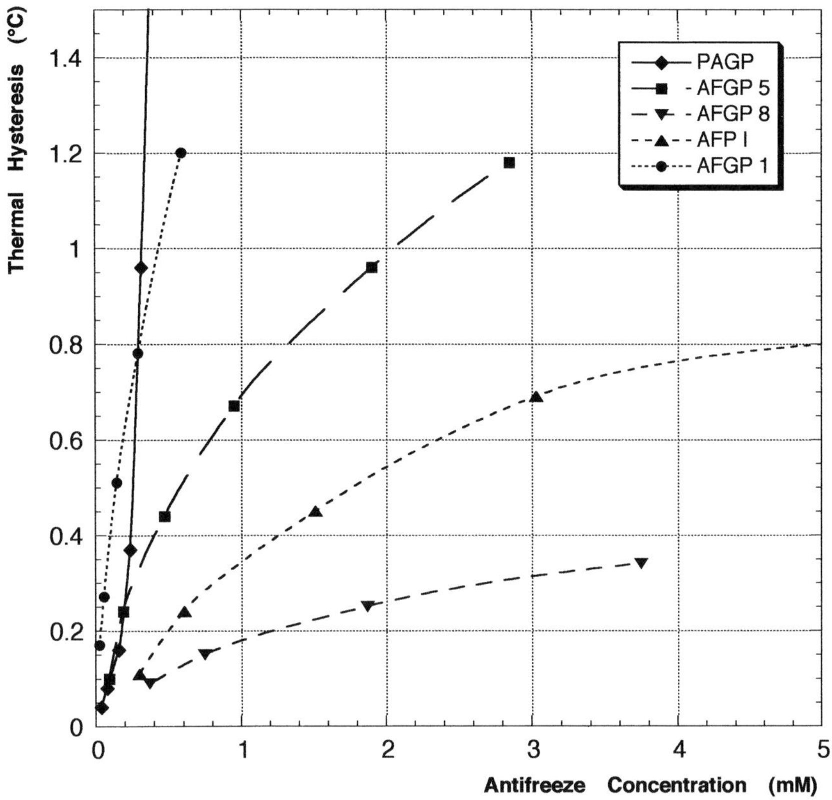

Fig. 4. Comparison of thermal hysteresis curves on a molar basis of HPLC-purified antifreeze glycopeptides obtained from *Pleuragramma antarcticum* (PAGP, MW 120 kDa) and *Trematomus pennellii* (AFGP 1, MW 33 700 Dalton; AFGP 5, MW 10 500 Da; AFGP 8, MW 2668 Da). The AFP I (Arctic shorthorn sculpin; MW 3300 Da) curve was adopted from Kao *et al.* (1986). Note the extreme increase of thermal hysteresis of the PAGP with a slight increase in concentration.

on a mass and a molar basis, the larger antifreeze components exhibited higher activity with the exception of the PAGP. In addition, the activity of the antifreeze peptides appear to require a minimum size and a defined configuration. Hew *et al.* (1985) has demonstrated that the two major peptides (MW 1400 and 2100) yielded by digestion of sculpin AFP were devoid of activity. This result is identical to that reported for AFGP where a molecular weight of greater than 2000 was necessary for activity (Geoghegan *et al.*, 1982). The smallest active AFGP found to date is the AFGP 8 with a MW of 2668 Da, determined by plasma desorption-mass spectrometry (Wöhrmann, 1995).

The AFGP from *Pleuragramma antarcticum*, *Dacodraco hunteri*, *Aethotaxis mitopteryx*, *Cryodraco antarcticus*, *Chaenodraco wilsoni*, and *Neopagetopsis ionah* are structurally homologous, having an amino acid composition of approximately 60% alanine, the disaccharide moiety Galβ1→3GalNAc, and a largely helical structure. Therefore, it is likely that their differences in activity can be attributed to both chain length as well as molecular weight. Hew *et al.* (1985) observed that within AFP I (Fletcher *et al.*, this volume) a decrease in chain length is associated with a reduction in α-helical content and the number of specific amino acid repeats within their primary structures. This amino acid repeat structure is the major functional domain for antifreeze activity. Kao *et al.* (1986) concluded that, within the different classes of AFP, activity increases with increasing molecular weight (3300 to 9700 Da). The specific antifreeze activity per mg protein decreases when the peptides have MW of 4000 Da and higher. Therefore, on a molecular weight basis, there appears to be no advantage in evolving large AFP molecules to achieve a freezing resistance required by the northern temperate fish species. In contrast, AFGPs and PAGP do not attain the antifreeze activity of the AFP until they reach a molecular weight greater than 8000 Da. Judging from the level of freezing resistance required by the Antarctic notothenioids, the large AFGP and PAGP in the special case of *P. antarcticum* molecules are essential for their survival in ice-laden water.

All AFGP components examined have a plateau in activity at high concentration; the actual value of the plateau activity, however, differs between the different AFGP components depending on their length. While the low molecular weight components lose activity at deep supercooling, very high concentrations can restore the activity (Burcham *et al.*, 1986). The activity data fit a reversible kinetic model of AFGP activity, and the coefficients obtained can be used to compare the activity differences between antifreeze components (Feeney, 1988). The model was also shown to describe the activity of the thermal hysteresis protein of the insect, *Tenebrio molitor*, and was useful for studying the mechanism of the *P. antarcticum*-specific PAGP.

Circular dichroism (CD) studies have shown that at low temperatures AFGP does possess an ordered structure and this is converted to a random coil structure at high temperatures or in the presence of high concentrations of $CaCl_2$ which is known to destroy ordered structures in polypeptides such as polyproline and in proteins (Hennessey & Johnson, 1981). The contribution of the sugar moiety to the observed CD spectra can be negligible. Differences in the activities of low and high molecular weight AFGP are due to the difference in their sizes and not due to variations in their conformation (Rao & Bush, 1987).

Furthermore, the loss of activity of the low molecular weight AFGP 8 on its own could be due to its inability to adequately cover ice growth sites. It is possible that co-operativity between high and low molecular weight AFGP may occur directly on the ice surface. In contrast, no co-operativity was observed at temperatures of nucleation <-3 °C between the high-alanine AFP of *Pseudopleuronectes americanus* and AFGP 8, even though additive effects were observed at temperatures of nucleation >-1 °C (Feeney, 1988).

Antifreeze and metabolism

There has been an evolutionary adaptation to overcome the rate-depressing effect of low temperature on the synthesis of proteins in notothenioids (Eastman, 1993). This is an important phenomenon as protein synthesis may constitute a significant portion of the maintenance metabolism and some enzymes, such as those involved in energy metabolism or synthesis of antifreeze peptides, require continuous replacement (Smith & Haschemeyer, 1980).

Although the metabolic costs required to synthesise and maintain active mature fish AFP and AFGP are unknown, some insight into relative energy costs can be gained by comparing the amounts of protein and peptide bonds (25 kcal/mol; Schulz & Schirmer, 1979) and peptide and carbohydrate bonds (4.4 – 5.8 kcal/mol; Toone, 1994) which are necessary to give the fish specific levels of antifreeze protection. In fact, synthesising continuously high amounts of antifreeze peptides (20 mg ml^{-1} blood serum and more) over the year implies high metabolic cost.

Zoarcids and some notothenioids (*Pogonophryne* spp. and *Bathydraco* spp.) are truly sluggish (Wöhrmann, unpublished observations) and have very low metabolic rates. Active notothenioid species such as *Trematomus bernacchii* show higher resting metabolism (Hubold, 1991). Scope for activity was therefore suggested to be one factor determining the resting metabolic rates of Antarctic notothenioid fish (Wells, 1987). All investigated species of both genera *Pogonophryne* and *Bathydraco* so far synthesise AFGP in relatively high amounts (Table 1). *Pogonophryne* spp. caught at 830 m depth and an ambient water temperature of +0.75 °C survived cooling to the freezing point in the presence of ice crystals as observed in captivity (Wöhrmann, 1990). However, more active species such as *Neopagetopsis ionah* living in the same environment (water depth of 800 m, water temperature above zero) possess lower amounts of AFGP (Fig. 4).

It should be noted that AFGP synthesis is not continuous throughout the year in all Antarctic fish species as has been reported in earlier studies (O'Grady & DeVries, 1982). The content of antifreezes in *P. antarcticum* depends on the fish's ontogenetic migration and hence is related to the ambient water temperature (Wöhrmann, 1996). Early post-larvae and

Table 1. *Antifreeze compounds of selected notothenioids and non-notothenioids found in the Weddell Sea and the Lazarev Sea*

Species	Antifreeze type	% FRG	TH (°C)	Antifreeze factor	Serum TH (°C)	Habitat zonation[a]	Distribution (m)[b]	Caught (m)	Temp. (°C)
Nototheniidae									
Lepidonotothen kempi *	AFGP	0.0636	0.52	0.033	n.d.	benthic	100–900	560	−1.40
Aethotaxis mitopteryx	AFGP	0.0356	0.89	0.032	0.97	mesopelagic	100–850	702	−0.16
*Pleuragramma antarcticum**	AFGP	0.0267	1.20	0.032	0.52	pelagic	0–900	450	−0.16
(adult specimen)[c]	PAGP	0.0143	2.45					to 630	to −1.84
Dissostichus mawsoni	AFGP	0.1053	1.10	0.116	1.05	mesopelagic	80–1600	626	−1.60
Trematomus bernacchii	AFGP	0.1021	1.01	0.103	1.08	benthic	100–700	626	−1.60
Trematomus eulepidotus	AFGP	0.1989	1.02	0.203	1.39	epibenthic	70–550	467	−1.60
Trematomus lepidorhinus	AFGP	0.1351	0.97	0.131	1.15	epibenthic	200–900	617	−1.40
Trematomus loennbergii	AFGP	0.1204	0.95	0.114	1.12	epibenthic	60–830	574	−1.92
Trematomus pennellii	AFGP	0.3337	1.06	0.354	1.61	benthic	0–730	405	−1.84
Artedidraconidae									
Artedidraco loennbergi	AFGP	0.0977	0.85	0.083	0.90	benthic	230–600	343	−0.50
Dolloidraco longedorsalis *	AFGP	0.0879	0.81	0.072	n.d.	benthic	200–2250	626	−1.60
Pogonophryne marmorata	AFGP	0.1595	0.87	0.139	0.86	benthic	140–1400	626	−1.00
Pogonophryne scotti	AFGP	0.1627	0.89	0.145	0.84	benthic	110–1200	830	+0.75
Pogonophryne barsukovi	AFGP	0.1544	0.86	0.133	n.d.	benthic	220–1120	800	+0.75
Pogonophryne permitini	AFGP	0.1509	0.87	0.131	n.d.	benthic	430–1120	830	+0.75
Pogonophryne macropogon	AFGP	0.1487	0.85	0.126	n.d.	benthic	570–840	830	+0.70
Bathydraconidae									
Bathydraco marri *	AFGP	0.0279	0.85	0.024	n.d.	benthic	300–1250	623	−1.00
Bathydraco macrolepis *	AFGP	0.0219	0.84	0.018	n.d.	benthic	450–2100	1400	+0.20
Bathydraco antarcticus *	AFGP	0.0231	0.85	0.020	0.49	benthic	320–2250	1400	+0.20
Gymnodraco acuticeps	AFGP	0.1973	0.90	0.178	1.34	benthic	0–550	197	−1.40
Cygnodraco mawsoni	AFGP	0.1794	0.89	0.160	1.27	benthic	110–300	197	−1.40

Gerlachea australis	AFGP	0.1643	0.84	0.138	1.16	benthic	200–670	407	−0.47
Racovitzia glacialis	AFGP	0.1147	0.84	0.096	0.96	benthic	220–610	574	−1.92
Channichthyidae									
Chaenodraco wilsoni	AFGP	0.2835	0.57	0.162	n.d.	benthopelagic	200–800	509	−1.60
Chionodraco hamatus	AFGP	0.2576	0.80	0.206	1.38	benthopelagic	0–600	407	−1.40
Chionodraco myersi	AFGP	0.1544	0.89	0.137	1.12	benthopelagic	200–800	623	−1.00
Cryodraco antarcticus	AFGP	0.0920	0.65	0.060	n.d.	epibenthic	250–800	623	−1.00
Dacodraco hunteri	AFGP	0.0809	1.00	0.081	0.87	benthopelagic	300–800	623	−1.00
Neopagetopsis ionah	AFGP	0.0621	0.52	0.032	0.48	benthopelagic	20–900	799	+0.20
Pagetopsis maculatus	AFGP	0.1796	0.94	0.169	1.31	epibenthic	200–800	453	−1.40
Pagetopsis macropterus	AFGP	0.2498	0.97	0.242	1.46	epibenthic	0–650	506	−1.89
Muraenolepididae									
Muraenolepis marmoratus	(AFGP)	0.0076	0.56	0.004	<0.1	benthic	20–1600	830	+0.75
Macrouridae									
Macrourus holotrachys	(AFGP)	0.0031	0.13	<0.000	<0.1	benthic	150–1100	742	+0.70
Liparididae									
Paraliparis somovi	AFP	0.0103	0.54	0.006	0.21	benthic	400–850	623	−1.00
Zoarcidae									
Lycenchelys hureaui	AFP	0.0041	0.18	<0.000	<0.1	benthic	560–940	830	+0.75
Myctophidae									
Gymnoscopelus opisthopterus	AFP	0.0070	≈0.1	<0.000	n.d.	mesopelagic	≥500	742	+0.70

Notes:
% FRG=level of HPLC-purified antifreeze peptides in % fresh weight of gutted fish; TH=thermal hysteresis of antifreeze compounds measured by differential scanning calorimetry (°C, standardised to 20 mg/ml); antifreeze factor=% FRG×TH; Serum TH=thermal hysteresis of blood serum (°C); Distribution=depth of distribution (m); Caught=depth of capture (m); Temp.=ambient water temperature (°C). For a detailed description of all procedures see Wöhrmann (1993, 1995).
* these species possess additional non-glycosylated antifreeze peptides.
[a] after Eastmann (1993).
[b] after Gon & Heemstra (1990).
[c] the TH (°C) of blood serum of *P. antarcticum* depends on the age of specimens.

maturing adults, abundant near the ice shelf in the southeastern Weddell Sea, possess higher amounts of antifreezes than juvenile fish feeding on krill in the East Wind Drift.

Pleuragramma antarcticum occurs in epi- and mesopelagic layers and has a very high mean lipid content, second to *A. mitopteryx*. The lipids are deposited predominantly as triacylglycerols, and the lipid stores greatly improve the buoyancy characteristics of *P. antarcticum* (Eastman, 1993). Morphological, ultrastructural and blood physiological studies suggest that pelagic Antarctic fishes like *P. antarcticum* and *A. mitopteryx* are rather sluggish with a low scope of activity and low metabolic requirements (Kunzmann, 1991; Wöhrmann, Hagen & Kunzmann, 1997*b*). In addition, the significantly increased viscosity of sea water at near-freezing temperatures elevate the energy cost of swimming (Hubold, 1992; Eastman, 1993). Reduced energy requirements due to low metabolic rates have also been indicated by starvation experiments with the epibenthic *Trematomus eulepidotus*, which can survive extended periods without food (Wöhrmann, 1988).

In conclusion, the more active and/or pelagic species may possess antifreeze peptides at a minimum concentration in order to minimise the metabolic cost for synthesising antifreezes. However, extremely sluggish species with a low resting metabolism can synthesise antifreezes in higher amounts than they would need in their natural habitat due to the 'sit and wait' foraging strategy and due to the year-round food supply. Furthermore, the synthesis of different antifreeze glycopeptides and peptides in relatively low concentrations and the high efficiency of these compounds is an extraordinary adaptation seen only in *P. antarcticum*, to minimise the energetic cost for a fully pelagic life in high-Antarctic waters.

The Weddell Sea environment

There is limited shallow water habitat in the Southern Ocean and this aspect has been an important factor influencing the evolution of the fish fauna. The shelf has been deep since at least the early Miocene (about 20 million years ago) and possibly much earlier in some areas. The shelf also contains inner shelf depressions, narrow trenches which reach depths of over 1000 m. Because of glacial isostasy the shelf in many areas has the most unusual characteristic of sloping or deepening toward the continent rather than offshore (Anderson, 1991).

The hydrographic regime of the Weddell Sea is governed by the slow moving Weddell Gyre, a large clockwise circulation driven by the East Wind Drift (EWD, Fig. 5). The gyre merges with circumpolar water from the Drake Passage and forms the eastward flowing Weddell–Scotia-Confluence northeast of the Antarctic Peninsula.

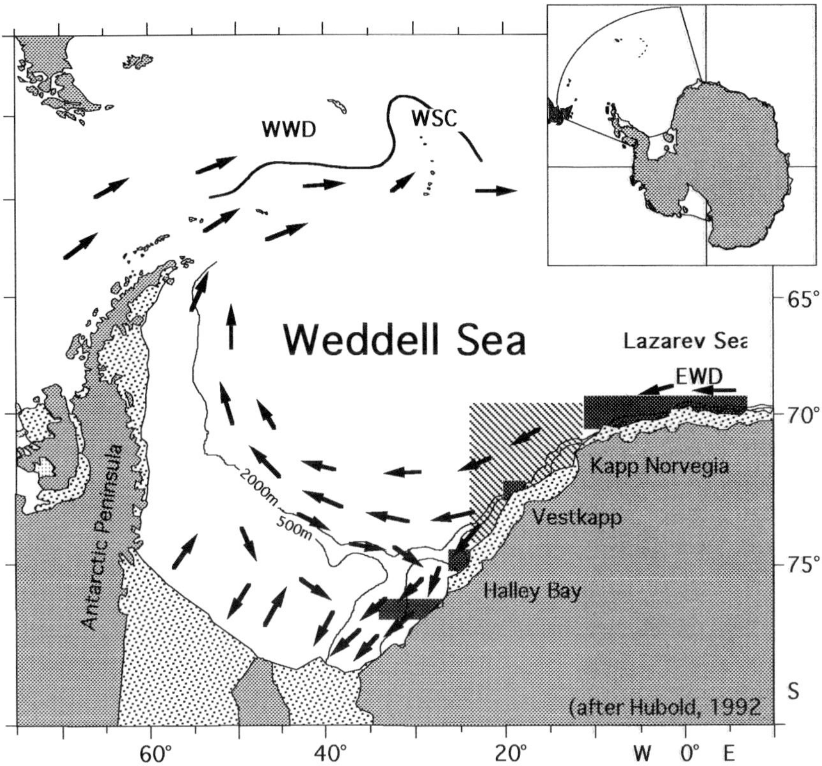

Fig. 5. Bathymetry and current patterns in the Weddell Sea region with the investigation areas during WWSP leg 3 (1986), EPOS leg 3 (1989) and ANT XI leg 3 (1991). Modified from Hubold (1992). WWD=West Wind Drift; WSC=Weddell-Scotia-Confluence; EWD=East West Drift.

The water masses of the Weddell Sea (Fig. 6) in the oceanic domain consist of the Ocean Surface Water, which is the colder and more saline Ocean Winter Water in 100–200 m depth (−1.6 to −2.0 °C), the Warm Deep Water (0 to 0.8 °C), the Antarctic Bottom Water below 2000 m (−0.8 to 0 °C) and the Weddell Sea Bottom Water (−1.4 to −0.8 °C) (Carmack & Foster, 1975; Hellmer & Bersch, 1985). Large portions of Antarctic Bottom Water are probably produced under the southern ice shelves (Foldvik, Kvinge & Torresen, 1985). The shelf domain is characterised by different colder water masses of the coastal current, which usually have temperatures between −1.6 and −2.0 °C with the exception of the more variable but warmer summer surface layer (Carmack & Foster, 1975; Hellmer & Bersch, 1985).

On a medium scale, the distribution of the principal water masses in the

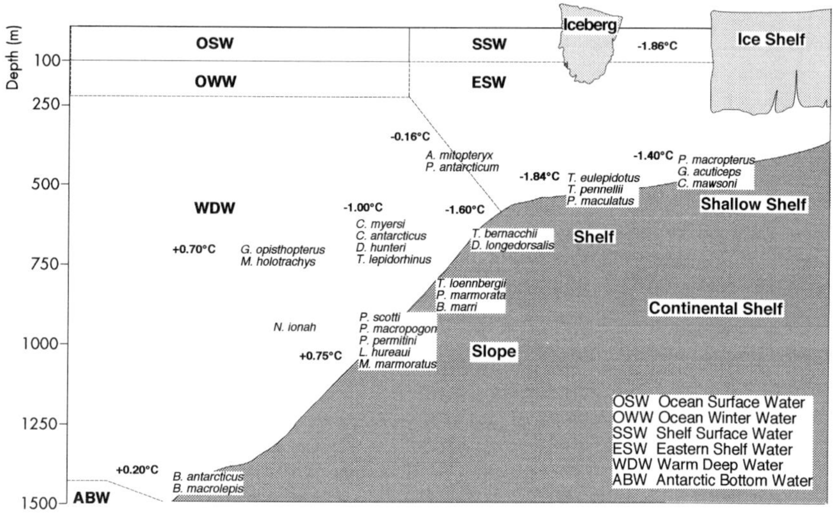

Fig. 6. Schematic cross-section of the water column in the Weddell Sea showing characteristic water masses and occurrence of investigated fish species. The water temperature shown would be measured by CTD at the different trawling stations. The definition of water masses was redrawn from Hubold (1991). For full species names see Table 1.

East Wind Drift and coastal current leads to differentiated temperature conditions on the shelf. Water temperature in the Eastern Shelf Water is constantly below −1.8 °C. Due to advection of warm meltwater from the north and east, temperatures can reach positive values during summer in a surface layer of 50–100 m (Rohardt, Ruhland & Schleif, 1990). At a depth below 500 m, the oceanic Warm Deep Water approaches the shelf and fills the deep trenches and inner shelf depressions of the eastern shelf, providing temperatures between 0.0° and +0.8 °C in the near-bottom layers. Interestingly, the Warm Deep Water is not present on the southern shelf or in the Filchner Depression, where the cold Antarctic Bottom Water is formed (Fahrbach *et al.*, 1987). The stable summer surface layer is another predictable feature of the Weddell Sea hydrography. All of these water masses are inhabited by notothenioid fish. The total temperature range experienced by the fish on the Weddell Sea shelf is no more than −2.2° to +1.0 °C.

Vertical distribution of fishes in the Weddell Sea

The Weddell Sea fish fauna is similar to that of the Ross Sea and Prydz Bay. It differs considerably from that of the subantarctic islands and Antarctic Peninsula, which is dominated by the genera *Notothenia* and

Champsocephalus (Kock, 1992). Some of the subantarctic species occur occasionally in the southern Weddell Sea where they live in the Warm Deep Water (Ekau, 1990).

The fish fauna of the eastern and southern Weddell Sea is diverse at the species level, in spite of the low biomass of the fishes and the dominance of individual species, consisting of more than 75 species from 8 orders, 14 families and 43 genera. Excluding oceanic species, the Weddell Sea fish fauna is richer in species (69) and genera (37) than the subantarctic island shelves (Hubold, 1991). Notothenioids contribute 70% of the species and dominate even more than in the Antarctic Ocean (Andriashev, 1987).

Typical species assemblages in the high-Antarctic Weddell Sea (Fig. 6) have been termed '*shallow shelf*' (200–350 m) with *Pagetopsis macropterus*, *Trematomus pennellii* and *Gymnodraco acuticeps* as dominant species, '*shelf*' (400–600 m) with *Pleuragramma antarcticum*, *Chionodraco myersi* and *Trematomus lepidorhinus* being the most abundant, and '*slope*' (500–800 m), dominated by *Macrourus holotrachys*, *Bathydraco marri* and *Muraenolepis* spp. A separate assemblage was found in the Gould Bay and in the Filchner Depression at 600–700 m which was made up mostly of nototheniids and bathydraconids, with *Trematomus loennbergii*, *Bathydraco* spp. and *Dacodraco hunteri* as the characteristic species (Schwarzbach, 1988; Ekau, 1990; Hubold, 1992).

The vertical distribution of mesopelagic fish is related to the Ocean Surface Water and the Warm Deep Water (Fig. 6). South of the Antarctic Convergence a number of myctophids (e.g. *Gymnoscopelus* spp.) may be found near the surface to more than 700 m depth (Lubimova *et al.*, 1983). Vertical migration in species, such as *Gymnoscopelus opisthopterus*, is probably confined by the Circumpolar Deep Water, as they have rarely been caught above 500–600 m depth (Duhamel, Hulley & Camus,1989).

Early life stages of notothenioids are mostly restricted to the upper 200 m of the water column, both in the Seasonal Pack-Ice Zone and in the High Antarctic Zone, although they may be found down to 600 m (Kellermann, 1989; Hureau *et al.*, 1990). Juveniles and adults of pelagic nototheniids, such as *Pleuragramma antarcticum* and *Aethotaxis mitopteryx*, have been caught at 800–1000 m (Hubold, 1992; Kunzmann & Zimmermann, 1992; Wöhrmann & Zimmermann, 1992), but they may occur at even greater depths.

Antifreeze and water depth

In the high-Antarctic shallow water species are abundant only above 400–450 m depth. The benthic/epibenthic *Cygnodraco mawsoni*, *Trematomus pennellii* and *Pagetopsis macropterus* are representatives of this group

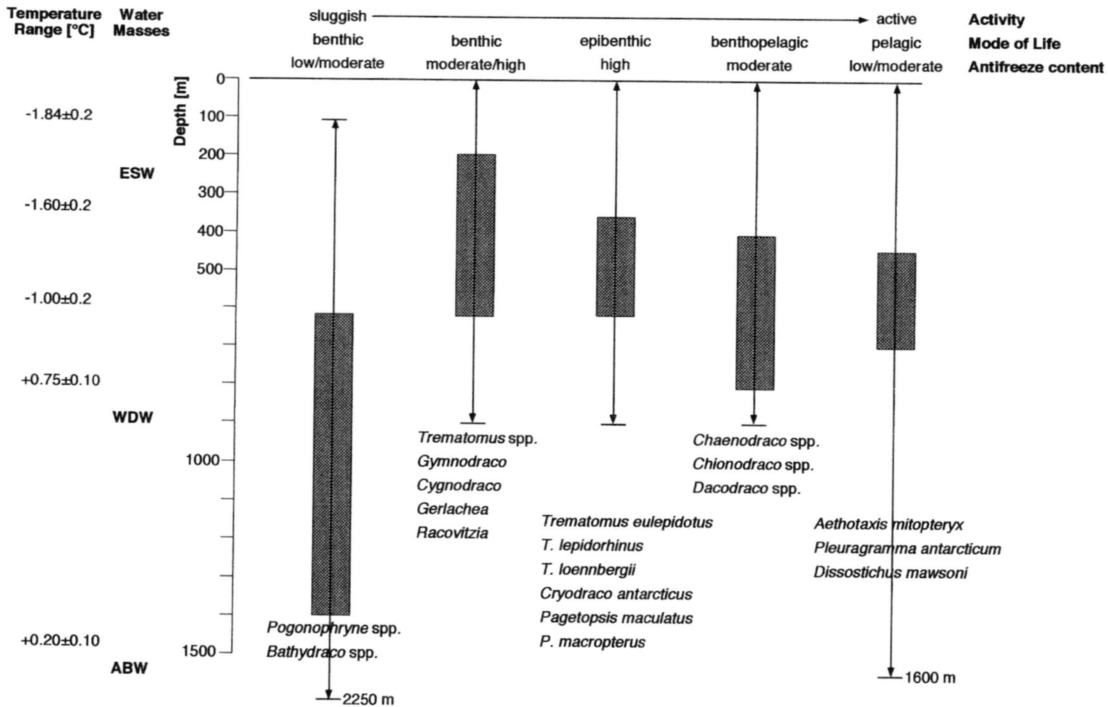

Fig. 7. Vertical distribution of investigated Antarctic fishes and their activity, mode of life and antifreeze content. Total range of distribution (after Gon & Heemstra, 1990) for each fish group is denoted by a vertical line, the bar represents the minimum and maximum depth of capture of the investigated species. ESW Eastern Shelf Water, WDW Warm Deep Water, ABW Antarctic Bottom Water.

(Figs. 6, 7), found to possess high amounts of both low and high molecular weight AFGP. Eurybathic species are abundant over most of the depth range down to 600–700 m. *Pogonophryne scotti*, *Racovitzia glacialis*, *Trematomus lepidorhinus* and *Cryodraco antarcticus* belong to this group. The antifreeze content in these species is moderate (Table 1), in active species (*C. antarcticus*) lower than in sluggish species (*P. scotti*). Deeper shelf and upper slope species occur only at depths below 400–450 m. *Lepidonotothen kempi*, *Dacodraco hunteri* and *Muraenolepis marmoratus* in the Lazarev Sea, *Bathydraco marri* and *Trematomus loennbergii* in the Weddell Sea are examples of this group. The Lazarev Sea species have lower amounts of antifreezes which is probably related to the higher water temperature (see also Fig. 8). In contrast, *D. hunteri*, caught in the Lazarev Sea, synthesises one of the most effective AFGP (Table 1: 1.0 °C at 20 mg/ml) with relatively high amounts of high molecular weight AFGP due to their distribution in colder areas such as the southern Weddell Sea.

Little is known about the bathymetric range of the deep-sea bottom and benthopelagic fish fauna. Some notothenioids, such as *Pogonophryne macropogon* and *Bathydraco antarcticus*, as well as a number of zoarcids and liparidids have been found as deep as 2000–2600 m and could be referred to as deep-sea rather than deeper shelf and upper slope species. They do not require any antifreeze protection owing to the water depth and temperature of their habitat. Nevertheless, these benthic and sluggish species still possess antifreezes in albeit small amounts. Antifreezes are the non-glycosylated AFP in the case of zoarcids and liparidids, and the AFGPs in *Pogonophryne* spp. and *Bathydraco* spp.. Moreover, *Bathydraco* spp., distributed in the cold water of the Filchner Depression, synthesize additional antifreeze peptides in the same concentration as the AFGP. Antifreeze activity is 0.84 °C at a peptide concentration of 40 mg ml^{-1}. So far, these peptides could not be further characterised (Wöhrmann, 1993).

The measurable thermal hysteresis (0.1 °C–0.5 °C) of blood serum of several deep-living species (e.g. *Muraenolepis marmoratus*, *Paraliparis somovi*, *Neopagetopsis ionah*, *Bathydraco* spp.) suggests that the serum contains modest amounts of antifreezes, but that this does not lower the freezing point of serum sufficiently to protect these species from freezing at the usual −1.9 °C surface temperature in the Shelf Surface Water (SSW; Fig. 6). Combined with elevated osmolality of body fluids, a widespread phenomenon in polar teleosts (O'Grady & DeVries, 1982), the freezing point depression of serum in deep water species is only 0.98 °C–1.38 °C below the freezing point of pure water.

However, the freezing behaviour of the serum of *Paraliparis somovi* is similar to that reported for some northern boreal species (Raymond, 1989; Wöhrmann, 1993). The freezing points are not well defined, and the ice

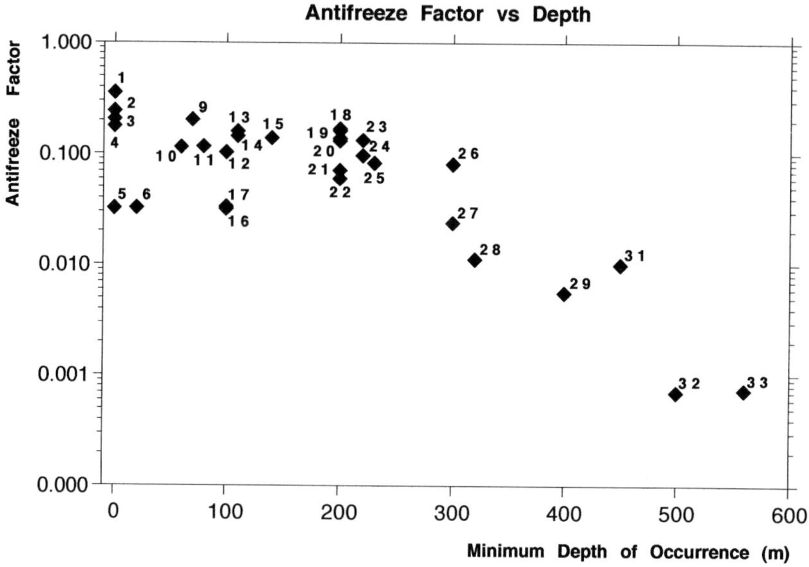

Fig. 8. Antifreeze factor (total body concentration of antifreeze compounds × thermal hysteresis of purified antifreezes standardised to 20 mg/ml) versus the minimum depth of occurrence. On the one hand in some species the antifreeze concentration in the blood serum is too low to measure a thermal hysteretic effect in the blood serum. On the other hand, only the skin mucus of some species contains significant amounts of antifreezes. The antifreeze factor compares these different kinds of freezing protection. 1 *Trematomus pennellii*, 2 *Pagetopsis macropterus*, 3 *Chionodraco hamatus*, 4 *Gymnodraco acuticeps*, 5 *Pleuragramma antarcticum*, 6 *Neopagetopsis ionah*, 9 *Trematomus eulepidotus*, 10 *T. loennbergii*, 11 *Dissostichus mawsoni*, 12 *Trematomus bernacchii*, 13 *Cygnodraco mawsoni*, 14 *Pogonophryne scotti*, 15 *P. marmorata*, 16 *Aethotaxis mitopteryx*, 17 *Lepidonotothen kempi*, 18 *Chaenodraco wilsoni* & *Pagetopsis maculatus*, 19 *Gerlachea australis* & *Chionodraco myersi*, 20 *Trematomus lepidorhinus*, 21 *Dolloidraco longedorsalis*, 22 *Cryodraco antarcticus*, 23 *Pogonophryne barsukovi*, 24 *Racovitzia glacialis*, 25 *Artedidraco loennbergi*, 26 *Dacodraco hunteri*, 27 *Bathydraco marri*, 28 *B. antarcticus*, 29 *Paraliparis somovi*, 31 *Bathydraco macrolepis*, 32 *Gymnoscopelus opisthopterus*, 33 *Lycenchelys hureaui*.

growth in the serum progresses through periods of stops and starts as the temperature is lowered. In other non-notothenioid species from the Lazarev Sea (e.g. *Macrourus holotrachys* and *Lycenchelys hureaui*) freezing point depression of serum could not be detected due to the extremely low concentrations of the antifreeze peptides (Table 1).

While antifreeze protection down to −2.2 °C is necessary for fishes inhabiting relatively shallow ice-laden water, most non-notothenioid species and some notothenioids (e.g. *Neopagetopsis ionah*) live at 600 m depth and deeper, a habitat with different physical parameters and less ice (with the exception of the southern Weddell Sea and Filchner Depression). Seawater at the surface of the Weddell Sea freezes at approx. −1.86 °C. Because of the pressure effect on the freezing point (0.00753 °C 10 m^{-1}; Lewis & Perkin, 1985), seawater at 500 m depth with a salinity of 34.7‰ freezes *in situ* at −2.24 °C. Hence, based on the freezing point of serum, the freezing point would be −1.50 °C in *P. somovi* (623 m), −2.28 °C in *P. scotti* (830 m), −1.90 °C in *N. ionah* (799 m), and −2.36 °C in *B. antarcticus* (1400 m).

Fishes living at these depths and temperatures in the eastern Weddell Sea and the Lazarev Sea are in little danger of freezing because there is no ice in the water. Some of these species were thought to lack antifreeze, remain in a supercooled state in this deep water habitat instead (DeVries & Lin, 1977). In the presence of ice, however, supercooling is not possible as the fish's gills and integument constitute no barrier to the propagation of ice in the body and even partial freezing leads to death (DeVries, 1988). In the southern Weddell Sea near the Filchner Ice Shelf the water at the undersurface of ice shelves is at its equilibrium freezing point. As this water flows out from under the shelf and rises, it becomes supercooled and potentially ice formation could occur at considerable depths. Large masses of single-crystal ice platelets were discovered in this area (Dieckmann *et al.*, 1986).

In addition to antifreeze glycopeptides and peptides in blood serum, some Antarctic species use trunk skin as a barrier to the entry of ice crystals into the body. Skin mucus of *Pogonophryne* spp. and *P. antarcticum* contains amounts of AFGP, and in the case of *P. antarcticum* a mixture of different antifreeze types protect against freezing (Wöhrmann, unpublished observations). Asakawa *et al.* (1989) have isolated glycoproteins similar to AFGPs from the skin mucus of *Trematomus bernacchii*. A high water content of the subdermal extracellular matrix of *Paraliparis devriesi* does not require additional protection against freezing (Jung *et al.*, 1995). Experiments *in vitro* on the skin of winter flounder (*Pseudopleuronectes americanus*) from Newfoundland support the hypothesis that the presence of epithelia excludes the entry of ice (Valerio, Kao & Flether, 1992). The antifreeze concentration in the blood of some of these species is too low to protect against freezing. It is likely that the modified skin surfaces act as a physical barrier (Fletcher *et al.*, this volume) to the propagation of ice into body fluids of sluggish species such as *Pogonophryne* spp., liparidids and probably in zoarcids and *P. antarcticum*.

In summary, antifreeze compounds are distributed in the blood serum and other body fluids. In some species they could be found in the skin mucus or in the subdermal extracellular matrix. In deeper living species the concentrations are too low to detect a thermal hysteresis in the blood serum. Furthermore, the thermal hysteresis depends on the relative composition of both high and low molecular weight AFGP. To take these facts into account and to compare the results of recent investigations of other authors with our own results, freezing resistance needs to be standardised. The 'antifreeze factor' (AF) has been defined as the content of HPLC-purified antifreeze peptides and glycopeptides in % of the gutted fish x the maximum hysteretic effect (°C; standardised to 20 mg/ml) of these antifreezes (Table 1).

The antifreeze factor versus the depth of minimum occurrence is shown in Fig. 8. As expected, the shallow water (epi)benthic species such as *Trematomus pennellii, Pagetopsis macropterus* and *Gymnodraco acuticeps*, with a moderate level of activity have high concentrations of AFGP with a high hysteretic effect due to the high molecular weight glycopeptides. This is resulting in the highest AF values. Species distributed in warmer waters of the Lazarev Sea (e.g. *Muraenolepis marmoratus, Lycenchelys hureaui*) possess lower amounts of antifreeze with less thermal hysteretic effect compared with species (e.g. *Pogonophryne* spp.) of the same depth (830 m) in the Weddell Sea, resulting in a much lower AF value. Non-notothenioid species of the Lazarev Sea do possess some antifreezes in the blood serum and probably in the skin mucus, even though in extremely low concentration. A thermal hysteretic effect is not measurable with conventional methods. Using the antifreeze factor it is possible to compare these low values with antifreeze concentrations in other species. On the other hand, *Bathydraco* species have unusually high AF values. These species living in the extremely cold waters of the southern Weddell Sea need high concentrations of highly functional antifreeze compounds. The antifreeze factor is a possibility to qualify the antifreeze avoidance of Antarctic and Arctic fishes.

Antifreeze formation and evolution

It is believed that notothenioids have evolved from a bottom-living ancestral group after the establishment of the Antarctic Circumpolar Current (Eastman & Grande, 1989). Monophyly was assumed (Eakin, 1981; Iwami, 1985). However, antifreeze compounds were not essential for survival of the notothenioid stock and other fish groups during the rapid cooling that took place 36 million years ago or prior to the formation of the Southern Ocean 25–22 million years ago (Clarke & Johnston, 1996). Most of the Southern Ocean was not cold enough to freeze fish (Eastman, 1993, 1995).

Antifreezes have been identified in all investigated high-Antarctic species (Eastman, 1993; Jung *et al.*, 1995; Wöhrmann, 1996). Studies have focused on the nototheniids, bathydraconids and channichthyids of the Weddell Sea and McMurdo Sound, with little attention devoted to other Antarctic families. Hence, antifreezes are probably a universal characteristic of notothenioids (and other fish groups), synthesised in significant concentrations only by those species which inhabit waters with subzero temperatures where ice is present or by species liable to encounter ice during latitudinal or vertical migrations.

The urinary loss of AFGPs in notothenioids (Eastman, 1993) and in the Arctic gadiform species *Boreogadus saida* (Christiansen, Dalmo & Ingebrigtsen, 1996) is effectively circumvented by the evolutionary alteration of renal morphology. Species with significant amounts of AFGPs have aglomerular nephrons. The evolution of this aglomerular nephron in a small group of teleost fishes is the most significant deviation from the basic renal pattern in vertebrates. In aglomerular nephrons, urine is produced by tubular secretion rather than by filtration and small molecules such as AFGPs are spared the filtration process (Eastman, 1993).

Eastman (1993) mapped the distribution of AFGPs and glomeruli on an area cladogram for the Notothenioidei. Together with the results reviewed in this chapter it is evident that all species which possess AFGPs exhibit the apomorphic state of kidney. The plesiomorphic condition of kidney is retained by the non-Antarctic bovichtids without AFGPs. Species of the Nototheniidae distributed in extremely cold water (e.g. *Pagothenia borchgrevinki*) and in warmer water of the subantarctic area (e.g. *Dissostichus eleginoides*) have mixed character states for both AFGPs and glomeruli. The high-Antarctic subfamilies Trematominae and Pleuragramminae all possess AFGPs, and all have aglomerular kidneys. On the other hand, the sub-Antarctic Eleginopinae and Nototheniidae are mixed with regard to these traits (Eastman, 1993).

It is suspected that the distribution of notothenioids to the south and of the polar cod (*Boreogadus saida*) and the arctic cod (*Arctogadus glacialis*) to the north has been primarily influenced by the acquisition of aglomerularism, and that this trend can be correlated with the phylogenetic diversification of notothenioids and gadiforms respectively, and with inhabitation of colder water. The adaptation to aglomerularism so as to spare small molecules such as AFGPs from filtration is a condition for species living in the high-Antarctic zone to preserve the energy costs of protein synthesis. Hence, it could be argued, that aglomerularism in the notothenioid families Artedidraconidae, Bathydraconidae and Channichthyidae is a key evolutionary innovation (Brooks & McLennon, 1991) to survive in cold high-Antarctic waters. However, the ability to synthesize antifreeze peptides is not correlated with water temperature and the synthesis of AFGPs is only one innovation in fishes living in the north and in the south.

The glacial–interglacial cycle

Organisms living in a stable environment have the opportunity to develop a physiology which is finely tuned to operate under these conditions. However, organisms living in a changing environment (e.g. seasonally) cannot afford such fine tuning (Clarke, 1987). The Antarctic environment has changed quite dramatically over the few million years and even over the last 750 000 years in a cyclic way (Quilty, 1990). Furthermore, the pelagic production in the Southern Ocean undergoes a marked seasonal oscillation (Clarke, 1988); the waters south of the Antarctic Convergence are productive only during summer and contain relatively few non-notothenioid fish.

Eastman and Grande (1989) discussed factors that might have contributed to the paucity of non-notothenioid species in the modern Antarctic ichthyofauna. They concluded that low water temperature was not of high priority and that factors in the realm of ecological constraints were probably at least as important in restricting diversity. Given that antifreeze compounds may have evolved independently several times in the course of evolution, as suggested by the presence of antifreeze peptides in ten families of five suborders, it seems unlikely that low temperature as such has caused the paucity of non-notothenioid species in the Antarctic fish fauna (Clarke, 1987, 1990). Rather, the limited shallow water habitat on the continental shelf and seasonal oscillation in the food supply are the plausible ecological factors.

Clarke (1983) and Clarke and Crame (1989) provide summaries of Southern Ocean palaeotemperatures for the biologist interested in the physiological adaptation and evolution of Antarctic marine organisms. A general cooling trend began in the late Paleocene/early Eocene and has continued into recent times. During this period of about 50 million years, water temperatures fell from about 15 °C to less than 0 °C (Eastman, 1993; Clarke & Johnston, 1996). This trend has been interspersed with periods of slight warming and there have been occasional drops in water temperature that deviated sharply from the general decline.

During periods of ice sheet retreat over the past five million years, sea levels rose and large marine embayments reached deep into the interior of the continent (Webb & Harwood, 1991). These seaways moderated inland temperatures and provided an ice-free environment for part of the year. It is conceivable that during glacial periods the shelf ice grounds on the shelf and thus may extend beyond the shelf edge. Therefore, the ice impact could cause extinction of the shelf fauna. When the shelf ice retreats during the subsequent interglacial, the defaunated shelf can be recolonised from other, less affected areas.

There is evidence that this almost regular cycle of shelf ice formation and retreat may have existed for the past two million years. In earlier times glaciation events also occurred, but at less regular intervals (Quilty, 1990). These glacial–interglacial cycles may have been the environmental driving force towards the evolutionary development of antifreeze synthesis in Antarctic fish. In contrast with the north polar sea, Antarctic shelf species cannot retreat from the advancing ice towards lower latitudes without moving into the deeper waters of the continental slope. Species without this ability are likely to become extinct, whilst those with a wider bathymetric range, an efficient freezing resistance and adapted physiology (e.g. aglomerularism) could survive in deep water and quickly recolonise the shelf during interglacial periods.

Summary and conclusions

The concentration and composition of antifreeze compounds appear to be dependent on the ambient water temperature, the depth of occurrence, the level of activity (resting metabolism), and the mode of life. Sluggish and benthic species, such as *Pogonophryne* spp. and *Bathydraco* spp., possess higher amounts of antifreezes than they would need in their natural habitat. Antifreeze concentration of more active benthopelagic or pelagic species, such as channichthyids, *P. antarcticum* and *A. mitopteryx*, is exactly tailored to the ambient water temperature and depth, thereby minimizing the metabolic cost for synthesising these glycopeptides and peptides. Freezing resistance is an important adaptation which all Antarctic fish species have successfully developed. Hence, the ability of species living in warm deep water to synthesise antifreeze indicates that antifreezes are ubiquitous. Furthermore, low water temperature as such has not caused the paucity of non-notothenioid species in the Antarctic fish fauna. Rather, the limited shallow water habitat on the continental shelf, the glacial–interglacial periods and seasonal oscillation in food supply are plausible, ecological factors. Finally, the present time which represents a point of interglacial warm period in the evolutionary history of Antarctic fish species reflects that some species do not need fully effective levels of antifreeze but synthesise only the essential minimum amounts.

Acknowledgements

I wish to express my gratitude to Professor R. Geyer, Professor R. Rudolph, and to Dr A. Haselbeck for their helpful discussions and advice. Special thanks are due to Dr K. Thomas for thoroughly checking the final version and improving the English. Part of this work was supported by the Deutsche Forschungs Gemeinschaft (DFG) and the European Science Foundation (ESF).

References

Ahlgren, J. A. & DeVries, A. L. (1984). Comparison of antifreeze glycopeptides from several Antarctic fishes. *Polar Biology*, **3**, 93–7.

Anderson, J. B. (1991). The Antarctic continental shelf: results from marine geological and geophysical investigations. In *The Geology of Antarctica*, ed. R. J. Tingey, pp. 285–334. Oxford: Oxford University Press.

Andriashev, A. P. (1987). A general review of the Antarctic bottom fish fauna. In *Proceedings of the 5th Congress of European Ichthyologists*, ed. S. O. Kullander & B. Fernholm, pp. 357–72. Stockholm: Museum of Natural History.

Asakawa, M., Nakagawa, H., Fukuda, Y. & Fukuchi, M. (1989). Characterization of glycoprotein obtained from the skin mucus of an Antarctic fish, *Trematomus bernacchii*. *Proceedings of the NIPR Symposium on Polar Biology*, Tokyo, No 2, pp. 131–8.

Brooks, D. R. & McLennan, D. A. (1991). *Phylogeny, Ecology, and Behavior: A Research Program in Comparative Biology*. Chicago: University of Chicago Press.

Burcham, T. S., Osuga, D. T., Chino, H. & Feeney, R. E. (1984). Analysis of antifreeze glycoproteins in fish serum. *Analytical Biochemistry*, **139**, 197–204.

Burcham, T. S., Osuga, D. T., Rao, B. N., Bush, C. A. & Feeney, R. E. (1986). Purification and primary sequences of the major arginine-containing antifreeze glycopeptides from the fish *Eleginus gracilis*. *Journal of Biological Chemistry*, **261**, 6384–9.

Cao, Y., Karsten U. R., Liebrich, W., Haensch, W., Springer, G. F. & Schlag, P. M. (1995). Expression of Thomsen–Friedenreich-related antigens in primary and metastatic colorectal carcinomas. *Cancer*, **76**, 1700–8.

Carmack, E. C. & Foster, T. D. (1975). Circulation and distribution of oceanographic properties near the Filchner Ice Shelf. *Deep-Sea Research*, **22**, 77–90.

Cheng, C. C. & DeVries, A. L. (1989). Structures of antifreeze peptides from the antarctic eel pout, *Austrolycichthys brachycephalus*. *Biochimica et Biophysica Acta*, **997**, 55–64.

Cheng, C. C. & DeVries, A. L. (1991). The role of antifreeze glycopeptides and peptides in the freezing avoidance of cold-water fish. In *Life under Extreme Conditions. Biochemical Adaptation* ed. G. di Prisco, pp. 1–15. Berlin: Springer Verlag.

Christiansen, J. S., Dalmo, R. A. & Ingebrigtsen K. (1996). Xenobiotic excretion in fish with aglomerular kidneys. *Marine Ecology Progress Series*, **136**, 303–4.

Clarke, A. (1983). Life in cold water: the physiological ecology of polar marine ectotherms. *Oceanography and Marine Biology: An Annual Review*, **21**, 341–53.

Clarke, A. (1987). The adaptation of aquatic animals to low temperatures.

In *The Effects of Low Temperatures on Biological Systems*, ed. B. W. W. Grout & G. J. Morris, pp. 313–48. London: Edward Arnold.

Clarke, A. (1988). Seasonality in the Antarctic environment. *Comparative Biochemistry and Physiology*, **90B**, 461–73 .

Clarke, A. (1990). Temperature and evolution: Southern Ocean cooling and the Antarctic marine fauna. In *Antarctic Ecosystems: Change and Conservation*, ed. K. R. Kerry & G. Hempel, pp. 9–22. Berlin: Springer Verlag.

Clarke, A. & Crame, J. A. (1989). The origin of the Southern Ocean marine fauna. In *Origins and Evolution of the Antarctic Biota*, ed. J. A. Crame, pp. 253–68. London: Geol. Soc. Spec. Publ. No. 47, The Geological Society.

Clarke, A. & Johnston, I. A. (1996). Evolution and adaptive radiation of Antarctic fishes. *Trends in Ecology and Evolution*, **11**, 212–18.

Davies, P. L., Hew, C. L. & Fletcher, G. L. (1988). Fish antifreeze proteins: physiology and evolutionary biology. *Canadian Journal of Zoology*, **66**, 2611–17.

Deacon, G. (1984). *The Antarctic Circumpolar Ocean*. Cambridge: Cambridge University Press.

DeVries, A. L. (1988). The role of antifreeze glycopeptides and peptides in the freezing avoidance of Antarctic fishes. *Comparative Biochemistry and Physiology*, **90B**, 611–21.

DeVries, A. L. & Eastman, J. T. (1981). Physiology and ecology of notothenioid fishes of the Ross Sea. *Journal of the Royal Society NZ*, **11**, 329–40.

DeVries, A. L. & Lin ,Y. (1977). The role of glycoprotein antifreezes in the survival of Antarctic fishes. In *Adaptations within Antarctic Ecosystems*, ed. G. A. Llano, pp. 439–58. Gulf: Gulf Publishing Co.

DeWitt, H. H. (1970). The character of the midwater fish fauna of the Ross Sea, Antarctica. In *Antarctic Ecology, Vol 1*, ed. M. W. Holdgate, pp. 305–14. London: Academic Press.

Dieckmann, G., Rohardt, H., Hellmer, H. & Kipfstuhl, J. (1986). The occurence of ice platelets at 250 m depth near the Filchner Ice Shelf and its significance for sea ice biology. *Deep-Sea Research*, **33**, 141–8.

Duhamel, G., Hulley, P. A. & Camus, P. (1989). Ichthyological results of cruise MD42/SIBEX II, Part I: Fishes from RMT-8 stations, with additional records of laternfishes (Myctophidae, Osteichthyes) from the Southern Ocean. *Cybium*, **13**, 83–99.

Eakin, R. R. (1981). Osteology and relationship of the fishes of the Anatarctic family Harpagiferidae (Pisces, Notothenioidei). In *Biology of the Antarctic Seas*, ed. L. S. Kornicker, pp. 81–147. Washington: Antarctic Research Series 31, American Geophysical Union.

Eastman, J. T. (1990). The biology and physiological ecology of notothenioid fishes. In *Fishes of the Southern Ocean*, ed. O. Gon & P. C. Heemstra, pp. 34–51. Grahamstown: J.L.B. Smith Institute of Ichthyology.

Eastman, J. T. (1991). Evolution and diversification of Antarctic notothenioid fishes. *American Zoologist*, **31**, 93–109.

Eastman, J. T. (1993). *AntarcticFish Biology*. San Diego: Academic Press.

Eastman, J. T. (1995). The evolution of Antarctic fishes: questions for consideration and avenues for research. *Cybium*, **19**, 371–89.

Eastman, J. T. & DeVries, A. L. (1982). Buoyancy studies of notothenioid fishes in McMurdo Sound, Antarctica. *Copeia*, **1982**, 385–93.

Eastman, J. T. and Grande, L. (1989). Evolution of the Antarctic fish fauna with emphasis on the recent notothenioids. In *Origins and Evolution of the Antarctic Biota*, ed. J. A. Crame, pp. 241–52. London: Special Publications of the Geological Society, No 47.

Ekau, W. (1990). Demersal fish fauna of the Weddell Sea, Antarctica. *Antarctic Science*, **2**, 129–37.

Ewart, K. V. & Fletcher, G. L. (1990). Isolation and characterization of antifreeze proteins from smelt (*Osmerus mordax*) and Atlantic herring (*Clupea harengus harengus*). *Canadian Journal of Zoology*, **68**, 1652–58.

Fahrbach, E., Klindt, H., Muus, D., Rohardt, G. & Salameh, P. (1987). Physical oceanography. *Berichte zur Polarforschung*, **39**, 156–69.

Fahrbach, E. (1995). Die Polarmeere – ein Überblick. In *Biologie der Polarmeere*, ed. I. Hempel & G. Hempel, pp. 24–44. Jena: Gustav Fischer.

Feeney, R. E. (1988). Inhibition and promotion of freezing: fish antifreeze proteins and ice-nucleating proteins. *Comments on Agricultural Food Chemistry*, **1**, 147–81.

Feeney, R. E. & Yeh, Y. (1978). Antifreeze proteins from fish bloods. *Advances in Protein Chemistry*, **32**, 191–282.

Foldvik, A., Kvinge, T. K. & Torresen, T. (1985). Bottom currents near the continental shelf break in the Weddell Sea. *Antarctic Research Series*, **43**, 21–4.

Geoghegan, K. F., Osuga, D. T., Ahmed, A. I., Yeh, Y. & Feeney, R. E. (1982). Antifreeze glycoproteins from polar fish. Structural requirements for function of glycopeptide 8. *Journal of Biological Chemistry*, **255**, 663–7.

Gon, O. & Heemstra, P. C. (1990). *Fishes of the Southern Ocean*. Grahamstown: JLB Smith Institute of Ichthyology.

Hellmer, H. H. & Bersch, M. (1985). The Southern Ocean. A survey of oceanographic and marine meteorological research work. *Berichte zur Polarforschung*, **26**, 1–115.

Hennessey, J. P. & Johnson, W. C. (1981). Information content in the circular dichroism of proteins. *Biochemistry*, **20**, 1085–94.

Hew, C. L., Joshi, S. B., Wang, N. C., Kao, M. H. & Ananthanarayanan, V. S. (1985). Structures of shorthorn sculpin antifreeze polypeptides. *European Journal of Biochemistry*, **151**, 167–72.

Hew, C. L., Slaughter, D., Joshi, S. B., Fletcher, G. L. & Ananthanarayanan, V. S. (1984). Antifreeze polypeptides from the Newfoundland ocean pout, *Macrozoarces americanus*: presence of multiple and compositionally diverse components. *Journal of Comparative Physiology*, **155B**, 81–8.

Hubold, G. (1990). Seasonal patterns of ichthyoplankton distribution and abundance in the southern Weddell Sea. In *Antarctic Ecosystems. Ecological Change and Conservation*, ed. K. R. Kerry & G. Hempel, pp. 149–58. Berlin: Springer Verlag.

Hubold, G. (1991). Ecology of notothenioid fishes in the Weddell Sea. In *Biology of Antarctic Fishes* ed. G. di Prisco, B. Maresca & B. Tota, pp. 3–22. Berlin: Springer-Verlag.

Hubold, G. (1992). Zur Ökologie der Fische im Weddellmeer. *Berichte zur Polarforschung*, **103**, 1–157.

Hubold, G. & Ekau, W. (1987). Midwater fish fauna of the Weddell Sea, Antarctica. In *Proceedings of the Fifth Congress of European Ichthyologists*, eds. S. O. Kullander & B. Fernholm, pp. 391–6. Swedish Museum of Natural History.

Hureau, J. C., Balguerias, E., Duhamel, G., Kock, K. H., Ozouf-Costaz, C. & White M. (1990). Fish fauna of the eastern Weddell Sea. *Berichte zur Polarforschung*, **68**, 130–8.

Iwami, T. (1985). Osteology and relationships of the family Channichthyidae. *Memoirs of the National Institute for Polar Research, Series E*, **36**, 1–69.

Jung, A., Johnson, P., Eastman, J. T. & DeVries, A. L. (1995). Protein content and freezing avoidance properties of the subdermal extracellular matrix and serum of the Antarctic snailfish, *Paraliparis devriesi*. *Fish Physiology and Biochemistry*, **14**, 71–80.

Kao, M. H., Fletcher, G. L., Wang, N. C. & Hew, C. L. (1986). The relationship between molecular weight and antifreeze polypeptide activity in marine fish. *Canadian Journal of Zoology*, **64**, 578–82.

Karsten, U., Butschak, G., Cao, Y., Goletz, S. & Hanisch, F.-G. (1995). A new monoclonal antibody (A78–G(A7) to the Thomsen-Friedenreich pan-tumor antigen. *Hybridoma*, **14**, 37–44.

Kellermann, A. (1989). The larval fish community in the zone of the seasonal pack-ice cover and its seasonal and interannual variability. *Archiv für Fischereiwissenschaft*, **39**, 81–109.

Kennett, J. P. (1977). Cenozoic evolution of Antarctic glaciation, the circumantarctic ocean and their impact on global paleoceanography. *Journal of Geophysical Research*, **82**, 3843–76.

Knight, C. A., DeVries, A. L. & Oolman, L. D. (1984). Fish antifreeze proteins and the freezing and recrystallization of ice. *Nature*, **308**, 295–6.

Knox, G. A. (1970). Antarctic marine ecosystems. In *Antarctic Ecology*, ed. M. W. Holdgate, vol. 1, pp. 69–96. London: Academic Press.

Kock, K. H. (1992). *Antarctic Fish and Fisheries*. Cambridge: Cambridge University Press.

Komatsu, S. K., DeVries, A. L. & Feeney, R. E. (1970). Studies of the structure of the freezing point-depressing glycoproteins from an Antarctic fish. *Journal of Biological Chemistry*, **245**, 2901–8.

Kunzmann, A. (1991). Blood physiology and ecological consequences in Weddell Sea fishes (Antarctica). *Berichte zur Polarforschung*, **91**, 1–79.

Kunzmann, A. & Zimmermann, C. (1992). *Aethotaxis mitopteryx*, a high-Antarctic fish with benthopelagic mode of life. *Marine Ecology Progress Series*, **88**, 33–40.

Lewis, E. L. & Perkin, R. G. (1985). The winter oceanography of McMurdo Sound, Antarctica. In *Antarctic Research Series,* Vol 43, *Oceanology of the Antarctic Continental Shelf*, ed. S. S. Jacobs, pp. 145–65. Washington: American Geophysical Union.

Lubimova, T. G., Shust, K. V., Troyanovskij, F. M. & Semenov, A. B. (1983). The ecology of mass species of myctophids from the Atlantic Antarctic. *Antarktika*, **22**, 99–106 (in Russian).

O'Grady, S. M., DeVries, A. L. (1982). Osmotic and ionic regulation in polar fishes. *Journal of Experimental Marine Biology and Ecology*, **57**, 219–28.

Quilty, P. G. (1990). Significance of evidence for changes in the Antarctic marine environment over the last 5 million years. In *Antarctic Ecosystems: Change and Conservation*, ed. K. R. Kerry & G. Hempel, pp. 3–8. Berlin: Springer Verlag.

Rao, B. N. & Bush, C. A. (1987). Comparison by ^{1}H-NMR spectroscopy of the conformation of the 2600 Dalton antifreeze glycopeptide of polar cod with that of the high molecular weight. *Biopolymers*, **26**, 1227–44.

Raymond, J. A. (1989). Freezing resistance in some northern populations of Pacific herring, *Clupea harengus pallasi*. *Canadian Journal of Fisheries and Aquatic Sciences*, **46**, 2104–7.

Raymond, J. A. & DeVries, A. L. (1977). Adsorption inhibition as a mechanism of freezing resistance in polar fishes. *Proceedings of the National Academy of Sciences, USA*, **74**, 2589–93.

Rohardt, G., Ruhland, G. & Schleif, U. (1990). The Halley Bay – Kapp Norvegia comparison. *Berichte zur Polarforschung*, **68**, 39–49.

Rudels, B. (1993). High latitude ocean convection. In *Flow and Creep in the Solar System: Observations, Modeling and Theory*, ed. D. B. Stone & S. K. Runcorn, pp. 323–356. Dordrecht: Kluwer Academic Publ.

Saint Paul, U., Hubold, G. & Ekau, W. (1988). Acclimation effects on routine oxygen consumption of the antarctic fish *Pogonophryne scotti* (Artedidraconidae). *Polar Biology*, **9**, 125–8.

Schrag, J. D., O'Grady, S. M. & DeVries, A. L. (1982). Relationship of amino acid composition and molecular weight of antifreeze glycopeptides to non-colligative freezing point depression. *Biochimica et Biophysica Acta*, **717**, 322–6.

Schrag, J. D., Cheng, C. C., Panico, M., Morris, H. R. & DeVries, A. L. (1987). Primary and secondary structure of antifreeze peptides from Arctic and Antarctic zoarcid fishes. *Biochimica et Biophysica Acta*, **915**, 357–70.

Schulz, G. E. & Schirmer, R. H. (1979). *Principle of Protein Structure*. New York: Springer-Verlag.

Schwarzbach, W. (1988). Die Fischfauna des östlichen und südlichen Weddellmeeres: geographische Verbreitung, Nahrung und trophische

Stellung der Fischarten. *Berichte zur Polarforschung*, **54**, 1–94.

Shier, W. T., Lin, Y. & DeVries, A. L. (1975). Structure of the carbohydrate of antifreeze glycoproteins from an Antarctic fish. *FEBS Letters*, **54**, 135–8.

Smith, M. A. K. & Haschemeyer, A. E. V. (1980). Protein metabolism and cold adaptation in Antarctic fishes. *Physiological Zoology*, **53**, 373–82.

Toone, E. J. (1994). Structure and energetics of protein – carbohydrate complexes. *Current Opinion in Structural Biology*, **4**, 719–28.

Valerio, P. F., Kao, M. H. & Fletcher, G. L. (1992). Fish skin: an effective barrier to ice crystal propagation. *Journal of Experimental Biology*, **164**, 135–51.

Webb, P.-N. & Harwood, D. M. (1991). Late Cenozoic glacial history of the Ross Embayment, Antarctica. *Quarternary Science Review*, **10**, 215–23.

Wells, R. M. G. (1987). Respiration of Antarctic fishes from McMurdo Sound. *Comparative Biochemistry and Physiology*, **88A**, 417–24.

Wöhrmann, A. P. A. (1988). Jahreszeitliche Unterschiede in der Ernährung antarktischer Fische. MSc thesis, pp. 1–109, Kiel University.

Wöhrmann, A. P. A. (1990). Maintenance of Antarctic fish (Notothenioidei). *Berichte zur Polarforschung*, **68**, 204–5.

Wöhrmann, A. P. A. (1993). Gefrierschutz bei Fischen der Polarmeere. *Berichte zur Polarforschung*, **119**, 1–99.

Wöhrmann, A. P. A. (1995). Antifreeze glycopeptides of the high-Antarctic silverfish *Pleuragramma antarcticum* (Notothenioidei). *Comparative Biochemistry and Physiology*, **111C**, 121–9.

Wöhrmann, A. P. A. (1996). Antifreeze glycopeptides and peptides in Antarctic fish species from the Weddell Sea and the Lazarev Sea. *Marine. Ecology Progress Series*, **130**, 47–59.

Wöhrmann, A. P. A. & Haselbeck, A. (1992). Characterization of antifreeze glycoproteins of *Pleuragramma antarcticum* (Pisces: Notothenioidei). *Biological Chemistry Hoppe-Seyler*, **373**, 854.

Wöhrmann, A. P. A. & Zimmermann, C. (1992). Comparative investigations on fishes of the Weddell Sea and the Lazarev Sea. *Berichte zur Polarforschung*, **100**, 208–22.

Wöhrmann, A. P. A., Geyer, R., Hösel, W. & Haselbeck, A. (1997*a*). Identification of an additional antifreeze substance in an Antarctic fish *Pleuragramma antarcticum* (Pisces: Nototheniidae): Preliminary characterization of a novel glycoconjugate. In press.

Wöhrmann, A. P. A., Hagen, W. & Kunzmann, A. (1997*b*). Adaptations of *Pleuragramma antarcticum* (Pisces: Nototheniidae) to pelagic life in high-Antarctic waters. *Marine Ecology Progress Series*, **151**, 205–18.

Zachariassen, K. E. (1985). Physiology of cold tolerance in insects. *Physiological Reviews*, **65**, 799–832.

PART IV Integrative approaches

S. EGGINTON and W. DAVISON

Effects of environmental and experimental stress on Antarctic fish

In order to investigate physiological adaptations, it is important to be aware of stress and its effects. Stress acts on homeostatic mechanisms of an animal, thus affecting many parameters, from minor perturbations in cellular metabolism to major factors such as growth, reproduction, infection and ultimately, survival of the animal itself. Effects can differ according to whether the stress is acute, lasting a few minutes or hours, or chronic, where the animal is subjected to altered conditions for many days or months. Fish are no exception to the effects of stress and there has been a great deal of research devoted to this very large topic. Much of the earlier literature can be found reviewed in Pickering (1981), while subsequent work has been summarised by Randall and Perry (1992). It is perhaps not surprising to find that many previous 'normal' values in the literature turn out to be from highly stressed fish. As experimental techniques have been refined over the past few decades, so our awareness of what stress means for a fish has, in some cases, led to quite a different picture to that based on earlier research. Nowhere has this been more evident than in the study of Antarctic fishes, where remoteness and difficulty in sampling led to much of the early data being collected in an opportunistic, rather than a systematic manner.

The effects of stress can be nominally divided into primary and secondary stress responses. Primary relates to the initial responses of an animal to a stressor and involves changes to the autonomic nervous system and release of stress hormones, in particular catecholamines and corticosteroids. Elasmobranch fish appear to use noradrenaline as the major catecholamine while teleosts generally use adrenaline (Perry & Reid, 1994), although exceptions are found (Gamperl, Vijayan & Boutilier, 1994). Cortisol is the major corticosteroid found in fish. Secondary stress responses occur as a consequence of the primary stress hormones acting on the cells of the body. Examples of these following acute stress include altered red blood cell metabolism and pH, fluid and ion shifts in plasma due to increased branchial permeability, changes to heart rate and stroke volume, and consequent systemic effects (e.g. blood pressure, vascular resistance) (Randall & Perry, 1992). There is usually an increased metabolic rate accompanied by hyperglycaemia, hypercapnia and a lactacidosis, reflecting the mobilisation of fuel reserves with the shortfall being met by anaerobic metabolism. Chronic stress continues these metabolic perturbations, and if the stress is severe enough can lead ultimately to the demise of the animal, often as a result of impaired ionoregulation or a decreased disease resistance due to suppression of the immune system.

There is a generally held belief that a major primary stress response in fish is release of catecholamines from chromaffin tissue into the circulation, and that increased plasma levels are mainly responsible for the observed changes. A wide range of stressors will cause release of catecholamines, although Randall and Perry (1992) caution that the stressor has to be sufficiently severe. These include hypoxia (Perry & Thomas, 1991), partial emersion (Wahlquist & Nilsson, 1980), surgery (Gingerich & Drottar, 1989), hypercapnia (Kinkead *et al.*, 1993), exercise (Butler *et al.*, 1989) and extreme temperatures or abrupt temperature change (Milligan, Graham & Farrell, 1989). Absolute values will depend on the method of sampling, being higher via cardiac puncture than from cannulated fish (Fløysand *et al.*, 1992). The release of catecholamines is primarily controlled by nervous stimulation of the chromaffin tissue (Nilsson, 1994), although oxygen content of the blood is also a major (indirect) controller and can result in extremely large changes in circulating levels (Perry & Reid, 1994). Butler *et al.* (1989) showed that exercise in Atlantic cod increased adrenaline levels from 4 to 46 nmol l^{-1}. Similar values were measured in brown trout exercised at Ucrit (Butler & Day, 1993), although rainbow trout exercised at maximal sustainable speeds showed no significant changes (Wilson & Egginton, 1994). Indeed, Ristori and Laurent (1985) showed that plasma catecholamine levels did not rise during modest exercise, only following bursts of violent activity. Extreme values of over 300 nmol l^{-1} have been reported for adrenaline in trout exposed to hypoxia (Thomas *et al.*, 1991; Perry & Reid, 1994), hypercapnia (Kinkead *et al.*, 1993) and for noradrenaline in dogfish exposed to hypoxia (Metcalfe & Butler, 1989). Some early work on catecholamines showed some quite remarkable rises in these hormones. For example, adrenaline values of 1000 nmol l^{-1} for sockeye salmon and 800 nmol l^{-1} for coho salmon were reported by Mazeaud, Mazeaud and Donaldson (1987).

While adrenaline is released into the circulation very quickly, release of cortisol is much slower, although its effects are likely to be much more prolonged (Gamperl *et al.*, 1994; Waring *et al.*, 1996). In contrast to the higher catecholamine levels, swimming at Ucrit did not result in release of cortisol in brown trout (Butler & Day, 1993). Chronic confinement stress produced large increases in circulating cortisol in striped bass (Davis & Parker, 1990) and in rainbow trout, with the elevated levels persisting for many days (Gamperl *et al.*, 1994).

Much work has been carried out on the effects of stress on blood composition and the cardiovascular system. Stress causes an increased haematocrit due to a combination of increased numbers of red blood cells and red cell swelling. The spleen is a major site for storage of erythrocytes in teleost fish, and stress causes catecholamine-mediated contraction leading to expulsion of erythrocytes into the circulation (Yamamoto, Itazawa & Kobayashi 1983; Wells & Weber, 1990; Pearson, van der Kraak & Stevens, 1992). The increased haemoglobin levels allow for increased oxygen transport via an increased carrying capacity, although there is a potential trade-off in that an elevated haematocrit increases blood viscosity, making more work (increased afterload) for a heart that may be already working hard (such as during intense exercise). However, Sorensen and Weber (1995) have suggested that increased circulating levels of adrenaline reduces blood viscosity at low shear rates, which the authors regard as physiologically important. This may occur in two ways. Red cell swelling is a product of a β adrenergic-mediated expulsion of protons in exchange for Na^+ ions, resulting in maintenance or even elevation of intracellular pH in the face of lowered plasma pH (Nikinmaa, 1983; Thomas & Perry, 1992; Berenbrink & Bridges, 1994). A modest swelling will lower the internal viscosity of red cells, thereby aiding deformation within the microcirculation (Nash & Egginton, 1993). Alternatively, as plasma viscosity essentially reflects protein concentration, any dilution would have the desired effect. However, this appears an unlikely mechanism for marine teleosts, as stress typically causes haemoconcentration due to loss of water.

Most stressors cause increased cardiac output as a consequence of a combination of increased heart rate and stroke volume, due to the well known chronotropic and inotropic effects of circulating catecholamines on myocytes. The major effects on cardiac afterload reflect the balance between β adrenoceptor- mediated dilatation in the branchial circulation, and α adrenoceptor-mediated vasoconstriction in the systemic circulation. This tends to result in an increased dorsal aortic blood pressure and increased perfusion of the gill secondary lamellae, allowing increased oxygen uptake (Randall & Daxboeck, 1984). The permeability of the gill epithelial cells may be increased, resulting in greater diffusion of oxygen, but also inorganic ions and water (Isaia, Maetz & Haywood, 1978; Perry, Booth & McDonald, 1985), such that increasing branchial oxygen uptake may incur additional costs in order to maintain osmo- and ionoregulation.

Antarctic fish

In terms of their physiology in general, and their response to stress in particular, Antarctic fish are a relatively unknown quantity. The present-day fauna is quite recent, appearing about 30 million years ago, but with no apparent link to the fish fauna existing before that time (Eastman, 1993). There are at least 274 species of teleost fishes regarded as Antarctic, although many of these are transient visitors to the northern extremities of the Southern Ocean. Many other species are better described as deep water fishes, such as the myctophids, liparids and zoarcids. The animals usually regarded as representative Antarctic fish are perciform species belonging to the suborder Notothenioidei. They comprise the bulk of the biomass of the region and are almost exclusively the only fish caught in shallow coastal waters. This has ensured that most research has been carried out on this one group of animals. The notothenioids are Antarctic specialists. Of 120 species, 95 are endemic to Antarctic waters, a consequence of rapid radiation following their initial invasion (Eastman, 1993). These fishes show some remarkable adaptations to life in subzero temperatures, including production of plasma glycopeptide antifreezes. However, the ability to produce antifreezes is not restricted to these animals, and both peptide and glycopeptide antifreezes can be found in Arctic fish (DeVries, 1983) and many Antarctic non-notothenioids (Wöhrmann, this volume).

Thus, in the following discussion it must be always kept in mind that virtually all stress work has been carried out on this one group of fishes. The Southern Ocean offers a uniquely stable thermal environment within which cold adaptation of fishes may be expected to have occurred, obviating the need to retain functional plasticity required by more variable ecosystems. For example, in the inshore marine environment the annual range is only around −1.5 to +1.5 °C while in deep water, which occurs closer than around other continents, temperature hardly varies from around the freezing point of seawater, −1.86 °C (Eastman, 1993). On the other hand, physical isolation caused by the opening of the Drake Passage has provided the potential for unique features resulting from endemic speciation. What might, at first, appear to be a cold adaptation may, instead, be a specialisation of the notothenioids. Unravelling the stress response of these fishes may help to clarify the situation a little, e.g. quantitative differences such as time required for recovery could reflect the degree of cold adaptation, while qualitative differences more likely reflect group characteristics. However, more detailed comparisons with other species are required in order to establish which adaptations are unique to the Antarctic ichthyofauna and which are merely extensions of those, for example, observed during winter acclimatisation of temperate water fishes.

Establishing a baseline

One of the earliest adaptations proposed for Antarctic fishes was that of metabolic cold adaptation, which suggested that the oxygen consumption ($\dot{V}O_2$) of indigenous species is higher than that of a comparable ecotype from outside Antarctica when measured at a similar temperature (Clarke, this volume). Many of the original high values of $\dot{V}O_2$ can be explained by the effects of anthropogenic stress (capture, tank-holding, etc.), such that it has been dismissed by several authors as an experimental artefact (Holeton, 1970; Clarke, 1991). Others are not convinced by this argument, pointing out that the apparently definitive experiments of Holeton used sluggish species, whereas more active fishes appear to show a true elevation of $\dot{V}O_2$ (Forster *et al.*, 1987; MacDonald, Montgomery & Wells, 1987). As experimental animals are subject to potentially severe levels of stress, it is important to first characterise the response to, and the time for recovery from, such interventions in order to establish true baseline values. This is clearly essential when attempting to quantify the magnitude of any stress response, which typically increases $\dot{V}O_2$. A further complication arises with the icefish (Channichthyidae) whose behavioural response to stress (immobility reflex on handling, rigidity in nets) will greatly attenuate any such increase in $\dot{V}O_2$ (Twelves, 1972; Egginton, 1994). Indeed, this sort of response may be widespread, e.g. the zoarcid *Lycodichthys dearboni* adopts a flaccid posture when handled (Montgomery & Wells, 1993).

The most immediate, and perhaps extensive, anthropogenic stress is that associated with initial capture of the fish. One of the more common, and very effective, methods is the use of trawl or trammel nets. The latter type in particular will restrict opercular movement and efficiency of the buccal pump, such that even in these oxygen-rich waters specimens may be expected to experience a relative hypoxia. In addition, an osmotic and ionic disturbance usually occurs when temperate water fish are subjected to handling stress, with marine teleosts experiencing an increased plasma osmolarity and ionic concentration following either an efflux of water or influx of ions. For nototheniids, this primarily reflects changes in branchial permeability, but for the channichthyids any physical damage to the scaleless integument will exacerbate the situation due to increased water loss. For example, following capture, renal handling of divalent cations in *Notothenia coriiceps* (previously *N. neglecta*) maintained plasma Mg^{2+} within normal levels (1–2 mM), while it appeared to be inadequate to cope with the increased gut absorption subsequent to presumed elevated drinking rates or increased integumental permeability in *Chaenocephalus aceratus* (range 1–18 mM). There are potentially two populations represented in Fig. 1, those with intact integuments and thus low [Mg^{2+}], and those with damaged integuments and thus high [Mg^{2+}].

Given the severe logistical problems of conducting experiments in

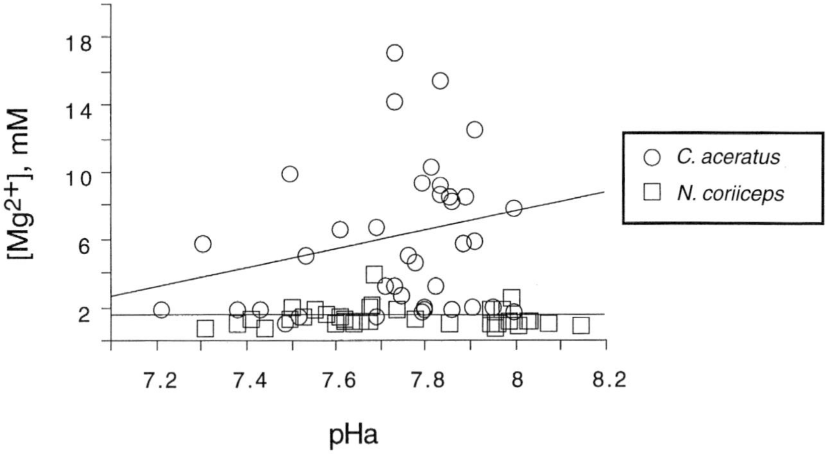

Fig. 1. Plasma Mg^{2+} concentration from fish of varying levels of stress, as indicated by the arterial blood pH (pHa) in *N. coriiceps* (□) and *C. aceratus* (○). The integument in some specimens of the latter, the scaleless icefish, had clearly been damaged during capture by netting, causing a loss of water and subsequent increased drinking rate and possibly permeability of the skin (Egginton, 1994, 1997).

$$[Mg^{2+}]=2.812-0.175 * pHa;\ r^2=0.004\ (N.\ coriiceps)$$

$$[Mg^{2+}]=-38.37+5.748 * pHa;\ r^2=0.06\ (C.\ aceratus)$$

Antarctica, there is increasing pressure to work on specimens shipped back to home research institutions. This is not without its problems. Contrasting haematological values among closely related *Notothenia* species was thought to reflect more the problems associated with extended transportation/holding (and all the factors associated with this, such as different diet, temperatures, water quality, photoperiod etc.) than true interspecific differences (Egginton *et al.*, 1991). Indeed, specimens of *N. coriiceps* sampled *in situ* had higher pH and plasma [chloride], and lower P_{CO_2} and blood [lactate], in blood taken at rest compared with those returned to the UK, while *N. rossii* had substantially higher haematocrit (Egginton, 1997). Similarly, Davison *et al.* (1995) found higher heart rate and mean corpuscular haemoglobin concentration, and significantly lower [Cl^-] in *Trematomus bernacchii* brought to New Zealand compared with freshly caught specimens (Table 1).

Viscometric considerations

There is a trend towards lowered numbers of circulating red blood cells in the nototheniids, which may be a specific adaptation to lower blood viscosity

Table 1. *Effect of transportation and prolonged captivity on selected blood values of nototheniids*

		pHa	Pa_{CO_2} (kPa)	lac⁻ (mmol l^{-1})	Hct (%)	MCHC (g dl^{-1})	Cl^- (mmol l^{-1})	Ref.
N. coriiceps	Signy Island	7.96±0.02	0.34±0.02	0.28±0.08	16.2±0.15	22.3±0.9	182±3	*a*
	UK	*7.81±0.03	*0.57±0.05	0.41±0.08	18.6±2.5	*31.6±2.2	175±10	*b*
N. rossii	Signy Island	7.90±0.04	0.34±0.02	0.03±0.07	18.4±0.11	—	157±11	*a*
	UK	8.02±0.07	*0.48±0.03	0.11±0.02	*11.7±0.75	—	172±6	*b*
T. bernacchii	Scott Base	—	—	—	11.1±0.9	15.4±0.6	236±4	*c*
	NZ	—	—	—	10.8±1.0	*18.0±1.2	*223±3	*c*

Notes:

*P<0.05 between sites.

Abbreviations : pHa, arterial blood pH; Pa_{CO_2}, arterial blood CO_2 tension; lac⁻, whole blood lactate concentration; Hct, haematocrit; MCHC, mean corpuscular haemoglobin concentration; Cl^-, plasma ion concentration.

[a] Egginton, 1997; [b] Egginton *et al.*, 1991; [c] Davison *et al.*, 1995.

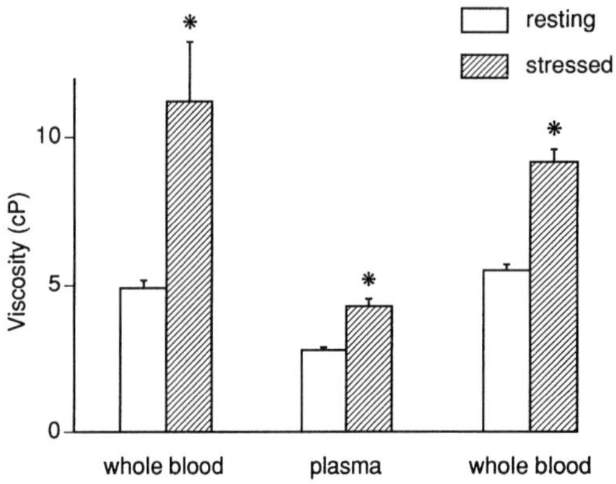

Fig. 2. Viscosity changes in whole blood of *P. borchgrevinki* (left), and plasma and whole blood of *N. coriiceps* (right), subjected to handling stress. The substantial rise in haematocrit and unspecified changes in plasma produce a significant rise in all cases (* $P<0.05$ v. resting). Parallel changes are not observed in icefish (Wells *et al.*, 1990; Egginton, 1996).

(MacDonald & Wells, 1991). As the afterload on the heart posed by viscous blood at subzero temperatures would be substantial, this clearly would reduce the energetic cost of perfusing the blood vascular system, and Wells, Macdonald & di Prisco (1990) favour the view that nototheniids are following the same evolutionary path as the channichthyids in reducing their dependency on respiratory pigments (see also Sidell, this volume). This may, however, be an example where apparent cold adaptation is masking an effect of endemic speciation. Several pieces of evidence suggest that viscometric considerations are not restrictive for nototheniid physiology. First, the generalised stress response is thought to involve an increased haematocrit (MacDonald *et al.*, 1987), more than doubling in response to forced exercise in *Pagothenia borchgrevinki* (Davison *et al.*, 1988). Indeed, handling stress increased blood viscosity in this species by 230% (Wells *et al.*, 1990), and both whole blood and plasma viscosity in *N. coriiceps* by around 150% (Egginton, 1996) (Fig. 2). These represent substantial increases in the cost of vascular perfusion which appear to be easily accommodated. In addition, while there is some evidence for a reduced plasma viscosity relative to other species (Egginton, 1996), the haematocrits found are not abnormally low, simply lying toward the lower end of normal teleost values. Indeed, a population of rainbow trout has been worked with that naturally had a lower haematocrit (S.E. Taylor, S. Egginton & E.W. Taylor, unpublished data). Furthermore, nototheniids have red cells with a very low mean corpuscular haemoglobin concentration (MCHC), which appears to be a latitudinal adaptation (Wells *et al.*, 1980). While a low MCHC may increase erythrocyte deformability by decreasing intracellular viscosity (Nash & Egginton, 1993), this effect on *Notothenia* red cell deformability is quite small (Lecklin, Nash & Egginton, 1994). If blood viscosity really were a limiting factor, then a more 'efficient' way to reduce haemoglobin content would be to further reduce haematocrit, rather than MCHC.

Viscosity arguments really fall down, however, when we consider the icefish. Any apparent reduction in red cell count has been taken to the extreme in the family Channichthyidae, where only a remnant of vestigial erythrocytes lacking haemoglobin are found, with a packed cell volume of $<1\%$. While this is usually taken to be the logical extension of the condition found in nototheniids and further proof of the importance of viscosity, and such a low 'haematocrit' certainly earns them the label of 'thin blooded' (Wells *et al.*, 1990), this is somewhat misleading. Whole blood, essentially plasma, viscosity is actually higher than that of nototheniid plasma, being almost identical to human plasma at 0 °C (Egginton, 1996). More importantly, though, is the presence of large-bore microvessels that are some four-fold larger than those of trout (Egginton & Rankin, 1991). Unlike nototheniids, there is no increase in plasma viscosity following handling stress and cardiac puncture (Wells *et al.*, 1990; Egginton, 1996). However, as vascular resistance increases directly with blood or plasma viscosity but with the fourth power of vessel radius (Poiseuille's Law), the vascular anatomy precludes any meaningful role for viscosity in determining either cardiac work load or tissue perfusion (Egginton & Rankin, 1997).

Oxygen transport

Fishes that are found exclusively in the waters around Antarctica are relatively anaemic, having around half the haemoglobin concentration of coastal species from more temperate latitudes. As haematocrit is not reduced to the same extent (see above), this primarily reflects a reduction in the haemoglobin content per erythrocyte (MCHC). It is usually assumed that the very high dissolved oxygen content in seawater, combined with low resting $\dot{V}O_2$, obviates the need for higher blood oxygen-carrying capacity. Indeed, haemoglobin is probably only of significant benefit in the face of environmental hypoxia (Holeton, 1970), which is unlikely to be experienced by fishes in this ecosystem, and recently di Prisco and MacDonald (1992) demonstrated that exposure to carbon monoxide has little effect on survival of these fish. Thus, circulating haemoglobin appears to be of little importance for transport of oxygen in resting animals. In view of the high oxygen affinity of haemoglobins from Antarctic fishes, and that intraerythrocytic oxygen affinity is dependent on haemoglobin concentration, it may be that a reduced MCHC avoids an unnecessary high affinity at such low temperatures (Wells *et al.*, 1980).

Higher levels of oxygen demand, however, do require the extra oxygen bound to haemoglobin. Franklin, Davison and McKenzie (1993) showed that restricting haemoglobin levels by isolating the spleen severely affected the ability of *Pagothenia borchgrevinki* to swim. The P_{50} of its haemoglobin was 2.7 kPa (Tetens, Wells & DeVries, 1984) while the venous Po_2 (ventral aorta) of resting specimens was approximately 5kPa, suggesting little or no desaturation of the haemoglobin (cf. 10 kPa in dorsal aorta blood). Exercise causes a fall in venous Po_2 to close to zero indicating a major role for the oxygen bound to haemoglobin during exercise (W. Davison, unpublished data), and perhaps any other stressor that produces an increase in metabolic rate.

The reduced importance of facilitated oxygen transport is most evident in the channichthyids which, until recently, were thought to be unique among vertebrates in lacking significant gene expression for the respiratory pigments myoglobin and haemoglobin (Sidell, this volume). A very large heart (Hemmingsen & Douglas, 1972) and impressive blood volume (Acierno *et al.*, 1995) accommodated in large blood vessels means that, despite a relatively low capillary density (Fitch, Johnston & Wood, 1984), these fish are able to extract enough oxygen that is transported in physical solution only (Hemmingsen & Douglas, 1970). This low pressure, low resistance cardiovascular system represents a perfusion, rather than the usual diffusion-limited situation. This is likely to be tenable only in the cold Antarctic waters with the associated low standard metabolic rates, although it should be remembered that most channichthyids are active predators and would often require oxygen at higher than resting rates, but presumably with a lower $\dot{V}o_2$ max than red-blooded nototheniids. This may explain their tonic immobility reflex in response to stress, mentioned earlier.

It has been suggested that the low metabolic rate and high oxygen solubility associated with this environment may also allow survival under pathological conditions likely to be lethal at warmer temperatures. The best example is the presence of so-called X-cell tumours of fishes from McMurdo Sound, where fusion of gill lamellae and tissue hyperplasia would otherwise decrease effective respiratory surface area and increase diffusion distances to a lethal degree (Davison, Franklin & Carey, 1990).

Primary stress response

While there is some variability in the magnitude of the response, and whether adrenaline or noradrenaline dominates, it is clear that teleosts have an immediate and vigorous response to induced stress. By contrast, the limited available data suggest that Antarctic notothenioids release little or no catecholamines in response to otherwise potent stressors. This has been shown for *Notothenia coriiceps* and *Chaenocephalus aceratus* from the Antarctic Peninsula subjected to net capture or surgery (Egginton, 1994), or forced exercise (Egginton, 1997). These samples were obtained via post-branchial (dorsal aorta) cannulae and, as gills are involved in biogenic amine metabolism (Randall & Perry, 1992), may represent minimum values. However, pre-branchial (ventral aorta) samples were within the range for post-exercise values taken from the dorsal aorta, suggesting this is unlikely to pose a serious problem (S. Egginton, unpublished data). Likewise, *Trematomus bernacchii* from McMurdo Sound subjected to the stress of either 5 min. handling or forced exercise had pre-branchial catecholamine concentrations that were similar to those of other teleosts at rest (Davison *et al.*, 1995). There is a single reference to increased catecholamines in an Antarctic fish. Ryan (1995) cites some unpublished data giving levels of 20 nmol l^{-1} for adrenaline and 40 nmol l^{-1} for noradrenaline in exercise-stressed *Pagothenia borchgrevinki*, values close to the stressed levels of some temperate water fishes (Axelsson & Nilsson, 1986). However, this has not been verified by the cited author (T. Lowe, pers. comm.). Until plasma samples from stressed *P. borchgrevinki* can be collected and analysed this must remain in doubt. It is unclear to what extent this trait reflects notothenioid radiation, or simply a relatively inactive lifestyle. For example, resting Gulf toadfish *Opsanus beta* has catecholamine levels of <1 nmol l^{-1}, with post-exercise increases in noradrenaline and adrenaline levels of only 1.7 and 8 nmol l^{-1}, respectively (Walsh, 1989). However, a few samples from pelagic icefish caught in midwater trawls (surely one of the most severe stressors) also had low catecholamine levels (S. Egginton & J.C. Rankin, unpublished data).

At present, it is unclear whether this strikingly weak primary stress response is a result of an inefficient chromaffin tissue (failure to release or synthesise) or down-regulated adrenergic control (receptor density or efficacy). Injection of adrenaline and noradrenaline into Antarctic fish produced a normal rise in systemic blood pressure (Axelsson *et al.*, 1994; Egginton, 1997) and fall in resistance of isolated gill arches (Rankin, 1989). As blood-borne catecholamines rise only a little during stress in these fish, it is interesting to note that in *P. borchgrevinki* exercise caused a marked decrease in dorsal aortic blood pressure which, coupled with a large increase in cardiac output, led to a decrease in systemic resistance (Axelsson *et al.*, 1994). Nevertheless, the blunted release may still account for a doubling of circulating noradrenaline in *N. coriiceps* (Fig. 3), and the resolution of their role must await data on receptor characteristics. Faced with low circulating titres it is reasonable to expect, if anything, an up-regulation of adrenoceptor populations. Also, rainbow trout acclimated to low temperature show an

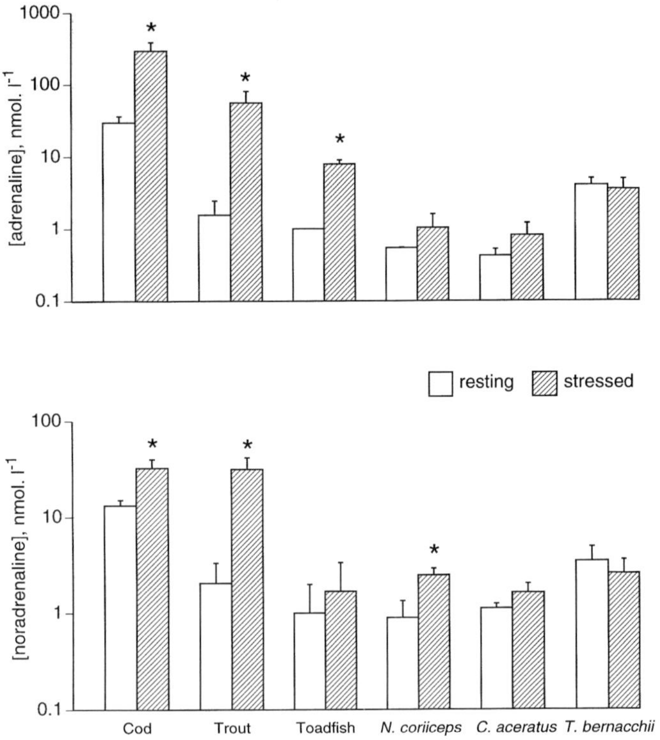

Fig. 3. Effect of induced stress on circulating catecholamine levels. Cod and trout represent the more usual teleost response, which is depressed in the sluggish toadfish, and suppressed in Antarctic fishes (* $P<0.05$ v. resting). Cod (emersion), Wahlqvist & Nilsson 1980; trout (surgery), Gingerich & Drottar, 1989; toadfish (exercise), Walsh, 1989; *N. coriiceps* and *C. aceratus* (netting), Egginton, 1994; *T. bernacchii* (handling), Davison *et al.*, 1995.

increased adrenergic sensitivity (Keen *et al.*, 1993), and it may be that Antarctic fishes represent an extension of this trend. However, using a perfused gill arch preparation Rankin (1989) demonstrated catecholamine-induced branchial vasodilatation in *C. aceratus* similar to that seen in gills of brown trout, and with an EC_{50} (median effective concentration) within the normal range for teleosts. Taken together, these data suggest that the potential for adrenoceptor response is present, but that adequate circulating levels are not produced. It is unclear what, if any, selective advantage this confers.

Stress also leads to changes in colouration of Antarctic fish, usually causing the fish to turn a darker colour. This is seen in response to elevated temperature in *P. borchgrevinki* (Franklin, Davison & Carey, 1991; Ryan, 1995), in several species of nototheniid subjected to high and low oxygen levels (Fanta, Lucchiari & Bacila, 1989), and to the water soluble fraction of fuel oil (Davison *et al.*, 1992). Melanocytes are usually under both hormonal (pituitary release of melanocyte stimulating hormone, MSH) and neural control (antagonistic sympathetic and parasympathetic innervation). The relative importance of these mechanisms in Antarctic fishes is not known.

Cortisol levels rose following heat stress from 15 to 70 ng ml^{-1} in *P. borchgrevinki* (Ryan, 1995). However, in another study using *P. borchgrevinki*, exercise produced a fall in cortisol, while levels in *T. bernacchii* did not change, although the authors do warn that control levels suggested that the animals were chronically stressed (Lowe & Wells, 1997). It is clear that much more work needs to be carried out on release of primary stress hormones in Antarctic fishes. While the available data are sparse and contradictory, it does appear that circulating catecholamines are not a major part of the stress response and this, of course, has consequences relating to the manifestation of secondary responses.

Secondary stress responses

If Antarctic fishes do not release catecholamines in response to stress, then it might be hypothesised that many of the secondary stress responses seen in temperate fishes as a consequence of increases in these hormones might not be seen. This depends very much on the species, although Wells, Tetens and DeVries (1984) and Davison *et al.* (1988) comment that nototheniids in general are very susceptible to stress, and that care needs to be taken in any physiological work with these fish. For example, the active nototheniid *P. borchgrevinki* shows massive increases in haematocrit in response to a range of stressors such as exercise (Davison *et al.*, 1988; Franklin *et al.*, 1993), high temperature (Franklin *et al.*, 1991) handling (Wells *et al.*, 1984), hypoxia (Wells *et al.*, 1989), anaesthesia (Ryan, 1992), gill disease (Davison *et al.*, 1990) and oil pollution (Davison *et al.*, 1992, 1993). Similarly, *Dissostichus mawsoni* can show substantial increases in haematocrit following capture stress (Wells *et al.*, 1984). The relationship between degree of haematocrit increase and the amount of measured cell swelling (Fig. 4a) may reflect the type or intensity of stressor, or lifestyle of different species. It appears that a substantial stress-induced rise in haematocrit is generally found in pelagic species, while benthic nototheniids are less sensitive. For example, while the benthic *T. bernacchii* also shows haematocrit increases in response to a range of stressors (Davison, Franklin & McKenzie, 1994) including exercise, high temperature and hypoxia, the response is modest in comparison to that shown by *P. borchgrevinki*. Indeed, two Antarctic Peninsula species (*Notothenia rossii* and *N. coriiceps*) appear to be acutely insensitive to the stress of imposed exercise (Egginton *et al.*, 1991; Egginton, 1997). However,

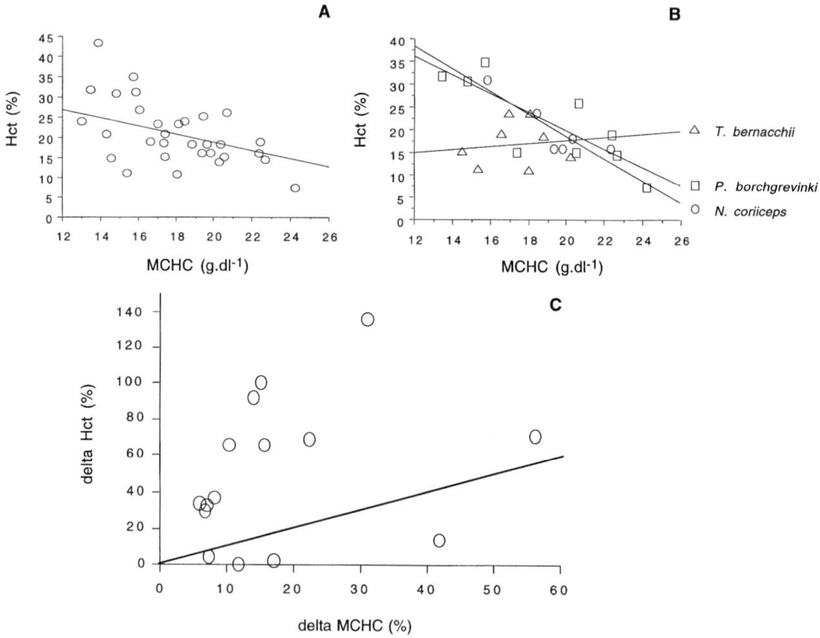

Fig. 4. A. Published haematological values for a range of Antarctic fishes suggest a weak correlation between haematocrit and degree of red cell swelling, indicated by lower MCHC values (means for individual species; $r^2=0.1$). B. Interspecific differences cannot be explained by lifestyle alone: a subset of the data in A shows that the pelagic *P. borchgrevinki* is very sensitive to stress, while the benthic *T. bernacchii* is not. However, *N. coriiceps* is also a benthic species and is sensitive to stressors other than induced exercise. (Mean values from different experiments; Franklin *et al.*, 1993; Davison *et al.*, 1994; Egginton, 1994.) C. Line of identity shows the expected response if changes in haematocrit were due solely to red cell swelling. By contrast, the relative change in haematocrit and MCHC shows that the general response to stress in nototheniids is polycythaemia.

Egginton (1994) did note a significant decrease in haematocrit during recovery from handling stress and anaesthesia in *N. coriiceps*, suggesting that other factors may dominate (Fig. 4b).

Stressed temperate water fish invariably show catecholamine-mediated red cell swelling. Available data from experiments with nototheniid fish have shown a range of responses. Exercise certainly produced swelling in *P. borchgrevinki* (Franklin *et al.*, 1993; Lowe & Wells, 1997) as did anaesthesia (Ryan, 1992) and a temperature of 8 °C (Ryan, 1995), although a lower temperature in the same study (5 °C) elicited no change in cell size, and stress from confinement actually produced cell shrinking. Wells *et al.* (1984) showed red cell swelling in *P. borchgrevinki* and *D. mawsoni* following capture, although Wells *et al.* (1989) showed no cell swelling in response to hypoxia. Chronic exposure of *P. borchgrevinki* to fuel oil did not result in red cell volume change (Davison *et al.*, 1993). Handling stress, exercise, a temperature of 10 °C and hypoxia produced no change to red cell volume in *T. bernacchii* (Davison *et al.*, 1994, 1995), while handling stress increased blood viscosity due to red cell swelling (Wells *et al.*, 1990), and exercise caused cell shrinkage (Lowe, T.E. and R.M.G. Wells, unpublished data). No change to red cell volume was apparent following exercise in *N. coriiceps* (Egginton *et al.*, 1991), although handling stress caused cell swelling in this species (Egginton, 1994). It thus appears that different stressors may produce different responses, and indeed the same stressor may produce opposite responses in different species. However, *P. borchgrevinki*, a very active pelagic fish, shows some of the biggest changes in both haematocrit and cell swelling and this actually confuses the picture. While the haematocrit in this fish can increase by over 150% following severe stress, there is the potential that what is being measured as cell swelling might actually reflect differences between cells in circulation and cells released from storage in the spleen. Certainly, there is not a single response and it may be that the size of the red cells is governed more by the osmotic state of the blood rather than any change to the red cell membrane.

While there is a broad correlation between haematocrit and MCHC (Fig. 4A), reminiscent of the adrenergic stress response of salmonids where an increase in haematocrit is largely, though not exclusively due to red cell swelling (i.e. a decrease in MCHC), in nototheniids this appears not to be the major source of the increase (Fig. 4C). Rather, there appears to be a substantial polycythaemia with increased red cell numbers as a result of splenic contraction (Fig. 5), which is affected mainly via cholinergic nerve fibres rather than circulating hormones (Wells *et al.*, 1989; Franklin *et al.*, 1993; Davison *et al.*, 1995; Nilsson *et al.*, 1996). Interestingly, although they appear to lack an exercise polycythaemia, the relative spleen mass (% body weight, BW) of *N. coriiceps* following the stress of handling or anaesthesia was similar to that of *P. borchgrevinki* (range 0.09–0.76% BW; S. Egginton, unpublished data), suggesting the degree of change in spleen mass depended on the type of stress experienced. Icefish, of course, do not have red blood cells and so stress will not affect an increased haematocrit. However, they do possess a spleen, and the wide range of relative spleen masses: 0.07 to 0.39% BW in several specimens of relatively stressed *Chionodraco hamatus* (W. Davison, unpublished data), and 0.17 to 0.75% BW in *Chaenocephalus aceratus* (S. Egginton, unpublished data), suggest that these fish may still have a contractile spleen, and that a similar degree of splenic contraction may occur in both red- and white-blooded species.

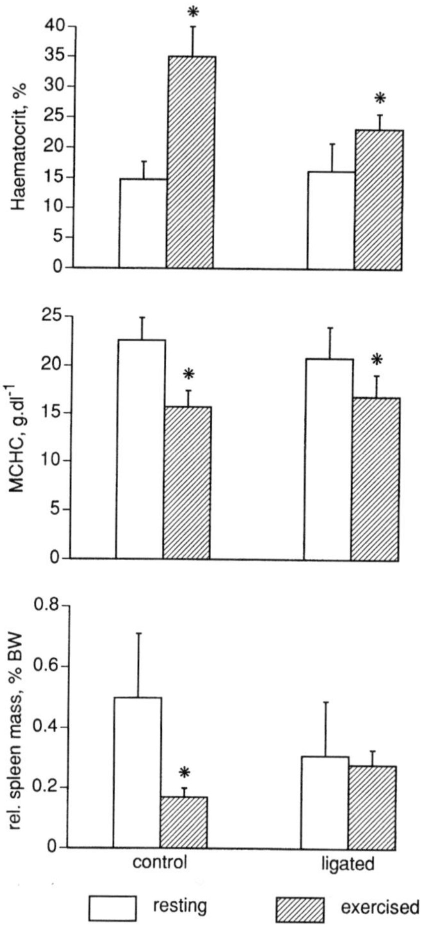

Fig. 5. Exercise induces a substantial increase in haematocrit of *P. borchgrevinki*, with a decrease in relative spleen mass (as a percentage of body weight) that suggests it is due to release of stored red cells rather than red cell swelling. Ligation of the spleen shows that polycythaemia is the major contributor to the elevated haematocrit (* $P<0.05$ vs. resting) (Franklin *et al.*, 1993).

Cholinergic control seems to be dominant in many control systems in Antarctic fish. As has already been stated, the spleen is almost entirely under cholinergic control (Nilsson *et al.*, 1996). The heart also appears to be mainly controlled by similar mechanisms, with resting heart rate determined by a very high inhibitory cholinergic tone of 55% in *P. borchgrevinki* and a massive 80% in *T. bernacchii* (Axelsson *et al.*, 1992). In this study it was shown that these species are unusual in that they do not display a hypoxic bradycardia, presumably because of the existing dominance of the cholinergic inhibition, and in fact often show a slight tachycardia. *T. bernacchii* shows no increase in heart rate over a temperature increase of three degrees, although it loses control after that, while atropine abolished this control (Axelsson *et al.*, 1992). Gills also appear to be under cholinergic control (Axelsson *et al.*, 1994). However, adrenergic control is still potentially important. Blood vessels are responsive to adrenaline rather than acetylcholine (Nilsson *et al.*, 1996), and both α and β responses can be seen in the gills (Rankin, 1989; Axelsson *et al.*, 1994). The balance of control may reflect lifestyle, with the benthic *T. bernacchii* having an excitatory adrenergic tone of around 30%, while in the cryopelagic *P. borchgrevinki* it is an order of magnitude less (Axelsson *et al.*, 1992).

The relatively long recovery times following stress recorded for blood values in temperate water teleosts may be expected to be further increased at very cold temperatures, unless significant adaptation has occurred. This has important implications for the behavioural ecology of the fishes. In fact, the time required for recovery of blood composition and standard oxygen consumption to resting values appears to be similar to that observed with temperate water species, and for most parameters this is essentially complete within 8–24 h. (Wells *et al.*, 1984; Forster *et al.*, 1987; Davison *et al.*, 1988; Saint-Paul *et al.*, 1988; Egginton *et al.*, 1991; Franklin *et al.*, 1991; Egginton, 1994, 1997). These data clearly demonstrate that an effective rate compensation has been achieved by the nototheniids during the process of gradual cold adaptation. By contrast, icefish show a prolonged recovery period which may reflect a limitation on aerobic metabolism imposed by the lack of respiratory pigments (Fig. 6). This is most clearly demonstrated by the response to catecholamine infusion, causing a 130% increase in blood pressure in *N. coriiceps* which returned to basal level around 70 min later, whereas a reduced pressor response of only a 90% increase in blood pressure in *C. aceratus* took nearly twice as long for full recovery (S. Egginton, unpublished data).

In addition to the blunted catecholamine response, there is also a severely attenuated lactacidosis in response to stress. For example, although the experiment of leaving *D. mawsoni* for 24 h on a set line was sufficient to decrease arterial pH from 8.2 to 7.7, the highest level of blood lactate recorded was a mere 2 mmol l^{-1} (Qvist *et al.*, 1977). Even lower levels were found in *P. borchgrevinki, N. rossii* and *N. coriiceps* when swam to exhaustion (Davison *et al.*, 1988; Egginton *et al.*, 1991). This is paralleled by a relatively constant intracellular concentration (Fig. 7), showing that it reflects low production rather than non- release, which has its origin in the limited anaerobic (glycolytic) capacity of skeletal muscle and a particularly heavy reliance on

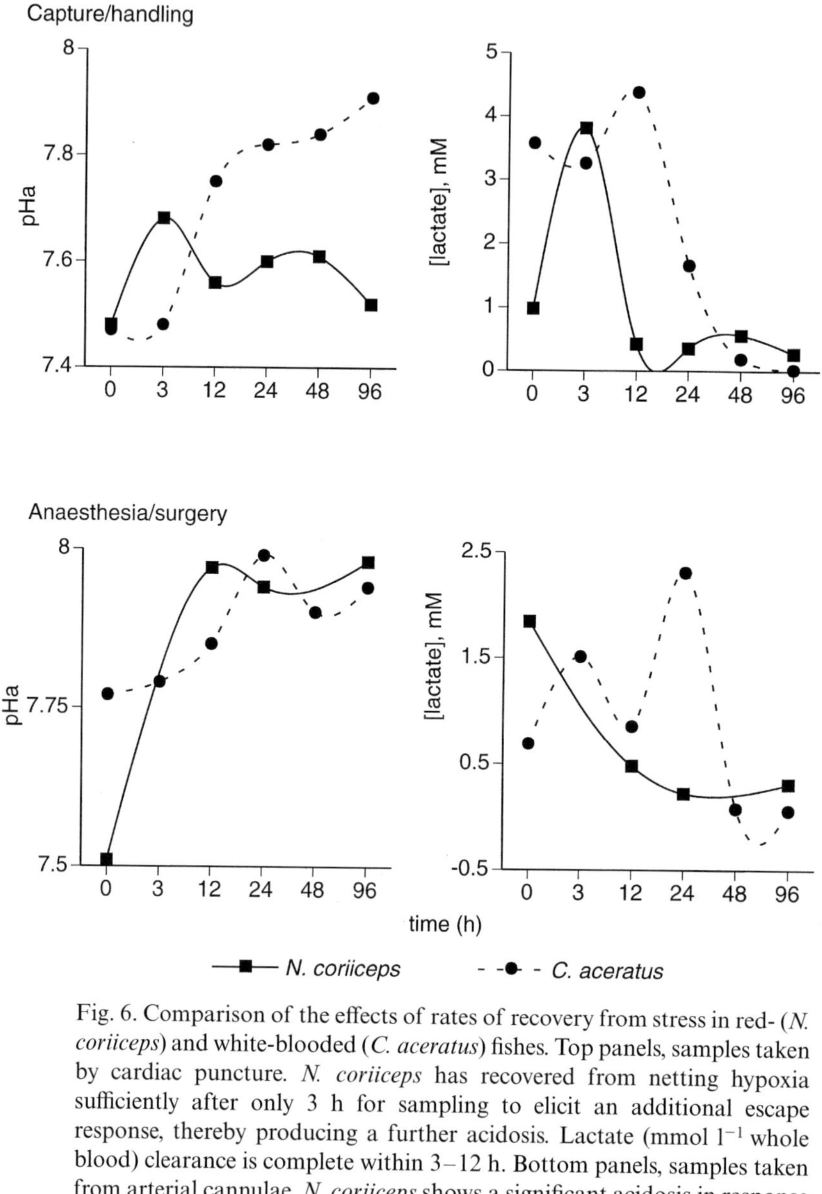

Fig. 6. Comparison of the effects of rates of recovery from stress in red- (*N. coriiceps*) and white-blooded (*C. aceratus*) fishes. Top panels, samples taken by cardiac puncture. *N. coriiceps* has recovered from netting hypoxia sufficiently after only 3 h for sampling to elicit an additional escape response, thereby producing a further acidosis. Lactate (mmol l^{-1} whole blood) clearance is complete within 3–12 h. Bottom panels, samples taken from arterial cannulae. *N. coriiceps* shows a significant acidosis in response to surgical hypoxaemia which recovers by 12 h, and is paralleled by lactate clearance. In all cases, the icefish shows a significantly prolonged recovery in pHa and [lactate], even when the initial acidosis is less. N=2–6, error bars omitted for clarity (Egginton, 1994).

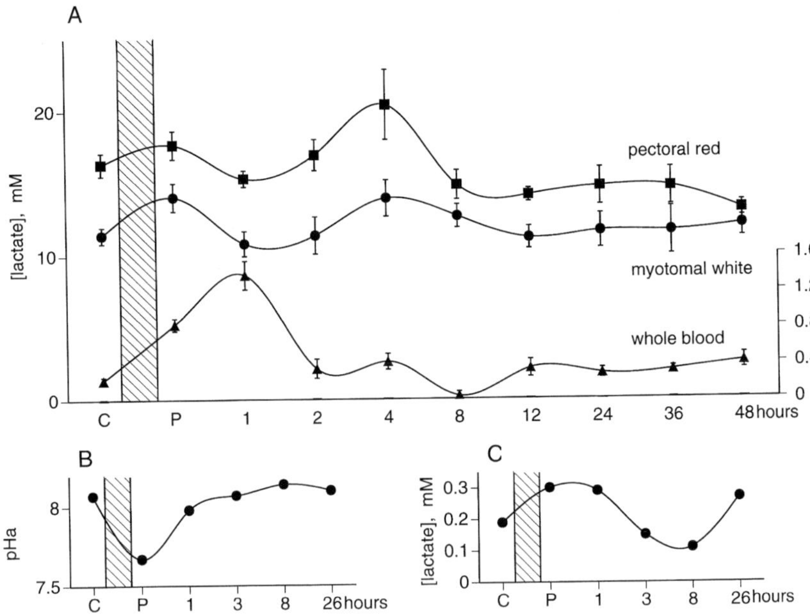

Fig. 7. Lactate handling in nototheniids. A. Whole blood and intracellular (mmol kg^{-1} wet weight) lactate concentrations in *P. borchgrevinki* following exercise. The modest lactacidosis follows the usual short delay, but similar values in muscle reflect a low glycolytic capacity. Time course of blood pH, B, and lactate, C in *N. coriiceps* after a similar increase in activity. C, Control (resting) values; P, values immediately following 5 min induced activity (hatched bar). N=3–6, error bars in B and C omitted for clarity (Davison *et al.*, 1988; Egginton, 1997).

phosphocreatine hydrolysis for burst activity (Dunn & Johnston, 1986; Davison *et al.*, 1988). Consequently, in *Notothenia* the major effect of induced stress is a respiratory, rather than a metabolic acidosis (Fig. 8). Interestingly, the icefish, *C. aceratus*, shows a modest hypercapnia and lactacidosis (Egginton, 1994). This may reflect both behaviour that minimises stress intensity, and the greater capacity for glycolysis and gluconeogenesis (Bacila *et al.*, 1989) than *Notothenia*, respectively. As a result, the accumulated fixed-acid load in all species during exercise stress is much lower than that observed in temperate water fishes, in keeping with their lower $\dot{V}O_2$ (Wells *et al.*, 1989; Egginton, 1997).

In temperate water fish, stress causes plasma osmotic disruption. Similar changes have been observed in Antarctic fish, although the recent report by Gonzalez-Cabrera *et al.* (1995) indicates that great care needs to be exercised in interpreting these data. Antarctic fishes are unusual in that plasma

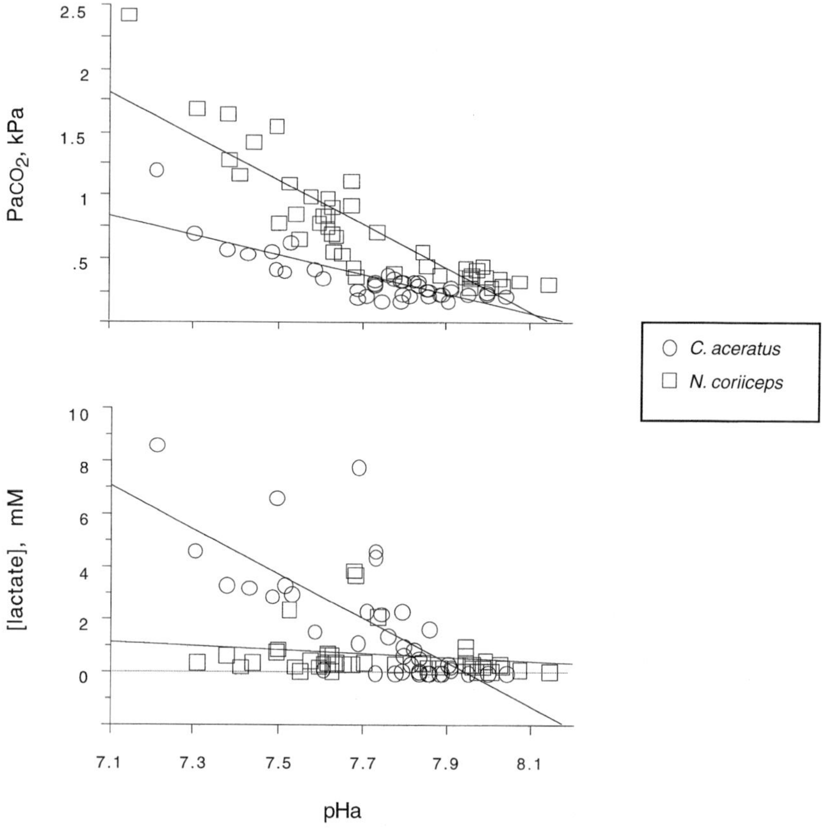

Fig. 8. Data gathered from red- and white-blooded species at rest (high pHa values) and following capture, anaesthesia or exercise. In the nototheniid respiratory, rather than the usual metabolic acidosis dominates, whereas in the channichthyid both a modest hypercapnia and lactacidosis (mmol l^{-1} whole blood) are evident. *Chaenocephalus aceratus* ○ and *Notothenia coriiceps* □. (Egginton, 1994, 1997). $Paco_2=14.01-1.72\cdot pHa$, $r^2=0.78$ (*N. coriiceps*); $Paco_2=6.27-0.77\cdot pHa$, $r^2=0.66$ (*C. aceratus*); $[lactate]=67.28-8.47\cdot pHa$, $r^2=0.56$ (*N. coriiceps*); $[lactate]=6.80-0.80\cdot pHa$, $r^2=0.05$ (*C. aceratus*)

osmolality is very high, often approaching 600 mosmol l^{-1}. Gonzalez-Cabrera *et al.* (1995) looked at osmotic regulation in warm-acclimated Antarctic fish (*Trematomus bernacchii* and *T. newnesi*). They discovered that 5 weeks at + 4 °C (which must be regarded as chronic stress, being so close to the upper lethal limit of these fish) produced a marked fall in plasma osmolality, coupled with similar changes to Na^+ and Cl^- concentrations (Fig 9). This

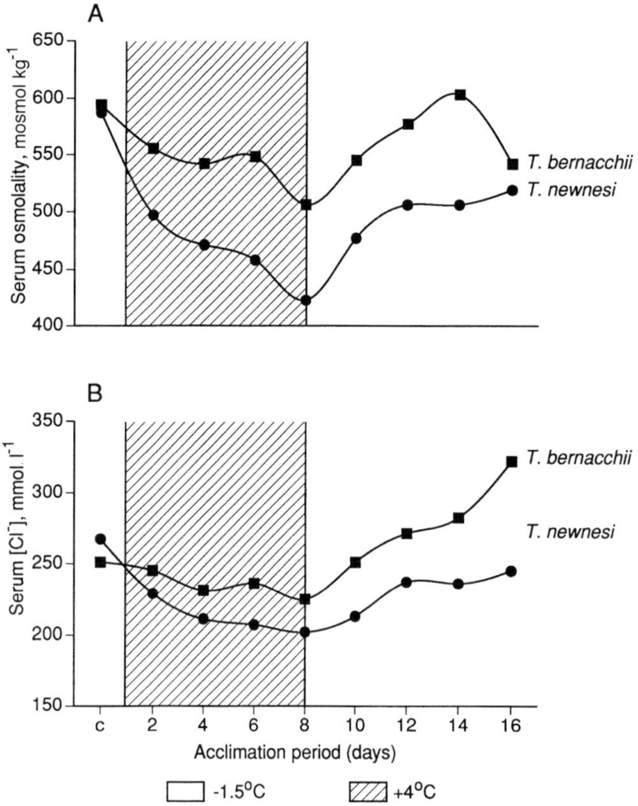

Fig. 9. Hypo-osmoregulation in nototheniids subjected to chronic thermal stress. Note the high control (C) values, the prolonged duration of the adaptive response, and the difference in magnitude of the response between species. (Mean values from Gonzalez-Cabrera *et al.*, 1996.)

was very marked in the more active *T. newnesi*, bringing osmolality down towards values recorded for temperate water fishes. This was a controlled change, rather than a general disruption of osmotic regulation, and of note is the fact that the change is opposite to that predicted for stress in marine teleosts. The authors suggest that the higher osmolality at low temperature is an energy saving cold adaptation, offering a decreased gradient between plasma and seawater, a suggestion also advanced by Eastman (1993). However, it seems odd that these fish should regulate osmolality downwards when chronically stressed, rather than the expected haemoconcentration. This is obviously a neglected area that requires more attention.

Davison *et al.* (1995) exercised *T. bernacchii* and showed no change to chloride levels, although they did provide data that suggested [chloride], and thus osmolality was influenced by the amount of time in captivity. No change to chloride levels or osmolality were seen after acute and chronic exposure to fuel oil in the same species (Davison *et al.*, 1992, 1993) or hypoxia (Davison *et al.*, 1994). However, chloride levels did rise following exercise and high temperature in this study. *P. borchgrevinki* showed large increases in osmolality and chloride following exposure to high temperatures, though there was no change following severe exercise (Franklin *et al.*, 1993) or as a result of gill disease (Davison *et al.*, 1990). Both these thermal stress experiments showed the opposite effect to that seen by Gonzalez-Cabrera *et al.* (1995). However, the former experiments were acute and, at 10 °C, were well past the lethal limit. Sodium and chloride levels did not change in *N. coriiceps* or *N. rossii* following exercise, though of note in this study was a large decrease in plasma protein concentration in *N. rossii* (Egginton *et al.*, 1991). Capture stress did not affect plasma protein in *N. coriiceps* and the icefish *C. aceratus*, while sodium and chloride levels gave opposite results: recovery from the stress led to a reduction in plasma ion levels in *N. coriiceps* and an increase in *C. aceratus* (Egginton, 1994), although the latter result may reflect chronic stress resulting from damage to the integument during initial capture (Fig. 1).

Conclusions

From the relatively sparse data available, and on the basis of information from one group of fishes largely endemic to the seas around Antarctica (the red-blooded nototheniids), the stress response would appear to differ from that seen in temperate water teleosts in both quantitative and qualitative aspects. There are unusually small changes in circulating catecholamine levels following induced stress. By contrast, there may be a substantial increase in haematocrit. In the absence of significant adrenergic stimulation this reflects an impressive polycythaemia due to release of stored red cells from the spleen, rather than red cell swelling. Unlike other teleosts, splenic contraction, and other aspects of cardiovascular physiology, would appear to be largely under cholinergic control. Recovery times are similar to those observed in other fishes, suggesting effective rate compensation for the extreme low temperatures. A reduced glycolytic capacity is reflected in a dominance of respiratory over metabolic acidosis. Unusually for marine teleosts, there seems to be an enhanced hypo-osmoregulation, although this occurs from a relatively hyperosmotic baseline. The few data available for the haemoglobinless icefish suggests a similar down-regulation of adrenergic control but with a greater degree of metabolic acidosis, and a prolonged recovery period.

Acknowledgements

The authors would like to thank the British Antarctic Survey, Natural Environment Research Council and the New Zealand Antarctic Programme for supporting their work.

References

Acierno, R., Macdonald, J.A., Agnisola, C. & Tota, B. (1995). Blood volume in the haemoglobinless Antarctic teleost *Chionodraco hamatus* (Lönnberg). *Journal of Experimental Zoology*, **272**, 407–9.

Axelsson, M. & Nilsson, S. (1986) Blood pressure control during exercise in the Atlantic cod, *Gadus morhua. Journal of Experimental Biology*, **126**, 225–36.

Axelsson, M., Davison, W., Forster, M.E. & Farrell, A.P. (1992). Cardiovascular responses of the red-blooded Antarctic fishes, *Pagothenia bernacchii* and *P. borchgrevinki. Journal of Experimental Biology*, **167**, 179–201.

Axelsson, M., Davison, W., Forster, M.E. & Nilsson, S. (1994). Blood pressure control in the Antarctic fish *Pagothenia borchgrevinki. Journal of Experimental Biology*, **190**, 265–79.

Bacila, M., Rosa, R., Rodrigues, E., Lucchiari, P.H. & Rosa, C.D. (1989). Tissue metabolism of the ice-fish *Chaenocephalus aceratus* Lönnberg. *Comparative Biochemistry and Physiology*, **92B**, 313–18.

Berenbrink, M. & Bridges, C.R. (1994). Catecholamine-activated sodium/proton exchange in the red blood cells of the marine teleost *Gadus morhua. Journal of Experimental Biology*, **192**, 253–67.

Butler, P.J. & Day, N. (1993). The relationship between intracellular pH and swimming performance of brown trout exposed to neutral and sublethal pH. *Journal of Experimental Biology*, **176**, 271–84.

Butler, P.J., Axelsson, M.M., Ehrenström, F., Metcalfe, J.D. & Nilsson, S. (1989). Circulating catecholamines and swimming performance in the Atlantic cod, *Gadus morhua. Journal of Experimental Biology*, **141**, 377–87.

Clarke, A. (1991). What is cold adaptation and how should we measure it? *American Zoologist*, **31**, 81–92.

Davis, K.B. & Parker, N.C. (1990). Physiological stress in striped bass: effect of acclimation temperature. *Aquaculture*, **91**, 349–58.

Davison, W., Forster, M.E., Franklin, C.E. & Taylor, H.H. (1988). Recovery from exhausting exercise in an Antarctic fish, *Pagothenia borchgrevinki. Polar Biology*, **8**, 167–72.

Davison, W., Franklin, C.E. & Carey, P.W. (1990). Oxygen uptake in the Antarctic teleost *Pagothenia borchgrevinki.* Limitations imposed by X-cell gill disease. *Fish Physiology and Biochemistry*, **8**, 69–77.

Davison, W., Franklin, C.E., McKenzie, J.C. & Dougan, M. (1992). The effects of acute exposure to the water soluble fraction of fuel oil on the survival and metabolic rate of an Antarctic teleost. *Comparative Biochemistry and Physiology*, **102C**, 185–8.

Davison, W., Franklin, C.E., McKenzie, J.C. & Carey, P.W. (1993). Effects of chronic exposure to the water soluble fraction of fuel oil on an Antarctic fish *Pagothenia borchgrevinki. Comparative Biochemistry and Physiology*, **104C**, 67–70.

Davison, W., Franklin, C.E. & McKenzie, J.C. (1994). Haematological changes in an Antarctic teleost *Tremotomus bernacchii* following stress. *Polar Biology*, **14**, 463–6

Davison, W., Axelsson, M., Forster, M.E. & Nilsson, S. (1995). Cardiovascular responses to acute handling stress in the Antarctic fish *Tremotomus bernacchii* are not mediated by circulatory catecholamines. *Fish Physiology and Biochemistry*, **14**, 253–7.

DeVries, A.L. (1983). Antifreeze peptides and glycopeptides in cold-water fishes. *Annual Review of Physiology*, **45**, 245–60.

di Prisco, G. & Macdonald, J.A. (1992). Antarctic fishes survive exposure to carbon monoxide. *Experientia*, **48**, 473–5.

Dunn, J.F. & Johnston, I.A. (1986). Metabolic constraints on burst-swimming in the Antarctic teleost *Notothenia neglecta. Marine Biology*, **91**, 433–40.

Eastman, J.T. (1993). *Antarctic Fish Biology*. San Diego: Academic Press. 322pp.

Egginton, S. (1994). Stress response in two Antarctic teleosts (*Notothenia coriiceps* Richardson and *Chaenocephalus aceratus* Lönnberg) following capture and surgery. *Journal of Comparative Physiology*, **164B**, 482–91.

Egginton, S. (1996). Blood rheology of Antarctic fishes: viscosity adaptations at very low temperatures *Journal of Fish Biology*, **48**, 513–21.

Egginton, S. (1997). Exercise-induced stress response in blood chemistry of three Antarctic teleosts. *Journal of Comparative Physiology*, **167B**, 129–34.

Egginton, S. & Rankin, J.C. (1991). The vascular supply to skeletal muscle in fishes with and without respiratory pigments. *International Journal of Microcirculation*, **10**, 396.

Egginton, S. & Rankin, J.C. (1997). Vascular adaptations for a low pressure / high flow blood supply to locomotory muscles of icefish. In *Fishes of Antarctica*, ed. G. di Prisco, A. Clarke & E. Pisano (in press).

Egginton, S., Taylor, E.W., Wilson, R.W., Johnston, I.A. & Moon, T.W. (1991). Stress response in the Antarctic teleosts *Notothenia neglecta* Nybelin and *N. rossii* Richardson. *Journal of Fish Biology*. **38**, 225–35.

Fanta, E., Lucchiari, P. & Bacila, M. (1989). The effect of environmental oxygen and carbon dioxide levels on the tissue oxygenation and the behavior of Antarctic fish. *Comparative Biochemistry and Physiology*, **93A**, 819–31.

Fitch, N.A., Johnston, I.A. & Wood, R.E. (1984). Skeletal muscle capillary supply in a fish that lacks respiratory pigments. *Respiration Physiology*, **57**, 201–11

Floysand, R., Ask, J.A., Serck-Hanssen,G. & Helle, K.B. (1992). Plasma catecholamines and accumulation of adrenaline in the atrial cardiac tissue of aquacultured Atlantic salmon (*Salmo salar*) during stress. *Journal of Fish Biology*, **41**,103–11.

Forster, M.E., Franklin, C.E., Taylor, H.H. & Davison, W. (1987). The aerobic scope of an Antarctic fish, *Pagothenia borchgrevinki* and its significance for metabolic cold adaptation. *Polar Biology*, **8**, 155–9.

Franklin, C.E., Davison, W. & Carey, P.W. (1991). The stress response of an antarctic teleost to an acute increase in temperature. *Journal of Thermal Biology*, **16**, 173–7.

Franklin, C.E., Davison, W. & McKenzie, J.C. (1993). The role of the spleen during exercise in the Antarctic teleost, *Pagothenia borchgrevinki*. *Journal of Experimental Biology*, **174**, 381–6.

Gamperl, A.K., Vijayan, M.M. & Boutilier, R.G. (1994). Epinephrine, norepinephrine, and cortisol concentrations in cannulated seawater-acclimated rainbow trout (*Oncorhynchus mykiss*) following black-box confinement and epinephrine injection. *Journal of Fish Biology*, **45**, 313–24.

Gingerich, W.H. & Drottar, K.R. (1989). Plasma catecholamine concentrations in rainbow trout (*Salmo gairdneri*) at rest and after anaesthesia and surgery. *General Comparative Endocrinology*, **73**, 390–7.

Gonzalez-Cabrera, P.J., Dowd, F., Pedibhotla, V.T., Rosario, R., Stanley-Samuelson, D. & Petzel, D. (1995). Enhanced hypo-osmoregulation induced by warm-acclimation in antarctic fish is mediated by increased gill and kidney Na^+/K^+-ATPase activities. *Journal of Experimental Biology*, **198**, 2279–91.

Hemmingsen, E.A. & Douglas, E.L. (1970). Respiratory characteristics of the haemoglobin-free fish *Chaenocephalus aceratus*. *Comparative Biochemistry and Physiology*, **33**, 733–44.

Hemmingsen, E.A. & Douglas, E.L. (1972). Respiratory and circulatory responses in a haemoglobin-free fish, *Chaenocephalus aceratus*, to changes in temperature and oxygen tension. *Comparative Biochemistry and Physiology*, **43A**, 1031–43.

Holeton, G.F. (1970). Oxygen uptake and circulation by a haemoglobinless Antarctic fish. *Comparative Biochemistry and Physiology*, **34**, 457–71.

Isaia, J., Maetz, J. & Haywood, G.P. (1978). Effects of epinephrine on branchial nonelectrolyte permeability in trout. *Journal of Experimental Biology*, **74**, 227–37.

Keen, J.E., Vianzon, D-M., Farrell, A.P. & Tibbits, G.F. (1993). Thermal acclimation alters both adrenergic sensitivity and adrenoceptor density in cardiac tissue of rainbow trout. *Journal of Experimental Biology*, **181**, 27–47.

Kinkead, R., Aota, S., Perry, S.F. & Randall, D.J. (1993). Propranolol impairs the hyperventilatory response to acute hypercapnia in rainbow trout. *Journal of Experimental Biology*, **75**, 115–26.

Lecklin, T., Nash, G.B. & Egginton S., (1994). Comparative effects of acute (*in vitro*) and chronic (*in vivo*) temperature changes in rheology of fish red blood cells. *Journal of Experimental Biology*, **198**, 1801–8.

Lowe, T.E. & Wells, R.M.G. (1997). Exercise challenge in Antarctic fishes: do haematology and muscle metabolite levels limit swimming performance? *Polar Biology*, **17**, 211–18.

Macdonald, J.A., Montgomery, J.C. & Wells, R.M.G. (1987). Comparative physiology of Antarctic fishes. *Advances in Marine Biology*, **24**, 321–88.

Macdonald, J.A. & Wells, R.M.G. (1991). Viscosity of body fluids from Antarctic notothenioid fish. In *Biology of Antarctic Fish*. ed. G. di Prisco, B. Maresca, & B. Tota, pp. 163–78. Berlin: Springer-Verlag.

Mazeaud, M.M., Mazeaud, F. & Donaldson, E.M. (1987). Primary and secondary effects of stress in fish: some new data with a general review. *Transactions of the American Fisheries Society*, **106**, 201–12.

Metcalfe, J.D., & Butler, P.J. (1989). The use of alpha-methyl-*p*-tyrosine to control circulating catecholamines in the dogfish *Scyliorhinus canucula*: the effects of gas exchange in normoxia and hypoxia. *Journal of Experimental Biology*, **141**, 21–32.

Milligan, C.L., Graham, M.S. & Farrell, A.P. (1989). The response of trout red cells to adrenaline during seasonal acclimation and changes in temperature. *Journal of Fish Biology*, **35**, 229–36.

Montgomery, J.C. & Wells, R.M.G. (1993). Recent advances in the ecophysiology of Antarctic notothenioid fishes: metabolic capacity and sensory performance, in *Fish Ecophysiology*, ed. J.C. Rankin & F.B. Jensen, pp. 341–74. London: Chapman & Hall.

Nash, G.B. & Egginton, S. (1993). Comparative rheology of human and trout red blood cells. *Journal of Experimental Biology*, **174**, 109–22.

Nikinmaa, M. (1983). Adrenergic regulation of haemoglobin oxygen affinity in rainbow trout red cells. *Journal of Comparative Physiology*, **152B**, 67–72.

Nilsson, S. (1994). The spleen. In *Comparative Physiology and Evolution of the Autonomic Nervous System*. S. Nilsson & S. Holmgren, Volume ed., G. Burnstock, Series ed. pp. 247–56. Chur, Switzerland: Harwood Academic.

Nilsson, S., Forster, M.E., Davison, W. & Axelsson, M. (1996). Nervous control of the spleen in the red-blooded Antarctic fish *Pagothenia borchgrevinki*. *American Journal of Physiology*, **270**, R599–604.

Pearson, M., van der Kraak, G. & Stevens, E.D. (1992). *In vivo* pharmacology of spleen concentration in rainbow trout. *Canadian Journal of Zoology*, **70**, 625–7.

Perry, S.F. & Reid, S.G. (1994). The effects of acclimation temperature on the dynamics of catecholamine release during acute hypoxia in the rainbow trout *Oncorhynchus mykiss*. *Journal of Experimental Biology*, **186**, 289–307.

Perry, S.F. & Thomas, S. (1991). The effect of endogenous or exogenous catecholamines on blood respiratory status during acute hypoxia in rainbow trout (*Oncorhynchus mykiss*). *Journal of Comparative Physiology B*, **161**, 489–97.

Perry, S.F., Booth, C.E. & McDonald, D.G. (1985). Isolated perfused head of rainbow trout. I. Gas transfer, acid–base balance and haemodynamics. *American Journal of Physiology*, **249**, R246–54.

Pickering, A.D. (1981). *Stress and Fish*. London: Academic Press, 367pp.

Qvist, J., Weber, R.E, DeVries, A.L. & Zapoul, W.M. (1977). pH and haemoglobin oxygen affinity in blood from the Antarctic cod *Dissostichus mawsoni*. *Journal of Experimental Biology*, **67**, 77–88.

Randall, D.J. & Daxboeck, C. (1984). Oxygen and carbon dioxide transfer across fish gills. In *Fish Physiology*. ed. W.S. Hoar & D.J. Randall, pp. 263–314. New York: Academic Press.

Randall, D.J. & Perry, S.F. (1992). Catecholamines. In *Fish Physiology*, vol. XIIb ed. W.S. Hoar, D.J. Randall & A.P. Farrell. pp. 255–300. San Diego: Academic Press.

Rankin, J.C. (1989). Blood circulation and gill water fluxes in the icefish, *Chaenocephalus aceratus* Lönnberg. *Proceedings of B.A.S. Antarctic Special Topic, University Research in Antarctica*, ed. R.B. Heywood.

Ristori, M.T. & Laurent, P. (1985). Plasma catecholamines and glucose during moderate exercise in the trout: comparison with bursts of violent activity. *Experimental Biology*, **44**, 247–53.

Ryan, S.N. (1992). Susceptibility of the Antarctic fish *Pagothenia borchgrevinki* to MS-222 anaesthesia. *Polar Biology*, **11**, 583–9.

Ryan, S.N. (1995). The effect of chronic heat stress on cortisol levels in the Antarctic fish *Pagothenia borchgrevinki*. *Experientia*, **51**, 769–74.

Saint-Paul, U., Hubold, G. & Ekau, W. (1988). Acclimation effects on routine oxygen consumption of the Antarctic fish *Pogonophryne scotti* . (Artedidraconidae). *Polar Biology*, **9**, 125–8.

Sorensen, B. & Weber, R.E. (1995). Effects of oxygenation and the stress hormones adrenaline and cortisol on the viscosity of blood from the trout *Oncorhynchus mykiss*. *Journal of Experimental Biology*, **198**, 953–9.

Tetens, V., Wells, R.M.G. & DeVries A.L. (1984). Antarctic fish blood. respiratory properties and the effects of thermal acclimation. *Journal of Experimental Biology*, **109**, 265–79.

Thomas, S., Kinkead, R., Wood, C.M., Walsh, P.J. & Perry, S.F. (1991). Densensitization of adrenaline-induced red blood cell H^+ extrusion *in vitro* after chronic exposure of rainbow trout (*Salmo gairdneri*) to moderate environmental hypoxia. *Journal of Experimental Biology*, **156**, 233–48.

Thomas, S. & Perry, S.F. (1992). Control and consequences of adrenergic activation of red blood cell Na^+/H^+ exchange on blood oxygen and carbon dioxide transport. *Journal of Experimental Zoology*, **263**, 160–75.

Twelves, E.L. (1972). Blood volumes of two Antarctic fishes. *British Antarctic Survey Bulletin*, **31**, 85–92.

Wahlquist, I. & Nilsson, S. (1980). Adrenergic control of the cardiovascular system of the Atlantic cod, *Gadus morhua*, during 'stress'. *Journal of Comparative Physiology*, **137B**, 145.

Walsh, P.J. (1989). An *in vitro* model of post-exercise hepatic gluconeogenesis in the Gulf toadfish *Opsanus beta. Journal of Experimental Biology*, **147**, 393–406.

Waring, C.P., Stagg, R.M. & Poxton, M.G. (1996). Physiological responses to handling in the turbot. *Journal of Fish Biology*, **48**, 161–73.

Wells, R.M.G., Tetens, V. & DeVries, A.L. (1984). Recovery from stress following capture and anaesthesia of Antarctic fish: haematology and blood chemistry. *Journal of Fish Biology*, **25**, 567–76.

Wells, R.M.G., Grigg, G.C., Beard, L.A. & Summers, G. (1989). Hypoxic responses in a fish from a stable environment: blood oxygen transport in the Antarctic fish *Pagothenia borchgrevinki. Journal of Experimental Biology*, **141**, 97–111.

Wells, R.M.G. & Weber, R.E. (1990). The spleen in hypoxic and exercised rainbow trout. *Journal of Experimental Biology*, **150**, 461–6.

Wells, R.M.G., Ashby, M.D., Duncan, S.J. & MacDonald, J.A. (1980). Comparative study of the erythrocytes and haemoglobins in nototheniid fishes from Antarctica. *Journal of Fish Biology*, **17**, 517–27.

Wells, R.M.G. Macdonald, J.A. & DiPrisco, G. (1990). Thin-blooded Antarctic fishes: a rheological comparison of the haemoglobin-free icefishes *Chionodraco katleenae* and *Cryodraco antarcticus* with a red blooded nototheniid, *Pagothenia bernacchii. Journal of Fish Biology*, **36**, 595–609.

Wilson, R.W. & Egginton, S. (1994). Assessment of maximum sustainable swimming performance in rainbow trout (*Oncorhynchus mykiss*). *Journal of Experimental Biology*, **192**, 299–305.

Yamamoto, K., Itazawa, Y. & Kobayashi, H. (1983). Erythrocyte supply from the spleen and hemoconcentration in hypoxic yellow-tail *Seriola quinqueradiata. Marine Biology*, **73**, 221–6.

M. AXELSSON, C. AGNISOLA, S. NILSSON
and B. TOTA

Fish cardio-circulatory function in the cold

Temperature influences the physiology of poikilotherm organisms such as fish in two general ways: (a) it determines the rates of chemical reactions (i.e. influences biochemical reaction rates by a factor of two to three for every change of 10 °C) ; and (b) it shifts the equilibria between the formation and disruption of the non-covalent interactions that stabilise biological structures. Accordingly, temperature can affect fish cardio-circulatory function at least at two biological levels: at the metabolic level influencing the rate of enzyme catalysis and direction of metabolic flow, and at the morpho-functional level influencing the composition and dynamic interrelationships of the various components of the cardiovascular apparatus.

The heart, like the myotomal musculature, is a chemo-mechanical converter of energy, made up of excitable and contractile units. Therefore, its rhythmical performance is particularly sensitive to temperature. However, as in other excitable tissues, adaptive mechanisms operate also in the cardiovascular system by reshaping it metabolically and structurally to confer a degree of independence from the intrinsic effects of temperature. Despite its limited and yet mostly descriptive nature, the available information indicates that the cardio-circulatory biology of fish mirrors, to some extent, the differences in the thermal tolerance limits experienced by the different teleost groups. Thus, eurythermal species in general show a remarkable ability to compensate cardiovascular performance for the direct consequences of seasonal changes in environmental temperature. Other species, such as the stenothermal Antarctic fishes that live in a nearly constant temperature, have developed cardio-circulatory adaptations for functioning within a very restricted thermal regime (Hazel & Prosser, 1974; Cossins & Bowler, 1987).

Such relative thermal independence of cardio-circulatory function in fish can be either a consequence of *evolutionary adaptation*, which results from the action of natural selection over evolutionary time spans, or the result of an *acclimatory compensation* due to mechanisms that lie within the genetic repertoire of an individual during its lifetime. Despite the differences in the temporal dimensions and in the mechanisms underlying these two kinds of responses, in both cases the global aim of cardiovascular adaptations in fish is (a) to maintain rates and directions at which chemical energy in the form of metabolic fuels is converted into mechanical work, so that (b) a certain volume of blood can be pumped against pressure to meet the changing demands of the organism. This in turn requires (c) the maintenance and integration of efficient informational loops (intrinsic and extrinsic cardiovascular control). The aspects of cardio-circulatory homeostasis or, as better termed by Mangum and Towle (1977), enantiostasis (conserved function), illustrated here will emphasise how these temperature limits and relations cannot be explained only by the temperature relations and limits of the cell components and their constituent molecules. Indeed, they also need to be analysed by taking into account the interactions between different types of cells at different levels of organisation.

This chapter will deal mainly with examples from these three levels of cardio-circulatory biology in Antarctic fish as a paradigm of cold adaptation. However, for comparative purposes some examples from other teleosts have been selected, including those obtained in the laboratory, with the intention to stimulate research to find coincident or causal links between evolutionary and acclimatory cardiovascular adjustments to low temperatures.

Structure of the fish heart

General anatomical aspects

The fish heart, posterior to the gills and protected by the shoulder girdle, is made up of four serially arranged chambers, enclosed in the membranous pericardial sac. The parietal or outer pericardium, adherent to surrounding tissues, puts the heart in a rigid 'box'. This may account for some features of the cardiac performance, i.e. *vis-à-fronte* filling, in fish like the trout and particularly the elasmobranchs, where a 'suctorial heart' has been characterised (Satchell, 1991; Farrell & Jones, 1992). This suctorial function of the pericardium can contribute to maintain maximum stroke volume at higher heart rates, such as those following warm acclimation. The visceral pericardium, also called epicardium, is inseparable from the heart and in intimate contact with the subjacent myocardium. It is often endowed with its own capillary network and plays a nutritional role for the subjacent myocardium, especially in teleosts with an avascular heart. The pericardial space is filled with an ultrafiltrate from plasma, the pericardial fluid. The pericardial fluid may contain antifreeze glycopeptides, as found in stenothermal Antarctic fish (DeVries, 1971), and thus can provide an 'antifreeze effect' protection for the heart of polar species.

In most teleost species, the ventricle, which can be tubular, sac-like or pyramidal in shape, is fully trabeculated, i.e. an array of thin myocardial bundles (trabeculae) form a complex meshwork to give the internal surface of the wall a spongy appearance, thence named 'spongiosa'. The trabeculae, lined by a monolayer of endothelial cells, are supplied by the venous blood from the inter-trabecular spaces (lacunae) of the ventricular lumen. This multi-chambered division of the ventricular wall ensures a higher ratio of surface area to ventricular volume, compared with the compact myocardium of the homoiotherm vertebrates. The designation as type I has been given to the fully trabeculated and avascular ventricle, while the designation of type II, III and IV of heart ventricles has been assigned to those cardiac designs that show increasing degrees of ventricular vascularisation and myo-architectures (Tota *et al.*, 1983; Farrell & Jones, 1992). In type III and IV, i.e. mixed types of hearts, the spongiosa is invested in an outer layer of densely and orderly arranged myocardial bundles, thence called 'compacta'. This is supplied by an arterio-capillary bed of either cephalic (i.e. coronary) or caudal origin.

Morpho-functional design of the myocytes

The fine structure of the fish myocardium has been reviewed extensively by Santer (1985) and more recently by Tibbits *et al.* (1992*a*,*b*). The contractile activity of the myocardial tissue reflects the summated and integrated working of its individual contractile elements, the sarcomeres. The force of myocardial contraction depends on the number of actin–myosin interactions, which is determined by cytosolic Ca^{2+} concentration and myofilament sensitivity to Ca^{2+}. The time-course of contraction is dependent on the duration of intracellular Ca^{2+} transients, and the rate of chemical turnover of interaction sites, which is in turn determined mainly by the properties of the myosin (Katz, 1992). Heart rate, contraction velocity and intracellular calcium regulation in the cardiac muscle are closely connected. Cardiac function is also critically dependent on the extracellular calcium ion concentration $[Ca^{2+}]_o$. $[Ca^{2+}]_o$ is known to affect force generation, time-course of contraction and duration of the action potential (AP). Therefore, one may expect that thermal adaptation can modify not only beat frequency, but also the protein composition of the sarcomere and the excitation-contraction coupling processes.

Dimensions of ventricular myocytes and diffusion distances

As in other ectothermic vertebrates, most fish myocytes are smaller than those of the homoiotherms. While mammalian ventricular myocytes range between 10–25 μm in diameter, those from most fish have an average diameter around 3 or 4 μm (Santer, 1985). Thus, the cell surface-to-volume ratio is higher in fish hearts than in the mammalian hearts. This is of relevance for the exchange of substances between the intra- and extracellular spaces, especially in relation to the absence of T-tubules in poikilotherm myocardia, and to the major role of sarcolemmal Ca^{2+} transport in the excitation–contraction coupling in the fish heart compared with the mammalian heart. Such high sarcolemmal surface to cytosolic volume ratio is of importance in view of the correspondingly higher area available for sarcolemmal transport proteins in general, and Ca^{2+} stores and exchange, in particular. These shorter ventricular intra-myocyte diffusion distances can be also of functional advantage in an avascular type of ventricle. In these ventricles, the myocardial trabeculae lack the capillary network that surrounds the muscle bundles of the compact myocardium, and must rely on the venous blood passing through the lacunae for oxygen supply. Thus, it is likely that, in an avascular fish heart, a dimensional constraint will keep within very narrow range any cold-induced hypertrophic myocyte remodelling of the ventricle.

Sarcolemma, sarcoplasmic reticulum (SR) and excitation-contraction coupling

As in other poikilothermic vertebrates, myocardial contraction in fish is activated mainly by sarcolemmal Ca^{2+} entry, such inflow being the primary source of Ca^{2+} which attaches to the binding sites of the troponin of the myofilaments. Therefore, the duration of sarcolemmal action potential (AP) is considered a major determinant of the contractile activity in most teleosts (Tibbits *et al.*, 1992*a,b*; Driedzic & Gesser, 1994). Recent evidence suggests that Ca^{2+} is bound both on extra- and intracellular leaflets of sarcolemma, the majority being localised on the intracellular phospholipids (Post & Langer, 1992). The external binding sites participate in the regulation of Ca^{2+} entry through sarcolemmal Ca^{2+} channels and the Na/ Ca^{2+} exchanger (Yee, Weiss & Langer, 1989). Although there is a lack of information on this aspect, the possibility exists that thermal responses in fish hearts can influence contractile force and the time-course of contraction through their modifications in sarcolemmal function, since cold-induced prolongation of the AP (flounder: Lennard & Huddart, 1991*a,b*; trout: Møller-Nielsen & Gesser, 1992) may enhance Ca^{2+} availability to the contractile myofilaments. The sarcolemma can be a target of thermal modulation at different levels, including the number and/ or affinity of Ca^{2+} binding sites, the density and properties of sarcolemmal Ca^{2+} channels (Tibbits & Kashihara, 1992), the Na/ Ca^{2+} exchange (guinea pig ventricular myocytes: Kimura, Miyamae & Noma, 1987; Vornanen, Shepherd & Isenberg, 1994), and the adrenoceptors. The latter two were investigated in the trout: Tibbits *et al.* (1992*a,b*) found that Na^+/ Ca^{2+} exchange, whose K_m is unaffected by temperature, is even less temperature dependent than that of the frog, without apparent breakpoint in the Arrhenius plot. Instead, an increased β-adrenoceptor density, that paralleled the different adrenergic sensitivities of the corresponding *in situ* and *in vitro* cardiac preparations, was found in the sarcolemmal fractions from heart ventricle tissues acclimated to 8 °C in comparison to their 18 °C-acclimated counterparts, suggesting an adrenergic compensatory response offsetting the reduced contractility at lower temperature (Keen *et al.*, 1993). The finding that antifreeze polypeptides from the winter flounder (*Pseudopleuronectes americanus*) inhibit Ca^{2+}- and K^{+}-currents (Rubinsky *et al.*, 1992) uncovers new properties of these molecules that may be important in cold adaptation.

The sarcoplasmic reticulum (SR) is less abundant in fish cardiomyocytes than it is in those of homoiotherms. Exceptions have been reported in some species with mixed type of heart, such as mackerel and trout (Santer, 1985), and in rainbow trout ventricle oxalate-dependent Ca^{2+} uptake data have suggested that the SR is well developed in comparison to other poikilotherm hearts (Dybro & Gesser, 1986). The suggestion from the majority of the ultrastructural and functional studies, including the use of inhibitors such as ryanodine to assess the relevance of SR Ca^{2+} release in contraction, is that delivery of Ca^{2+} to the contractile apparatus from SR binding sites may be less important in fish heart (for references, see Driedzic & Gesser, 1994; Tibbits *et al.*, 1992*a*,*b*). However, the role of SR in excitation–contraction coupling could vary with temperature, as well as during increased cardiac demand (acidosis, hypoxia, physical activity). The putative selective thermal sensitivity of the ryanodine-sensitive Ca^{2+} release channel was explored in trout ventricle strips. At 25 °C the force was correlated with the maximal rate of force development, and the SR contributed significantly to excitation–contraction coupling, while at 5 °C force was correlated with time-to-peak tension and appeared regulated by a ryanodine-insensitive mechanism (Hove-Madsen, 1992). Work of this kind has led to the hypothesis that SR function in some fish, like the trout, becomes progressively less important with lower temperature (Tibbits, Hove-Madsen & Bers, 1991), while in other species it can play a significant role. In fact, a substantial increase of SR was detected in the heart ventricle of cold-acclimated perch (*Perca fluviatilis*) exposed to 5 °C for at least three weeks in comparison with the warm acclimated counterparts, while both myoglobin or mitochondrial volume densities were unchanged (Bowler & Tirri, 1990). The study of Vornanen (1996) suggests that thermal acclimation alters Ca^{2+} activation of contraction at the level of sarcolemma without any significant contribution by SR Ca^{2+} release. In the ventricular myocardium of the crucian carp (*Carassius carassius* L.) acclimated to 2 °C and 22 °C and assayed at 12 °C, cold-acclimated fish exhibited contraction activated at much lower external Ca^{2+} concentration and longer duration of contraction and relaxation than in the warm-acclimated counterparts. A marked prolongation of contraction duration by low Ca^{2+} levels was present only in the cold-acclimated fish.

In contrast to the *lengthening* of contraction duration in cold-acclimated crucian carp heart, Bailey and Driedzic (1990) reported *shortening* of relaxation time and better ability to maintain contractility at higher frequencies in the cold-acclimated yellow perch, *Perca flavescens.* Such species differences in cardiac acclimatory responses may reflect ecophysiological characteristics. The crucian carp becomes torpid during the winter, while the yellow perch remains active at low temperature. On the other hand, the differences stress the difficulty of discerning which particular basic mechanism is involved in the adjustment of the fish heart to cold.

Myofibrils

In the ventricular myocytes of both teleosts and elasmobranchs, the myofibrils are typically located in the periphery as a single sub-sarcolemmal ring, surrounding the centrally located mitochondria. This arrangement, and the short intra-myocyte diffusion distances and the lack of T-tubules, conceivably makes the contractile apparatus less dependent on structures like the SR for rapid exchanges of cytoplasmic Ca^{2+} and other inotropic modulators. As in homoiotherms, ventricular myocytes of athletic species have higher myofibrillar content than more sedentary species, the lowest values being found in the icefish.

Mitochondria

In many species, positive thermal compensation in response to cold acclimation is achieved by increasing the levels of some oxidative mitochondrial enzymes, particularly those necessary for aerobic fatty acid catabolism. Increased volume density of mitochondria is a frequent response to low-temperature acclimation and is viewed as an adaptive enhancement of the functional capacity of the tissue at low temperature (Egginton & Sidell, 1989). The enlargement of the mitochondrial compartment of the myocytes, particularly in the ventricle, is responsible for the hypertrophic changes observed in these cells and often is associated with the increased heart mass frequently reported for cold-adapted hearts.

The heart of polar fish

Within Notothenioidei, cardiac mass is higher in the haemoglobin-less species (icefish), and among the red-blooded species, it is higher in the most active fish (pelagic and cryopelagic: Montgomery & Wells, 1993). The ventricle weight/ body weight ratio of icefish is higher than that of other fish, including Notothenioidei, at 3–4 times the typical values found in poikilotherm vertebrates and coincident with that of tunas (Johnston *et al.*, 1983; Tota *et al.*, 1991). All Notothenioids examined so far have a fully trabeculated type of heart ventricle (type I). It has been suggested that this might be one of the morphological features reflecting the ecophysiological characteristics (sedentary benthic habitat) of the ancestral stock from which they have radiated (Eastman, 1993). In the icefish the cardiac enlargement is attained without changing the basic myoarchitecture of the spongiosa (cardiomegaly of the 'spongy type'), i.e. without any development of an outer compact layer, which leads to very high surface area-to-cavity volume ratios. This may help diffusion processes between the myocardium and the superfusing blood in the absence of haemoglobin and myoglobin in a cold environment. Moreover, the extensive compartmentalisation of the icefish ventricle appears specifically adapted to the task of pumping large volumes of blood against very low output pressure with relatively low tension (stress) of the myocardial trabeculae (Harrison *et al.*, 1991; Zummo *et al.*, 1995). The importance of this kind of ventricular myoarchitecture will be discussed later when comparing the mechanical behaviour of different fish heart types in terms of their relative capability to generate volume and pressure work.

The ventricular myocytes of both red-blooded and haemoglobinless Notothenioidei have diameters ranging from 4 to 6 μm, and thus approach the upper end of the range reported for the majority of teleosts (Fig. 1). Such cardio-myocyte enlargement appears to be caused by a dramatic increase in the content of mitochondria, the volume densities of which are among the largest in vertebrate hearts (Johnston *et al.*, 1983; Tota, Acierno & Agnisola, 1991; Zummo *et al.*, 1995). This is achieved at the expense of the density of myofibrils. These are arranged in columns, between one or several fibrils, at the cell periphery and show volume densities in the lower end of the range reported for fish hearts (Fig. 1). Consequently, a potential trade-off in space economy and organelle logistics, i.e. the concept of accommodating increased mass of intracellular organelles (mitochondria) without impeding cell dimension (Londraville & Sidell, 1990), may occur in icefish ventricular myocytes. This suggests that physical constraints, such as diffusional distances, can prevent further myocyte enlargement (Tota, Acierno & Agnisola, 1994; Zummo *et al.*, 1995).

It remains to be experimentally tested whether the very high mitochondrial content already matches the maximum aerobic capacity required by the myofibril compartment for maintaining myocardial performance in the icefish. The icefish heart is also characterised by an interstitial space very rich

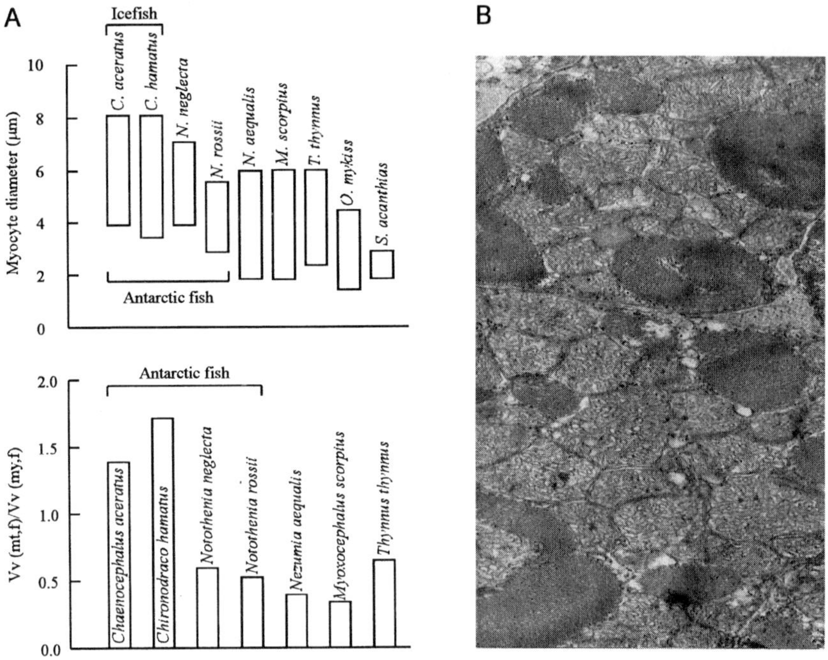

Fig. 1. Ventricular myocytes in fish. A. Range of fibre diameters (top left) and mitochondrial/myofibrillar density ratio (bottom left); Vv (mt,f)= mitochondrial volume density; Vv (my,f)=myofibrillar volume density. From Zummo *et al.* (1995). B. Transmission electron micrograph showing the large mitochondrial content in the cardiomyocytes of the icefish *C. hamatus*.(× 8000).

in acellular, electron-dense amorphous material and microfibrils that may increase the compliance of the ventricular wall, possibly improving its capacity to act as volume pump (Zummo *et al.*, 1995).

Biochemical and metabolic aspects of cardiac performance

The adaptive 'strategies' include alterations in intracellular ionic and membrane phospholipid composition, functional isozyme and allozyme differences such as the expression of temperature-specific isoenzymes of energy metabolism in polyploid fish (Somero, 1975, 1995), and alterations of protein composition. The latter may reflect compensatory changes in many enzymatic activities (aerobic energy metabolism), and/ or changes in the quantity of organelles (mitochondria, sarcoplasmatic reticulum) (Cossins & Bowler, 1987; Kent, Koban & Prosser,1988; Torres & Somero, 1988).

At the membrane level, temperature influences both the rate of molecular motion of membrane constituents, and the phase state and order of membrane lipids. Many studies on adaptive changes in cellular biochemistry and enzyme activity have emphasised the remodelling of the lipid composition of cellular membranes. In particular, the extensive remodelling by phospholipids, and the proportions of saturated fatty acids present, have often been interpreted as producing a degree of homeoviscous adaptation that offsets changes in the physical state of the membrane imposed by the temperature change (Hazel, 1995).

Biochemical and metabolic profiles

Myofibrillar proteins

The largest amount of information on thermal adaptation of myofibrillar systems in fish comes from studies on myotomal musculature. However, it is reasonable to expect that the fundamental thermal adaptations described for the contractile proteins of the myotomal musculature very probably will also be valid for the heart. Somero (1983) reviewed, from a comparative perspective, the key properties of proteins, i.e. their structural integrity, their catalytic functions, and their regulatory characteristics. It was shown how the flexibility in structure, necessary for efficient catalytic function, creates the drawback of an increased susceptibility to thermal denaturation. This is illustrated by the myofibrillar ATPase activity from polar teleosts, that is substantially higher than those from tropical species when assayed at the same (low) temperature. These proteins from polar fish become much more rapidly denaturated by heat than those from their warmer counterparts (Johnston *et al.*, 1975; Johnston & Walesby, 1977). Clearly, there is a need to include the heart in such comparative analyses. Cardiac myosin isoforms have been reported in some teleosts by Karasinski (1988) and Martinez *et al.* (1991), but only one study (Vornanen, 1994) seems to have documented a temperature-induced switch between teleost cardiac myosin isoforms. In the ventricular myocardium of the crucian carp (*Carassius carassius* L.), only one myosin heavy chain was present during the winter, when the fish inhabits ice-covered shallow ponds. It was partly replaced by a heavier isoform during summer (water temperature around 22 °C). Acclimation experiments in the laboratory under constant photoperiod and oxygen tension reproduced the above pattern, with cold-acclimated (2 °C) fish showing only winter myosin, and warm-acclimated (22 °C) fish showing in equal amounts both summer and winter myosin. Myosin ATPase activity was higher in the warm-acclimated carps than in the cold-acclimated counterparts. The former group of fish also show higher heart rate *in vitro* and shorter contraction duration. This kind of cardiac response mirrors the type of acclimation response so far described in the skeletal musculature of cyprinid fishes. This group belongs to polyploid taxa, and polyploid fishes are characterised by the unusual ability to respond to temperature challenge by expressing temperature-specific isoforms (for references see Somero, 1975, 1995; Sidell & Moerland, 1989).

Metabolic aspects

According to the concept of metabolic independence (homeokinesis), Hazel and Prosser (1974) suggested that oxidative enzymes are more thermally compensated than glycolytic enzymes. The hearts of *Notothenia gibberifrons* and *Trematomus newnesi* at 1 °C exhibited roughly the same activities in the majority of mitochondrial oxidative enzymes compared to the hearts from temperate teleosts assayed at 15 °C (Sidell *et al.*, 1987), which speaks in favour of cardiac metabolic compensation in the Antarctic fish. Further evidence (Crockett & Sidell, 1990) indeed suggested that Antarctic fish are metabolically cold adapted in central pathways of aerobic metabolism (markers: citrate-synthase and cytochrome oxidase) and β-oxidation of lipid fuels (markers: carnitine palmitoyltransferase and 3-hydroxyacyl-CoA dehydrogenase). In fact, when comparisons of metabolic capacities were made between Antarctic fishes (*Notothenia gibberifrons* and *Trematomus newnesi*) and ecotypically similar species from temperate latitudes (*Myoxocephalus octodecimspinosus* and *Tautoga onitis*), maximal enzyme activities assayed at 1 °C were higher in heart ventricles from polar species than from their temperate counterparts.

In contrast, prevailing evidence suggests that the hearts of most Antarctic fish have not compensated metabolically for both anaerobic (markers: 6-phosphofructokinase, pyruvate kinase, and lactate dehydrogenase) and aerobic (hexokinase) carbohydrate pathways. However, the cardiac ventricles of *N. gibberifrons* and the icefish *Channichthys rhinoceratus* seem to deviate from this pattern, since the former displays activities of glycolytic markers (PFK, PK, and LDH) that are from 1.3 to 1.8 times higher than those of its temperate latitude counterpart (*M. octodecimspinosus*) (Crockett & Sidell, 1990). Icefish show high potential for anaerobic metabolism (Feller & Gerday, 1987).

The low activities of the glycolytic enzymes are indicative of a minor role of carbohydrate catabolism in Antarctic fish and may reflect their preference for fuelling aerobic metabolism with lipids. Lipids, predominantly triacylglycerols, the primary depot lipid of most of the prey organisms of many Antarctic teleosts, reach high levels in serum and tissues of the Antarctic fish. In oxidative muscle tissue from *Dissostichus mawsoni* and *Pagothenia borchgrevinki*, for example, they may account for up to 37% of the dry weight (Lin, Dobbs & deVries, 1973). With the exception of salmonids, temperate species have considerably lower serum lipid content. In addition to conferring static lift to the swimbladder-less notothenioids, this high lipid content could also constitute a primary fuel for aerobic energy metabolism. In agreement with this, isolated paced heart ventricle and pectoral adductor muscle (made up of oxidative fibres) of *Gobionotothen gibberifrons* have shown clear preference for fatty acids, and particularly monounsaturates, over glucose or lactate as substrates for energy metabolism (Sidell, Crockett & Driedzic, 1995).

Extrinsic cardiac control mechanisms

Innervation patterns

All teleost fish species studied have a cholinergic inhibitory innervation of the heart, and most teleosts also possess an adrenergic excitatory innervation of the heart. The cholinergic nerve fibres reach the heart via the cardiac branch of the vagus, which may also contain adrenergic nerve fibres that enter the vagus trunk from the sympathetic chains (Morris & Nilsson, 1994). It has been shown that, in most species, there is a tonic influence on the heart, the cholinergic vagal fibres mediating an inhibitory effect, while the adrenergic system (adrenergic nerves and circulating catecholamines) produce positive chronotropic and inotropic effects on the heart. The relative tonic influence of the cholinergic and adrenergic systems varies between species. The cholinergic control of the heart in Antarctic fish species appears to be well developed, and in *T. bernacchii* a cholinergic tone of about 80% has been recorded, which is the highest described in any teleost fish (Axelsson *et al.*, 1992). In the cryopelagic *P. borchgrevinki,* the cholinergic tone is smaller (54%), but still high compared to most other teleosts species. The intrinsic heart rates of these species are not dramatically different from those reported for other species (see p. 344), and the lower 'resting' heart rate is a function of a large cholinergic tone and a much smaller adrenergic influence. During periods of 'stress' or exercise (see below) heart rate can change rapidly, but it seems that most of the change is due to changes in the cholinergic tone and less to an increase in the adrenergic influence on the heart. This dependence on cholinergic mechanisms for the control of the heart (and other organs, see below) is in line with the observations made by Wood, Pieprzack & Trott (1979). These authors found a higher cholinergic tone in rainbow trout acclimated for two weeks to 5 °C, compared to animals acclimated to 20 °C.

Humoral control of the heart

Measurements of plasma levels of circulating catecholamines of Antarctic fish show that the basal levels are comparable to the levels reported in studies of temperate teleost species (Davison *et al.*, 1995;

Randall & Perry, 1992). There is no conclusive evidence for any temperature-induced changes in the levels of circulating catecholamines in fish, but some studies have shown seasonal changes in the sensitivity to adrenergic stimulation of red blood cells (see Randall & Perry, 1992, for references). Using the β-adrenoceptor antagonist sotalol, it was shown that there is a small adrenergic tone on the heart in *P. borchgrevinki*, but no further attempts have been made to find out to what extent this tone is mediated via adrenergic nerves or circulating catecholamines from chromaffin tissue (Axelsson *et al.*, 1994). Egginton (1994) studied the effects of 'stress' on the plasma levels of catecholamines in two Antarctic teleosts *Notothenia coriiceps* and *Chaenocephalus aceratus*, and found no changes in the plasma concentration of either adrenaline or noradrenaline. Davison *et al.* (1995) noted a similar lack of increase in the concentrations of circulating catecholamines during 'stress' in *T. bernacchii*. During acute 'stress' in *T. bernacchii,* the heart rate did not increase by more than a few beats and this change could have been mediated by a decrease in cholinergic tone on the heart (Davison *et al.*, 1995). These studies indicate a low reliance on adrenergic mechanisms in the control of the heart in Antarctic fish species.

Mechanical properties of the heart

Cardiac output, the product of heart rate and stroke volume, depends on a combination of myocardial and mechanical factors that lets the heart pump blood to meet the changing metabolic needs of the organism. The myocardial factors concern changes in contractility, i.e. changes in mechanical performance independent of resting length or end-diastolic volume, that reflect changes in the physical and chemical interactions of the contractile proteins in the myofibrils. An example is the temperature-induced shift in cardiac myosin isoform pattern. The mechanical factors include various adaptive re-modellings, primarily of chamber dilatation or hypertrophy. In contrast to the mammalian literature, mechanical aspects of heart performance in fish have been poorly studied. A better understanding would be obtained with the wider application in fish cardiology of techniques utilising, e.g. cine-angiocardiography, for defining time-related simultaneous changes in chamber anatomy and function in an integrative way. Despite the limits of conventional flow and pressure measurements both in cold-adapted species and from acclimation studies in assessing the effectiveness of the heart as a muscle and as a pulsatile pump, they may be of value as a basic description of the effects of cold on the mechanical behaviour of fish heart.

Performance of polar fish hearts

The heart of Antarctic species is characterised by relatively large stroke volume values compared with temperate fish. In Fig. 2, a number of fish species are ranked on the basis of the relative contribution of stroke volume and blood pressure to the stroke work (=the product of stroke volume and mean driving pressure, i.e. ventral aortic pressure minus venous pressure). This is an approximation of real heart work calculated from pressure–volume loops (Farrell & Jones, 1992). It allows the distinction between species characterised by high pressure development, placed on the left side of the bar graph, such as tunas, living in warm waters, and species characterised by low pressure–high volume displacement, on the right side of the bar graph. Interestingly, all Antarctic species are found on the right side of the figure. The extreme is represented by the icefish *C. hamatus* and *C. aceratus*, whose hearts may be considered specialised volume pumps (Tota *et al.*, 1991). The specialisation of the heart as a pump (volume *vs.* pressure pump) has been related with the myoarchitectural arrangements of the ventricular wall (Agnisola & Tota, 1994). This distinction is reinforced by the fact that in the species on the left side of the graph, the maximal stroke work is mainly attained by a pressure increase, while in those on the right side it is mainly attained by a volume increase.

Fig. 2 also shows a trend for heart rate to be lower in the species on the right side of the graph. This could be due to a high cholinergic inhibitory tone on the heart (see Table 1) but when comparing the intrinsic heart rate between Antarctic and non-Antarctic species the difference is less pronounced. There may be a link between the low heart rate typical of Antarctic fish, and the high stroke volume they show, as heart rate and stroke volume are strictly related as will be emphasised below.

Heart rate and stroke volume

The force of contraction, as measured *in vitro* on ventricular strips, is dependent on the length of contraction and relaxation periods and on the rate of contraction. Changes in heart rate, while altering the ratio between calcium release and calcium uptake per unit time, may change the duration of the contraction–relaxation cycles and then affect contractility of the ventricle muscle and, ultimately, the stroke volume and pressure generation at the whole organ level. This effect is dependent on the actual heart rate: at low heart rate, an increase in heart rate is usually associated with an increase in contraction force (positive staircase effect) , while at higher frequency the relationship between heart rate and force is negative (negative staircase,

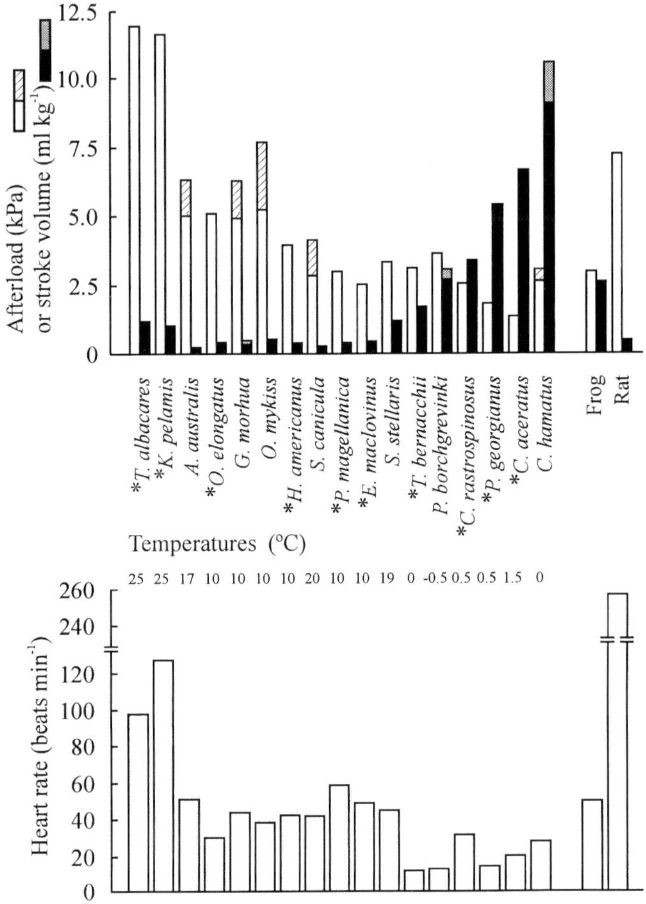

Fig. 2. Afterload, stroke volume and heart rate in fish. Top panel shows resting afterload (mean ventral aorta pressure minus mean sinus venosus pressure, open bars) and stroke volume (closed bars) values. The corresponding values of heart rate are reported in the low panel together with the temperature of determination. Increases in afterload (cross-hatched columns) and stroke volume (shaded columns) under conditions of maximum stroke work (measured *in vivo* during exercise or *in vitro* under conditions of maximum sustainable output pressure) are also reported in the top panel. Asterisks: species for which maximum stroke work data are not available. References: *Thunnus albacares* and *Katsuwomis pelamis*: Bushnell & Brill, 1992; *Anguilla australis*: Hipkins, 1985; *Ophiodon elongatus*: Farrell, 1982; *Gadus morhua*: Axelsson & Nilsson, 1986; *Onchorynchus mykiss*: Kiceniuk & Jones, 1977; Farrell *et al.*, 1996; *Hemitripterus americanus*: Farrell *et al.*, 1985; *Scyliorhinus canicula*: Short *et al.*, 1977; *Paranotothenia magellanica* and

particularly evident in those mammalian species that are highly dependent on intracellular calcium for contractile activation).

Heart rate is determined by the intrinsic rate of the pacemaker in the sinus venosus or in the sino-atrial ring, and in its extrinsic control by the autonomic nervous system (Nilsson, 1983; Santer, 1985; Taylor, 1992) and humoral agents (Nilsson & Holmgren, 1992). In teleosts the heart rate is scarcely affected by preload and afterload changes (Randall, 1968), but is remarkably influenced by temperature (Driedzic & Gesser,1994). Electro cardiogram (ECG) measurements of intact fish, and especially telemetry records from animals in the natural environment, allow evaluation of the contribution of extrinsic components to temperature-induced changes in heart rate, while isolated cardiac preparations give information on the intrinsic components.

The general and intuitive view that an increase in temperature increases heart rate, often in parallel with contractile activity and cardiac metabolism, is in accord with predictable higher oxygen demands during warming. This is beautifully illustrated by an important, yet scarcely explored short-term type of response represented by behavioural hypothermia, elicited by the exposure of fish to hypoxia. Goldfish exposed to anoxia for 5 h select a body temperature about 5 °C lower than the corresponding controls (Crawshaw, Wollmulh & O'Conner, 1989), thus reducing oxygen demand and costly processes of increasing cardiac output and ventilation. In many acclimation studies, non-compensatory or inverse acclimation has been observed, with the cold-acclimated fish exhibiting lower heart rate, contractile performance and Q_{10} than warm-acclimated fish. For example, in the crucian carp (*Carassius carassius* L.), measurements of heart rate both *in vivo* and *in vitro* showed that winter fish (compared to summer fish) exhibited higher heart rate, a Q_{10} value slightly higher, and absence of arrhythmic disturbance when temperature was increased up to 30 °C (Matikainen & Vornanen, 1992). In contrast, positive compensation in heart rate (and heart size) at low temperatures has been found in *Anguilla anguilla* (Seibert, 1979), in the goldfish *Carassius auratus* (Tsukuda, Liu & Fujii, 1985), in the sole *Solea vulgaris* (Sureau, Lagardere & Pennec, 1989), and in the trout (Graham & Farrell,

Eloginops maclovinus: Agnisola, Acierno & Tota, unpublished data; *Scyliorhinus stellaris*: Farrell & Jones, 1992; *Trematomus bernacchii* and *Pagothenia borchgrevinki*: Axelsson *et al.*, 1992; *Chionodraco rastrospinosus*: Acierno & Tota, unpublished data; *Pseudochaenichthys georgianus*: Hemmingsen & Douglas, 1977; *Chaenocephalus aceratus*: Hemmingsen *et al.*, 1972; *Chionodraco hamatus*: Tota *et al.*, 1991. For comparison basal data for the frog heart (volume pump) (Acierno *et al.,* 1994) and for rat heart (pressure pump) (Neely *et al.*, 1967) are also reported.

Table 1. *A summary of the effects of various agonists and antagonists on some cardiovascular variables in* Pagothenia borchgrevinki *and* Trematomus bernacchii

	Pagothenia borchgrevinki			(Axelsson *et al.*, 1994)
	Adrenaline (5 nmol kg^{-1})	Prazosin	Atropine	Angiotensin (1 nmol kg^{-1})
Pda (kPa)	+0.9			+1.1
(%)	+38%	−26%	−13%	+39%
Pva (kPa)	+1.1			+1.0
(%)	32%	−10%	−7%	+26%
fH (bpm)	−3		+10	−6
(%)	−13%	−10%	+51%	−30%
Q (%)	+50%	+47%	+39%	−10%
Vs (%)	+50%	+73%	−4%	+60%
Rsys (%)	+35%	−45%	**−36%**	+50%
Rgill (%)	−15%	+25%	**+23%**	−20%

	Trematomus bernacchii	(Axelsson *et al.*, 1992)
	Adrenaline (10 nmol kg^{-1})	Atropine
Pva (kPa) (%)	+0.9	+0.1
	+31%	+3%
fH (bpm) (%)	+8	**+16.9**
	+72%	**+161%**
Q (%)	+42%	+6%
Vs (%)	+25%	−59%
Rtot (%)	+25%	−6%

Note:
Some of the more notable effects are shown in **bold**. *Abbreviations:* Pda, dorsal aortic blood pressure (kilopascals); Pva, ventral aortic blood pressure (kilopascals); fH, heart rate in beats per minute; Q, cardiac output, percentage change; Vs, cardiac stroke volume, percentage change; Rsys, systemic vascular resistance, percentage change; Rgill, branchial (gill) vascular resistance, percentage change, Rtot, total vascular resistance, percentage change.

1989). In the yellow perch, Bowler and Tirri (1990) detected heart rate *in vitro* to be higher in cold acclimated animals over the physiological range of temperature (from about 20 °C to 5 °C) than in warm acclimated fish. The compensation is, however, only partial (Precht type 3), while the oxygen consumption of the non-beating heart also showed an acclimation effect. In view of these different chronotropic responses, we must be cautious in seeking generalised adaptive explanations.

Different from mammals, heart rate in teleosts is a poor indicator of integrated cardiac performance. In most species, increases in cardiac output are associated with a simultaneous increase in both heart rate and stroke volume (Farrell & Jones, 1992). The relative contribution of heart rate and stroke volume to changes in cardiac output in fish varies between species and is a function of temperature (Kolok & Farrell, 1994). Based on acclimation studies, temperature is considered to have negative inotropic effects on the teleost myocardium (Lennard & Huddart, 1992; Matikainen & Vornanen, 1992). This may be the indirect consequence of the positive chronotropic effect of temperature combined with an inverse relationship between heart rate and stroke volume, both for reduced filling time and negative staircase. In contrast to other vertebrates, including elasmobranchs in which the heart shows a positive correlation between frequency and peak force, i.e. positive staircase, teleost myocardia tend to display a negative staircase, at least in a lower frequency range (Driedzic & Gesser, 1985). Conceivably, the positive staircase might be due to an enhanced Ca^{2+} influx during activation, secondary to a frequency-mediated increase of intracellular Na^+. In the absence of a plausible hypotheses available for the negative staircase, Johansen and Gesser (1986) have postulated that the relatively higher Na^+ concentration of the teleost cell leaves little room for affecting the frequency. A stimulation of the Na^+/K^+ pump, caused by increasing extracellular K^+ concentration, would result in decreased intracellular Na^+ concentration, and thus change the staircase phenomenon. This was indeed the case in the trout heart, in which the staircase became less negative or even reversed to positive at an elevated extracellular K^+ concentration (Hove-Madsen & Gesser, 1989).

An alternative hypothesis has been proposed to explain the negative inotropic effect of temperature by a reduction of the positive inotropic adrenergic influence (Farrell *et al.*, 1996). In the absence of adrenaline, the beat frequency of the spontaneously beating atrium of the trout was more than doubled (in parallel with contractile force and total tension generated during one minute), by raising the temperature from 2 °C to 17 °C. Adrenaline caused a similar chronotropic response at low as well at high temperatures, the chronotropic and inotropic effects both being mediated by a β_2-adrenoceptor (Ask, Stene-Larsen & Helle, 1981). In the same work, more pronounced inotropic effects of adrenaline at 2 °C indicated that adrenergic tone can be especially important in adjusting maximal cardiac performance at lower temperature. In ventricular strips, Keen *et al.* (1993) demonstrated that the higher amount of adrenaline necessary at 18 °C to give the same level of tension measured at 8 °C was associated with a reduction in sarcolemmal β-adrenoreceptor densities.

As shown in Fig. 2, Antarctic stenothermal fish have relatively low heart rates, for example around 15–18 beats per minute (bpm) in the icefish *C. aceratus* and in *C. hamatus*. However, their intrinsic chronotropism is higher when compared with that of some cold-acclimated temperate species (sole, perch, and sea raven) (Montgomery and Wells, 1993). In two red-blooded nototheniids, Axelsson *et al.* (1992), have uncovered a very high inhibitory cholinergic tone, the highest value found in resting fish (Table 1).

By sequentially injecting a cholinoceptor antagonist (e.g. atropine) and a β-adrenoceptor antagonist (e.g. propranolol or sotalol) *in vivo*, it is possible to assess the cholinergic and adrenergic influences on the heart, and to express the percentage change in heart rate as a 'cholinergic tone' and 'adrenergic tone', respectively. In the cryopelagic *P. borchgrevinki* at 0 °C the resting heart rate of 10–12 bpm resulted from an inhibitory cholinergic tone of 54.5% and an excitatory adrenergic tone of 3.2%, the intrinsic heart rate being 23.3 bpm. In the benthic *T. bernacchii* at 0 °C the resting heart rate of 10.5 bpm resulted from an inhibitory cholinergic tone of 80.4% and an excitatory adrenergic tone of 27.5%, the intrinsic heart rate being 21.7 bpm. By increasing temperature acutely from 0 °C to 5 °C, *T. bernacchii* increased heart rate (with a corresponding decrease in stroke volume) and the intrinsic heart rate over this temperature range had a Q_{10} of 1.96; the chronotropic inhibition persisted up to 2.5–3.0 °C, but was lost at 5 °C. Other remarkable findings of this study were the lack of hypoxic bradycardia, particularly evident in *T. bernacchii*, as also observed in *C. aceratus* (Hemmingsen *et al.*, 1972), and the ability of the two nototheniids, unlike most teleosts, to alter heart rate to a greater degree than they alter stroke volume.

Oxygen consumption

Temperature acclimated poikilotherms may show a partial compensation in oxygen consumption (Kent *et al.*, 1988). Despite this, tissues like liver, heart or red muscle have been shown to present perfect acclimation or over-compensation.

Fish ventricles show mass and protein hypertrophy during cold acclimation (Kent *et al.*, 1988; Graham & Farrell, 1989, 1990). Compensation at the level of the heart may be different from the compensation at whole animal level, and this will alter the ratio between myocardial Vo_2 and total Vo_2. As myocardial Vo_2 is related to myocardial power output, compensation at the level of power output could involve adaptation in heart rate, stroke volume (ventricular inotropism) and peripheral resistance. For example, Graham and Farrell (1985) described the seasonal differences in the heart performance *in vitro* of sea raven. Hearts from winter fish (T=2–3 °C) displayed lower power output values than those from summer animals (T=12–14 °C) when perfused under the same load and temperature conditions. However, there was no significant difference in stroke work, and the higher power output of summer hearts was due to a higher heart rate (Q_{10} of 1.7 for power output and 1.9 for heart rate). An acute change in perfusion temperature (from 3 to 13 °C for winter animals, and from 13 to 3 °C for the summer animals) affected heart rate but not stroke work. So, there was apparently full compensation in stroke work (possibly dependent on the heart rate increase, and confirmed in trout by Graham and Farrell, 1989) and no compensation in heart rate and power output. Farrell *et al.* (1985) reported a higher myocardial Vo_2 at a given power output in the summer, suggesting a higher efficiency of winter acclimated heart muscle. Still this may be linked to the lower intrinsic heart rate of winter animals, as heart efficiency is usually inversely related to heart rate (Schipke, 1994). Interestingly, summer fish showed a higher sensitivity to pre-load increases (higher Starling response).

In trout the slope of the Vo_2-power output relationship has been reported to be lower for 5 °C acclimated animals with respect to 15 °C acclimated animals. However, the Vo_2 per mW values were similar: 0.25 $\mu l\ O_2\ s^{-1}\ mW^{-1}$ for the 5 °C and 0.27 $\mu l\ O_2\ s^{-1}\ mW^{-1}$ for the 15 °C acclimated animals. As the cold acclimated animals had a bigger heart (0.110% compared to 0.084% for 5° and 15 °C acclimated animals, respectively), the power produced and the corresponding oxygen consumption by the heart of a 1 kg trout acclimated to 5 °C was 20% higher than by the heart of a trout of the same size acclimated to 15 °C. Both in the chain pickerel (*E. niger*) and American eel (*A. rostrata*) acclimated to 10–12 °C, the perfusion of the isolated heart at 5 °C reduced power output and Vo_2, but increased efficiency with respect to the perfusion at 15 °C (Bailey, Sephton & Driedzic, 1991). It should be noted that in these experiments, the hearts were paced at different rates at the two temperatures (lower at 5 °C).

Table 2 reports cardiac power output *in vivo* in fish expressed per g of ventricle and per kg of body weight. Usually cardiac power output per kg of animal is similar or slightly lower than cardiac output per g of ventricle. Tuna and icefish represent exceptions because of the high ventricular mass typical of these fish. Although the myocardial oxygen consumption *in vivo* is not known, and the power output calculated from flow and ventral aorta is only an approximation of the real power produced by the heart (Farrell & Jones, 1992), we can see that there is an aspect which may distinguish icefish from all the other species so far examined: the total myocardial oxygen

Table 2. *Resting cardiac power output and oxygen consumption in fish*

	Temperature (°C)	Power output of the heart (mW g^{-1} heart weight)	Power output of the heart (mW kg^{-1} body weight)	Myocardial $\dot{V}o_2$ ($\mu l\ min^{-1}\ g^{-1}$)	Myocardial $\dot{V}o_2$ ($\mu l\ s^{-1}\ mW^{-1}$)	Animal $\dot{V}o_2$ ($\mu l\ min^{-1}\ kg^{-1}$)	O_2 cost of cardiac pump (%)	Ref.
Tropical fish								
Katsuwomis pelamis	25	6.31	29.2	–	–	18200	2.3*	*a*
Thunnus albacares	25	5.65	27.2	–	–	10500	–	*a*
Temperate water fish								
Anguilla anguilla	10	0.96	0.67	–	–	–	–	*b*
Hemitripterus americanus	10	1.08	0.82	18.96	0.29	–	0.6	*c*
Oncorhynchus mykiss	10	1.27–1.56	1.23–1.52	19.9	0.23	582–645	2.7–3.6	*d,e,f*
Oncorhynchus mykiss	5	0.86	0.94	13.74	0.27	–	–	*g*
Oncorhynchus mykiss	15	1.91	1.62	28.27	0.25	–	–	*g*
Squalus acanthias	15	2.27	0.92	31.2	0.23	–	–	*h*
Sub-Antarctic fish								
Eleginops maclovinus	10	1.37	0.84	24.1	0.29	–	–	*i*
Paranotothenia magellanica	10	0.67	0.64	16.32	0.41	–	–	*i*
Patagonotothenia tessellata	10	0.76	0.71	25.59	0.56	–	–	*i*
Antarctic fish								
Trematomus bernacchii	−0.5	2.01	0.85	–	–	–	–	*j*
Chionodraco rastospinosus	0.5	0.84	3.94	20.0	0.40	–	–	*k*
Chaenocephalus aceratus	1	1.54	4.62	37.8	0.41	333–533	21–34	*k,l, m*

Note:
[a] Bushnell & Brill, 1992; [b] Agnisola & Tota, unpublished observation; [c] Farrell *et al.*, 1985; the % cost of cardiac pump has been estimated by these as on the basis of the Vo_2 of the lingcod; [d] Kiceniuk & Jones, 1977; [e] Houlihan *et al.*, 1988; [f] Wood *et al.*, 1979; [g] Graham & Farrell, 1990; animals acclimated at the reported temperature; [h] Davie & Franklin, 1992; [i] Agnisola, Acierno & Tota, unpublished observation; [j] Acierno, Agnisola & Tota, unpublished observation; [k] Acierno & Sidell, unpublished observation; [l] Holeton, 1970; [m] Hemmingsen *et al.*, 1972.
*As estimated by Farrell and Jones, 1992.

consumption is very high in relation to the resting metabolism (21–34%). So far, this can be calculated only for two species (*C. aceratus* and *O. mykiss*), and has been estimated in two other species (tuna, 2.3%, Farrell & Jones, 1992; and sea raven, 0.6%, Farrell *et al.*, 1985). Interestingly, the ratio between cardiac $\dot{V}O_2$ and power output is higher in all notothenioids studied so far (Antarctic and Sub-Antarctic), suggesting the possibility of peculiar adaptations of heart metabolism in this group of animals independent of the environmental temperature.

Clearly, more studies are necessary to identify specific temperature adaptations in the relation between heart metabolism and heart mechanics in Antarctic fish.

Control of the vasculature

Innervation patterns

Branchial vasculature

In fish, the heart and gills are connected in series, and de-oxygenated blood leaving the heart enters the gills via the four pairs of afferent branchial arteries. The blood may leave the branchial circulation either via the arterio-arterial pathway into the systemic circulation, or flow back to the heart via the arterio-venous, or nutritive pathway. In resting teleosts more than 90% of the blood entering the branchial circulation is directed through the arterio-arterial pathway where it is oxygenated (Ishimatsu, Iwama & Heisler 1988; Sundin & Nilsson, 1992). The regulation of the vascular resistances is complex, and involves α- and β-adrenergic, cholinergic and serotonergic (5-HT) mechanisms (Nilsson, 1984; Morris & Nilsson, 1994). There is also growing evidence for an involvement of different neuropeptides in the regulation of the vascular resistance of the branchial vasculature. Changes in the branchial vascular resistance, especially in the arterio-arterial pathway, affect the post-branchial pressure, which is the driving pressure for the perfusion of the entire systemic circulation. It is also important to understand where in the branchial circulation the regulation occurs, since this may affect the perfusion pressure not only of the systemic circulation but also of the respiratory part of the gill, altering the oxygenation of the blood.

In the Atlantic cod (*Gadus morhua*) swimming at sustained swimming speed there is a small decrease in gill vascular resistance, while in the rainbow trout (*Oncorhynchus mykiss*) swimming at U_{crit} there is a small increase (Table 3) (Kiceniuk & Jones, 1977; Axelsson & Nilsson, 1986). At sustained swimming speed there is a 23% fall in the branchial vascular resistance also in the rainbow trout (Kiceniuk & Jones, 1977). In a study on the Antarctic cryopelagic species *P. borchgrevinki*, it was shown that during sub-maximal

Table 3. *Effects of three different stimuli on the cardiovascular system in two Notothenioid species,* P. borchgrevinki *and* T. bernacchii

	f_H (bpm)	P_{VA}	P_{DA}	Q	SV	R_{sys}	R_{gill}	Reference
Hypoxia								
T. bernacchii (7 kPa, 0 °C)	**+2** **+27%**	+0.7 kPa +22%		+3%	−10%	±0%, (R_{tot})		Axelsson *et al.*, 1992
Gadus morhua (6 kPa, 10 °C)	−13 −32%	+1.5 kPa +31%	+1.8 kPa +60%	(±0%	+70%	+59%	±0%	Fritsche & Nilsson,1989
Oncorhynchus mykiss (9 kPa)	−33 −39%	+27%	+0.9 kPa	+20%	+89%			Wood & Shelton, 1980
Temperature								
T. bernacchii (0–5 °C)	+17 +138%	±0 kPa ±0%		+16%	−37%	−20%, (R_{tot})		Axelsson *et al.*, 1992
Hemitripterus americanus (5–10 °C)	+17 +121%	+2.4 +133%						Bailey & Driedzic, 1989
Exercise								
P. borchgrevinki (0.8–1.0 bls s^{-1},10 °C)	+7 33%	+0.8 kPa +10%	**−0.3 kPa** **−18 %**	+78%	+36%	−50%	+**30%**	Axelsson *et al.*, 1994
Gadus morhua (2/3 bls s^{-1},10 °C)	+8 +19%	+1.3 +26%	+0.8 +25%	+47%	26%	−7%	−7%	Axelsson & Nilsson, 1986
Oncorhynchus mykiss (U_{crit}, 9–10 °C)	+20 +64%	+3.1 +59%	+0.8 +19%	+35 ml min^{-1} kg^{-1} +198 %	+124%	−141 Pa min kg ml^{-1} −60%	+6%	Kiceniuk & Jones, 1977

Note:

f_H, heart rate; bpm, beats per minute; P_{VA}, P_{DA} ventral and dorsal aortic blood pressure; Q, cardiac output; SV, stroke volume, R_{sys}, R_{gill}, systemic and branchial vascular resistance. Values in **bold** point out major differences in the response of Antarctic species compared to other teleosts studied. Abbreviations: bls s^{-1}, swimming velocity in body lengths per second; U_{crit}, critical swimming velocity; see also abbreviations in Fig. 1.

exercise there was a 30% increase in branchial resistance leading to a fall in systemic blood pressure (Axelsson *et al.*, 1994). The regulation of the branchial vascular resistance in this species is complex and it was shown that atropine, a muscarinic receptor antagonist, blocked the increase in branchial resistance seen during exercise. The β-adrenoceptor antagonist sotalol restored the branchial vasoconstrictor response to exercise that had been lost in atropinised fish, which indicates a β-adrenoceptor mediated vasodilation (Axelsson *et al.*, 1994). An α-adrenergic vasoconstrictor component was also found in these experiments, and it was concluded that the α-adrenergic tonus was dominant over the β-adrenergic vasodilator tonus in the gill. Rankin (1989) found evidence for a noradrenaline-mediated vasodilatation of the branchial vasculature in the icefish, *Chaenocephalus aceratus.* It is not known whether this effect was mediated by α- or β-adrenoceptors. Recent studies on gill blood vessels indicate that 5-hydroxytryptamine is a more potent vasoconstrictor than adrenaline in the presence of sotalol. *In vivo*, there is a profound effect on the branchial vascular resistance of intra-vascular injections of 5-hydroxytryptamine, mimicking the response seen during exercise. The control mechanisms found in the two Antarctic species studied so far are not different from what has been described for other teleost species. However, the cholinergic increase in branchial vascular resistance during exercise is unusual (Table 3).

Systemic vasculature

There is a limited amount of information about the control of the systemic vasculature in Antarctic fish species, and the information is derived from only a few of the around 270 species found in the area. The calculated systemic vascular resistance tends to be lower compared to other temperate species and blood flows are high (Hemmingsen *et al.*, 1972; Rankin, 1989; Axelsson *et al.*, 1994). In *Pagothenia borchgrevinki*, injection of adrenaline produced a transient increase in systemic vascular resistance, and a long lasting increase in both pre- and postbranchial blood pressure (Axelsson *et al.*, 1994). Prazosin, an α-adrenoceptor antagonist, produced a marked fall in systemic resistance, indicating the presence of α-adrenoceptors in the systemic circulation and an adrenergic tone on the vasculature.

Two studies of the stress response in Antarctic fish have shown that there are minor, if any, changes in the plasma levels of catecholamines during or after stress (Egginton, 1994; Davison *et al.*, 1995). This may indicate that the α-adrenoceptor mediated vascular tone may be mediated via adrenergic nerves rather than circulating catecholamines. However, so far no studies on the innervation pattern of the systemic vasculature exist.

Humoral control of the circulatory system

Catecholamines

An adrenergic control of the circulatory system occurs in most vertebrates. Such control may originate from three systems (i) endocrine cells (=chromaffin cells) that release their stored catecholamines into the bloodstream, (ii) paracrine cells, such as the chromaffin cells of the cyclostome and dipnoan heart, that modulate the cardiac activity by local release (leakage?) of catecholamines, and (iii) an adrenergic innervation of the heart and vasculature (for references, see Morris & Nilsson, 1994). From studies of teleosts, there is good evidence for an adrenergic innervation of both the heart (in many, but not all, species studied), the branchial vasculature and the systemic vasculature. The evidence for an involvement of humoral (circulating) catecholamines in any of the fish groups is less substantial. In both elasmobranchs (Davies & Rankin, 1973) and teleosts (*cf.* Nilsson, 1984), a control of the branchial vascular resistance, notably of the arterio-arterial pathway, by humoral catecholamines seems probable. However, humoral adrenaline and noradrenaline may be less important in the control of the systemic vasculature.

Adrenaline injected into the blood of *Pagothenia borchgrevinki* or *Trematomus bernacchii* produced marked increase in arterial blood pressure. The hypertensive effect of 5–10 $nmol \cdot kg^{-1}$ was mainly due to an increase in cardiac output due to a marked (>40%) increase in stroke volume, but effects on the vascular resistance were small and transient (Axelsson *et al.*, 1992, 1994).

In the Antarctic nototheniid fish *T. bernacchii*, handling stress causes a rise of heart rate and ventral aortic blood pressure (Davison *et al.*, 1995). However, measurements of the plasma levels of catecholamines (adrenaline and noradrenaline) in *Notothenia coriiceps*, *Chaenocephalus aceratus* and *Trematomus bernacchii* at rest and after stressful stimuli show no rise in these levels (Egginton, 1994; Davison *et al.*, 1995). The observations thus suggest that if an adrenergic control of the circulatory systems indeed occurs in these Antarctic species, such control must be due to adrenergic nerves.

The renin-angiotensin system

The renin–angiotensin system (RAS) of teleost fish appears well established as a control mechanism also affecting the arterial blood pressure. In fact, studies on temperate fish such as the Atlantic cod, suggest that the RAS is more important than the adrenergic system in maintaining a tonic influence on the resting blood pressure (Olson, 1992; Platzack, Axelsson & Nilsson, 1993). An even more important role for the RAS may be to serve as an 'anti-

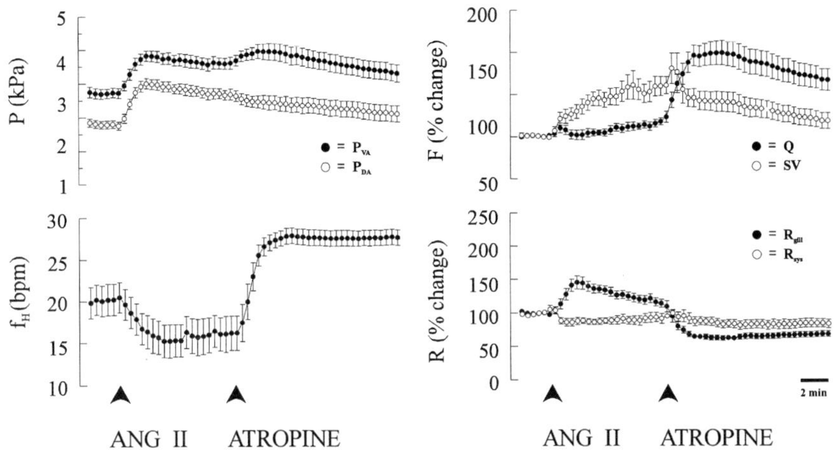

Fig. 3. Effects of angiotensin II (ANG II) followed by the muscarinic cholinoceptor antagonist atropine (ATROPINE) on ventral (P_{VA}) and dorsal (P_{DA}) aortic blood pressure, heart rate (f_H), cardiac output (Q), cardiac stroke volume (SV) and the vascular resistances of the branchial (R_{Gill}) and systemic (R_{Sys}) vascular beds. Note the increase in blood pressures accompanied by a (reflexogenic) decrease of heart rate. Atropine abolishes the heart rate decrease, demonstrating the cholinergic inhibition of the heart. (Modified from Axelsson *et al.*, 1994.)

drop factor' under certain circumstances (Olson, 1992). Part of the effect of angiotensin II (AngII) on the vasculature may be indirect, due to secondary release of catecholamines from nervous stores (Olson, 1992).

Injection of AngII consistently produced an increase of both the ventral and dorsal aortic blood pressures in *Pagothenia borchgrevinki*, due to a major increase in the systemic vascular resistance (Fig. 3) (Axelsson *et al.*, 1994). Similarly, injection of angiotensin I (AngI) slowly increased the blood pressure, suggesting the presence of the angiotensin converting enzyme (ACE) which acts on AngI to produce AngII. The ACE inhibitor enalapril abolished the effect of AngI, confirming the presence of a functional RAS in *P. borchgrevinki* (Axelsson *et al.*, 1994).

Blood pressure regulation

Effects of exercise

Exercise is known to elicit profound changes in the cardiovascular system of most vertebrates studied, including different fish species (Kiceniuk and Jones, 1977; Jones and Randall, 1978; Axelsson & Nilsson, 1986; Axelsson and Fritsche, 1991). The information on the cardiovascular adjustments to exercise in Antarctic fish is limited to one study by Axelsson and co-workers on *P. borchgrevinki* (Axelsson *et al.*, 1994). In this study it was shown that during non-exhaustive short term exercise there was an increase in cardiac output, heart rate, stroke volume and ventral aortic pressure similar to the changes recorded in other teleosts (Table 3). Unlike the situation in the Atlantic cod, *Gadus morhua,* and rainbow trout, *Oncorhynchus mykiss*, there was a decrease in dorsal aortic blood pressure of *P. borchgrevinki* during exercise due a massive increase in the branchial vascular resistance (Table 3). Calculations of systemic vascular resistance showed a 50% decrease in this during swimming, a change that was unaffected by atropine, prazosin, or sotalol pre-treatment. In the Atlantic cod and rainbow trout it has been shown that the decrease in systemic vascular resistance during exercise is counteracted by an increase in the adrenergic vasomotor tone. After pre-treatment with α-adrenoceptor antagonists there is a massive fall in the systemic vascular resistance (Kiceniuk & Jones, 1977; Axelsson & Nilsson, 1986). In the Atlantic cod the large fall in systemic vascular resistance after α-adrenoceptor blockade results in a decrease in both the dorsal and ventral aortic blood pressure. This drop in pressure activates the RAS system which counter-balances the local metabolite-induced vasodilatation (see p. 351) (Platzack *et al.*, 1993).

Effects of hypoxia

The water around the Antarctic continent is cold and well oxygenated, and it is unlikely that Antarctic fish are exposed to hypoxia. In temperate species, living under conditions of fluctuating water temperature and oxygen concentration, a complex cardiovascular response to hypoxia exists, including a reduction in heart rate (hypoxic bradycardia) and an increase in dorsal aortic blood pressure due to an increase in systemic vascular resistance (Axelsson, Farrell & Nilsson, 1990; Farrell, 1991; Fritsche & Nilsson, 1990; Fritsche, 1990; Wood & Shelton, 1980). In *T. bernacchii* the hypoxic bradycardia is absent and replaced by a small tachycardia (Table 3), although in these experiments some animals with higher initial heart rate showed occasional phasic decreases in heart rate (Axelsson *et al.*, 1992). The 'classical' increase in blood pressure seen in other teleosts was also found in *T. bernacchii,* but it was small and occurred only late in the hypoxic exposure.

Effects of temperature

In trout there is no evidence for temperature-dependence of blood pressure (Holeton & Randall, 1967). Arterial blood pressure does not change following acute temperature transitions in winter flounder (*Pseudopleuronectes americanus*), eel (*Anguilla anguilla*) and lingcod (*O. elongatus*) (Cech *et al.*, 1976)

Barron, Tari and Hayton (1987) showed a temperature dependence of *in vivo* cardiac output (measured with the dilution method) in trout acclimated to 6, 12 and 18 °C. Heart rate and blood pressure were measured in two animals only (acclimated at 12 °C). The most dramatic change seen during acute temperature challenge in the Antarctic species *T. bernacchii*, is the increase in heart rate. This is offset by a decrease in cardiac stroke volume leading to a modest 20% increase in cardiac output. During this acute exposure the dorsal aortic blood pressure and total vascular resistance is increased up to about 4 °C, when the cardiovascular system starts to fail (Axelsson *et al.*, 1992). There are no data available about long-term effects on the cardiovascular system of temperature changes in any Antarctic fish species.

It can be concluded that blood pressure is scarcely dependent of temperature. Intrinsic heart rate is highly dependent of temperature (Q_{10} up to 2), and explains the temperature-dependence of cardiac output and power output. Stroke work is scarcely affected by acclimation temperature. The $\dot{V}O_2$-power output relationship may also be affected by acclimation temperature. The power output per kg of fish is higher in cold-acclimated animals.

Control of the haematocrit

Fish belonging to the family Nototheniidae generally have a low resting haematocrit (Hct), but changes in Hct are substantial in some of the red-blooded nototheniids (Macdonald, Montgomery & Wells, 1987; Davison, Franklin & McKenzie, 1994). In the borch (*Pagothenia borchgrevinki*) the resting Hct is about 15%, and may rise to about 35% during exercise due to erythrocyte release from the spleen. The Hct is controlled by several factors, notably the release of erythrocytes from the spleen (Davison *et al.*, 1988; Franklin, Davison & McKenzie, 1993; Nilsson *et al.*, 1996).

Control of spleno-somatic index and haematocrit

In many teleost species, the spleen acts as a reservoir of sequestered erythrocytes. During hypoxia or other types of 'stress' that cause an increased adrenergic nerve activity, these erythrocytes may be released into the bloodstream, increasing the haematocrit. Pearson and Stevens (1991) demonstrated a stress-induced release of red cells from the spleen of the rainbow trout, that added about one-third of the observed haemo-concentration. The release was concluded to be due to an adrenergic mechanism, mediated via α-adrenoceptors. No cholinergic influence could be seen (Pearson *et al.*, 1992).

In the studies by Axelsson *et al.* (1994) and Nilsson *et al.* (1996), a major increase in the spleno-somatic index (from 0.60 to 0.89) was observed in *Pagothenia borchgrevinki* after treatment with the cholinergic antagonist atropine. In addition, it could be shown that the Hct decreased substantially during atropine treatment (from 18.6±1.4% to 6.6±0.8%). The results show that when the cholinergic innervation of the spleen is blocked, splenodilation occurs together with an increased trapping of the erythrocytes. This suggests that modulation of the activity of a cholinergic innervation of the spleen is responsible for the regulation of splenic erythrocyte sequestering and release (Nilsson *et al.*, 1996).

Splenic innervation

The 'normal' pattern of innervation of the vertebrate spleen is one of a mainly, if not solely, adrenergic excitatory control (see Nilsson, 1994). Activation of the adrenergic fibres innervating the spleen will produce erythrocyte release in several mammals (Davies & Withrington, 1973), the toad (*Bufo marinus*) (Nilsson, 1978) and elasmobranchs (*Squalus acanthias, Scyliorhinus canicula*) (Nilsson, Holmgren & Grove, 1975). In the Atlantic cod *(Gadus morhua)*, the spleen is controlled by autonomic nerve fibres, electrical stimulation of which cause erythrocyte release (Nilsson & Grove, 1974). The innervation exhibited both adrenergic and cholinergic properties (Nilsson & Grove, 1974; Winberg, Holmgren & Nilsson, 1981). In fact, from 20% up to 80% of the splenic response to nerve stimulation of a perfused spleen preparation *in situ* can be abolished by atropine treatment, and pharmacological analysis of the splenic innervation clearly shows a functional cholinergic component in the innervation (Holmgren & Nilsson, 1975; Winberg *et al.*, 1981).

Thus the cholinergic innervation of the borch (*Pagothenia borchgrevinki*) spleen could represent the cholinergic extreme on the adrenergic/cholinergic innervation pattern observed in the cod. Nilsson *et al.* (1996) speculated, based also on observations by Wood *et al.* (1979) on the adrenergic and cholinergic influence on the rainbow trout heart at different temperatures, that cholinergic control (of some organs) could increase in importance at low temperature. The body temperature of the borch is, quite obviously, at the very low end of vertebrate body temperatures.

Summary and conclusions

The cardiovascular system of ectothermic vertebrates such as fish can be expected to be greatly affected by environmental temperatures. The systems show adaptive responses that provide a degree of independence from the direct temperature effects. Such thermal independence is the result of short-term mechanisms, i.e. acclimation, or long-term changes, i.e. evolutionary adaptation. Some examples of cardio-vascular cold-adaptation in fish have been indicated on molecular, tissue, organ and systemic levels. A comparison of acclimation is made between eurythermal fish species and the stenothermal Antarctic fish.

Biochemical evidence from the ventricle of the crucian carp (*C. carassius*) suggests that the effect of seasonal temperature variations on cardiac performance may produce switches in myosin isoforms. There are also data to indicate an important role of homeoviscous adaptation in the thermal reshaping of the cardiocyte membrane, involving the sarcolemma, the mitochondria and the sarcoplasmatic reticulum. The adaptations reach an extreme in the haemoglobin-less Antarctic icefish. In these the heart is very large, and there is an impressive mitochondrial proliferation with a consequent decrease of myofibrils. This conflict in space economy severely constrains the myocardial capability to meet mechanical challenges, e.g. increases in afterload.

Quantitative analysis of temperature dependence of the mechanical factors that control cardiac output is further needed. Conventional flow and pressure measurements indicate the heart of polar fish as a low rate, low pressure and high volume pump. Low heart rate at low temperature is a universal feature affecting both myocardial inotropy and heart efficiency.

The control of cardiovascular functions in the examined Antarctic species show some differences from that in temperate species. For instance, there is a remarkably high resting inhibitory vagal (cholinergic) tone to the heart and also an, as yet unexplained, cholinergic influence on systemic vascular resistance. Even the splenic innervation seems to be largely cholinergic in nature, which is not the case in most other vertebrates. During exercise the branchial vascular resistance increases substantially, and the dorsal aortic blood pressure drops. This is in contrast to other teleost species, where both ventral and dorsal aortic blood pressure usually increase during exercise. Whether or not these observed differences are due to true evolutionary adaptation of the Antarctic fish remains to be determined.

Acknowledgements

Our own work on Antarctic fish has been supported by the Italian National Research Program in Antarctica (PNRA) (B.T. and C.A.) and the Swedish Natural Science Research Council (NFR), Forestry and Agriculture Research Council (SJFR) and the New Zealand Antarctic Research Programme (NZAP) (M.A. and S.N.).

References

Acierno, R., Gattuso, A., Cerra, M. C., Pellegrino, D., Angola, C. & Tota, B. (1994). The isolated and perfused working heart of the frog, *Rana esculenta*: an improved preparation. *General Pharmacology*, **25,** 521–6.

Agnisola, C. & Tota, B. (1994). Structure and function of the fish cardiac ventricle: flexibility and limitations. *Cardioscience*, **5**, 145–53.

Ask, J.A., Stene-Larsen, G. & Helle, K.B. (1981). Temperature effects on the β_2-adrenoceptors of the trout atrium. *Journal of Comparative Physiology*, **143B**, 161–8.

Axelsson, M. & Fritsche, R. (1991). Effects of exercise, hypoxia and feeding on the gastrointestinal blood flow in the Atlantic cod, *Gadus morhua. Journal of Experimental Biology*, **158,** 181–98.

Axelsson, M. & Nilsson, S. (1986). Blood pressure control during exercise in Atlantic cod, *Gadus morhua. Journal of Experimental Biology*, **126**, 225–36.

Axelsson, M., Farrell, A.P. & Nilsson, S. (1990). Effects of hypoxia and drugs on the cardiovascular dynamics of the Atlantic hagfish *Myxine glutinosa. Journal of Experimental Biology*, **151**, 297–316.

Axelsson, M., Davison, W., Forster, M.E. & Farrell, A.P. (1992). Cardiovascular responses of the red-blooded Antarctic fishes, *Pagothenia bernacchii* and *Pagothenia borchgrevinki. Journal of Experimental Biology*, **167**, 179–201.

Axelsson, M., Davison, B., Forster, M. & Nilsson, S. (1994). Blood pressure control in the Antarctic fish *Pagothenia borchgrevinki. Journal of Experimental Biology*, **190**, 265–79.

Bailey, J.R. & Driedzic, W.R. (1989). Effects of acute temperature change on cardiac performance and oxygen consumption of a marine fish, the sea raven (*Hemitripterus americanus*). *Physiological Zoology*, **62,** 1089–101.

Bailey, J.R. & Driedzic, W.R. (1990). Enhanced maximum frequency and force development of fish hearts following temperature acclimation. *Journal of Experimental Biology*, **149,** 239–54.

Bailey, J., Sephton, D. & Driedzic, W.R. (1991). Impact of an acute temperature change on performance and metabolism of pickerel (*Esox niger*) and eel (*Anguilla rostrata*) hearts. *Physiological Zoology*, **64**, 697–716.

Barron, M.G., Tarr, B.D. & Hayton, W.L. (1987). Temperature-dependence of cardiac output and regional blood flow in rainbow trout, *Salmo gairdneri* Richardson. *Journal of Fish Biology*, **31**, 735–44.

Bowler, K. & Tirri, R. (1990). Temperature dependence of the heart isolated from the cold or warm acclimated perch (*Perca fluviatilis*). *Comparative Biochemistry and Physiology*, **96A,** 177–80.

Bushnell, P.G. & Brill, R.W. (1992). Oxygen transport and cardiovascular responses in skipjack tuna (*Katsuwonus pelamis*) and yellowfin tuna (*Thunnus albacares*) exposed to acute hypoxia. *Journal of Comparative Physiology*, **162B**, 131–43.

Cech, J.J.Jr, Bridges, D.W., Rowell, D.M. & Balzer, P.J. (1976). Cardiovascular responses of the winter flounder, *Pseudopleuronectes americanus*, (Walbaum), to acute temperature increase. *Canadian Journal of Zoology*, **54**, 1383–8.

Chan, D.K.O. (1986). Cardiovascular, respiratory and blood adjustments to hypoxia in the Japanese eel, *Anguilla japonica. Fish Physiology and Biochemistry*, **2**, 179–93.

Cossins, A. R. & Bowler, K. (1987). *Temperature Biology of Animals*. London: Chapman & Hall.

Cossins, A. R., Behan, M. K., Jones, G. & Bowler, K. (1987) Lipid-protein interactions in the adaptive regulation of membrane function. *Biochemical Society Transactions*, **15**, 77–81.

Crawshaw, L. I., Wollmuth, L. P. & O'Conner, C. S. (1989). Intracranial ethanol and ambient anoxia elicit selection of cooler water by goldfish. *American Journal of Physiology,* **256**, R133–7.

Crockett, E.L. & Sidell, B.D. (1990). Some pathways of energy metabolism are cold adapted in Antarctic fishes. *Physiological Zoology*, **63**, 472–88.

Davie, P. S. & Franklin, C. E. (1992). Myocardial oxygen consumption and mechanical efficiency of a perfused dogfish heat preparation. *Journal of Comparative Physiology*, **162**, 256–62.

Davies, D. T. & Rankin, J. C. (1973). Adrenergic receptors and vascular responses to catecholamines of perfused dogfish gills. *Comparative and General Pharmacology*, **4,** 139–47.

Davies, B. N. & Withrington, P. G. (1973). The actions of drugs on the smooth muscle of the capsule and blood vessels of the spleen. *Pharmacology Review*, **25,** 373–414.

Davison, W., Forster, M.E., Franklin, G.E. & Taylor, H.H. (1988). Recovery from exhausting exercise in an Antarctic fish, *Pagothenia borchgrevinki*. *Polar Biology*, **8**, 167–71.

Davison, W., Franklin, C. E. & McKenzie, J. C. (1994). Haematological changes in an Antarctic teleost, *Trematomus bernacchii*, following stress. *Polar Biology*, **14**, 463–6.

Davison, W., Axelsson, M., Forster, M. & Nilsson, S. (1995). Cardiovascular responses to acute handling stress in the Antarctic fish *Trematomus bernacchii* are not mediated by circulatory catecholamines. *Fish Physiology and Biochemistry*, **14**, 253–7.

DeVries, A. L. (1971). Glycoproteins as biological antifreeze agents in Antarctic fishes. *Science*, **172**, 1152–5.

Driedzic, W.R. & Gesser, H. (1985). Ca^{2+} protection from the negative inotropic effect of contraction frequency on teleost hearts. *Journal of Comparative Physiology*, **156B**, 135–42.

Driedzic, W.R. & Gesser, H. (1994). Energy metabolism and contractility in hearts of ectothermic vertebrates: impact of hypoxia, acidosis and low temperature. *Physiological Reviews*, **74**, 221–58.

Dybro, L. & Gesser H. (1986). Capacity of the sarcoplasmic reticulum and contractility in heart of 5 vertebrate species. *Acta Physiologia Scandinavia*, **128**, 74.

Eastman, J. T. (1993). *Antarctic Fish Biology*. New York: Academic Press.

Egginton, S. (1994). Stress response in two Antarctic teleosts (*Notothenia coriiceps* Richardson and *Chaenocephalus aceratus* Lönnberg) following capture and surgery. *Journal of Comparative Physiology*, **164B**, 482–91.

Egginton, S. & Sidell, B.D. (1989). Thermal acclimation induces adaptive changes in subcellular structure of fish skeletal muscle. *American Journal of Physiology*, **256**, R1–9.

Farrell, A.P. (1982). Cardiovascular changes in the unanaesthetized lingcod (*Ophiodon elongatus*) during short-term, progressive hypoxia and spontaneous activity. *Canadian Journal of Zoology*, **60**, 933–41.

Farrell, A.P. (1991). From hagfish to tuna: a perspective on cardiac function in fish. *Physiological Zoology*, **64**, 1137–64.

Farrell, A.P. & Jones, D.R. (1992). The heart. In *Fish Physiology*, Vol. XIIA, ed. W.S. Hoar, D.J. Randall & A.P. Farrell, pp. 1–88. San Diego: Academic Press.

Farrell, A.P., Wood, S., Hart, T. & Driedzic, W.R. (1985). Myocardial oxygen consumption in the sea raven, *Hemitripterus americanus*: the effects of volume loading, pressure loading and progressive hypoxia. *Journal of Experimental Biology*, **117**, 237–50.

Farrell, A.P., Gamperl, A.K., Hicks, J.M.T., Shiels H.A. & Jain, K.E. (1996). Maximum cardiac performance of rainbow trout (*Oncorhynchus mykiss*) at temperatures approaching their upper lethal limit. *Journal of Experimental Biology*, **199**, 663–72.

Feller, G., & Gerday, C. (1987). Metabolic pattern of the heart of hemoglobin- and myoglobin-free Antarctic fish *Channichthys rhinoceratus*. *Polar Biology*, **7**, 225–9.

Franklin, G.E., Davison, W. & McKenzie, J.C. (1993). The role of the spleen during exercise in the Antarctic teleost, *Pagothenia borchgrevinki*. *Journal of Experimental Biology*, **174**, 381–6.

Fritsche, R. (1990). Effects of hypoxia on blood pressure and heart rate in three marine teleosts. *Fish Physiology and Biochemistry*, **8,** 85–92.

Fritsche, R. & Nilsson, S. (1990). Autonomic nervous control of blood pressure and heart rate during hypoxia in the cod, *Gadus morhua*. *Journal of Comparative Physiology*, **160B,** 287–92.

Fritsche, R. & Nilsson, S. (1989). Cardiovascular responses to hypoxia in the Atlantic cod, *Gadus morhua*. *Experimental Biology*, **48,** 153–60.

Graham, M. & Farrell, A. (1985). The seasonal intrinsic cardiac performance of a marine teleost. *Journal of Experimental Biology*, **118**, 173–83.

Graham, M.S. & Farrell, A.P. (1989). The effect of temperature acclimation and adrenaline on the performance of a perfused trout heart. *Physiological Zoology*, **62**, 38–61.

Graham, M.S. & Farrell, A.P. (1990). Myocardial oxygen consumption in trout acclimated to 5 °C and 15 °C. *Physiological Zoology*, **63**, 536–54.

Harrison, P., Zummo, G., Farina, F., Tota, B. & Johnston, I. A. (1991). Gross anatomy, myoarchitecture, and ultrastructure of the heart ventricle in the haemoglobinless icefish *Chaenocephalus aceratus*. *Canadian Journal of Zoology*, **69**, 1339–47.

Hazel, J. R. (1995). Thermal adaptation in biological membranes: Is homeoviscous adaptation the explanation? *Annual Review of Physiology*, **57**, 19–42.

Hazel, J. R. & Prosser, C.L. (1974). Molecular mechanisms of temperature compensation in poikilotherms. *Physiological Reviews*, **54**, 620–77.

Hemmingsen, E. A. & Douglas, E. L. (1977). Respiratory and circulatory adaptations to the absence of haemoglobin in Chaenichthyid fishes. In *Adaptations within Antarctic Ecosystems*, ed. G. A. Llano, pp. 479–87. Washington: Smithsonian Institution.

Hemmingsen, E.A., Douglas, E.L., Johansen, K. & Millard, R.W. (1972). Aortic blood flow and cardiac output in the hemoglobin-free fish *Chaenocephalus aceratus*. *Comparative Biochemistry and Physiology*, **43A**, 1045–51.

Hipkins, S.F. (1985). Adrenergic responses of the cardiovascular system of the eel, *Anguilla australis, in vivo*. *Journal of Experimental Biology*, **235**, 7–20.

Holeton, G. F. (1970). Oxygen uptake and circulation by a haemoglobin-less Antarctic fish. *Comparative Biochemistry and Physiology*, **34**, 457–71.

Holeton, G. F. & Randall, D. J. (1967). Changes in blood pressure in the rainbow trout during hypoxia. *Journal of Experimental Biology*, **46**, 297–305.

Holmgren, S. & Nilsson, S. (1975). Effects of some adrenergic and cholinergic drugs on isolated spleen strips from the cod. *Gadus morhua*. *European Journal of Pharmacology*, **32**, 163–69.

Houlihan, D.F., Agnisola, C., Lyndon, A.R., Gray, C. & Hamilton, N.M. (1988). Protein synthesis in a fish heart: responses to increased power output. *Journal of Experimental Biology*, **137**, 565–87.

Hove-Madsen, L. (1992). The influence of temperature on ryanodine sensitivity and the force-frequency relationship in the myocardium of rainbow trout. *Journal of Experimental Biology*, **167**, 47–60.

Hove-Madsen, L. & Gesser, H. (1989). Force frequency relation in the myocardium of rainbow trout. Effect of K^+ and adrenaline. *Journal of Comparative Physiology*, **159B**, 61–9.

Ishimatsu, A., Iwama, G. K. & Heisler, N. (1988). *In vivo* analysis of partitioning of cardiac output between systemic and central venous sinus circuits in rainbow trout: a new approach using chronic cannulation of the branchial vein. *Journal of Experimental Biology*, **137**, 75–88.

Johansen K., & Gesser, H. (1986). Fish cardiology: structural, haemodynamic, electromechanical and metabolic aspects. In *Fish Physiology: Recent Advances*, ed. S. Nilsson & S. Holmgren, pp.71–85. London: Croom Helm.

Johnston, I. A., & Walesby, N. J. (1977). Molecular mechanisms of temperature adaptation in fish myofibrillar adenosine triphosphatases. *Journal of Comparative Physiol*ogy, **119B**, 195–206.

Johnston, I. A., Walesby, N. J., Davison, W. & Goldspink, G. (1975). Temperature adaptation of myosin in Antarctic fish. *Nature (London)*, **254**, 74–5.

Johnston, I. A., Fitch, N., Zummo, G., Wood, R. E., Harrison, P. & Tota, B. (1983). Morphometric and ultrastructural features of the ventricular myocardium of the haemoglobinless icefish *Chaenocephalus aceratus. Comparative Biochemistry and Physiology*, **76A**, 475–80.

Jones, D. R. & Randall, D. J. (1978). The respiratory and circulatory systems during exercise. In *Fish Physiology*. ed. W.S. Hoar & D.J. Randall, Vol. 7, pp. 425–501. New York: Academic Press.

Karasinski, J. (1988). Myosin isoenzymes of fish hearts. *Comparative Biochemistry and Physiology*, **91B**, 359–63.

Katz, A. M. (1992). *Physiology of the Heart*. pp. 303–8. New York: Raven Press.

Keen, J.E., Vianzon, D-M., Farrell, A.P. & Tibbits, G.F. (1993). Thermal acclimation alters both adrenergic sensitivity and adrenoceptor density in cardiac tissue of rainbow trout. *Journal of Experimental Biology*, **181**, 27–47.

Kent, J., Koban, M. & Prosser, C.L. (1988). Cold-acclimation-induced protein hypertrophy in channel catfish and green sunfish. *Journal of Comparative Physiology*, **158B**, 185–98.

Kiceniuk, J.W. & Jones, D.R. (1977). The oxygen transport system in trout (*Salmo gairdneri*) during sustained exercise. *Journal of Experimental Biology*, **69**, 247–60.

Kimura, J., Miyamae, S., & Noma, A. (1987). Identification of sodium-calcium exchange current in single ventricular cells of guinea-pig. *Journal of Physiology London*, **384**, 199–222.

Kolok, A.S. & Farrell, A.P. (1994). Individual variations in the swimming performance and cardiac performance of northern squawfish, *Ptychocheilus oregonensis*. *Physiological Zoology*, **67**, 706–22.

Lennard, R., & Huddart, H. (1991*a*). The effect of thermal stress on electrical and mechanical responses and associated calcium movements of flounder heart and gut. *Comparative Biochemistry and Physiology*, **98A**, 221–8.

Lennard, R. & Huddart, H. (1991*b*). Anomalous thermal stress effects on calcium-dependent activity of flounder (*Platichthys flesus*) intestinal smooth muscle. *Comparative Biochemistry and Physiology*, **98A**, 229–234.

Lennard, R. &Huddart, H. (1992). The effect of thermal stress and adrenaline on the dynamics of the flounder (*Platichthys flesus*) heart. *Comparative Biochemistry and Physiology*, **101C**, 307–10.

Lin, Y., Dobbs, G. H., & DeVries, A. L. (1973). Oxygen consumption and lipid content in red and white muscles of Antarctic fishes. *Journal of Experimental Zoology*, **189**, 379–86.

Londraville, R. L.& Sidell, B. D. (1990). Ultrastructure of aerobic muscle in Antarctic fishes may contribute to maintenance of diffusive fluxes. *Journal of Experimental Biology*, **150**, 205–20.

Macdonald, J. A., Montgomery, J. C. & Wells, R. M. G. (1987). Comparative physiology of Antarctic fishes. In *Advances in Marine Biology*. ed. J.H.S. Blaxter & A.J. Southward Vol. 24. pp. 321–88. London: Academic Press.

Mangum, C. P., & Towle, D. W. (1977). Physiological adaptation to unstable environments. *American Scientist*, **65**, 67–75.

Martinez, I., Christiansen, J.S., Ofstad, R. & Olsen, R.L. (1991). Comparison of myosin isoenzymes present in skeletal and cardiac muscles of the Arctic charr *Salvelinus alpinus* (L.): sequential expression of different myosin heavy chains during development of the fast white skeletal muscle. *European Journal of Biochemistry*, **195,** 743–54.

Matikainen, N. & Vornanen, M. (1992). Effect of season and temperature acclimation on the function of crucian carp (*Carassius carassius*) heart. *Journal of Experimental Biology*, **167**, 203–20.

Møller-Nielsen T. & Gesser, H. (1992). Sarcoplasmic reticulum and excitation-contraction coupling at 20° and 10 °C in rainbow trout myocardium. *Journal of Comparative Physiology*, **162B,** 526–34.

Montgomery, J.C. & Wells, R.M.G. (1993). Recent advances in the ecophysiology of Antarctic notohenioid fishes: metabolic capacity and sensory performance. In *Fish Ecophysiology* ed. J.C. Rankin & F.B. Jensen, pp. 341–374. London: Chapman & Hall.

Morris, J. L. & Nilsson, S. (1994). The circulatory system. In *Comparative Physiology and Evolution of the Autonomic Nervous System*. series ed. G. Burnstock, ed. S. Nilsson & S. Holmgren, pp. 193–246. Chur, Switzerland: Harwood Academic.

Neely, J.R., Liebermeister, H., Battersby, E.J. & Morgan, H.E. (1967). Effect of pressure development on oxygen consumption by isolated rat heart. *American Journal of Physiology*, **212**, 804–14.

Nilsson, S. (1978). Sympathetic innervation of the spleen of the cane toad, *Bufo marinus*. *Comparative Biochemistry and Physiology*, **61C**, 133–49.

Nilsson, S. (1983). Innervation and pharmacology of the gills. In *Fish Physiology*, ed. W.S. Hoar & D.J. Randall, vol. XA, pp. 185–227. Academic Press: Orlando.

Nilsson, S. (1984). Review: Adrenergic control systems in fish. *Marine Biology Letters*, **5**, 131–8.

Nilsson, S. (1994). The spleen. In: *Comparative physiology and evolution of the autonomic nervous system*, series ed. G. Burnstock, ed. S. Nilsson, & S. Holmgren, pp. 247–56. Chur, Switzerland: Harwood Academic.

Nilsson, S. & Grove, D. J. (1974). Adrenergic and cholinergic innervation of the spleen of the cod, *Gadus morhua*. *European Journal of Pharmacology*, **28**, 135–43.

Nilsson, S. & Holmgren, S. (1992). Cardiovascular control by purines, 5–hydroxytryptamine, and neuropeptides. In *Fish Physiology*, ed. W.S. Hoar, D.J. Randall & A.P. Farrell, vol. 12B, pp. 301–41. San Diego: Academic Press.

Nilsson, S., Holmgren, S. & Grove, D. J. (1975). Effects of drugs and nerve stimulation on the spleen and arteries of two species of dogfish, *Scyliorhinus canicula* and *Squalus acanthias*. *Acta Physiologia Scandinavia*, **95**, 219–30.

Nilsson S., Forster M., Davison W. & Axelsson M. (1996). Nervous control of the spleen in the red-blooded Antarctic fish, *Pagothenia borchgrevinki*. *American Journal of Physiology*, **270,** R599–604.

Olson, K.R. (1992). Blood and extracellular fluid volume regulation: role of the renin angiotensin system, kallikrein–kinin system, and atrial natriuretic peptides. In *Fish physiology*, ed. W.S. Hoar, D.J. Randall & A.P. Farrell, Vol. XIIB pp. 135–254. San Diego, New York, London: Academic Press.

Pearson, M. P. & Stevens E.D. (1991). Size and hematological impact of the splenic erythrocyte reservoir in rainbow trout, *Oncorhynchus mykiss*. *Fish Physiology and Biochemistry*, **9**, 39–50.

Pearson, M., van der Kraak, G. & Stevens, E. D. (1992). *In vivo* pharmacology of spleen contraction in rainbow trout. *Canadian Journal of Zoology*, **70,** 625–27.

Platzack, B., Axelsson, M. & Nilsson, S. (1993). The renin angiotensin system in blood pressure control during exercise in the cod, *Gadus morhua*. *Journal of Experimental Biology*, **180**, 253–62.

Post, J.A. & Langer, G.A. (1992). Sarcolemmal calcium binding sites in heart: I. Molecular origin in 'gas-dissected' sarcolemma. *Journal of Membrane Biology*, **129,** 49–57.

Randall, D. J. & Perry, S. F. (1992). Catecholamines. In *Fish Physiology*. ed. W.S. Hoar, D.J. Randall & A.P. Farrell Vol. 12B, pp. 255–300. San Diego: Academic Press.

Randall, D. J. (1968). Functional morphology of the heart in fishes. *American Zoologist*, **8**, 179–89.

Rankin, J. C. (1989). Blood circulation and gill water fluxes in the icefish, *Chaenocephalus aceratus* Lönnberg. *Antarctic Special Topic*, **1989**, 87–91.

Rubinsky, B., Mattioli, M., Arav, A., Barboni, B. & Fletcher, G. L. (1992). Inhibition of Ca^+ and K^+ currents by 'antifreeze' proteins. *American Journal of Physiol*ogy, **262**, R542–5.

Santer, R.M. (1985). Morphology and innervation of the fish heart. *Advances in Anatomy Embryology and Cell Biology*, **89**, 1–102.

Satchell, G. H. (1991). *Physiology and Form of Fish Circulation*. Cambridge: Cambridge University Press.

Schipke, J.D. (1994) Cardiac efficiency. *Basic Research in Cardiology*, **89**: 207–40.

Seibert, H. (1979). Thermal adaptation of heart rate and its parasympathetic control in the European eel *Anguilla anguilla*. *Comparative Biochemistry and Physiology*, **64C**, 275–8.

Short, S., Butler, P.J. & Taylor, E.W. (1977). The relative importance of nervous, humoral and intrinsic mechanisms in the regulation of heart rate and stroke volume in the dogfish, *Scyliorhinus canicula* L. *Journal of Experimental Biology*, **70**, 77–92.

Sidell, B. D., & Moerland, T. S. (1989). Effects of temperature on muscular function and locomotory performance in teleost fish. In *Advances in Comparative and Environmental Physiology*, pp. 116–56. Berlin: Springer-Verlag.

Sidell, B.D., Driedzic, W.R., Stowe, D.B. & Johnston, I.A. (1987). Biochemical correlations of power development and metabolic fuel preferenda in fish hearts. *Physiological Zoology*, **60**, 221–32.

Sidell, B.D., Crockett, E.L. & Driedzic, W.R. (1995). Antarctic fish tissues preferentially catabolize monoenoic fatty acids. *Journal of Experimental Zoology*, **271,** 73–81.

Somero, G. N. (1975). The role of isozymes in adaptation to varying temperatures. In *Isozymes II: Physiological Function*, ed. C. L. Markert, pp. 221–34. New York: Academic Press.

Somero, G. N. (1983). Environmental adaptation of proteins: strategies for the conservation of critical functional and structural traits. *Comparative Biochemistry and Physiology*, **76A**, 621–33.

Somero, G. N. (1995). Proteins and temperature. *Annual Review of Physiology*, **57**, 43–68.

Sundin, L. & Nilsson, S. (1992). Arterio-venous branchial blood flow in the Atlantic cod, *Gadus morhua*. *Journal of Experimental Biology*, **165**, 73–84.

Sureau, D., Lagardere, J.P. & Pennec, J.P. (1989). Heart rate and its cholinergic control in the sole (*Solea vulgaris*) acclimatized to different temperatures. *Comparative Biochemistry and Physiology*, **92A**, 49–51.

Taylor, E. W. (1992). Nervous control of the heart and cardiorespiratory interactions. In *Fish Physiology*, ed. W. S. Hoar, D. J. Randall, & A. P. Farrell, vol. XII B, pp. 343–87. San Diego: Academic Press.

Tibbits, G. F., Moyes, C. D. & Hove-Madsen, L. (1992*a*). Excitation–contraction coupling in the teleost heart. In *Fish Physiology*, ed. W. S. Hoar, D. J. Randall & A. P. Farrell Vol. XII A, pp.267–304. San Diego: Academic Press.

Tibbits, G.F., Philipson, K.D. & Kashihara, H. (1992*b*). Characterization of myocardial sodium–calcium exchange in rainbow trout. *American Journal of Physiology*, **262,** C411–17.

Tibbits, G.F., Hove-Madsen, L. & Bers, D.M. (1991). Calcium transport and the regulation of cardiac contractility in teleosts: a comparison with higher vertebrates. *Canadian Journal of Zoology*, **69**, 2014–19.

Tibbits, G. F. & Kashihara, H. (1992). Myocardial sarcolemma isolated from skipjack tuna, *Katsuwonus pelamis*. *Canadian Journal of Zoology*, **70**, 1240–45.

Torres, J. J. & Somero, G. N. (1988). Metabolism, enzymic activities and cold adaptation in Antarctic mesopelagic fishes. *Marine Biology*, **98,** 169–80.

Tota, B., Cimini, V., Salvatore, G. & Zummo, G. (1983). Comparative study of the arterial and lacunary systems of the ventricular myocardium of elasmobranch and teleost fishes. *American Journal of Anatomy*, **167**, 15–32.

Tota, B., Acierno, R. & Agnisola, C. (1991). Mechanical performance of the isolated and perfused heart of the haemoglobinless Antarctic icefish *Chionodraco hamatus* (Lönnberg): effects of loading conditions and temperature. *Philosophical Transactions of the Royal Society of London, B*, **332**, 191–8.

Tota, B., Acierno, R. and Agnisola, C. (1994). Morphodynamic analysis of *C. hamatus*, a hemoglobinless Antarctic fish. *Physiologist*, **37**, A93.

Tota, B., Agnisola, C., Schioppa, M., Acierno, R., Harrison, P. & Zummo, G. (1991). Structural and mechanical characteristics of the heart of the icefish *Chionodraco hamatus*. In *Biology of Antarctic Fish*, ed. G. DiPrisco, B. Maresca & B. Tota, pp. 204–19. Berlin: Springer-Verlag.

Tsukuda, H., Liu, B. & Fujii, K. I. (1985). Pulsation rate and oxygen consumption of isolated hearts of goldfish, *Carassius auratus*, acclimated to different temperatures. *Comparative Biochemistry and Physiology*, **82A**, 281–3.

Vornanen, M. (1994). Seasonal and temperature-induced changes in myosin heavy chain composition of the crucian carp hearts. *American Journal of Physiology*, **267**, R1567–73.

Vornanen, M. (1996). Effect of extracellular calcium on the contractility of warm and cold-acclimated crucian carp heart. *Journal of Comparative Physiology*, **165B**, 507–17.

Vornanen, M., Shepherd, N. & Isenberg, G. (1994). Tension-voltage relations of single myocytes reflect Ca release triggered by Na/Ca exchange at 35 °C but not 23 °C. *American Journal of Physiology*, **267,** C623–32.

Winberg, M., Holmgren, S. & Nilsson, S. (1981). Effects of denervation and 6–hydroxydopamine on the activity of choline acetyltransferase in the spleen of the cod, *Gadus morhua*. *Comparative Biochemistry and Physiology*, **69C**, 141–3.

Wood, C.M. & Shelton, G. (1980). Cardiovascular dynamics and adrenergic responses of the rainbow trout *in vivo*. *Journal of Experimental Biology*, **87**, 247–70.

Wood, C. M., Pieprzak, P. & Trott, J. N. (1979). The influence of temperature and anaemia on the adrenergic and cholinergic mechanisms controlling heart rate in the rainbow trout. *Canadian Journal of Zoology*, **57**, 2440–7.

Yee, H.F. Jr., Weiss, J.N. & Langer, G.A. (1989). Neuraminidase selectively enhances transient Ca^{++} current in cardiac myocytes. *American Journal of Physiology*, **256,** C1267–2.

Zummo, G., Acierno, R., Angola, C. & Tota, B. (1995). The heart of the icefish: bioconstruction and adaptation. *Brazilian Journal of Medical and Biological Research*, **28**, 1265–76.

L.S. PECK

Feeding, metabolism and metabolic scope in Antarctic marine ectotherms

The rise in metabolism following feeding exhibited by animals, known as the heat increment or specific dynamic action (SDA) of feeding has been well known since the 1930s when it was described in domesticated mammals (Brody, 1945). The effect is usually assessed in terms of oxygen consumed following a feeding event, and is characterised by a rapid rise in metabolism which peaks within a few hours of the consumption of a meal. In marine ectotherms, peaks are typically two to four times higher than prefeeding standard or basal metabolic rates, where basal metabolism is defined as the cost of maintaining the body tissues in the absence of factors such as activity, growth or SDA which may cause elevations. Some species are, however, capable of elevations much greater than two to four times basal metabolism, and recently Burmese pythons have been shown to have SDA peaks 45 times higher than prefeeding levels (Secor & Diamond, 1995), a metabolic scope which is as high as the greatest aerobic scopes exhibited by highly trained active species, such as racehorses (Brody, 1945; Birlenbach & Leith, 1994), under maximal workloads.

It is now generally accepted that SDA is composed of the physical costs of processing food and both the anabolic (transformation of absorbed material and growth) and catabolic (formation and excretion of wastes) processes associated with the feeding event (Pandian, 1987; Wieser, 1994). The contribution of many of these components have been well studied in fish during the last 20 years (Jobling, 1981,1994; Pandian, 1987; Lyndon, Houlihan & Hall, 1992), and more recently protein synthesis has been shown to be a major component of SDA in temperate species (Houlihan, 1991; Houlihan *et al.*, 1990, 1995*a*; Houlihan, Carter & McCarthy, 1995*b*; Lyndon *et al.*, 1992). Metabolic elevation due to excitement produced by the presence of food has also been indicated as a potentially important SDA component in the supralittoral isopod *Ligia pallasii* (Carefoot, 1990*a*). It might also be expected that temperature would affect the SDA response. Recently the effects of temperature on post-prandial metabolism have received attention, and data have been obtained on SDA responses in Antarctic species.

In the following discussion, two terms will be used: SDA factorial scope, which means the peak post-prandial elevation expressed relative to prefeeding levels, and the absolute increment, which is purely the maximum numerical increase in metabolism after a feeding event. Factorial scopes are always expressed as multiples of prefeeding metabolic rates, they are dimensionless and, as seen above, are usually in the range ×2 to ×4 for marine ectotherms. Absolute increments, on the other hand, are usually quoted numerically in molar, mass or volume units.

Temperature and scaling effects on SDA peaks

Recent studies, mainly on fish have shown that temperature and animal size have little effect on the size of post-prandial factorial scopes (Jobling & Davies, 1980; Johnston & Battram, 1993). However, assessments of the elevations in metabolism in terms of absolute increments, instead of relative to prefeeding levels produce very different results. Metabolic rate in marine ectotherms rises in a non-linear fashion with temperature, and although factorial scope may be the same at different temperatures, absolute increments can vary dramatically (Fig. 1). Consider an idealised crustacean living at a temperature near to 0 °C with an oxygen uptake of around 0.02 mol h^{-1}. A post-prandial maximum factorial scope of 2 means it can raise its metabolism by 0.02 mol h^{-1}, which converts to 2.7 watts of power output. At 20 °C metabolic rates are around 0.13 mol h^{-1}, and a factorial scope of 2 gives a rise of 0.13 mol h^{-1} (16 watts), and at approaching 30 °C the rate is around 0.27 mol h^{-1} giving a power output of 32 watts. Thus, although the factorial scope is constant the absolute increment of the SDA is 12 times greater at 30 °C than at 0 °C. Clearly, the animals at higher temperatures are capable of more work and greater metabolic power output than those at lower temperatures, and these differences may be highly significant. This effect is purely due to the higher resting or standard metabolic rates of species living at higher temperatures. It should also be noted, that the higher the factorial scope the greater is this temperature effect in absolute terms. This has bearing on the comparison made earlier between maximum post-prandial factorial scopes in snakes and maximum factorial aerobic scopes exhibited by trained mammals. Both Burmese pythons and trained racehorses had factorial scopes of 45. However, the absolute power generated by a racehorse at full speed is much greater than that of the python following a large meal because of the large difference in their prefeeding resting metabolic rates.

A similar argument is applicable to the effects of animal size on metabolic scope or power generation. Factorial SDA scopes have been shown to be

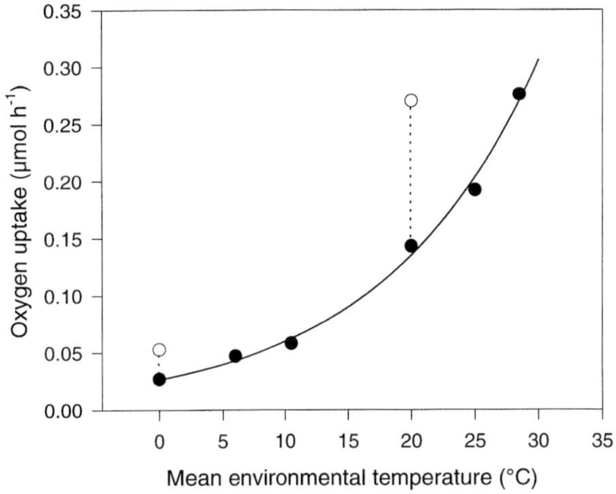

Fig. 1. The relationship between respiration rate and mean environmental temperature for marine crustaceans (adapted from Clarke, 1987). The effects of a doubling of metabolic rate following feeding for idealised species at 0 °C and 20 °C are shown. The species at 20 °C generates nearly six times as much metabolic power during the SDA as a species at 0 °C, even though the factorial rise in metabolic rate (×2) is the same.

constant in relation to animal size for dogfish (Sims & Davies, 1994), the Antarctic plunderfish (Boyce and Clarke, submitted) and the isopod *Ligia pallasii* (Carefoot, 1990*a*). However, standard or basal metabolic rates increase with animal size, usually to the 0.75 power of mass, in inter-specific comparisons (Hemmingsen, 1960; Heusner, 1982; Prothero, 1984, Wieser, 1984; LaBarbera, 1986). They also increase in a similar fashion in intra-specific assessments, although the scaling coefficient often varies from 0.75. In the Antarctic brachiopod *Liothyrella uva* starved (basal) metabolic rate ($\dot{M}O_2$, mol h^{-1}) is related to animal ash-free dry mass (AFDM, g) by the equation:

$\log_e\dot{M}O_2=0.41 + 0.40 \log_e AFDM$ ($r^2=0.93$, $F=571$, $P < 0.001$, $n=45$; Peck *et al.*, 1987)

L. uva has a post-prandial factorial scope of 1.6 (Fig. 2, Table 1). Larger animals, therefore, have greater post-prandial absolute increments and are capable of producing more metabolic power than smaller ones, because of their higher prefeeding standard metabolic rates. A 100 mg AFDM individual has a basal metabolic rate of 0.18 µmol h^{-1} and an absolute post-prandial increment of 0.11 µmol (12.5 µwatts of power), whereas in a 1 g AFDM individual these figures are 1.13 µmol h^{-1} and 141 µwatts, respectively. On a

Table 1. *SDA peak elevations (factorial rise over prefeeding starved levels) and duration (days) in oxygen consumption for Antarctic ectotherms*

Species	Peak elevation	Time to reach peak (days)	Total SDA duration (days)	Source
Antarctic species				
Invertebrates				
Liothyrella uva (Brachiopoda)	1.6	1–5	20	Peck (1996)
Parborlasia corrugatus (Nemertea)	1.8		20–25	Boyce & Clarke (1997)
Nacella concinna (Gastropoda)	2.3	6–7	8–12	Peck & Veal (in prep.)
Glyptonotus antarcticus (Isopoda)	2.2	3–4	7	Robertson (pers. comm.)
Waldeckia obesa (Amphipoda)	4.0	1	4	Chapelle *et al.* (1994)
Fish				
Notothenia coriiceps	2.0–2.3	2–3	9	Johnston & Battram (1993)
Harpagifer antarcticus	2.1–2.5	2–3	10–15	Boyce & Clarke (in press)
Temperate species				
Invertebrates				
Mytilus edulis (Bivalvia)	2.2	<0.05	<1	Thompson & Bayne (1972)
Nassarius reticulatus (Gastropoda)	2.2–5.2	<0.05	2–3	Crisp *et al.* (1978)

Octopus vulgaris (Cephalopoda)	2.7	<0.125	0.5	Wells *et al.* (1983)
Ligia pallasii (Isopoda)	2.5	<0.05	<1	Carefoot (1990*a*,*b*,*c*)
Ocypode quadrata (Decapoda)	2.7	<0.5	1.75	Burggren *et al.* (1993)
Goniopsis cruentata (Decapoda)	2.5	<0.5	2	Burggren *et al.* (1993)
Fish				
Myoxocephalus scorpius	2.6–3.5	<1	6–8	Johnston & Battram (1993)
Kuhlia sandvicensis	2.5	0.5–0.75	2–2.5	Muir & Niimi (1972)
Gadus morhua	2–3	<0.5	2–7	Soofiani & Hawkins (1982)

Note:
Data for a range of temperate marine ectotherms are also included for comparison. Data for temperate species are from studies where large meals were fed, and both peak elevations and SDA durations were recorded.

mass-specific basis, however, power utilisation is not as different, with the 100 mg individual producing 125 μwatts per gram of AFDM and the 1 g animal producing 141 μwatts per gram AFDM. The difference between mass specific power utilisation by small and large animals in a given species will depend on how metabolism scales to size in that species, and the magnitude of its SDA factorial scope. How significant the consequences of such differences as those calculated for *L. uva* are in terms of physiological capabilities is yet to be determined.

The implications of temperature effects on metabolic power generation are that species at lower temperatures are restricted in the absolute amounts of power they can generate compared to warmer water species, even though their factorial scopes are the same. This is also true of the effects of animal size on power generation with larger individuals being capable of producing more absolute power than smaller ones. The consequences of these relations on lifestyles and organismal distributions have yet to be elucidated, but they may be important in dictating which groups are successful in cold but seasonal Antarctic marine ecosystems.

SDA in Antarctic ectotherms

Recent studies have shown that the elevation of metabolism after feeding in Antarctic species is characterised by relatively low peaks and a highly extended SDA duration. Peck (1996) starved specimens of the brachiopod *Liothyrella uva* at Signy Island for 25 days prior to allowing them to feed on naturally occurring phytoplankton for 24 h. Following feeding, oxygen consumption rose to a peak 1.6 times the prefeeding basal level, and it took between 1 and 5 days for the peak to be attained (Fig. 2). The total duration of the elevation in metabolism lasted for 18–20 days. Peak factorial scopes for other Antarctic species are in the range 1.9 to 2.5 times prefeeding levels (Table 1), with the exception of the amphipod *Waldeckia obesa*, which was assessed following an extended period of starvation of 64 days. Durations of elevated oxygen consumption in the SDA of these species of up to 25 days were all dramatically extended compared with the same responses in temperate ectotherms, where durations of between a few hours and up to 3 days are typical (Bayne & Scullard, 1977; Wells *et al.*, 1983; Pandian, 1987; Jobling, 1992). In studies of SDA in temperate species, peaks are usually reached immediately, or within a few hours of the feeding event. In cold-water species, the time needed for oxygen consumption to peak is again extended to a period greater than 24 h, but usually occurs within the 1–5 days seen for *Liothyrella uva*.

The factorial scopes in oxygen consumption exhibited by the Antarctic species so far studied may seem slightly reduced in comparison with data

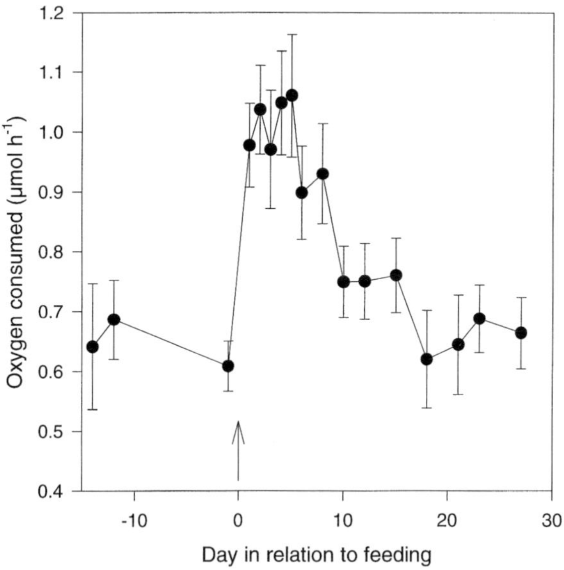

Fig. 2. Oxygen consumption in relation to feeding in the Antarctic brachiopod *Liothyrella uva*. Data presented are mean values (±SE, n=6) for a standard animal size of 290 mg, ash-free dry mass (37 mm length) which was close to the average size of the group. The arrow indicates the period when the brachiopods were fed for 24 h. (From Peck, 1996.)

from temperate and tropical species (Table 1). There has, however, only been one study directly comparing species living with similar lifestyles from similar habitats in polar, temperate and tropical localities. Johnston and Battram (1993) studied demersal fish from the Antarctic, the North Sea and the Indian Ocean. Prefeeding rates of oxygen consumption were positively correlated with habitat temperature and factorial scopes (peak rate/prefed starved oxygen consumption) were independent of habitat temperature and all were in the range two to three. SDA duration, however, increased markedly with reduced environmental temperature and was three to four times longer in cold-water fish than species from warm environments. Times taken to reach SDA peaks were between two and ten times longer in the Antarctic species than those from temperate or tropical regions. A similar effect was found in a within-species investigation comparing SDA and temperature on a temperate fish, *Pleuronectes platessa*, where reduced temperature increased the duration of the SDA, with a decline in temperature from 20 °C to 10 °C causing an approximate doubling of the time required for a return to prefed oxygen consumption levels (Jobling & Davies, 1980). Thus the effect of reduced temperature on the SDA appears to be one where factorial rises to peak levels are maintained at constant levels, but durations are extended. Also, because prefeeding basal metabolic rates are correlated with temperature, the overall size of the SDA (the total energy used) remains constant at different temperatures.

The slightly lower SDA factorial scopes for Antarctic marine invertebrates seen in Table 1, appears superficially to conflict with the consistent factorial scopes found at different environmental temperatures both within fish species (Jobling & Davies, 1980) and between species of similar lifestyles (Johnston & Battram, 1993). However, most of the Antarctic invertebrates studied are characterised by generally low levels of activity, a point that has been made powerfully for brachiopods (James *et al.*, 1992; Peck *et al.*, 1996), and this is a general trait for most of the common marine benthic species from Antarctica (Clarke & Peck, 1991). There is therefore, a strong phylogenetic component in any general reduction of factorial scopes seen in the SDA assessments of Antarctic species to date. This probably reflects the success of species with low energy lifestyles in polar oceans. It has been argued that this success is due to the low basal metabolic rates exhibited by cold-water benthos which provide advantages in resource limited, or highly seasonal, environments (Clarke, 1983, 1988, 1993; Clarke & Peck, 1991). The constant relationship found between basal metabolism and peak metabolic rise (i.e. a constant factorial scope) across latitudinal zones would therefore, not conflict with the observed slight reduction in factorial SDA scopes in oxygen consumption in Antarctic species because the successful groups would be ones with phylogenetic restrictions to metabolic elevation during an SDA, and good comparisons for these species with temperate and tropical equivalents are rare or absent.

Animal size, meal size and the SDA coefficient

In the above discussion of the form of the SDA in Antarctic species, it was seen that temperature had little or no effect on the factorial rise in metabolism, where comparisons were made between species with similar lifestyles. However, the time taken to attain peak levels and the overall duration of the response were both extended. The total energy used in the SDA for a given meal size was also independent of temperature. These three factors may seem paradoxical, but are balanced by the lower standard prefeeding metabolic rates of species at lower temperatures producing lower absolute increments for the same factorial scope. In temperate animals other factors than temperature have been shown to alter the characteristics of the metabolic response to feeding. These include meal size and composition. Animal size, on the other hand, is found to have little effect on the SDA factorial scope. For example Sims and Davies (1994) found that, after being fed a meal of

~7% body mass, small and large dogfish both had SDA factorial scopes of 2.8 ± 0.2 (times prefeeding oxygen consumption rate), and in both groups peak metabolism was attained 4–10 h post feeding. Large fish, however, took nearly twice as long as small fish (96 h vs 50 h) for oxygen consumption to return to prefeeding levels. Similar results to this were found for the Antarctic plunderfish *Harpagifer antarcticus*, with animal size not affecting peak factorial scope, but significantly increasing the duration of the SDA from around 250 h for small individuals to between 320 and 390 h for large fish (Boyce & Clarke, 1997). The above results showing consistent factorial scopes and enhanced SDA durations with increased animal size would tend to suggest that the overall size of the SDA as total oxygen consumed or total energy expended should increase in larger animals. However, because specimens were fed constant proportions of their own body mass absolute meal sizes increased in larger specimens. Carefoot (1990*b*) working on the isopod *Ligia pallasii* used ANOVA and multiple comparison tests to separate the various factors affecting the rise in metabolism. He found that, although total oxygen consumed rose with animal size, the rises were small and the effect was only significant in one of his six treatments. Clearly, body size has only a very small effect, if any, on overall energy utilisation during an SDA.

There has been much work over many years on the effects of meal size and composition on post-prandial metabolism, initially on domesticated mammals, and later mainly with fish (Brody, 1945; Kleiber, 1961; Pandian, 1987; Jobling, 1994; Houlihan *et al.*, 1995*a*). This showed dietary protein to have a much larger effect on the SDA than lipids or carbohydrates. Increasing ration increases the overall magnitude of the SDA, and the general finding is that the duration of the effect is increased at larger meal sizes. The increase in total energy utilised during the SDA (the area under the curve) was found to be linearly related to meal energy content for juvenile cod, *Gadus morhua* by Soofiani and Hawkins (1982). For the isopod *Ligia pallasii*, the relationship between total energy used and meal size was exponential (Carefoot, 1990*b*) and there was also a strong positive correlation between the overall size of the SDA and dietary amino acid content (Carefoot, 1990*a*).

The factorial scope of the rise in metabolism, or the ratio of peak to prefed oxygen consumption rates is often not found to be affected by changing meal size. This was the case for the Antarctic plunderfish, *Harpagifer antarcticus* where juveniles and adults were both studied and fed a high protein diet of *Euphausia superba* (Antarctic krill) muscle. They were fed at either low rations (meal size=2–3% wet body mass), or to satiation (12–14% wet body mass). For both sizes of fish, factorial scope declined slightly, but non-significantly, from small to large meals. Unexpectedly, the duration of the SDA response was also not affected by meal size, where times taken for

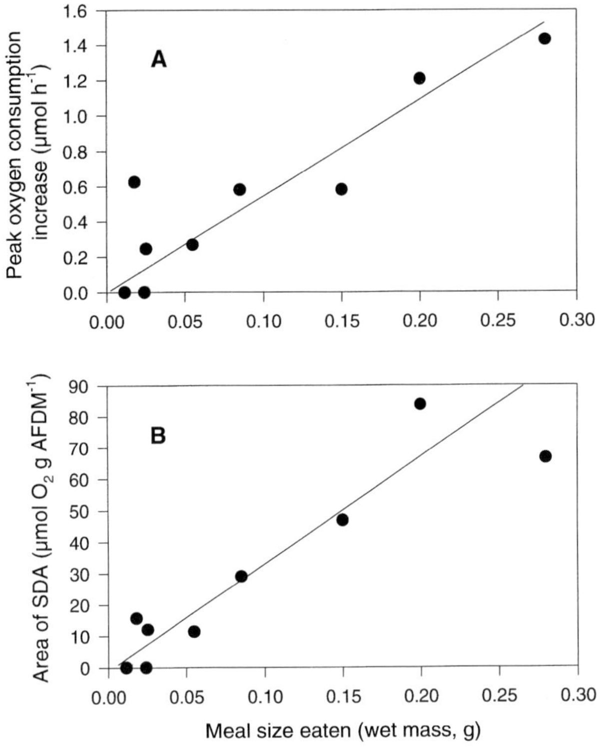

Fig. 3. SDA responses in the Antarctic starfish *Odontaster validus* following meals of different sizes. A. peak elevations. B. total magnitude of the SDA. Note both are linearly related to meal size. Mean starfish mass was approximately 10 g live weight. Meal sizes were, therefore in the range 0.25% to 2.75% body mass, and the food used was foot muscle of the limpet *Nacella concinna*. (Figures produced from data provided by Omana, 1995.)

metabolism to return to prefeeding levels were longer for large meals, but differences again were not significant (Boyce & Clarke, 1997).

This picture is not maintained, however, at very low ration levels, as indicated by data for the starfish *Odontaster validus*, from Signy Island, Antarctica. Specimens were fed on pieces of foot of the limpet *Nacella concinna*, a known natural prey item, and meal size varied between 0.25% and 2.75% wet body mass. Both maximum peak rise and overall magnitude of the SDA as area under the curve increased linearly with meal size (Fig. 3). Clearly, factorial scope rises with increasing meal size for very small rations. The effect does, however, saturate at small meal sizes. The overall SDA magnitude, as total extra energy used appears to be directly related to meal size for all ration levels.

The SDA coefficient is defined as the proportion of the ingested meal that is accounted for by the overall summed elevation in metabolism during the SDA, and it is usually quoted in energetic terms (e.g. Johnston & Battram, 1993). Thus, if the post-prandial rise in metabolism lasts for 20 days the SDA coefficient is the sum of metabolic costs above the prefeeding standard level over the whole 20 days divided by the energy content of the food ingested. In domesticated mammals and rodents it can be as high as 80% (Kleiber, 1961). These values are generally much lower for marine ectotherms. In a wide range of fish species the SDA coefficient ranged between 6% and 21% (Pandian, 1987), and in juvenile salmon the value was as high as 45% (Averett, 1969). Measures in invertebrates are rare, but in *Mytilus edulis* it was around 24% (Bayne & Scullard, 1977). In the isopod *Ligia pallasii* the SDA coefficient was 12–17% on natural diets (Carefoot, 1989), but was only 2.7% on an optimised artificial diet (Carefoot, 1990*a*). In Antarctic species the SDA coefficient was not affected by animal size, but varied markedly with ration level in the plunderfish *Harpagifer antarcticus* (Boyce & Clarke, 1997). At low ration levels it was 55% of ingested energy, but when fed larger meals the figure fell to around 9%. SDA coefficients of 50–60% were found for the giant isopod *Glyptonotus antarcticus* (R. Robertson, pers. comm.) fed Antarctic krill muscle at rations of around 5% wet body mass. The Antarctic nemertean, on the other hand had SDA coefficients of 1% or less when fed tissues of the limpet *Nacella concinna*, at ration levels between 20% and 110% wet body mass. Johnston and Battram (1993) measured SDA coefficients in tropical, temperate and polar fish of similar lifestyles, fed meals of the shrimp *Crangon crangon* between 5% and 15% body mass. They found a constant SDA coefficient of between 15 and 23% for *Cirrhytichys bleekeri, Myoxocephalus scorpius* and *Notothenia neglecta* (now *coriiceps*), from tropical, temperate and Antarctic latitudes respectively. These studies appear to show a decline of SDA coefficient with increasing meal size, with those studies utilising low rations (5% body mass, or less) producing coefficients up to 60%, moderate ration investigations (5–15% body mass) giving coefficients between 5% and 25%, and high ration studies (20%-110% body mass) producing values less than 5%. These data are still clearly limited, and factors such as diet composition will have powerful effects here, as seen for temperate species (Pandian, 1987; Carefoot, 1990*c*; Houlihan, 1991; Jobling, 1994). However, an inverse relationship between meal size and magnitude of SDA coefficient is emerging for Antarctic species.

Nitrogen excretion and the O:N ratio

Increased ammonia excretion following feeding has been observed previously in temperate species. In the mussel *Mytilus edulis* a pulse of ammonia was produced much later than the oxygen peak, and the duration of the

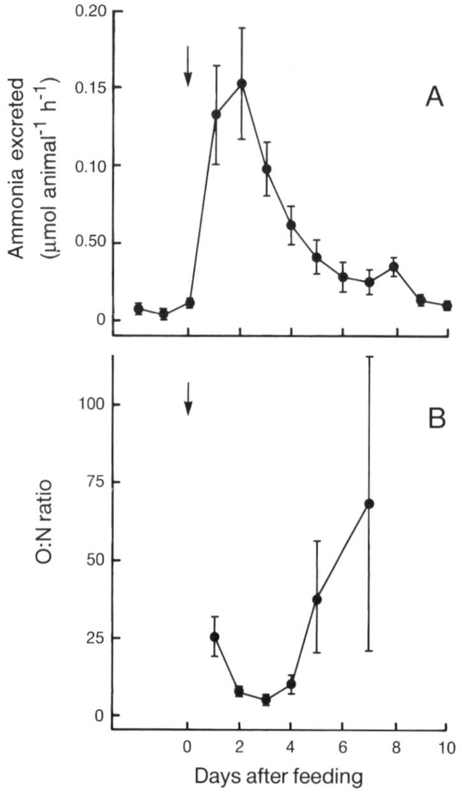

Fig. 4. Post-prandial responses in ammonia excretion (A) and O:N ratio (B) in the scavenging Antarctic amphipod *Waldeckia obesa*, after Chapelle *et al.*, 1994. Arrows indicate period when animals were fed.

period of increased excretion was around 10–15 hours (Bayne & Scullard, 1977). However, this aspect of metabolic response to a feeding event has received little attention. Ammonia excretion following feeding in Antarctic species shows a wide range of responses. Some species exhibit increased ammonia excretion rates which have a greater factorial scope and are of even longer duration than the SDA measured using oxygen consumption. In the amphipod *Waldeckia obesa* peaks and durations of enhanced ammonia excretion were two to three times the equivalent responses in the oxygen SDA (Chapelle, Peck & Clarke, 1994). The peak in ammonia production was also later than the oxygen peak and, because of this, O:N ratios declined from levels in excess of 20 to low values around 6 on days 3–5 after feeding. They then returned to high levels by the 7th post-prandial day (Fig. 4). Similar responses in ammonia excretion were found in the isopod *Glyptonotus antarcticus* where peak rises were around six times prefed levels and the duration was for 8–10 days, compared with a rise of 2.2 times and 7 days for oxygen (R. Robertson, pers. comm). A much longer response has been found in the nemertean *Parborlasia corrugatus,* where ammonia excretion stayed high for over 30 days (Clarke & Prothero-Thomas, 1997). However, in that study meal sizes were extremely large and, in some trials, were larger than the worms in terms of mass consumed. The Antarctic plunderfish, *Harpagifer antarcticus* showed different responses depending on meal size (Boyce & Clarke, 1997, Fig. 5). Small meals elicited a small transient increase in ammonia excretion lasting for 2–5 days followed by an extended period of 10–12 days when excretion rates were less than prefeeding levels. When fed to satiation, however, ammonia excretion rose to a peak over ten times prefeeding levels (4–6 days post-prandially), and the overall response lasted 15–20 days. O:N ratios were the reverse of this (Fig. 5B). Prefeeding ratios were between 20 and 40. Low rations produced enhanced O:N ratios, up to ~160, and ratios stayed high for over 20 days. High rations caused ratios to fall to values around 5, before returning to between 20 and 40 after 15–20 days.

In the brachiopod *Liothyrella uva* the response was very different again (Fig. 6A). Ammonia excretion was little affected by feeding except for a short sharp rise in excretion 4–5 days after feeding (Peck, 1996). Under prefeeding starved conditions O:N ratios were very low, in the region of three to five (Fig. 6B). Post-prandially these rose rapidly to high values around 7, and declined very slowly, such that they had only returned to prefeeding levels around 20 days after the feeding event. Similar low O:N ratios during an SDA have been found in the Antarctic nemertean *Parborlasia corrugatus*, where values were between 2 and 7 throughout (Clarke & Prothero-Thomas, 1997).

These data show that the balance of substrates being burnt during the SDA changes dramatically during the course of the response. The O:N ratio can be used as an indicator of the proportion of protein being used to fuel metabolic requirements. The lowest ratios possible are around two to three, and indicate the sole use of protein (Mayzaud & Conover, 1988), and ratios around 25 indicate a protein utilisation of around 50% (Ikeda, 1974). These figures are based on utilisation of substrates within the body tissues and, therefore refer to absorbed or stored proteins. The amphipod *W. obesa,* when fed meals of Antarctic krill (*Euphausia superba*), moved from using protein for less than half of its metabolic requirements in prefed conditions (the amphipods were starved for 64 days) to utilising solely protein during the SDA (Chapelle *et al.*, 1994). It therefore appears that this species relies mainly on lipid or carbohydrate stores to survive periods of reduced resource

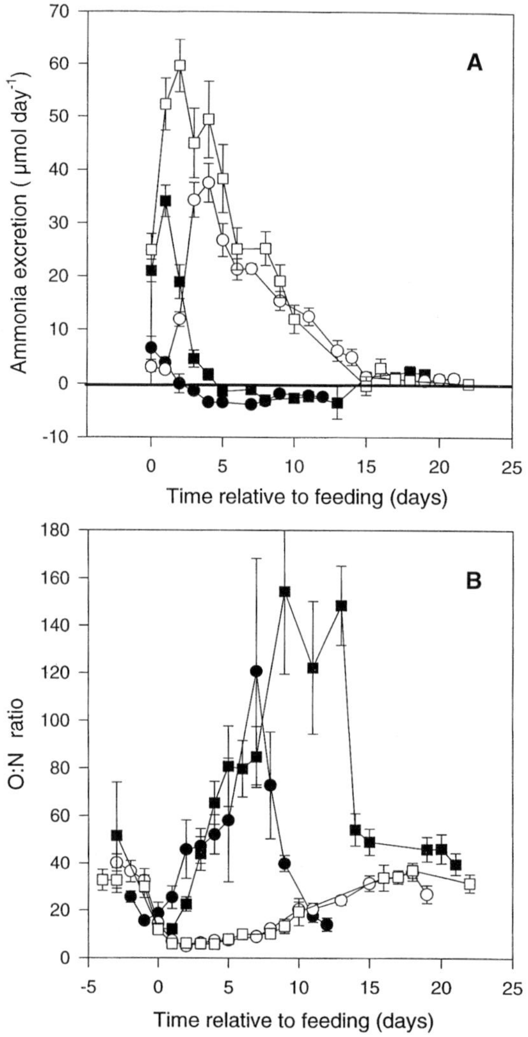

Fig. 5. Post-prandial responses in ammonia excretion (A) and O:N ratio (B) in the Antarctic plunderfish *Harpagifer antarcticus.* Data shown are for juvenile fish of around 1.7 g wet mass (circles) and large adult specimens of around 3.9 g wet mass (squares), and for small meals of 2–3% wet body mass (closed symbols) and animals fed to satiation at 12–14% wet body mass (open symbols). Note that low rations elicit increased ammonia production, but that high meals cause a reduction in ammonia excreted. As expected, O:N ratios showed an opposite trend to ammonia excretion, with small meals producing enhanced O:N ratios and large meals reduced ratios. In all cases the effect lasted for between 15 and 20 days. (From Boyce & Clarke, 1997.)

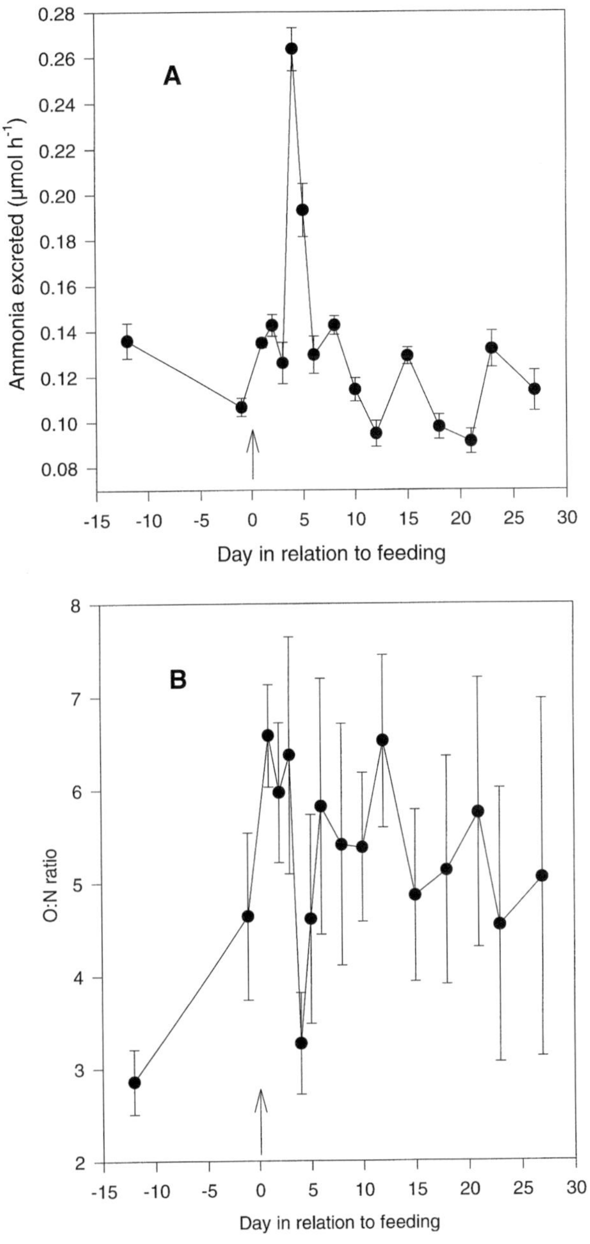

Fig. 6. Post-prandial responses in ammonia excretion (A) and O:N ratios (B) in the Antarctic brachiopod *Liothyrella uva*. Arrows indicate the day of feeding. (From Peck, 1996.)

supply. The brachiopod *L. uva* (fed on naturally occurring phytoplankton) and the nemertean *P. corrugatus* (fed foot muscle of the limpet *Nacella concinna*), on the other hand had O:N ratios close to the minimum possible at all times, indicating an almost sole use of protein to fuel metabolic requirements before, during and after the feeding event. Finally, when fed meals of Antarctic krill (*E. superba*) the plunderfish (*H. antarcticus*) moved from ratios indicating a significant reliance on protein (approaching 50%) to either using virtually no protein at all (on low rations), or using almost exclusively protein for metabolic requirements (satiation diet). Clearly, there is a need for more work to elucidate the mechanisms controlling which substrates are important for fuelling metabolism during an SDA, and also whether pulses of ammonia excretion are associated with amino acid deamination associated with enhanced protein turnover, growth or storage functions, or associated with an enhanced energy requirement for processing food. The data on Antarctic species, however, suggest that meal size and composition have important roles in determining which metabolic fuels are used during the SDA. There is also some evidence that phylogenetic or lifestyle traits may again be influencing the response, with the low energy lifestyle brachiopods and nemerteans utilising protein for metabolic requirements under all conditions; whereas the scavenging amphipod *Waldeckia obesa*, which has to maintain a more dynamic lifestyle in response to patchy resource availability, survives low food supply periods utilising mainly lipid and/or carbohydrate stores. The plunderfish has a more complex response, relying on a mix of reserves during protracted periods of low resource supply, and fuelling metabolism with dramatically different balances of substrates during the first 10–20 days after feeding, depending on meal size.

Urea excretion has been measured in association with a feeding event in some Antarctic species. In the brachiopod *L. uva* urea accounted for between 1 and 16% of total nitrogen excreted. It was measured for a period extending from 12 days before the meal to 30 days after feeding, and there was no clear pattern in urea excretion associated with the SDA. In the plunderfish *H. antarcticus* urea accounted for 10–50% of total nitrogen excreted. It did not vary with feeding at low rations, but did show a small increase in excretion when specimens were fed to satiation (Boyce & Clarke, 1997). In the nemertean *Parborlasia corrugatus* urea excretion did not change following feeding for approximately 35 days, and then a relatively large pulse of urea was produced over a period of ~5 days (Clarke & Prothero-Thomas, 1997). Such long-term responses as this have not been seen before, and delayed urea pulses following a feeding event may be present in marine ectotherms in general. However, they would not have been noted in previous investigations, because few studies have measured urea production, and only the nemertean investigation has continued measures over such a long duration. However, this is clearly a phenomenon of considerable interest because, although surveys are far from comprehensive, the ornithine cycle is not thought to be present in invertebrates (Baldwin, 1967; Nicol, 1967; Regnault, 1987; Clarke, Prothero-Thomas & Whitehouse, 1994). If this is the case urea excretion in the above species could only come from the deamination of arginine and/or the excretion of purines, pyrimidines or their breakdown products. This could signify an important role for nucleic acid metabolism generally in some Antarctic marine invertebrates. It would also indicate a large increase in nucleic acid metabolism in the nemertean *P. corrugatus,* when not fed for 30–40 days after a previous meal. Whether the dramatically delayed pulse of urea excreted should be viewed as a normal response to feeding or a change associated with starvation is also yet to be elucidated. In the plunderfish, the urea excretion is presumably associated with maintenance of pH balance via the ornithine cycle (Atkinson, 1992).

Growth and the SDA

The current understanding of the major components of elevated post-prandial metabolism is that there are two major components: the physical costs of processing food, including digestion, absorption and nutrient storage (catabolic processes); and secondly anabolic processes such as deamination of amino acids, synthesis of excretory products, increased synthesis and turnover of proteins and deposition of tissues (Pandian, 1987; McMillan & Houlihan, 1988; Jobling, 1994; Wieser, 1994; Houlihan *et al.*, 1995*a*,*b*). Krebs (1960), mainly based on considerations of mammalian responses to feeding, contended that the SDA was totally accounted for by catabolic processes. However, since then a large body of data has shown that, in most species the anabolic costs (usually measured as a whole and described as growth), form a large component of the SDA and that enhanced protein synthesis is often a major anabolic cost (Houlihan *et al.*, 1990, 1995*a*,*b*). Studies in fish and isopods manipulating diets by diluting them with α-cellulose suggested that the mechanical component accounted for 5–30% of the rise in metabolism (Tandler & Beamish, 1979; Carefoot, 1990*a*,*c*), whereas other investigations (e.g. Cho & Slinger, 1979) have found mechanical contributions as low as 1–3%. Interestingly, Wells *et al.* (1983) interpreted SDA responses in *Octopus vulgaris* as having two components, a short sharp peak lasting around 6 hours which they attributed to the costs of processing food, and a longer-term elevation of metabolic rate lasting 3–4 days associated mainly with growth.

Studies assessing whole animal oxygen consumption following injected amino acids have found SDA responses similar to those elicited by meals of equivalent size in alligators (Coulson & Hernandez, 1979) and channel

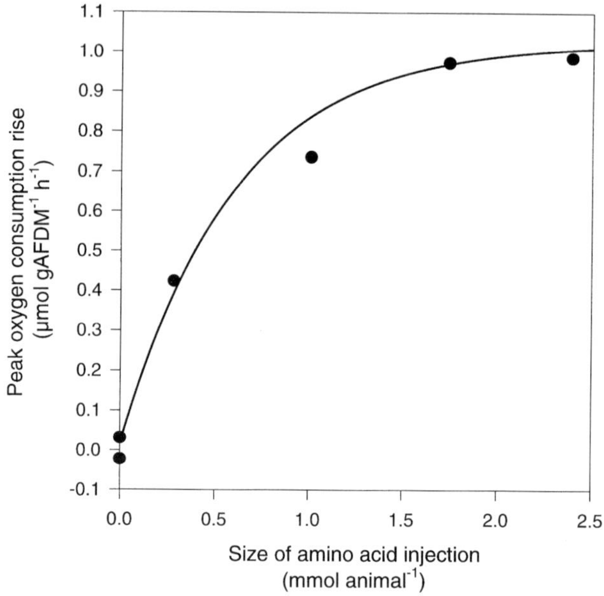

Fig. 7. Peak oxygen consumption rises following injection of amino acids into the coelom of the starfish *Odontaster validus*. Injection volumes were maintained at the same size, but amounts of amino acid injected were varied to be equivalent to meals of limpet foot (*Nacella concinna)* between 0 and 4% of starfish wet body mass. Peak rises increased with percentage of body weight injected, but the effect appeared to become saturated at injections between 2.5% and 4% body weight. (Figure drawn from data provided by Omana, 1995.)

catfish (Brown & Cameron, 1991). This also suggests that mechanical processing costs are small. In the Antarctic starfish *Odontaster validus* injecting a mixture of 17 amino acids induced elevated metabolic rates similar to SDA responses (Omana, 1995), and peak elevations (absolute increments) were related to size of injection (Fig. 7). The response was curvilinear, compared with the linear fit of peak height to meal size from the same study (Fig. 3). Absolute increments following injections were between 60% and 100% of maximum responses from the same-sized meal, with smaller injections giving closer responses to meals than larger injections. It should be stressed here that, in the *O. validus* investigation, small meal sizes were used, so the figure of 60% is likely to be more representative of normal feeding conditions than higher values.

The close relationship between feeding and growth seen in temperate ectotherms has also been found in some investigations of Antarctic invertebrates. In a study of the limpet *Nacella concinna* specimens were firstly starved for 20 days to reduce metabolic rates close to basal levels (Peck and Veal in prep.). Limpets were then allowed to feed *ad libitum* for 24 h on epilithic algae, and the ensuing SDA was measured. The algae used to feed the limpets were cultured in an enhanced strontium medium (~10 mM), and the experimental trials run in seawater with the same enhanced Sr conditions. Growth increments during trials were clearly visible in shell sections viewed under back-scattered scanning electron microscope images, and daily growth bands discernible in some specimens (Fig. 8). In this study the duration of the period of shell growth was estimated as 60–150% of the time needed for post-prandial rise in oxygen consumption to return to prefeeding levels.

Recent work on the seasonality of growth of the brachiopod *Liothyrella uva* at Signy Island has found that this species grows much faster in shell length during the winter period, but increases in tissue mass during the summer, when food availability is high because of the intense phytoplankton bloom (Peck, Brockington & Brey, 1997). This pattern was explained as a period of feeding where resources are diverted to a predominantly protein storage area between the mantle and the shell valves, followed by a winter period of utilisation of stored material in growth and gonad maturation. It was also thought that this could be an adaptation for efficient annual resource utilisation in an environment where food supplies are either temporally limited (e.g. for suspension feeders exploiting phytoplankton in the >20 μm size range) or only present at low standing stocks during the year (species feeding on particles in the 2–20 μm size range).

In a small pilot study, oxygen consumption of isolated guts of *L. uva* was found to increase five-fold when the brachiopods were fed (L. Peck, unpublished observations). For a 40 mm length individual the absolute increase in gut oxygen consumption was 0.86 μmol O_2 h^{-1}. This compares with an absolute rise of 0.59 μmol O_2 h^{-1} for whole animals of 37 mm length (Fig. 2). The implication here is that for Antarctic brachiopods the SDA response may predominantly be composed of the costs of pre-absorptive processes. In this context it is interesting that *L. uva* has the smallest peak post-prandial rise of all Antarctic species so far studied at ×1.6 prefeeding levels (Table 1). The reduced SDA factorial scope could be interpreted as a greatly reduced growth component in this species, and it may be the case that the interaction of storage cycles with growth and feeding events affect the SDA produced. This may also be a more important effect in highly seasonal polar environments where storage might be expected to be a more common response to feeding than in other areas. However, data on the interaction of growth, metabolism and storage cycles are absent for Antarctic marine ectotherms.

Fig. 8. Scanning electron micrograph of a section through a *Nacella concinna* shell. The light area indicates the zone of high Sr content laid down during an SDA experiment. Daily growth bands are visible in the area of enhanced Sr content. (From Peck and Veal, in prep.)

Metabolic scope and power generation

The concept of maximum metabolic capacity or scope is a relatively easy one to grasp in relation to active species, such as running mammals, or fast swimming fish, where it is routinely measured as the energy utilised under maximum locomotory activity. Factorial aerobic scopes (the maximum levels animals attain over basal or standard rates) are usually in the range five to ten for active vertebrate species. However, highly active mammals can improve on this, with human sprinters achieving a factorial aerobic scope of 15–20 (Margaria *et al.*, 1963), and this value is just over 20 for foxes (Weibel *et al.*, 1983). The highest aerobic scope recorded for a vertebrate is 45 for racehorses (Brody, 1945; Birlenbach & Leith, 1994). Insects have even higher values than this, with several species increasing metabolism 100 fold between rest and flight (Sacktor, 1975; Ziegler, 1985).

Maximum levels of activity are harder to assess in sessile or slow-moving species. In some snakes there are dramatic elevations of metabolism following feeding, and Burmese pythons have a factorial metabolic rise of 45 following feeding, which is as large as the maximum scope exhibited by any active vertebrate species under maximum work. Also studies on fish have indicated that the rise in metabolism following feeding reduces maximum aerobic capacity under exercise on a pro-rata basis (Boutilier, this volume; Lucas & Priede, 1992)

In many Antarctic marine invertebrates peak post-prandial metabolic rates may also indicate maximum metabolic (aerobic) scope. The brachiopod *Liothyrella uva* and the limpet *Nacella concinna* have been subject to SDA investigations conducted at temperatures of around 0 °C. They had factorial scopes of 1.6 and 2.3, respectively (Table 1). In both species the effects of temperature on basal metabolic rate have also been assessed (Peck, 1989). Following increases in temperature, oxygen consumption was shown to increase to acute peak levels before declining to steady acclimated rates. Maximum oxygen consumption rates in the temperature and metabolism investigation were 1.6 times acclimated rates at 0 °C for *L. uva*, and 2.4 times greater for *N. concinna* (Peck, 1989). The close correlation of these metabolic scopes obtained under markedly different conditions suggests that the SDA peaks do constitute a measure of maximum metabolic scope, and that this may be a widespread phenomenon for cold-water low activity benthos. It also suggests that in Antarctic marine ectotherms limits to SDA peaks may be due to constraints on maximum aerobic capacity, and not to limitations on rates at which resources can be processed or translocated.

Summary

Factorial rises in oxygen consumption following a feeding event in marine ectotherms appear to be consistent across latitudinal zones, whereas absolute peak rises in polar ectotherms are smaller and of longer duration than those for warmer water species. Post-prandial metabolic rates are intrinsically linked to basal metabolic rates and factorial scopes are, therefore, phylogenetically dictated. Animal size has no effect on SDA factorial scope, and has little or no effect on the overall size of the SDA. However, absolute peak height, duration and total energy utilised are all affected by meal size. The absolute increment is reduced at small meal sizes and duration and total energy used are usually related to amounts eaten by linear or exponential relationships. Although the data are limited, SDA coefficients measured so far for Antarctic species are similar to those from other latitudes, with coefficients mainly in the range 5–24%, but with some values are as high as 60%. The coefficients also appear to be inversely related to meal size.

Ammonia excretion following feeding is variable in Antarctic species. Generally the response is of longer duration and greater amplitude than for oxygen consumption. O:N ratios indicate different utilisation of substrates by different species, and also that meal size plays a role in dictating which substrates are used.

Anabolic processes are the major component of SDA in most species. Storage cycles and lifestyle in some low energy organisms such as brachiopods may result in the growth component being dramatically reduced, and there is evidence that in such species growth may be separated from feeding events.

Absolute metabolic increments in Antarctic species appear to be reduced compared with temperate and tropical groups. Temperature effects, and ecological and phylogenetic constraints are all factors involved in the control of the level to which metabolism rises post-prandially.

References

Atkinson, D.E., (1992). Functional roles of urea synthesis in vertebrates. *Physiological Zoology,* **65**, 243–67.

Averett, R.C. (1969). The influence of temperature on energy and material utilization by juvenile coho salmon. PhD thesis, Oregon State University, Corvallis, Oregon.

Baldwin, E. (1967). *Dynamic Aspects of Biochemistry*. 5th edn. Cambridge: Cambridge University Press.

Bayne, B.L. & Scullard, C. (1977). An apparent specific dynamic action in *Mytilus edulis* L. *Journal of the Marine Biological Association, UK*, **57**, 371–8.

Birlenbach, U.S. & Leith, D.E. (1994). Intraspecific adaptations for different aerobic capacities: thoroughbred and draft horses. *FASEB Journal*, **8**, A2.

Boyce, S.J. & Clarke, A. (1997). Effect of body size and ration on specific dynamic action in the plunderfish, *Harpagifer antarcticus* Nybelin 1947. *Physiological Zoology*, in press.

Brody, S. (1945). *Bioenergetics and Growth*. New York: Reinhold Publ. C.

Brown, C.R. & Cameron, J.N. (1991). The relationship between specific dynamic action (SDA) and protein synthesis rates in the channel catfish. *Physiological Zoology*, **64**, 298–309.

Burggren, W.W., Moreira, G.S. & Santos, M. doC. F. (1993). Specific dynamic action and the metabolism of the brachyuran land crabs *Ocypode quadrata* (Fabricius, 1787), *Goniopsis cruentata* (Latreille, 1803) and *Cardisoma guanhumi* Latreille, 1825. *Journal of Experimental Marine Biology and Ecology*, **169**, 117–30.

Carefoot, T.H. (1989). Diet and its effect on oxygen consumption in the semiterrestrial isopod *Ligia. Monitore zool. ital.* (N.S.) Monograph **4**, 193–210.

Carefoot, T.H. (1990*a*). Specific dynamic action (SDA) in the supralittoral isopod *Ligia pallasii*: identification of components of apparent SDA and effects of amino acid quality and content on SDA. *Comparative Biochemistry and Physiology,* **95A**, 309–16.

Carefoot, T.H. (1990*b*). Specific dynamic action (SDA) in the supralittoral isopod *Ligia pallasii*: effect of ration and body size on SDA. *Comparative Biochemistry and Physiology,* **95A**, 317–20.

Carefoot, T.H. (1990*c*). Specific dynamic action (SDA) in the supralittoral isopod *Ligia pallasii*: relationship of growth to SDA. *Comparative Biochemistry and Physiology,* **95A**, 553–7.

Chapelle, G, Peck, L.S. & Clarke, A. (1994). Effects of feeding and starvation on the metabolic rate of the necrophagous Antarctic amphipod *Waldeckia obesa* (Chevreux, 1905). *Journal of Experimental Marine Biology and Ecology*, **183**, 63–76.

Cho, C.Y. & Slinger, S.J. (1979). Apparent digestibility measurements in feedstuffs for rainbow trout. *Proceedings of the World Symposium on Finfish Nutrition and Fishfeed Technology*, **2**, 239–47.

Clarke, A. (1983). Life in coldwater: the physiological ecology of polar marine ectotherms. *Oceanography and Marine Biology: An Annual Review*, **21**, 341–453.

Clarke, A. (1987). Temperature, latitude and reproductive effort. *Marine Ecology Progress Series,* **38**, 89–99.

Clarke, A. (1988). Seasonality in the Antarctic marine environment. *Comparative Biochemistry and Physiology*, **90B**, 461–73.

Clarke, A. (1993). Seasonal acclimatization and latitudinal compensation in metabolism: do they exist? *Functional Ecology*, **7**, 139–49.

Clarke, A. & Prothero-Thomas, E. (1997). The effect of feeding on the oxygen consumption and nitrogen excretion of the antarctic nemertean *Parborlasia corrugatus*. *Physiological Zoology*, in press.

Clarke, A. & Peck, L.S. (1991). The physiology of polar marine zooplankton. *Polar Research,* **10**, 355–69.

Clarke, A., Prothero-Thomas, E. & Whitehouse, M.J. (1994). Nitrogen excretion in the Antarctic limpet *Nacella concinna* (Strebel, 1908). *Journal of Molluscan Studies*, **60**, 141–7.

Coulson, R.A. & Hernandez, T. (1979). Increase in metabolic rate of the alligator fed protein or amino acids. *Journal of Nutrition,* **109**, 538–50.

Crisp, M., Davenport, J. & Shumway, S.E. (1978). Effects of feeding and of chemical stimulation on the oxygen uptake of *Nassarius reticulatus* (Gasteropoda: Prosobranchia). *Journal of the Marine Biological Association, UK,* **58**, 387–99.

Hemmingsen, A.M. (1960). Energy metabolism as related to body size and respiratory surfaces, and its evolution. *Report of the Steno Memorial Hospital,* **9**, 7–110.

Heusner, A.A. (1982). Energy metabolism and body size. I. Is the 0.75 mass exponent of Kleiber's equation a statistical artefact? *Respiration Physiology,* **48**, 1–12.

Houlihan, D.F. (1991). Protein turnover in ectotherms and its relationships to energetics. In *Advances in Comparative and Environmental Physiology*, ed. R. Gilles, Vol. 1, pp. 1–43. Springer-Verlag.

Houlihan, D.E., Waring, C.P., Mathers, E. & Gray, C. (1990). Protein synthesis and oxygen consumption of the shore crab *Carcinus maenas* after a meal. *Physiological Zoology,* **63**, 735– 56.

Houlihan, D.F., Pedersen, B.H., Steffensen, J.F. & Brechin, J. (1995*a*). Protein synthesis, growth and energetics in larval herring (*Clupea harengus*) at different feeding regimes. *Fish Physiology and Biochemistry*, **14**, 195–208.

Houlihan, D.F., Carter, C.G. & McCarthy, I.D. (1995*b*). Protein turnover in animals. In *Nitrogen Metabolism and Excretion* ed P.J. Walsh, & P. Wright. pp. 1–32. Boca Raton: CRC Press.

Ikeda, T. (1974). Nutritional ecology of marine zooplankton. *Memoirs of the Faculty of Fisheries of Hokkaido University*, **22**. 1–97.

James, M.A., Ansell, A.D., Collins, M.J., Curry, G.B., Peck, L.S. & Rhodes, M.C. (1992 Biology of living brachiopods. *Advances in Marine Biology*, **28**, 175–387.

Jobling, M. (1981). The influences of feeding on the metabolic rate of fishes: a short review. *Journal of Fish Biology*, **18**, 385–400.

Jobling, M. (1992). Bioenergetics: feed intake and energy partitioning. In *Fish Ecophysiology*, ed. J.C. Rankin & F.B. Jensen, Chap 1, pp. 1–44. London: Chapman & Hall.

Jobling, M. (1994). *Fish Bioenergetics*. 309 pp. London: Chapman & Hall.

Jobling, M. & Davies, P.S. (1980). Effects of feeding on the metabolic rate and the Specific Dynamic Action in plaice, *Pleuronectes platessa* L. *Journal of Fish Biology*, **16**, 629–38.

Johnston, I.A. & Battram, J. (1993). Feeding energetics and metabolism in demersal fish species from Antarctic, temperate and tropical environments. *Marine Biology*, **115**, 7–14.

Kleiber, M. (1961). *The Fire of Life*. New York: Wiley.

Krebs, H.A. (1960). The cause of the specific dynamic action of foodstuffs. *Arzneimittel- Forschung*, **10**, 369–73.

LaBarbera, M. (1986). The evolution and ecology of body size. In *Patterns and Processes in the History of Life. Dahlem Konferenzen*, ed. D.M. Raup, & D. Jablonski, pp. 69–98. Berlin: Springer-Verlag.

Lucas, M.C. & Priede, I.G. (1992). Utilisation of metabolic scope in relation to feeding and activity by individual and grouped zebrafish, *Brachydania rerio* (Hamilton-Buchanan). *Journal of Fish Biology*, **41**, 175–90.

Lyndon, A.R., Houlihan, D.F. & Hall, S.J. (1992). The effect of short term fasting and a single meal on protein synthesis and oxygen consumption in cod, *Gadus morhua*. *Journal of Comparative Physiology*, **162B**, 209–15.

Margaria, R., Cerretelli, P., Aghemo, P. & Sassi, G. (1963). Energy cost of running. *Journal of Applied Physiology*, **18**, 367–70.

Mayzaud, P.& Conover, R.J. (1988). O:N atomic ratio as a tool to describe zooplankton metabolism. *Marine Ecology Progress Series*, **45**, 289–302.

McMillan, D.N. & Houlihan, D.F. (1988). The effect of refeeding on tissue protein synthesis in rainbow trout. *Physiological Zoology*, **6(5)**, 429–41.

Muir, B.S. & Niimi, A.J. (1972). Oxygen consumption of the euryhaline fish alehole (*Kuhlia sandvicensis*) with reference to salinity, swimming and food consumption. *Journal of the Fisheries Research Board of Canada*, **29**, 67–77.

Nicol, J.A. (1967). *The Biology of Marine Animals.* 2nd edn. New York: John Wiley.

Omana, C. (1995). Investigation into oxygen consumption of the Antarctic asteroid *Odontaster validus*, following feeding and injection of L-amino acid solution. Project report, Cambridge University Department of Zoology, 48 pp.

Pandian, T.J. (1987). Fish. In *Animal Energetics*, ed. T.J. Pandian, & F.J. Vernberg, Chap. 7 pp. 358–465. San Diego: Academic Press.

Peck, L.S. (1989). Temperature and basal metabolism in two Antarctic marine herbivores. *Journal of Experimental Marine Biology and Ecology*, **127**, 1–12.

Peck, L.S. (1996). Metabolism and feeding in the Antarctic brachiopod *Liothyrella uva*: a low energy lifestyle species with restricted metabolic scope. *Proceedings of the Royal Society of London B*, **236**, 223–8.

Peck, L.S., Clarke, A. & Holmes, L.J. (1987). Summer metabolism and seasonal changes in biochemical composition of the Antarctic brachiopod *Liothyrella uva* (Broderip, 1833). *Journal of Experimental Marine Biology and Ecology*, **114**, 85–97.

Peck, L.S., Ansell, A.D., Curry, G.B. & Rhodes, M.C. (1996). Physiology. Chapter III. In, *Treatise on Invertebrate Palaeontology, H, Brachiopoda*, ed. A. Williams, pp. 178–261. Lawrence, Kansas: Geological Society of America and the University of Kansas.

Peck, L.S., Brockington, S. & Brey, T. (1997). Growth and metabolism in the Antarctic brachiopod *Liothyrella uva*. *Philosophical Transactions of the Royal Society B*, **352**, 851–8.

Prothero, J. (1984). Scaling of standard energy metabolism in mammals: I Neglect of circadian rhythms. *Journal of Theoretical Biology*, **106**, 1–8.

Regnault, M. (1987). Nitrogen excretion in marine and freshwater crustacea. *Biological Reviews*, **62**, 1–24.

Sacktor, B. (1975). Biochemistry of insect flight. Part 1. Utilisation of fuels by muscle. In *Insect Biochemistry and Function*, ed. D.J. Candy & B.A. Kilby, pp. 1–88. London: Chapman & Hall.

Secor, S.M. & Diamond, J. (1995). Adaptive responses to feeding in Burmese pythons: pay before pumping. *Journal of Experimental Biology*, **198**, 1313–25.

Sims, D.W. & Davies, S.J. (1994). Does specific dynamic action (SDA) regulate return appetite in the lesser spotted dogfish *Scyliorhinus canicula*? *Journal of Fish Biology*, **45**, 341– 8.

Soofiani, N.M. & Hawkins, A.D. (1982). Energetic costs at different levels of feeding in juvenile cod , *Gadus morhua* L. *Journal of Fish Biology*, **21**, 577–92.

Tandler, A. & Beamish, F.W.H. (1979). Mechanical and biochemical components of apparent specific dynamic action in largemouth bass, *Micropterus salmoides* Lacepede. *Journal of Fish Biology*, **14**, 343–50.

Thompson, R.J. & Bayne, B.L. (1972). Active metabolism associated with feeding in the mussel *Mytilus edulis* L. *Journal of Experimental Marine Biology and Ecology*, **9**, 111–24.

Weibel, E.R., Taylor, C.R., O'Neil, J.J., Keith, D.E., Gehr, P., Hoppeler, H., Longman, V. & Baudinette, R.V. (1983). Maximum oxygen consumption and pulmonary diffusing capacity: a direct comparison of physiologic and morphometric measurements in canids. *Respiration Physiology*, **54**, 173–88.

Wieser, W. (1984). A distinction must be made between the ontogeny and the phylogeny of metabolism in order to understand the mass exponent of energy metabolism. *Respiration Physiology*, **55**, 1–9.

Wieser, W. (1994). Cost of growth in cells and organisms: general rules and comparative aspects. *Biological Reviews*, **68**, 1–33.

Wells, M.J., O'Dor, R.K., Mangold, K. & Wells, J. (1983). Feeding and metabolic rate in *Octopus*. *Marine Behaviour and Physiology*, **9**, 305–17.

Ziegler, R. (1985). Metabolic energy expenditure and its hormonal regulation. Chapter 4. I *Environmental Physiology and Biochemistry of Insects*, ed. K.H. Hoffman, pp. 95–118. Heidelberg: Springer-Verlag.

P.W. HOCHACHKA and P.D. MOTTISHAW

Evolution and adaptation of the diving response: phocids and otariids

Although biologists have been intrigued by diving mammals and birds for well over a century, the physiological and metabolic mechanisms now known to permit an air breathing animal to operate successfully deep into the water column were first exposed in the 1930s and 1940s through the pioneering work of Scholander, Irving, and their colleagues (Irving *et al.*, 1935; Scholander, 1940). The basis of their work provided the fundamental foundations of diving physiology, which are now known to include three key physiological 'reflexes': (i) apnea, (ii) bradycardia, and (iii) vasoconstriction and thus hypoperfusion of most peripheral tissues. Scholander referred to these physiological reflexes in combination as the 'diving response', and, in simulated diving under controlled laboratory conditions, he imagined the marine mammal reducing itself to a 'heart, lung, brain machine'. The metabolic representation of this response included the gradual development of oxygen-limiting conditions in hypoperfused (ischaemic) tissues, with attendant accumulation of end products of anaerobic metabolism (especially lactate and H^+ ions). Scholander realised that because peripheral tissues were hypoperfused, most of the lactate would remain at sites of formation during the course of a (simulated) dive, and that most of it would not be 'washed out' of the tissues into the circulation until perfusion was restored at the end of diving. By explaining why only a small lactate accumulation is observed in blood plasma during diving, while a peak of lactate is usually seen early in recovery from a dive, Scholander used the lactate data as indirect evidence for hypoperfusion and vasoconstriction of peripheral tissues.

Initially, Scholander and Irving expected that any energy deficit due to oxygen lack during diving would be made up by anaerobic glycolysis (with concomitant lactate accumulation). However, in their seminal studies of the harbour seal, they observed that the post-diving oxygen debt was frequently less than the expected oxygen deficit during diving. Additionally, the amount of lactate accumulated often was substantially less than would be expected if the energy deficit were to be made up by anaerobic glycolysis. So, as early as 1940, the idea was introduced of a relative hypometabolism during diving to account for the missing lactate and the missing oxygen debt.

Over the subsequent four decades, many laboratory studies were performed which essentially confirmed the earlier framework developed by Scholander and Irving (see Zapol *et al.*, 1979; Butler & Jones, 1982; Elsner & Gooden, 1983). The key features which were observed in simulated diving studies over and over again were (i) apnea, (ii) bradycardia, (iii) peripheral vasoconstriction, (iv) low lactate accumulation during diving *per se*, and (v) a post-dive lactate washout peak appearing in the plasma within 1–4 min of recovery; the term for this combination of processes became fixed in the literature as 'the diving response'.

Scholander was an adventurer as well as a creative scientist; he and many students following in his path observed that the key features of the diving response were evident in many different kinds of animals. Diving bradycardia was often used as a kind of index or indicator of the diving response and it was so seemingly universal among the vertebrates that Scholander (1963) referred to it as the 'master switch of life'. In his day, this 'master switch', or the diving response, was viewed as an obvious 'physiological adaptation' to diving, even if there was little indication as to how the response evolved through any particular lineage. At this time, little attention was paid to the criteria of evolutionary biology: that to be defined as adaptive a character either had to have arisen by natural selection or to be maintained by selective forces.

The advent of modern field study technologies, especially of microprocessor-assisted monitoring of aquatic animals while diving voluntarily in their natural environment (Guppy *et al.*, 1986; Hill *et al.*, 1987), has confirmed over the last two decades the validity and plasticity of the overall 'diving response' first elucidated in the 1930s and 1940s and has greatly extended our understanding of how the diving response is used under natural diving conditions (Kooyman *et al.*, 1980; Qvist *et al.*, 1986; Le Boeuf *et al.*, 1989; DeLong & Stewart, 1991; Castellini, Kooyman & Ponganis, 1992; DeLong, Stewart & Hill, 1992; Hindell *et al.*, 1992; Thompson & Fedak, 1992; Guyton *et al.*, 1995; Hochachka *et al.*, 1995; Hurford *et al.*, 1995). Studies have proposed life history causes for why phocids are better divers than otariids (Costa, 1991, 1993), and extended physiological mechanisms for the differences in the diving abilities of pinnipeds (Fedak, 1986; Hochachka & Foreman, 1993). Yet, just as in Scholander's time, these approaches have added little insight into how the diving response evolved in any given lineage, or into how physiological characters may be adapted for different kinds of life styles and environments.

It is these latter questions that will be addressed in this chapter, restricting the analysis to the evolution of diving physiology within the pinnipeds, and testing the hypothesis that the above well-established physiological characters of the diving response should correlate positively with diving ability in these species.

Pinniped phylogeny

The evolutionary origins and relationships of the pinnipeds have been topics of debate among biologists for the better part of a century. However, one of the main arguments, whether or not the pinnipeds are monophyletic, has been gradually resolved. A typical phylogenetic tree (Fig. 1) depicts a monophyletic origin for the pinnipeds with bears as their nearest relatives. Assuming a monophyletic origin, the placement of the walrus is ambiguous. In some studies, the walrus is paired with the otariids (sea lions and fur seals) while in some it is paired with the true seals. One possible phylogeny, incorporating molecular and morphological information (Mottishaw, 1997), is presented for illustrative purposes in Fig. 2. A major implication of the phylogenetic information encapsulated in Figs 1 and 2 is that the pinnipeds are broadly represented by two evolutionary branches – the otariid and the phocid lineages. Do the physiological traits of the diving response differ along these two major phylogenetic branches? The answer turns out to be affirmative. Close examination of available data indicates two very different evolutionary physiologies for the two lineages: in the phocids, the diving strategy is energy conserving, while in the otariids, it is much more energy dissipative.

The 'slow-lane' phocid strategy of diving

To illustrate the phocid strategy of diving, consider the elephant and Weddell seals for which there are good data bases. First of all, the two species of elephant seals currently hold the dive duration record, 2 hours (Hindell *et al.*, 1992), and dive depth record, 1.5 km (DeLong *et al.*, 1992), among marine mammals. Northern elephant seals go to sea for months at a time and migrate thousands of km per year, during which they spend approximately 90% of the time submerged (Le Boeuf *et al.*, 1989, 1992). One of the major food items for these phocids appears to be midwater vertically migrating squid. The food is pelagic and so the northern elephant seal too becomes a pelagic, indeed a mesopelagic, mammal. While the natural histories of the southern elephant seals (McConnell, Chambers & Fedak, 1992) are different, their diving capabilities are similarly spectacular (Hindell *et al.*, 1992). The natural history of the Weddell seal differs even more from that of the northern elephant seal. Weddell seals operate in seas that freeze overwinter. By maintaining breathing holes in these areas, these seals are able to

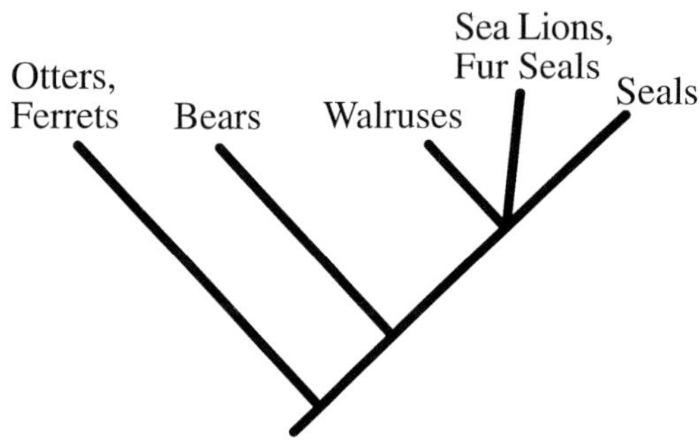

Fig. 1. A phylogeny for the pinnipeds. The monophyletic origin for the phocids and otariids is consistent with recent molecular data (Lento *et al.*, 1995), but the relationship between them and the walrus lineage is not yet clear.

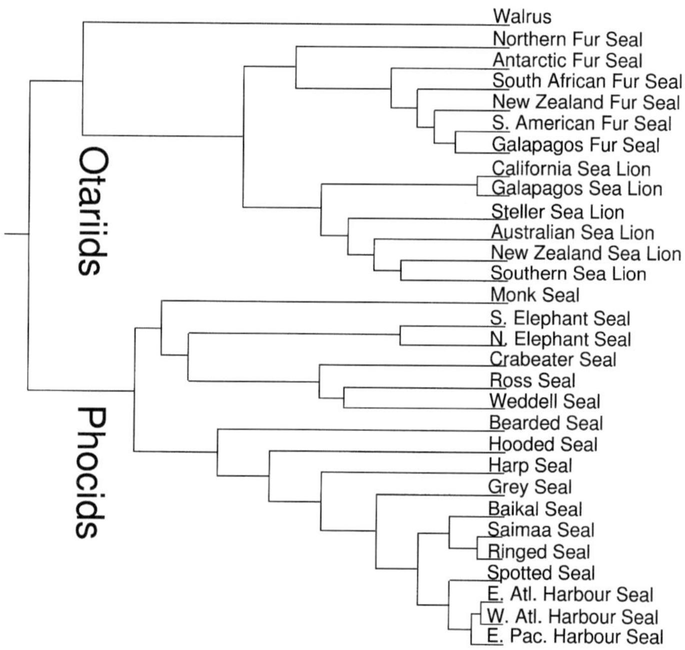

Fig. 2. One composite phylogenetic hypothesis for the pinnipeds describing phocid and otariid relationships and based on molecular and morphological data (compiled from Arnason *et al.*, 1995; Berta & Demere, 1986; Berta & Wyss, 1994; Burns & Fay, 1970; Lento *et al.*, 1995; Muizon, 1982; Perry *et al.*, 1995; additional literature in Mottishaw, 1997). Some branches of the tree are supported by both molecular and morphological evidence, while other areas, such as the fur seal species, are not fully resolved.

exploit a niche and food resource, again well into the pelagic water column, that would otherwise be unavailable. Not surprisingly, the diving capacities of the Weddell seal (in terms of depth and duration) are similarly awesome and this species easily qualifies as another of nature's champion divers (Castellini *et al.*, 1992; Hochachka, 1992).

Many, and probably most, of the biological and physiological adjustments underpinning the making of a mesopelagic marine mammal are now well understood and are reviewed elsewhere (Hochachka, 1986, 1992; Kooyman, 1985). Excluding obvious morphological adaptations for pelagic performance (flippers vs. feet, collapsible pulmonary system, etc.), the major functional characteristics for such sensational diving capabilities include:

(i) apnea, with exhalation upon initiation of diving (for minimising buoyancy and other pressure-related problems),
(ii) bradycardia (in 1:1 proportion with changes in cardiac output),
(iii) peripheral vasoconstriction and hypoperfusion (in order to conserve oxygen for the central nervous system (CNS) and heart),
(iv) hypometabolism of (vasoconstricted) ischemic tissues (also in order to conserve oxygen and plasma borne fuels for the CNS and heart),
(v) an enhanced oxygen carrying capacity (enlarged blood volume, expanded red blood cell mass within the blood volume, i.e. higher haematocrit (Hct), higher haemoglobin concentration ([Hb]) in red blood cells, and possibly higher myoglobin concentration ([Mb]) in muscles and heart), and
(vi) an enlarged spleen (for regulating the Hct so that a very high % of RBCs need not be circulated under all physiological conditions). Additionally
(vii) it should be noted that, for really outstanding diving, all of the above characteristics (i) are incorporated with a large body weight (in order to maximise the amount of oxygen that can be carried while minimising mass-specific energy demands during diving by allometric effects), and (ii) are coupled with slow swimming speed while at sea (to minimise the cost of locomotion while maximising submergence, hence foraging, time).

The evidence for these overall patterns arises from studies of several phocid species (for example, see Kooyman *et al.*, 1980; Guppy *et al.*, 1986; Qvist *et al.*, 1986; Castellini *et al.*, 1992; Hindell *et al.*, 1992; Thompson &

Fedak, 1992; Reed *et al.*, 1994; Hochachka *et al.*, 1995; Hurford *et al.*, 1995; Guyton *et al.*, 1995). It seems that this combination of characters enables phocids to dive longer than otariids.

The 'fast-lane' otariid strategy of diving

Several sea lions and fur seals, for which there are good data bases (see Gentry & Kooyman, 1986; Ponganis *et al.*, 1990; Costa, 1991, 1993), can be used to illustrate the situation in this lineage. First of all, these are animals that do not spend as much time diving, either expressed as a fraction of each year or as a fraction of each day when they are foraging. Average dive durations are on the order of a few minutes. Maximum duration dives observed so far are only 15–20 min and maximum depths are only in the 300–400 m range. Diet consists largely of inshore prey species. Swim speeds in terms of body lengths per second indicate that substantially higher swim speeds are used by otariids than by phocids (see Williams, Kooyman & Croll 1991; Ponganis *et al.*, 1990).

Again, many of the physiological characteristics underpinning this otariid diving behaviour are now well understood (for literature in this area, see Butler & Jones, 1982; Gentry & Kooyman, 1986; Costa, 1991, 1993; Ponganis *et al.* 1990; Williams *et al.*, 1991). They include:

(i) apnea (initiated on inhalation, in contrast to phocids) and a gas exchange system that does not completely collapse (so the lung probably continues to function as a gas exchange organ during most diving),

(ii) bradycardia, again presumably in 1:1 proportion to changes in cardiac output,

(iii) peripheral vasoconstriction, with propulsive muscle microvasculature presumably more relaxed than in phocids (because of the higher oxygen and fuel demands of swimming) relative to other less-energy demanding peripheral tissues,

(iv) hypometabolism of ischaemic tissues (the longer the dive, the lower the metabolic rate (Hurley, 1996), presumably to conserve oxygen for the CNS and the heart, as in diving phocids),

(v) an oxygen carrying capacity intermediate between that of the large phocids and terrestrial mammals (this includes blood volume, red blood cell mass or Hct, [Hb] in RBCs, and [Mb] in muscles and heart),

(vi) a modest-sized spleen, not much larger as a percentage of body weight than found in terrestrial animals, and, additionally,

(vii) a relatively small and hydrodynamically sleek body consistent with a higher speed predator life style. The operational trade-off seems to require sacrificing extended foraging time for energy needed for higher speed swimming (Costa, 1991).

Based on the above, one could on first glance conclude that the phocids are much 'better adapted' divers than are the otariids. However, even if the diving strategy of sea lions is relatively much more energy expensive than that of the seals, it may be the only one that could work for a predator targeting fast swimming inshore fish as a major food source. On second glance, therefore, the energy dissipative diving strategy of the sea lion is just as 'adaptive' for its ecology as is the energy conserving diving strategy of the seal for its midwater, pelagic zone of operation.

Tracing the evolution of diving capacity

To investigate these questions, what is required is a more quantitative comparison of diving strategies within the phocids and otariids. When our attention was turned towards this end, two unexpected situations were encountered. The first was that numerous physiological characteristics of diving animals, instead of systematically varying with diving capacities, were similar in all diving species for which could be found data. For example, values of maximum bradycardia during diving periods show no consistent phylogenetic patterns or relationship with diving duration (Fig. 3). This could mean that heart rate is too crude a measure of circulatory control during diving and recovery (and its measure is consistent with what Scholander knew all along: the basic reflex is nearly universally present in some form in all vertebrates). Or, it could mean that heart rate is selectively constrained due to roles in so many different biological settings that any adaptational changes for diving are too modest to detect with the crude physiological criteria thus far utilised.

Interestingly, a similar situation seems to hold for another physiological character, diving hypometabolism, which we initially expected to vary with dive time. In earlier studies, it was assumed (Guppy *et al.*, 1986; Hochachka & Guppy, 1987; Hochachka & Foreman, 1993; Le Boeuf *et al.*, 1989; Costa, 1991; Hindell *et al.* 1992) explicitly or implicitly that the impressive diving performance of large seals depended in large part on an 'energy conserving' physiology and diving strategy. Central to this was some concept of diving hypometabolism. Subsequent research has uncovered two potential underlying mechanisms. (i) One hypothesis is that hypoperfusion (vasoconstriction) of nonworking peripheral muscles and other tissues is the proximate cause of hypometabolism, with reduction in tissue metabolic rate being a direct function of the reduction in oxygen delivery, a relationship

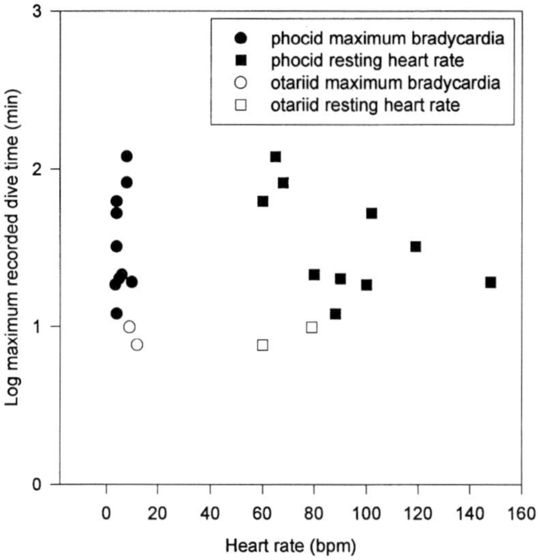

Fig. 3. Relationship between lowest observed diving heart rate (maximum bradycardia) (bpm) and $\log_{10}$ maximum recorded dive time (min) for nine species of phocids (closed circles) and for two species of otariids (open circles). The correlation of lowest heart rate and dive time for all pinnipeds was not significant (r=−0.44; *P*=0.09). The correlation was also not significant within phocids (r=−0.03; *P*=0.47). Any correlation (presumably within phocids) between resting heart rate and dive time would be due to allometric effects of body weight. The data illustrate that a small seal with a high resting heart rate and a large seal with a low resting heart rate are able to lower heart rate similarly during diving. Otariids are also able to lower their heart rate to a comparable degree, and differences seen between phocids and otariids are due to differences in maximum recorded dive times, not heart rates. Data compiled by Mottishaw (1997) and available in the Appendix tables. Maximum recorded diving duration here and in all subsequent figures is based on data obtained by remote sensing technology (microcomputer monitors).

also observed in terrestrial mammals (Hochachka, 1992; Guyton *et al.*, 1995; Hochachka *et al.*, 1995). (ii) An alternate postulate is that regional hypothermia contributes to low metabolic rates, with metabolic suppression being a function of tissue cooling (Hill *et al.*, 1987; Andrews *et al.*, 1994). While these two mechanisms are not mutually exclusive, it was at first thought that the physiological characteristic of hypometabolism would be largely restricted to the large seals, or at least to phocids. However, recent careful experimental studies with sea lions (animals trained to remain submerged and relatively inactive for defined time periods) indicate that the metabolic rate declines as a direct function of diving duration; the metabolic rate for 7-minute diving periods (water temperature at about 15 °C) falls to about 50% of resting metabolic rate (RMR). Since the times involved are so short, it is unlikely that regional hypothermia plays a significant role in this metabolic suppression (Hurley, 1996). This is interpreted to mean that activation of the diving response automatically leads to hypoperfusion of some tissues/organs and subsequently to diving hypometabolism. On its own, then, this physiological character, like bradycardia, would appear to be general among pinnipeds and thus could not be expected to vary (and indeed did not vary) in any systematic way with diving capacity.

The second unexpected situation encountered was purely practical. Ideally, for tracing the evolution of the diving response in these lineages, one would like to be able to compare all of the above metabolic and physiological characters in numerous phocid and otariid species. However, in reality, detailed information on complex characters such as tissue specific regional hypoperfusion is available for only a few species, so for multiple species comparisons, analysis had to be restricted to only a few data sets (or characters) that have been quantified for many species. The ones focused on here are (i) body weight, (ii) spleen weight, and (iii) whole body haemoglobin (defined as the content of Hb in the entire blood volume of the organism). It is implicit in much of the literature on diving that all three characters play an adaptive role in diving.

In terms of diving capacity, body weight influences the total on-board oxygen supplies as well as mass-specific energy demands; thus it would be expected to vary with diving capacity on purely theoretical considerations (see Butler & Jones, 1982). This trend has been previously demonstrated in the pinnipeds and discussed in the context of life-history parameters (Costa, 1991, 1993). The spleen acts as a SCUBA tank (oxygenated red blood cell storage) (Qvist *et al.*, 1986; Hurford *et al.*, 1995). The spleen is controlled by a catecholamine-based regulatory circuitry which also controls several other metabolic and physiological functions during diving-recovery cycles (Hurford *et al.*, 1995; Hochachka *et al.*, 1995; Lacombe & Jones, 1991). It was anticipated that spleen weight might also vary with diving requirements. Finally, whole body haemoglobin is a direct measure of oxygen carrying capacity; since maximum diving duration is presumably set by some complex balance between oxygen availability and oxygen demand, this too may respond to selection based on diving behaviour. Therefore, it was tested whether these variables differed between phocids and otariids, and body weight, spleen weight, and whole body oxygen stores were correlated with maximum recorded dive time within these groups (when enough data were available).

Statistical methods

To remove the influence of body weight, dependent variables were regressed on body weight using ordinary least squares linear regression (as in Bennett, 1987). Variables were $\log_{10}$ transformed to meet the assumptions of normality required for analysis. The residuals generated from the regression lines were correlated using Pearson product moment correlations. Correlations and regressions within each group were analysed with one tailed tests. Many studies attempt to remove body weight from biological characters by dividing by the weight of the animal and expressing the character as a ratio or percentage of weight. Metabolic rate, for instance, is expressed in the literature as ml oxygen per kg per minute, and is consequently presented as such in Table 1. Studies of spleen weights have demonstrated that the slope of the allometric relationship between spleen weight and body weight within a species is 1.0 (Bryden, 1971). As the percentage of body weight occupied by the spleen did not change in animals of differing weights, the measured value of spleen weight was divided by the weight of the animals, and this relative spleen size was compared between phocids and otariids. Differences between phocid and otariid characters were analysed with one tailed t-tests, or in the case of dive time and spleen weight, Mann–Whitney rank sum tests. All variables compared in this study, as well as a list of sources for diving and physiological information found for each pinniped species, are given in the Appendix.

Body size and long diving duration

Based on species for which data were available (Appendix 1: Tables 1 and 2), phocids are significantly larger than otariids (phocid mean weight: 151 kg, otariid mean weight: 66 kg; $P < 0.001$), and phocid maximum recorded dive duration is significantly longer than otariid maximum recorded dive duration (phocid median duration: 20 min; otariid median duration: 8 min; $P < 0.001$). The relationship between body weight and maximum recorded diving time is shown in Fig. 4 for phocids (left panel) and otariids (right panel). The general trend is essentially the same in both lineages: bigger animals tend to be able to dive for longer time periods, but the trend is statistically significant only for the phocids ($R^2=0.37$; $P < 0.01$).

Spleen size and long diving duration

The phocid spleen weights used in this study are significantly larger than otariid spleen weights (phocid mean spleen weight: 457 g; otariid mean spleen weight: 158 g; $P=0.043$). The measures of spleen weight

Table 1. *Net effects of putative physiological and biochemical adaptations for diving in phocids compared with otariids expressed as the ratios of diving metabolic rates/resting metabolic rates*

Species	Weight	RMR	DMR	DMR/RMR	Source
Phocids					
Weddell seal	355	4.1	5.0 (S)	1.20	1
			3.4 (L)	0.82	1
			4.5 all	1.10	1
Elephant seal	156	7.6	8.2	1.13	2
Grey seal	189	7.6	7.9 (S)	1.04	3
			3.5 (L)	0.46	3
			5.2 all	0.68	3
Harp seal	140	3.7	3.5 (S)	0.95	4
			2.9 (L)	0.78	4
			3.2 all	0.88	4
Harbour seal	42	6.1	6.75 all	1.16	5
Otariids					
Sea lion	85			4.80	6
Northern fur seal	37			6.0	6
Antarctic fur seal	39			7.0	6

Note:
Body weight in kg; for phocids, resting metabolic rates (RMR) and diving metabolic rates (DMR) are in ml oxygen$\times$kg$^{-1}\times$min^{-1}. Elephant seal DMRs estimated using doubly labelled water for animals at sea, while RMRs were for seals on shore; all other phocid metabolic rates are based on studies with free diving seals using indirect calorimetry. RMR values are not the same as basal metabolic rates or BMRs; by definition, the latter have to be taken at specified times and conditions (post-absorptive, quiescent, thermoneutral) and are rarely available for phocids. The RMR of fasting (about 345 kg) elephant seals on shore averaged over a 32-day period was 4.64 ml oxygen per kg per min (Worthy *et al.*, 1992) slightly lower than the value used above for younger seals; in this case, the DMR/RMR would increase to about 1.7. Otariid metabolic data in Costa (1991) are all given in kJ$\times$min^{-1} and are based on 'at sea' estimates using the doubly labelled water technique. S, data for short dives; L, data for long dives; all, data for all dives studied.
Source: (1) Castellini *et al.* (1992). (2) Andrews R.D. *et al.* (1994). (3) Reed *et al.* (1994). (4) Gallivan (1981). (5) Craig & Pasche (1980). (6) Costa (1991).

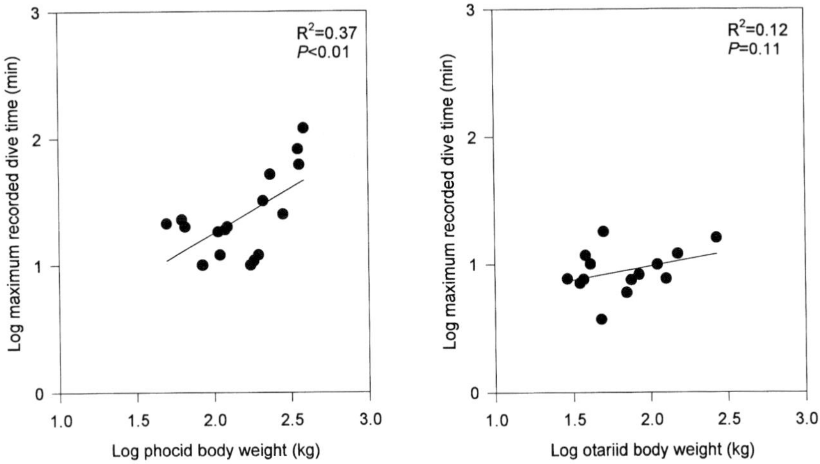

Fig. 4. Relationship between log body weight (kg) and log maximum recorded dive time (min) for 17 species of phocid (left panel) and for 15 species of otariid (right panel). Data compiled by Mottishaw (1997). For phocids $R^2=0.37$, $P<0.01$. For otariids the relationship is not statistically significant: $R^2=0.12$, $P=0.11$.

were conducted on animals of all ages and sizes. Spleen weight is a function of body weight for both groups (Fig. 5). If the relative spleen weight is estimated as percentage of body weight in both lineages, the difference in spleen size is also significant (median phocid spleen: 0.33%; median otariid spleen: 0.21%; $P=0.013$). Residuals of log phocid spleen vs log weight are plotted against log maximum recorded dive vs log body weight (Fig. 6). With the effect of body weight removed, these data clearly show that for phocids, diving capacity is strongly correlated with spleen size ($r=0.81$; $P < 0.0005$). Species with large spleens for their body weight have longer dive times. This relationship is not seen in the data on otariids.

Whole body haemoglobin and long diving duration

The third parameter examined in this manner was the total or whole body haemoglobin content (calculated from data on haemoglobin concentration in g per 100 ml of blood and total blood volume in litres). Phocids have a significantly higher [Hb] than otariids (phocid mean [Hb]: 20.8 g/100 ml; otariid mean [Hb]: 16.4 g/100 ml; $P=0.003$) (Appendix 1: Table 5). As data for blood volume were available in only three otariid species, the power of comparing means of blood volume and whole body Hb was greatly reduced

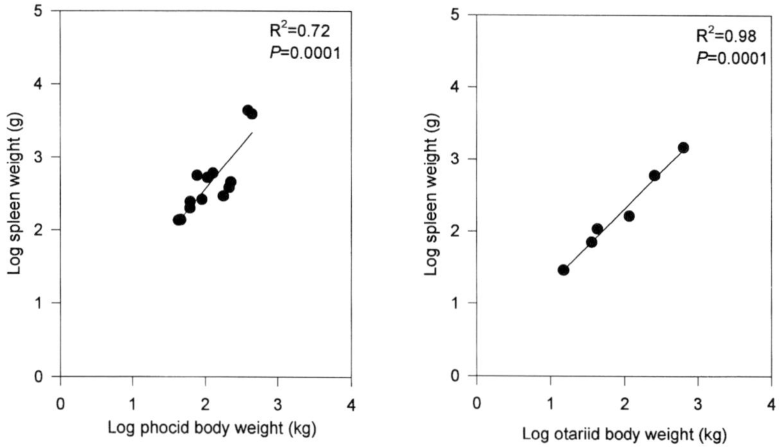

Fig. 5. Relationship between log body weight (kg) and log spleen weight (g) for phocids (left panel) and otariids (right panel). For phocids R^2=0.72, P<0.001, and for otariids R^2=0.98, P<0.001. For both lineages, spleen weight and body weight are clearly closely correlated; similar statistically significant relationships between body weight and total blood volume and between body weight and whole body haemoglobin (not shown) were also observed. (Data compiled by Mottishaw, 1997.)

(phocid mean estimated whole body Hb: 4898 g; otariid mean: 1698 g; P=0.04).

It was found that in phocids, blood volume significantly correlated with body weight (r=0.97; $P < 0.0001$). Haemoglobin concentration and blood volume were combined as whole body haemoglobin content, since this parameter is proportional to the total blood oxygen availability during diving. This parameter also correlated directly with body weight (r=0.95; $P < 0.0001$). As before, plotting residuals indicates that for phocids, independent of body weight effects, there is a statistically significant relationship between whole body blood volume and maximum diving duration (Fig. 7). Similarly, plotting residuals shows that whole body Hb content and maximum diving duration are significantly correlated (Fig. 8). This means that for phocids, independent of body size, the higher the whole body blood volume, the greater the maximum diving capacity; similarly, the greater the whole body Hb content, the greater the maximum diving duration that is possible, presumably because of the greater oxygen carrying capacity of the blood. Similar data are available for only three species of otariids, so an analysis cannot be made with confidence for this lineage.

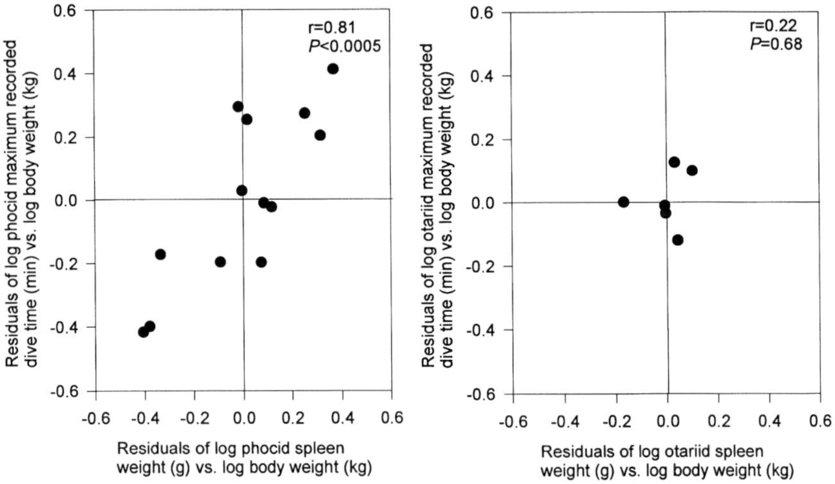

Fig. 6. Residuals of log phocid body weight (kg) vs log spleen weight (g) plotted as the independent parameter; residuals of log maximum dive time (min) vs log body weight (kg) as dependent parameter. Data for phocids in left panel; for otariids, in right panel. For phocids r=0.815; $P<0.001$, indicating a highly significant correlation. Since the relationship between body weight and maximum dive time for otariids is not significant (Fig. 4), the data for residuals are presented only for comparative purposes; no significant trends are observed. (From Mottishaw, 1997.)

Diving and phylogeny

Taken together, the characters analysed reflect most of the known components of the diving response (either directly, as in the case of bradycardia, or indirectly, as in the case of spleen weight). It is clear that a number of diving 'characters' show little variation in phocids and otariids, such as diving apnea, bradycardia, and regulated redistribution of cardiac output. The universality of some diving response traits, the fact that they can be elicited in terrestrial mammals including man, in a sense raises the possibility that their occurrence in diving animals is less 'an adaptation' for diving than it is a plesiomorphic or ancestral trait that simply preadapted air breathing vertebrates for dealing with a variety of stressful situations, including diving.

In contrast, the physiological characters which do vary, not only differ between the pinniped lineages in the expected direction, but also correlate with dive time in phocids. The analyses do not expose causes, they expose correlations. These traits likely allow the differences in dive times observed in pinnipeds, but proving this, and identifying which physiological characters explain different portions of dive time variation is not possible.

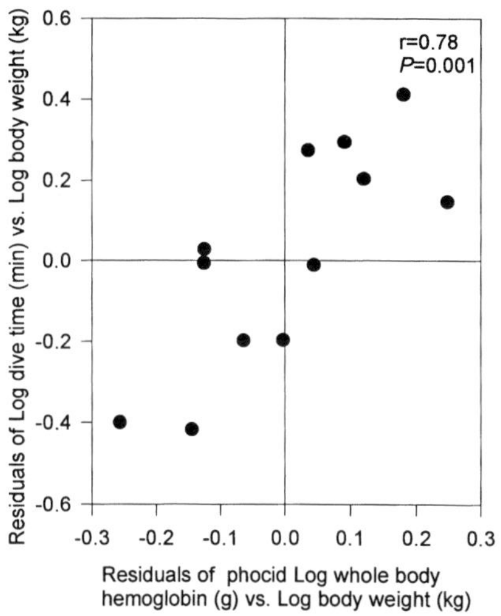

Fig. 7. Residuals of log whole body blood volume vs log body weight plotted as the independent parameter; residuals of log maximum recorded dive time vs log body weight as the dependent parameter. For phocids r=0.811, $P<0.01$, indicating a highly significant correlation. For otariids, these relationships were not statistically significant. (From Mottishaw, 1997.)

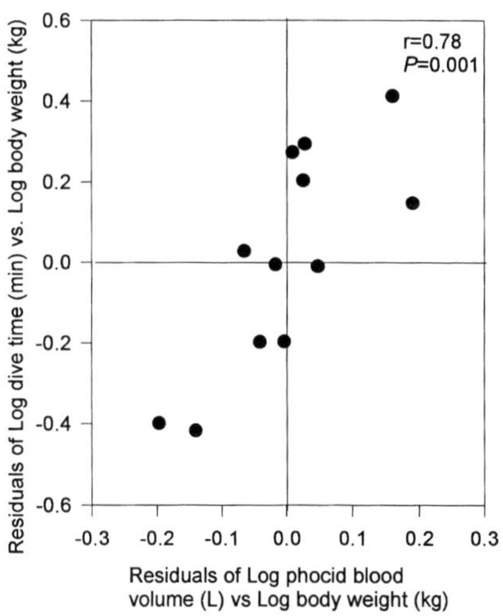

Fig. 8. Residuals of log whole body haemoglobin vs log body weight plotted as the independent parameter; residuals of log maximum recorded dive time vs log body weight plotted as the dependent parameter. For phocids r=0.815, $P<0.01$, again indicating a very strong correlation. For otariids, these analyses showed no statistically significant trends. (From Mottishaw, 1997.)

Nevertheless, the analysis demonstrates that at least three factors (i) body weight, (ii) spleen weight (independent of body weight) and (iii) whole body Hb (also independent of body weight) are consistent with the hypothesis that increased expression is an adaptation for extending diving duration in phocids.

The apparent lack of significant correlations within otariids needs explanation and may be a simple artefact of the available data. Tentative conclusions are (i) that there were not enough data for otariid species to reach the statistical power required, (as is probably the case in body weight vs. dive time), (ii) that variation in otariid maximum diving duration is not large enough for relatively insensitive diving response characters to decipher any adaptive trends, (otariids are more closely related than phocids (see Fig. 2) and may be less variable as a consequence), or (iii) that the evolution of the otariids has been 'driven' by factors other than requirements for long duration diving (such as reproductive requirements (Costa, 1991)). The next task (Mottishaw, 1997) will be to further evaluate using phylogenetically independent contrasts the evolution of the diving response in these pinnipeds.

Net effects of physiological/biochemical adaptations for diving

One way to evaluate the biological meaning of these putative adaptations, is to compare known differences in energetic costs in phocids and otariids. In phocids, the net effect of these characters is an 'at sea' or 'free diving' metabolic rate (DMR) which is similar to the 'routine' or resting metabolic rate (RMR) on land, or, in the case of other seals, RMR at the breathing hole (Table 1). In well-controlled experimental studies of the grey seal, DMR/RMR ratios, determined by direct oxygen consumption measurements, were found to be less than 0.5 for long dives (indicating significant diving hypometabolism) but these values were near 1 for short dives (Table 1). Field oxygen consumption studies of the Weddell seal also indicate values of DMR/RMR of close to 1 (Table 1). Using doubly labelled water technique, field studies of northern elephant seals indicate that the ratios of DMR/RMR range from fractionally less than 1 to about 1.3 (Andrews *et al.*, 1994), similar to values estimated from time-depth diving data for elephant seals at sea (Costa, 1991). (Boyd *et al.* (1995) reported that direct gas exchange estimates of average 24-hour metabolic rates were somewhat lower than estimates based on the doubly labelled water measurements; although both methods measure the same biochemical reactions, making it unclear which is the more reliable, the differences were small enough not to affect our current analysis.) Thus, it appears that for these species, the diving response is so effective that foraging at sea is no more costly than staying relatively inactive on shore. The same conclusions derive from independent studies (Table 1) of harp seals (Gallivan, 1981) and of harbour seals (Craig & Pasche, 1980).

Another way of evaluating the net effects of these diving characteristics is to compare diving and interdive intervals at sea. In phocids, the overall diving response is so effective that diving duration can greatly exceed interdive periods of recovery and recharging of oxygen stores. For elephant seals, these surface intervals are in the 2–4 min range (Hindell *et al.*, 1992; Le Boeuf *et al.*, 1989). In these large seals, ratios of diving duration / interdive periods can thus range from about 15 during routine diving to as high as 40 – 50 during maximum observed diving times (Le Boeuf *et al.* 1989, 1992; Hindell *et al.* 1992). Similar data are available for several other phocids (see Kooyman *et al.* 1980; Fedak, 1986; Reed *et al.*, 1994).

Although qualitatively some of the diving response characteristics are similar in sea lions and in seals, they function differently in each lineage. The most striking way in which to contrast the two diving strategies is to compare DMR/RMR for the two kinds of pinnipeds. While in the phocids this ratio is about 1, or even less than 1, in otariids it is in the 4–7 range (Table 1), either because of the high cost of swimming or thermogenesis, or both (Butler *et al.*, 1995). Not surprisingly, the ratios of diving duration / interdive time for recharging oxygen stores are much lower in sea lions than in elephant seals. Instead of values between 15 and 50 noted for the large seals, in sea lions the values are about 3 for routine dives and approach 10 in maximum duration diving.

Emerging principles of evolution of physiological systems

From this inquiry into the variability of the diving response in pinnipeds, the following conclusions seem to be valid:

(i) A number of physiological/biochemical characters considered necessary in diving animals are conserved in all pinnipeds; these traits, which are necessarily similar in phocids and otariids, include diving apnea, bradycardia, tissue hypoperfusion, and hypometabolism of hypoperfused tissues. At this stage in understanding of diving physiology and biochemistry, it is not possible to detect any correlation between these characters and diving capacity.

(ii) A number of physiological/biochemical characters are more malleable and are clearly correlated with long duration diving and prolonged foraging at sea. These characters are more lineage specific, and, for the phocids include body weight, spleen weight, and whole body oxygen carrying capacity. Within the phocids, the larger these are, the greater the diving capacity (defined as diving duration). Since the relationships between diving capacity and spleen weight or between diving capacity and whole body oxygen carrying capacities are evident even when corrected for body weight, it is reasonable to suggest that the two traits, large spleens and large whole body oxygen carrying capacities, extend diving duration. In contrast to conserved traits such as bradycardia, these characters (and presumably other similar ones, such as tissue specific metabolic organisation (Hochachka & Foreman, 1993)) have evolved to enable prolonged dive times. We conclude that increased spleen size and O_2 carrying capacity are likely to be physiological adaptations for increased diving duration.

Acknowledgements

This work was supported by NSERC (Canada). Special thanks are extended to our colleagues in marine mammal research (especially D.P. Costa, B.J. Le Boeuf, D.R. Jones, J.A. Hurley, and R.D. Andrews) for many useful discussions and, periodically, for releasing (unpublished) information in advance of its publication. Also thanks to Arne Mooers, Sally Otto, and Dolph Schluter for insightful discussions about evolutionary biology.

References

Andrews, R.D., Jones, D.R., Williams, J.D., Crocker, D.E., Costa, D.P. & Le Boeuf, B.J. (1994). Thermoregulation and metabolism in freely diving northern elephant seals. *FASEB Journal*, **8**, A2.

Arnason, O., Bodin, K., Gullberg, A., Ledje, C. & Mouchaty, S. (1995). A molecular view of pinniped relationships with particular emphasis on the true seals. *Journal of Molecular Evolution*, **40**, 78–85.

Bennett, A. F. (1987). Interindividual variability: an underutilized resource. In *New Directions in Ecological Physiology*, ed. M.E. Feder, A.F. Bennett, W.W. Burggren & R.B. Huey, pp. 147–69. Cambridge: Cambridge University Press.

Berta, A. & Demere, T.A. (1986). *Callorhinus gilmorei* n. sp., (Carnivora: Otariidae) from the San Diego formation (Blancan) and its implications for Otariid phylogeny. *Transactions of the San Diego Society for Natural History*, **21**, 111–26.

Berta, A. & Wyss, A.R. (1994). Pinniped phylogeny. *Proceedings of the San Diego Society for Natural History*, **29**, 35–56.

Boyd, I.L, Woakes, A.J., Butler, P.J., Davis, R.W. & Williams, T.M. (1995). Validation of heart rate and doubly labelled water as measures of metabolic rate during swimming in California sea lions. *Functional Ecology*, **9**, 151–60.

Bryden, M. M. (1971). Size and growth of viscera in the southern elephant seal, *Mirounga leonina* (L.). *Australian Journal of Zoology*, **19**, 103–20.

Burns J.J. & Fay, F.H. (1970). Comparative morphology of the skull of the Ribbon seal, *Histriophoca fasciata*, with remarks on systematics of Phocidae. *Journal of Zoology London*, **161**, 363–94.

Butler, P.J. & Jones, D.R. (1982). The comparative physiology of diving in vertebrates. *Advances in Comparative Physiology and Biochemistry* **8**, 179–364.

Butler, P.J., Bevan, R.M., Woakes, A.J., Croxall, J.P. & Boyd, I.L. (1995). The use of data loggers to determine the energetics and physiology of aquatic birds and mammals. *Brazilian Journal of Medical and Biological Research*, **28**, 1307–17.

Castellini, M.A., Kooyman, G.L. & Ponganis, P.J. (1992). Metabolic rates of freely diving Weddell seals: correlations with oxygen stores, swim velocity, and diving duration. *Journal of Experimental Biology*, **165**, 181–94.

Costa, D.P. (1991). Reproductive and foraging energetics of pinnipeds: implications for life history patterns. In *The Behaviour of Pinnipeds*, ed. D. Renouf, pp. 300–44. London: Chapman & Hall.

Costa, D.P. (1993). The relationship between reproductive and foraging energetics and the evolution of the Pinnipedia. *Symposia of the Zoological Society London*, **66**, 293–314.

Craig, A.B, Jr. & Pasche, A. (1980). Respiratory physiology of freely diving harbor seals (*Phoca vitulina*). *Physiological Zoology*, **53**, 419–32.

DeLong R.L. & Stewart, B.S. (1991). Diving patterns of northern elephant seal bulls. *Marine Mammal Science*, **7**, 369–84.

DeLong, R.L., Stewart, B.S. & Hill, R.D. (1992). Documenting migrations of northern elephant seals using day length. *Marine Mammal Science*, **8**, 155–9.

Elsner, R. & Gooden, B. (1983). *Diving and Asphyxia – A Comparative Study of Animals and Man.* pp. 1–177. Cambridge: Cambridge University Press.

Fedak, M. (1986). Diving and exercise in seals – a benthic perspective. In *Diving in Animals and Man*, ed. A. Brubakk, J.W. Kanwisher & G. Sundnes, pp. 11–32. Trondheim: Tapir Publ.

Gallivan, G.J. (1981). Ventilation and gas exchange in unrestrained harp seals (*Phoca groenlandica*). *Comparative Biochemistry and Physiology*, **69A**, 809–13.

Gentry, R.L. & Kooyman,G.L. (1986). *Fur Seals*. pp. 1–291. Princeton, NJ: Princeton Univ. Press.

Guppy, M., Hill, R.D., Schneider, R.C., Qvist, J. Liggins, G.C., Zapol W.M. & Hochachka, P.W. (1986). Microcomputer assisted metabolic studies of voluntary diving of Weddell seals. *American Journal of Physiology*, **250**, R175–87.

Guyton, G.P., Stanek, K.S., Schneider, R.C., Hochachka, P.W., Hurford, W.E., Zapol, D.K. & Zapol, W.M. (1995). Myoglobin saturation in free diving Weddell seals. *Journal of Applied Physiol*ogy, **79**, 1148–55.

Hill, R.D., Schneider, R.C., Liggins, G.C., Schuette, A.H., Elliott, R.L., Guppy, M., Hochachka, P.W., Qvist, J., Falke, K.J. & Zapol, W.M. (1987). Heart rate and body temperature during free diving of Weddell seals. *American Journal of Physiology*, **253**, R344–51.

Hindell, M.A., Slip, D.J., Burton, H.R. & Bryden, M.M. (1992). Physiological implications of continuous and deep dives of the southern elephant seal (*Mirounga leonina*). *Canadian Journal of Zoology*, **70**, 370–9.

Hochachka, P.W. (1986). Balancing the conflicting demands of diving and exercise. *Federation Proceedings*, **45**, 2949–54.

Hochachka, P.W. (1992). Metabolic biochemistry and the making of a mesopelagic mammal. *Experientia*, 48: 570–5.

Hochachka, P.W. & Guppy, M. (1987). *Metabolic Arrest and the Control of Biological Time*, pp. 1–237 Cambridge, USA: Harvard University Press.

Hochachka, P.W. & Foreman, R.A. III. (1993). Phocid and cetacean blueprints of muscle metabolism. *Canadian Journal of Zoology*, **71**, 2089–98.

Hochachka, P.W., Liggins, G.C., Guyton, G.P., Schneider, R.C., Stanek, K.S., Hurford, W.E., Creasy, R.K., Zapol, D.G. & Zapol, W.M. (1995). Hormonal regulatory adjustments during voluntary diving in Weddell seals. *Comparative Biochemistry and Physiology*, **112B**, 361–75.

Hurford, W.E., Hochachka, P.W., Schneider, R.C., Guyton, G.P., Stanek, K., Zapol, D.G., Liggins, G.C. & Zapol, W.M. (1995). Splenic contraction, catecholamine release and blood volume redistribution during voluntary diving in the Weddell seal. *Journal of Applied Physiology*, **80**, 298–306.

Hurley, J.A. (1996). Metabolic rate and heart rate during trained dives in adult California sea lions. PhD Thesis, University of California, Santa Cruz.

Irving, L., Solandt, O.M., Solandt, D.Y. & Fisher, K.C. (1935). The respiratory metabolism of the seal and its adjustment to diving. *Journal of Cellular and Comparative Physiology*, **7**, 137–51.

Kooyman, G.L. (1985). Physiology without restraint in diving mammals. *Marine Mammal Science*, **1**, 166–78.

Kooyman, G.L., Wahrenbrock, E.H. Castellini, M.A. Davis, R.W. &

Sinnett, E.E. (1980). Aerobic and anaerobic metabolism during voluntary diving in Weddell seals: evidence of preferred pathways from blood chemistry and behaviour. *Journal of Comparative Physiology*, **138B**, 335–46.

Lacombe, A.M. & Jones, D.R. (1991). Role of adrenal catecholamines during forced submergence in ducks. *American Journal of Physiology*, **261**, R1364–72.

Le Boeuf, B.J., Naito, Y., Huntley, A.C. & Asaga, T. (1989). Prolonged, continuous, deep diving by northern elephant seals. *Canadian Journal of Zoology*, **67**, 2514–19.

Le Boeuf, B.J., Naito, Y., Asaga, T., Crocker, D. & Costa, D. (1992). Swim velocity and dive patterns in a northern elephant seal, *Mirounga angustirostris*. *Canadian Journal of Zoology*, **70**, 786–95.

Lento, G.M., Hickson, R.E., Chambers, G.K. & Penny, D. (1995). Use of spectral analysis to test hypotheses on the origin of pinnipeds. *Molecular Biology and Evolution*, **12**, 28–52.

McConnell, D.J., Chambers, C., & Fedak, M.A. (1992). Foraging ecology of southern elephant seals in relation to the bathymetry and productivity of the Southern Ocean. *Antarctic Science*, **4**, 393–8.

Mottishaw, P.D. (1997). The diving physiology of pinnipeds: an evolutionary inquiry. MSc Thesis, University of British Columbia, Vancouver.

Muizon, Ch. de (1982). Phocid phylogeny and dispersal. *Annals of South Africa Museums*, **89**, 175–213.

Perry, E.A., Carr, S.M., Bartlett, S.E. & Davidson, W.S. (1995). A phylogenetic perspective on the evolution of reproductive behaviour in pagophilic seals of the northwest Atlantic as indicated by mitochondrial DNA sequences. *Journal of Mammalogy*, **76**, 22–31.

Ponganis, P., Ponganis, E.P., Gentry, R.L. & Trillmich, F. (1990). Swimming velocities in otariids. *Canadian Journal of Zoology*, **68**, 2105–12.

Qvist, J., Hill, R.D., Schneider, R.C., Falke, K.J., Guppy, M., Elliott, R.L., Hochachka, P.W. & Zapol, W.M. (1986). Hemoglobin concentrations and blood gas tensions of free diving Weddell seals. *Journal of Applied Physiology*, **61**, 1560–9.

Reed, J.Z., Chambers, C., Fedak, M.A. & Butler, P.J. (1994). Gas exchange of captive freely diving grey seals *(Halichoerus grypus)*. *Journal of Experimental Biology*, **191**, 1–18.

Scholander, P.F. (1940). Experimental investigations in diving mammals and birds. *Hvalradets Skrifter*, **22**, 1–131.

Scholander, P.F. (1963). The master switch of life. *Scientific American*, **209**, 92–106.

Thompson, D. & Fedak, M.A. (1992). Cardiac responses of grey seals during diving at sea. *Journal of Experimental Biology*, **174**, 139–64.

Williams, T.M., Kooyman, G.L. & Croll, D.A. (1991). The effect of submergence on heart rate and oxygen consumption of swimming seals and sea lions. *Journal of Comparative Physiology*, **160B**, 637–44.

Worthy G.A.J., Morris, P.A. Costa, D.P. & Le Boeuf, B.J. (1992). Moult energetics of the northern elephant seal *(Mirounga angastirostris). Journal of Zoology, London*, **227**, 257–65.

Zapol, W.M., Liggins, G.C. Schneider, R.C. Qvist, J. Snider, M.T. Creasy, R.K. & Hochachka, P.W. (1979). Regional blood flow during simulated diving in the conscious Weddell seal. *Journal of Applied Physiology*, **47**, 968–73.

Appendix

Table 1. *Maximum recorded and average dive durations of free diving pinniped species*

Common name	Latin name	Abrv	Max min	Avg min	N	Method	Source
Northern elephant seal	*Mirounga angustirostris*	Ma	62	19.2	7,6	TDR	Le Boeuf *et al.*, 1988, 1989
Southern elephant seal	*Mirounga leonina*	Ml	120	27.1	10	TDR	Hindell, Slip & Burton, 1991
Weddell seal	*Leptonychotes weddellii*	Lw	82	11.5	5,69	TDR	Castellini, Kooyman & Ponganis, 1992a; Castellini, Davis & Kooyman, 1992b
Ross seal	*Ommatophoca rossii*	Or	9.8	6.4	1	TDR	Bengston & Stewart, 1997
Crabeater seal	*Lobodon carcinophagus*	Lc	10.8	4.5	6,8	TDR/SLTDR	Bengston & Stewart, 1992; Nordøy, Folkow & Blix, 1995a
Hawaiian monk seal	*Monachus schauinslandi*	Ms	12	7	11	TDR	T. Ragen, pers. comm.
Bearded seal	*Erignatus barbatus*	Eb	25	8	4	SLTDR	Gjertz *et al.*, 1995; I. Gjertz, pers. comm.
Hooded seal	*Cystophora cristata*	Cc	52	10	19,14	SLTDR	Folkow & Blix, 1995; L.P. Folkow, pers. comm.; Stenson *et al.*, 1993
Harp seal	*Phoca groenlandica*	Pg	20	3.2	8,4	SLTDR/TDR	Nordøy *et al.*, 1995b; E.S. Nordøy, pers comm.; Lydersen & Kovacs, 1993
Grey seal	*Halichoerus grypus*	Hg	32	4.5	3	RFT	Thompson & Fedak, 1993
Baikal seal	*Phoca sibirica*	Ps	*40	4	4	TDR	Stewart *et al.*, 1996
Ringed seal	*Phoca hispida*	Phi	*26.4	5.7	14	RFT	Kelly & Wartzok, 1996

Table 1. *(cont.)*

Common name	Latin name	Abrv	Max min	Avg min	N	Method	Source
Saimaa seal	*Phoca h. siamensis*	Phs	22.8	5.6	4,2	RFT	Hyvarinen, Hamalainen & Kunnasranta, 1995; Koskela, Hyvarinen & Kunnasranta, 1995
Spotted seal	*Phoca largha*	Pla	10	4	8	SLTDR	K. Frost, pers. comm.
Harbour seal E Atl	*Phoca vitulina vitulina*	Pvv	12	2.5	2	RFT	Fedak, Pullen & Kanwisher, 1988
Harbour seal E Pac	*Phoca v. richardsii*	Pvr	*28.3	4.6	30,8	TDR	P. Olesiuk, pers. comm.; Eguchi & Harvey 1995, T. Eguchi, pers. comm.
Harbour seal W Atl	*Phoca v. concolor*	Pvc	18.3	3.3	19	TDR	Coltman *et al.*, 1995, Coltman, Bowen, Boness, & Iverson unpub. data
Northern fur seal	*Callorhinus ursinus*	Cu	7.6	2.6	7	TDR	Gentry *et al* 1986
Antarctic fur seal	*Arctocephalus gazella*	Ag	10	0.9	11	TDR	Boyd & Croxall, 1992
South African fur seal	*Arctocephalus pusillus*	Ap	7.5	1.9	2	TDR	Kooyman & Gentry, 1986
Guadalupe fur seal	*Arctocephalus townsendi*	At	18	2.6	1	SLTDR	Gallo-Reynoso *et al.*, 1995
Juan Fernandez fur seal	*Arctocephalus philippii*	Aph	3.7	1.8	9	TDR	Francis & Boness in prep
New Zealand fur seal	*Arctocephalus forsteri*	Af	11.7	1.6	17,13	TDR	Harcourt *et al.*, 1995, Harcourt unpubl. data, R.H. Mattlin 1993, pers. comm.
South American fur seal	*Arctocephalus australis*	Aa	7.1	2.8	1	TDR	Trillmich *et al.*, 1986
Galapagos fur seal	*Arctocephalus galapagoensis*	Aga	7.7	1	20	TDR	Kooyman & Trillmich 1986*a*
Southern sea lion	*Otaria flavescens*	Of	7.7	3	6	TDR	Werner & Campagna, 1995

Australian sea lion	*Neophoca cinerea*	Nci	8.3	3.3	7	TDR	Costa, Kretzman & Thorson, 1989; D.P. Costa, pers. comm.
New Zealand sea lion	*Phocarctos hookeri*	Pho	12	5.8	15	TDR	Gentry, Roberts & Cawthorn, 1987; R.L. Gentry pers. comm.; N. Gales pers. comm.
Steller sea lion	*Eumatopias jubatus*	Ej	16	2.4	14,2	SLTDR	Calkins & Swain, 1995; D.G. Calkins unpubl. data; Merrick *et al.*, 1994,
California sea lion	*Zalophus californianus*	Zc	9.9	2.3	10	TDR	Feldkamp, Delong & Antonelis, 1989
Galapagos sea lion	*Zalophus c. wollebaeki*	Zcw	6	2	4	TDR	Kooyman & Trillmich, 1986*b*
Walrus	*Odobenus rosmarus*	Oro	12.7	5.2	1	TDR	Wiig *et al* 1993

Notes:
Species are listed by common name, Latin name, and an abbreviation. Maximum (max) and average (avg) dive times are given in minutes. The number of animals (N), methods used, and the source of the values are given. Average dive times are the mean or modal dive times given in each study. A minor caveat is that time-depth recorders differ in data saving systems and minimum and maximum settings, which may introduce noise into a comparison of average dive times. Dive depth data are also available from most of these sources. Methods of recording dive times are: TDR, time-depth recorder; SLTDR, satellite linked time-depth recorder; RFT, Radio frequency transmitter.
*These values were published or made available to the authors after completion of the analysis for this paper, but were included, as this resource may be useful for future studies. The maximum recorded dive times used in analysis for this paper and their sources were: Ps (20 min. Reeves, Stewart & Leatherwood, 1992), Phi (21.2 min. C. Lydersen pers. comm.), Pvr (19 min. Eguchi & Harvey 1995).

Table 2. *Body weights of pinniped species used in the diving studies*

Common name	Abrv	Weight (kg)	Source
Northern elephant seal	Ma	361	Le Boeuf *et al.*, 1989
Southern elephant seal	Ml	387	Hindell *et al.*, 1991
Weddell seal	Lw	355	Castellini *et al.*, 1992a
Ross seal	Or	173	Ray, 1981
Crabeater seal	Lc	182	Nordøy *et al.*, 1995a
Hawaiian monk seal	Ms	195	T.J. Ragen, pers. comm.
Bearded seal	Eb	283	I. Gjertz, pers. comm.
Hooded seal	Cc	235	L.P. Folkow, pers. comm.
Harp seal	Pg	123	Lydersen & Kovacs, 1993
Grey seal	Hg	210	Thompson & Fedak, 1993
Baikal seal	Ps	*29	Stewart *et al.*, 1996
Ringed seal	Phi	*73	Kelly & Wartzok, 1996
Saimaa seal	Phs	63	Hyvarinen *et al.*, 1995
Spotted seal	Pla	85	K.J. Frost, pers. comm.
Harbour seal E Atl	Pvv	110	Fedak *et al.*, 1988
Harbour seal E Pac	Pvr	*93	P. Olesiuk, pers. comm.
Harbour seal W Atl	Pvc	107	Coltman *et al.*, 1995
Northern fur seal	Cu	37	Kooyman, Gentry & Urquhart, 1976
Antarctic fur seal F	Ag	41	Boyd & Croxall, 1992
South African fur seal	Ap	75	Kooyman & Gentry, 1986
Guadalupe fur seal	At	50	J.P. Gallo-Reynoso, pers. comm.
Juan Fernandez fur seal	Aph	48	J.M. Francis & D.J. Boness in prep
New Zealand fur seal	Af	38	R.G. Harcourt *et al.*, 1995, R.H. Mattlin, 1993
South American fur seal	Aa	35	Trillmich *et al.*, 1986
Galapagos fur seal	Aga	29	Kooyman & Trillmich, 1986a
Southern sea lion	Of	126	Werner & Campagna, 1995
Australian sea lion	Nci	84	Costa *et al.*, 1989
New Zealand sea lion	Pho	150	Gentry *et al.*, 1987
Steller sea lion	Ej	268	D.G. Calkins, pers. comm.
California sea lion	Zc	111	Odell, 1981
Galapagos sea lion	Zcw	70	Kooyman & Trillmich, 1986b
Walrus	Oro	1500	Wiig *et al.*, 1993

Notes:
Common names and abbreviation of Latin names for each species are given. Body weight (kg) was taken from the same sources as the diving data. If body weight was not given, data were requested from the authors (given as pers. comm.), or mean weight was used from other sources.
*These values were published or made available to the authors after completion of the analysis for this paper. The whole animal body weights used in analysis for this paper and the sources were: Ps (66 kg. Reeves *et al.*, 1992), Phi (50 kg. Lydersen, 1991), Pvr (119 kg. T. Eguchi pers. comm.).

Table 3. *Spleen weights of pinniped species*

Common name	Abrv	Spleen (g)	Weight (kg)	ESW (g)	% bw	N	Source
S. elephant seal	Ml	4373	387	4373	1.13	19	Bryden, 1971
Weddell seal	Lw	3916	440	3160	0.89	13	Zwillenberg, 1959 (in Qvist *et al.*, 1986)
Ross seal	Or	290	179	276.8	0.16	2	Bryden & Erickson, 1976
Crabeater seal	Lc	381	214	327.6	0.18	7	Bryden & Erickson, 1976
Bearded seal	Eb	454	227	566	0.20	6	Sokolov, Kozygin & Tikhomirov, 1966
Hooded seal	Cc	558	77.3	1692	0.72	1	Blessing *et al.*, 1972
Harp seal	Pg	600	128	578.1	0.47	4	Shustov & Yablokov, 1968
Grey seal	Hg	200	61	693	0.33	2	Bonner, 1981
Baikal seal	Ps*	163	65	73	0.25	3	Petrov, Elagin & Shoshenko, 1987
Ringed seal	Phi	138	46	150	0.30	8	Frost & Lowry, 1981
Ladoga seal	Phl**	136	42.4	201.6	0.32	5	Sokolov *et al.*, 1966
Spotted seal	Pla	260	90	246.5	0.29	64	Kosygin & Gol'tsev, 1971
Harbour seal E Atl	Pvv	244	62	429	0.39	2	Slijper, 1958
Harbour seal E Pac	Pvr	520	107	583.1	0.49	1	Crile & Quiring, 1940
Northern fur seal	Cu	70.6	36	70.3	0.19	75	Scheffer, 1960
Antarctic fur seal	Ag	108.3	43	102.5	0.25	3	Payne, 1979
S. American fur seal	Aa	29	15	66.5	0.19	1	Blessing, Ligensa & Winner, 1972
Southern sea lion	Of	600	260	289.8	0.23	1	Vaz-Ferreira, 1981
Steller sea lion	Ej	1472	640	616.4	0.23	5	Sokolov *et al.*, 1966
California sea lion	Zc	163	117	155.4	0.14	2	Blessing *et al.*, 1972
Walrus	Oro	4070	1100	5550	0.37	5	Sokolov *et al.*, 1966

Notes;

Common names and abbreviation of Latin names for each species are given. Mean spleen weight (g) and mean weight of the animals (kg) used in each study are given, along with the percent of body weight occupied by the spleen (% bw), number of animals (N), and the source of the given values. Spleen weight measures were taken from autopsy data, as in vivo studies did not give comparable results. Evidence from previous pinniped studies suggests that the slope of the allometric relationship between spleen weight and body weight within a species is 1.0 (Bryden, 1971; a similar trend can be seen in data from Scheffer, 1960 and Slijper, 1958). As the percentage of body weight occupied by the spleen did not change in animals of differing weights, the measured value of spleen weight was divided by the weight of the animals and multiplied by the weight of the animals from the diving measurements (Table 2). This measure was called estimated spleen weight (ESW) in grams.

*This value was made available to the authors after completion of the analysis for this paper.

**The Ladoga ringed seal spleen weight values were correlated with the Saimaa ringed seal dive times. As these are closely related subspecies, this should not affect the analysis.

Table 4. *Blood volumes of pinniped species*

Common name	Abrv	Blvol (l)	Weight (kg)	Eblvol (l)	N	Method	Source
N. elephant seal	Ma	22.7	105.4	78.0	7	^{131}I dilution	Simpson *et al.*, 1970
S. elephant seal	Ml	59.4	287	79.1	1	bled out	Bryden & Lim, 1969
Weddell seal	Lw	55.7	378	52.5	2	^{51}Cr cell dilution	Lenfant *et al.*, 1969
Ross seal	Or	18.3	179	17.6	2	bled out	Bryden & Erickson, 1976
Crabeater seal	Lc	18.1	201	16.4	7	bled out	Bryden & Erickson, 1976
Hooded seal	Cc	3.75	26.5	35.25	2	bled out	Scholander, 1940
Harp seal	Pg	12.3	80	18.9	1	^{51}Cr cell dilution	Ronald, 1970
Grey seal	Hg	15.6	129.8	21.6	4	Dye dilution	Castellini *et al.*, 1985
Ribbon seal	Hf	6.87	51.7	13.3	4	^{51}Cr cell dilution	Lenfant, Johansen & Torrance, 1970
Baikal seal	Ps*	11.8	65	5.25	3	^{51}Cr cell, ^{131}I dil.	Petrov *et al.*, 1987
Ringed seal	Phi	3.39	23.9	7.1	3	^{51}Cr cell, ^{125}I dil.	St. Aubin *et al.*, 1978
Spotted seal	Pla	3.14	23.8	11.22	5	^{51}Cr cell dilution	Lenfant *et al.*, 1970
Harbour seal E Atl	Pvv	1.02	8.7	12.9	1	Dye dilution	Harrison & Tomlinson, 1956
Harbour seal W Atl	Pvc	2.96	23.2	13.64	4	^{51}Cr cell dilution	Packer *et al.*, 1969
Northern fur seal	Cu	17.83	153.5	4.03	4	^{51}Cr cell dilution	Lenfant *et al.*, 1970
Steller sea lion	Ej	3.19	36.3	24.2	4	^{51}Cr cell dilution	Lenfant *et al.*, 1970
California sea lion	Zc			13.88		^{131}I dilution	Ridgway, 1972
Walrus	Oro	6.93	64.5	159	3	^{51}Cr cell dilution	Lenfant *et al.*, 1970

Notes:
Common names and abbreviation of Latin names are given for each species. The published measure of blood volume (Blvol) in litres is given along with the weight of the animal(s) and number (N) it was measured in, as well as the method used and source. As with spleen weights, the measured value of blood volume in litres (MBV) was divided by weight of the animals it was calculated in (kg), and multiplied by the weight of the animals from the diving measurements (Table 2). This measure was called estimated blood volume (Eblvol) in litres.
* This value was made available to the authors after completion of the analysis for this paper.

Table 5. *Blood oxygen carrying capacity of pinniped species*

Common name	Abrv	Hb (g/0.1l)	Hb WB (g (MBV))	E Hb WB (g (EBV))	N	Source
N. elephant seal	Ma	24.9	5652.3	19417	4	Hedrick, Duffield & Cornell, 1986
S. elephant seal	Ml	23.3	13840.2	18430.3	43	Lane, Morris & Sheedy,1972
Weddell seal	Lw	23.7	13200.9	12442.5	4	Lenfant *et al.*, 1969
Ross seal	Or	21.8	3989.4	3836.8	*	Kandinskii, 1974
Crabeater seal	Lc	19.2	3475.2	3148.8	*	Kandinskii, 1974
Leopard seal	Hl	17.0			*	Kandinskii, 1974
Hawaiian monk seal	Ms	18.5			15	Banish & Gilmartin, 1988
Bearded seal	Eb	17.4			11	Sokolov, 1965
Hooded seal	Cc	26.4	990	9306	2	Clausen & Ersland, 1969
Harp seal	Pg	21.5	2644.5	4063.5	22	Geraci, 1971
Grey seal	Hg	19.1	2980	4125.6	6	Greenwood, Ridgway & Harrison, 1971
Ribbon seal	Hf	24.5	1683.2	3269.8	5	Lenfant *et al.*, 1970
Baikal seal	Ps**	22.0	2596	1155	3	Petrov *et al.*, 1987
Ringed seal	Phi	24.5	830.6	1739.5	6	Geraci & Smith, 1975
Spotted seal	Pla	21.2	665.7	2378.6	5	Lenfant *et al.*, 1970
Harbour seal E Atl	Pvv	19.7	200.9	2541.3	23	McConnell & Vaughan, 1983
Harbour seal E Pac	Pvr	20.2			5	Lenfant, 1969
Harbour seal W Atl	Pvc	16.5	488.1	2250.6	8	Irving *et al.*, 1935
Northern fur seal	Cu	17.0	3025.8	683.9	4	Lenfant *et al.*, 1970
New Zealand fur seal	Af	16.9			1	Wells, 1978
Australian sea lion	Nci	18.4			19	Needham, Cargill & Sheriff, 1980

Steller sea lion	Ej	14.7	468.9	3557.4	5	Lenfant *et al.*, 1970
California sea lion	Zc	15.0	2164.5	2081.3	8	Ridgway 1972
Walrus	Oro	16.2	1122.7	25758	5	Lenfant *et al.*, 1970

Notes:
Common names and abbreviation of Latin names are given for each species. The published measure of haemoglobin (Hb) per 100 ml (g/0.1 l) of blood is given along with the number (N) it was measured in and sources of the values. Whole body haemoglobin content in grams (Hb WB) was calculated by multiplying the measured blood volume (MBV) in litres (Table 4) by the measured [Hb] in the blood. Estimated whole body haemoglobin (E Hb WB) was calculated using the estimated value for blood volume (EBV) of the diving pinniped.
* Kandinskii 1974 sampled 71 seals of four species (Or, Lc, Hl, Lw).
** These values were calculated using the blood volumes made available to the authors after completion of the analysis for this paper. The previous value for Hb and source was: Ps (18 g/100 ml. Hawkey, 1975)

Table 6. *Lowest recorded heart rate of pinniped species*

Common name	Abrv	LowHR (bpm)	rHR (bpm)	N	Condition	Source
Northern elephant seal	Ma	4	60	7	free	R.D. Andrews *et al.*, 1995, pers. comm.
Southern elephant seal	Ml	8	65	1	free	Hindell & Lea, 1995
Weddell seal	Lw	8	68	14	free	Hill *et al.*, 1987
Hooded seal	Cc	4	102	1	captive	Pasche & Krog, 1980
Harp seal	Pg	5	90	1	captive	Casson & Ronald, 1975
Grey seal	Hg	4	119	3	free	Thompson & Fedak, 1993
Ringed seal	Phi	6	80	3	forced	Elsner *et al.*, 1989
Harbour seal E Atl	Pvv	4	88	>2	forced	Harrison & Tomlinson, 1960
Harbour seal E Pac	Pvr	10	148	2	trained	Elsner, 1965
Harbour seal W Atl	Pvc	3.5	100	>2	forced	Irving, Scholander & Grinnell, 1941
Northern fur seal	Cu	12	60	3	forced	Irving *et al.*, 1963
California sea lion	Zc	9	79	4	trained	Hurley, 1996

Note:
Species are listed by common names and Latin name abbreviation. Lowest heart rate of animals (lowHR) and resting heart rate (rHR) in each study is given in beats per minute (bpm). The numbers of animals (N), methods used (condition), and the sources of the values are given. The state of the animals during recording of heart rate is given as: free, recorder attached to a freely diving animal; forced, EKG of an animal that is forced to dive; captive, EKG attached to an animal that is allowed to dive of its own volition in a captive tank; trained, EKG of an animal that has been trained to dive in response to a signal, and for a set duration.

Appendix references

Andrews, R.D., Jones, D.R., Williams, J.D., Crocker, D.E., Costa, D.P. & Le Boeuf, B.J. (1995). Metabolic and cardiovascular adjustments to diving in northern elephant seals (*Mirounga angustirostris*). *Physiological Zoology*, **68**, 105.

Banish, L. D. & Gilmartin, W. G. (1988). Hematology and serum chemistry of the young Hawaiian monk seal (*monachus schauinslandi*). *Journal of Wildlife Diseases*, **24**, 225–30.

Bengston, J. L. & Stewart, B. S. (1992). Diving and haulout behavior of crabeater seals in the Weddell Sea, Antarctica, during March 1986. *Polar Biology*, **12**, 635–44.

Bengston, J. L. & Stewart, B. S. (1997). Diving patterns of a Ross seal (*Ommatophoca rossii*) near the eastern coast of the Antarctic Peninsula. *Polar Biology*, In Press.

Blessing, M. H., Ligensa, K. & Winner, R. (1972). Zur Morphologie der Milz einiger im Wasser lebender Säugetiere. *Zeitschrift für Wissenschafliche Zoologie*, **184**, 164–204.

Bonner, W. N. (1981). Grey seal *Halichoerus grypus* Fabricius, 1791. In *Handbook of Marine Mammals Volume 2: Seals*, ed. S. H. Ridgway & J. H. Harrison, pp. 111–44. London: Academic Press.

Boyd, I. L. & Croxall, J. P. (1992). Diving behaviour of lactating Antarctic fur seals. *Canadian Journal of Zoology*, **70**, 919–28.

Bryden, M. M. (1971). Size and growth of viscera in the southern elephant seal, *Mirounga leonina* (L.). *Australian Journal of Zoology*, **19**, 103–20.

Bryden, M. M. & Erickson, A. W. (1976). Body size and composition of crabeater seals (*Lobodon carcinophagus*), with observations on tissue and organ size in Ross seals (*Ommatophoca rossi*). *Journal of Zoology, London*, **179**, 235–47.

Bryden, M. M. & Lim, G. H. K. (1969). Blood parameters of the southern elephant seal (*Mirounga leonina*, Linn.) in relation to diving. *Comparative Biochemistry and Physiology*, **28**, 139–48.

Calkins, D. G., & Swain, U. (1995). Movements and diving behavior of female Steller sea lions in southeastern Alaska. *Abstracts of the 11th Biennial Conference on the Biology of Marine Mammals*, 19.

Casson, D. M. & Ronald, K. (1975). The harp seal, *Pagophilus groenlandicus* (Erxleben, 1777)-XIV. Cardiac arrythmias. *Comparative Biochemistry and Physiology*, **50A**, 307–14.

Castellini, M. A., Murphy, B. J., Fedak, M., Ronald, K., Gofton, N. & Hochachka, P. W. (1985). Potentially conflicting metabolic demands of diving and exercise in seals. *Journal of Applied Physiology*, **58**, 392–9.

Castellini, M. A., Kooyman, G. L. & Ponganis, P. J. (1992*a*). Metabolic rates of freely diving Weddell seals: correlations with oxygen stores, swim velocity, and dive duration. *Journal of Experimental Biology*, **165**, 181–94.

Castellini, M. A., Davis, R.W. & Kooyman, G. L. (1992*b*). Diving behavior and ecology of the Weddell seal: annual cycles. *Bulletin of the Scripps Institute of Oceanography*, **28**, 1–54.

Clausen, G. & Ersland, A. (1969). The respiratory properties of the blood of the bladdernose seal (*Cystophora cristata*). *Respiration Physiology*, **7**, 1–6.

Coltman, D. W., Bowen, W. D. & Boness, D. J. (1995). Does body mass affect aquatic behaviour of harbour seal (*Phoca vitulina*) males during the mating season? *Abstracts of the 11th Biennial Conference on the Biology of Marine Mammals*, 24.

Costa, D. P., Kretzman, M. & Thorson, P. H. (1989). Diving pattern and energetics of the Australian sea lion, *Neophoca cinerea. American Zoologist*, **29**, 71A.

Crile, G. & Quiring, D. P. (1940). A record of the body weight and certain organ and gland weights of 3690 animals. *Ohio Journal of Science*, **15**, 219–59.

Eguchi, T. & Harvey, J. T. (1995). Diving behavior and movements of the Pacific harbor seal (Phoca vitulina richardsi) in Monterey Bay, California. *Abstracts of the 11th Biennial Conference on the Biology of Marine Mammals*, 34.

Elsner, R. (1965). Heart rate response in forced versus trained experimental dives in pinnipeds. *Hvalradets Skrifter*, **48**, 24–9.

Elsner, R., Wartzok, D., Sonafrank, N. B. & Kelly, B. P. (1989). Behavioral and physiological reactions of arctic seals during under-ice pilotage. *Canadian Journal of Zoo*logy, **67**, 2506–13.

Fedak, M. A., Pullen, M. R. & Kanwisher, J. (1988). Circulatory responses of seals to periodic breathing: heart rate and breathing during exercise and diving in the laboratory and open sea. *Canadian Journal of Zoology*, **66**, 53–60.

Feldkamp, S. D., Delong, R. L. & Antonelis, G. A. (1989). Diving patterns of California sea lions, *Zalophus californianus. Canadian Journal of Zoology*, **67**, 872–83.

Folkow, L. P. & Blix, A. S. (1995). Diving and feeding of hooded seals, *Cystophora cristata. Abstracts of the 11th Biennial Conference on the Biology of Marine Mammals*, 38.

Frost, K. J. & Lowry, L. F. (1981). Ringed, Baikal, and Caspian seals *Phoca hispida* Schreber, 1775; *Phoca sibirica* Gmelin, 1788 and *Phoca caspica* Gmelin, 1788. In *Handbook of Marine Mammals Volume 2: Seals*, ed. S. H. Ridgway & J. H. Harrison, pp. 29–53. London: Academic Press.

Gallo-Reynoso, J. P., Le Boeuf, B. J. & Figueroa, A. L. (1995). Track, location, duration and diving behavior during foraging trips of Guadelupe fur seal females. *Abstracts of the 11th Biennial Conference on the Biology of Marine Mammals*, 41.

Gentry, R. L., Kooyman, G. L. & Goebel, M. E. (1986). Feeding and diving behavior of Northern Fur Seals. In *Fur Seals: Maternal Strategies on Land and at Sea*, ed. R. L. Gentry, G. L. Kooyman & M. E. Goebel, pp. 61–78. Princeton: Princeton University Press.

Gentry, R. L., Roberts, W. E. & Cawthorn, M. W. (1987). Diving behavior of the hooker's sea lion. *Abstracts of the 7th Biennial Conference on the Biology of Marine Mammals*, 10.

Geraci, J. R. (1971). Functional hematology of the harp seal *Pagophilus groenlandicus*. *Physiological Zoology*, **44**, 162–70.

Geraci, J. R. & Smith, T. G. (1975). Functional hematology of ringed seals (*Phoca hispida*) in the Canadian arctic. *Journal of the Fisheries Research Board of Canada*, **32**, 2559–64.

Gjertz, I., Kovacs, K., Lydersen, C. & Wiig, Ø. (1995). Satellite tracking of bearded seals from Svalbard. *Abstracts of the 11th Biennial Conference on the Biology of Marine Mammals*, 44.

Greenwood, A. G., Ridgway, S. H. & Harrison, R. J. (1971). Blood values in young gray seals. *Journal of the American Veterinary Association*, **159**, 571–4.

Harcourt, R. G., Schulman, A. M., Davis, L. S. & Trillmich, F. (1995). Summer foraging by lactating female New Zealand fur seals (*Arctocephalus forsteri*) off Otago Peninsula, New Zealand. *Canadian Journal of Zoology*, **73**, 678–90.

Harrison, R. J. & Tomlinson, J. D. W. (1956). Observations on the venous system in certain pinnipedia and cetacea. *Proceedings of the Zoological Society London*, **126**, 205–33.

Harrison, R. J. & Tomlinson, J. D. W. (1960). Normal and experimental diving in the common seal (*Phoca vitulina*). *Mammalia*, **24**, 386–99.

Hawkey, C. M. (1975). *Comparative Mammalian Hematology*. London: William Heinemann Medical Books Ltd.

Hedrick, M. S., Duffield, D. A. & Cornell, L. H. (1986). Blood viscosity and optimal hematocrit in a deep-diving mammal, the northern elephant seal (*Mirounga angustirostris*). *Canadian Journal of Zoology*, **64**, 2081–5.

Hill, R. D., Schneider, R. C., Liggins, G. C., Schuette, A. H., Elliott, R. L., Guppy, M., Hochachka, P. W., Qvist, J., Falke, K. J. & Zapol, W. M. (1987). Heart rate and body temperature during free diving of weddell seals. *American Journal of Physiology*, **253**, R344–51.

Hindell, M. A. & Lea, M. (1995). Heart rate and swimming velocity in an adult female southern elephant seal. *Abstracts of the 11th Biennial Conference on the Biology of Marine Mammals*, 54.

Hindell, M. A., Slip, D. J. & Burton, H. R. (1991). The diving behaviour of adult male and female southern elephant seals, *Mirounga leonina* (Pinnipedia: Phocidae). *Australian Journal of Zoology*, **39**, 595–619.

Hurley, J. A. (1996) *Metabolic rate and heart rate during trained dives in adult California sea lions*. PhD Thesis, University of California, Santa Cruz.

Hyvarinen, H., Hamalainen, E. & Kunnasranta, M. (1995). Diving behavior of the Saimaa ringed seal (*Phoca hispida siamensis* Nordq). *Marine Mammal Science*, **11**, 324–34.

Irving, L., Solandt, O. M., Solandt, D. Y., & Fisher, K. C. (1935). Respiratory characteristics of the blood of the seal. *Journal of Cellular and Comparative Physiology*, **6**, 393–403.

Irving, L., Scholander, P. F. & Grinnell, S. W. (1941). Significance of the heart rate to the diving ability of seals. *Journal of Cellular and Comparative Physiology*, **18**, 283–97.

Irving, L., Peyton, L. J., Bahn, C. H. & Peterson, R. S. (1963). Action of the heart and breathing during the development of fur seals (*Callorhinus ursinus*). *Physiology Zoology*, **36**, 1–20.

Kandinskii, P. A. (1974). Viscosity, specific gravity, and hematocrit values of blood in Antarctic pinnipeds. *Proceedings of the Pacific Ocean Research Institute of Fisheries and Oceanography (TINRO)*, **87**, 216–18.

Kelly, B. P. & Wartzok, D. (1996). Ringed seal diving behavior in the breeding season. *Canadian Journal of Zoology*, **74**, 1547–55.

Kooyman, G. L., Gentry, R. L. & Urquhart, D. L. (1976). Northern fur seal diving behavior: a new approach to its study. *Science*, **193**, 411–12.

Kooyman, G. L., & Gentry, R. L. (1986). Diving behavior of South African Fur Seals. In *Fur Seals: Maternal Strategies on Land and at Sea*, ed. R. L. Gentry & G. L. Kooyman, pp. 142–52. Princeton: Princeton University Press.

Kooyman, G. L. & Trillmich, F. (1986*a*). Diving behavior of Galapagos Fur Seals. In *Fur Seals: Maternal Strategies on Land and at Sea*, ed. R. L. Gentry & G. L. Kooyman, pp. 186–95. Princeton: Princeton University Press.

Kooyman, G. L. & Trillmich, F. (1986*b*). Diving behavior of Galapagos Sea Lions. In *Fur Seals: Maternal Strategies on Land and at Sea*, ed. R. L. Gentry & G. L. Kooyman, pp. 209–19. Princeton: Princeton University Press.

Koskela, J. T., Hyvarinen, H. & Kunnasranta, M. (1995). Movements and habitat use of radio-tagged Saimaa ringed seals. *Abstracts of the 11th Biennial Conference on the Biology of Marine Mammals*, 63.

Kosygin, G. M. & Gol'tsev, V. N. (1971). Data on the morphology and ecology of the harbour seal of the tartar strait. *Transactions Series of the Canadian Fisheries Marine Services*, **3185**, 384–410.

Lane, R. A. B., Morris, R. J. H. & Sheedy, J. W. (1972). A haematological study of the southern elephant seal, *Mirounga leonina* (Linn.). *Comparative Biochemistry and Physiology*, **42A**, 841–50.

Le Boeuf, B. J., Costa, D. P., Huntley, A. C. & Feldkamp, S. D. (1988). Continuous, deep diving in female northern elephant seals, *Mirounga angustirostris*. *Canadian Journal of Zoology*, **66**, 446–58.

Le Boeuf, B. J., Naito, Y., Huntley, A. C. & Asaga, T. (1989). Prolonged, continuous, deep diving by northern elephant seals. *Canadian Journal of Zoology*, **67**, 2514–19.

Lenfant, C. (1969). Physiological properties of blood of marine mammals. In *The Biology of Marine Mammals*, ed. H. T. Andersen, pp. 95–116. New York: Academic Press.

Lenfant, C., Elsner, R., Kooyman, G. L. & Drabek, C. M. (1969). Respiratory function of blood of the adult and fetus Weddell seal *Leponychotes Weddelli*. *American Journal of Physiology*, **216**, 1595–7.

Lenfant, C., Johansen, K. & Torrance, J. D. (1970). Gas transport and oxygen storage capacity in some pinnipeds and the sea otter. *Respiration Physiology*, **9**, 277–86.

Lydersen, C. (1991). Monitoring ringed seal (*Phoca hispida*) activity by means of acoustic telemetry. *Canadian Journal of Zoology*, **69**, 1178–82.

Lydersen, C. & Kovacs, K. M. (1993). Diving behaviour of lactating harp seal, *Phoca groenlandica*, females from the gulf of St. Lawrence, Canada. *Animal Behavior*, **46**, 1213–21.

Mattlin, R. H. (1993). Seasonal behavior of the New Zealand fur seal, *Arctocephalus forsteri*. *Abstracts of the 10th Biennial Conference on the Biology of Marine Mammals*, 74.

McConnell, L. C. & Vaughan, R. W. (1983). Some blood values in captive and free-living common seals (*Phoca vitulina*). *Aquatic Mammals*, **10**, 9–13.

Merrick, R. L. Loughlin, T. R., Antonelis, G. A. & Hill, R. (1994). Use of satellite-linked telemetry to study Steller sea lion and northern fur seal foraging. *Polar Research*, **13**, 105–14.

Needham, D. J., Cargill, C. F. & Sheriff, D. (1980). Haematology of the Australian sea lion, *Neophoca cinerea*. *Journal of Wildlife Diseases*, **16**, 103–7.

Nordøy, E. S., Folkow, L. & Blix, A. S. (1995*a*). Distribution and diving behavior of crabeater seals (*Lobodon carcinophagus*) off Queen Maud Land. *Polar Biology*, **15**, 261–8.

Nordøy, E. S., Folkow, L. P., Potelov, V., Prichtchemikhine, V. & Blix, A. S. (1995b). Distribution and dive behavior of white sea harp seals, between breeding and moulting. *Abstracts of the 11th biennial conference on the biology of marine mammals*, 83.

Odell, D. K. (1981). California Sea Lion, *Zalophus californianus* (Lesson, 1828). In *Handbook of Marine Mammals Volume 1: The Walrus, Seal Lions, Fur Seals, and Sea Otter* ed. S. H. Ridgway & J. H. Harrison, pp. 67–97. London: Academic Press.

Pasche, A., & Krog, J. (1980). Heart rate in resting seals on land and in water. *Comparative Biochemistry and Physiology*, **67A**, 77–83.

Packer, B. S., Altman, M., Cross, C. E., Murdaugh, H. V., Jr., Linta, J. & Robin, E. D. (1969). Adaptations to diving in the harbor seal: oxygen stores and supply. *American Journal of Physiology*, **217**, 903–6.

Payne, M. R. (1979). Growth in the Antarctic fur seal *Arctocephalus gazella*. *Journal of Zoology., London*, **187**, 1–20.

Petrov, E. A., Elagin, O. K. & Shoshenko, K. A. (1987). The volume of circulatory blood in the seal, Pusa sibirica. *Journal of Evolutionary and Biochemical Physiology*, **23**, 390–2.

Qvist, J., Hill, R. D., Schneider, R. C., Falke, K. J., Liggins, G. C., Guppy, M., Elliott, R. L., Hochachka, P. W. & Zapol, W. M. (1986). Hemoglobin concentrations and blood gas tensions of free diving Weddell seals. *Journal of Applied Physiology*, **61**, 1560–9.

Ray, G. C. (1981). Ross seal *Ommatophoca rossii* Gray, 1844. In *Handbook of Marine Mammals Volume 2: Seals*, ed. S. H. Ridgway & J. H. Harrison, pp. 237–60. London: Academic Press.

Reeves, R. R., Stewart, B. S. & Leatherwood, S. (1992). *The Sierra Club Handbook of Seals and Sirenians*. San Francisco: Sierra Club Books.

Ridgway, S. H. (ed.). (1972). *Mammals of the Sea, Biology and Medicine*. Springfield: Charles C. Thomas.

Ronald, K. (1970). Physical blood properties of neonatal and mature harp seals. *International Council for the Exploration of the Sea CM*, **5**, 1–4.

Scheffer, V. B. (1960). Weights of organs and glands in the northern fur seal. *Mammalia*, **24**, 477–81.

Scholander, P. F. (1940). Experimental investigations on the respiratory function in diving mammals and birds. *Hvalradets Skrifter*, **22**, 1–131.

Shustov, A. P. & Yablokov, A. V. (1968). Comparative morphological characteristics of the harp and ribbon seals. *Transactions Series of the Fisheries Research Board of Canada*, **1084**, 1–19.

Simpson, J. G., Gilmartin, W. G. & Ridgway, S. H. (1970). Blood volume and other hematologic values in young elephant seals (*Mirounga angustirostris*). *American Journal of the Veterinary Research*, **31**, 1449–52.

Slijper, E. J. (1958). Organ weights and symmetry problems in porpoises and seals. *Arch Neer. Zool.*, **13**: 97–113.

Sokolov, A. S. (1965). Ecologo-functional and age characteristics of the blood of north Pacific pinnipeds. *Doklady Akademii Nauk SSSR*, **169**, 683–4.

Sokolov, A. S., Kozygin, G. M. & Tikhomirov, E. A. (1966). Information on the weight of the internal organs of the Bering sea pinnipeds. *Bulletin of the Pacific Scientific Institute of Fisheries and Oceanography*, **58**, 137–47.

St. Aubin, D. J., Geraci, J. R., Smith, T. G. & Smith, V. I. (1978). Blood volume determination in the ringed seal, *Phoca hispida. Canadian Journal of Zoology*, **56**, 1885–7.

Stenson, G. B., Hammill, M. O., Fedak, M. A. & McConnell, B. J. (1993). The diving behaviour and seasonal migration of adult hooded seals. *Abstracts of the 10th Biennial Conference on the Biology of Marine Mammals*, 99.

Stewart, B. S., Petrov, E. A., Baranov, E. A., Timonin, A. & Ivanov, M. (1996). Seasonal movements and dive patterns of juvenal Baikal seals, *Phoca sibirica. Marine Mammal Science*, **12**, 528–42.

Thompson, D. & Fedak, M. A. (1993). Cardiac responses of grey seals during diving at sea. *Journal of Experimental Biology*, **174**, 139–54.

Trillmich, F., Kooyman, G. L., Majluf, P. & Sanchez-Grinan, M. (1986). Attendance and diving behavior of South American Fur Seals during el Nino in 1983. In *Fur Seals: Maternal Strategies on Land and at Sea*, ed. R. L. Gentry & G. L. Kooyman, pp. 153–67. Princeton: Princeton University Press.

Vaz-Ferriera (1981). South American Sea Lion *Otaria flavescens* (Shaw, 1800). In *Handbook of Marine Mammals Volume 1: the Walrus, Sea*

Lions, Fur Seals, and Sea Otter, ed. S. H. Ridgway & J. H. Harrison, pp. 39–65. London: Academic Press.

Wells, R. M. G. (1978). Observations on the haematology and oxygen transport of the New Zealand fur seal, *Arctocephalus forsteri*. *New Zealand Journal of Zoology*, **5**, 421–4.

Werner, R. & Campagna, C. (1995). Diving behaviour of lactating southern sea lions (*Otaria flavescens*). *Canadian Journal of Zoology*, **73**, 1975–83.

Wiig, Ø., Gjertz, I., Griffiths, D. & Lydersen, C. (1993). Diving patterns of an Atlantic walrus *Odobenus rosmarus rosmarus* near Svalbard. *Polar Biology*, **13**, 71–2.

R.M.BEVAN, C.M.BISHOP and P.J.BUTLER

The physiology of polar birds

Polar birds have always held a fascination for man, e.g. the movements of birds to and from the Arctic have heralded the start of spring and the end of summer, respectively, for many northern cultures. Yet, it is only within recent years that research has started to unfold the complex interaction between the physiology of the birds and the polar environment. The early studies concentrated on the breeding behaviour and ecology of the birds. Increasingly, researchers are investigating how the physiology of the birds has evolved and how it enables them to survive the polar environment. Coupled with this, the advent of new technology has meant that the physiology of many species can now be studied remotely in the wild.

Few bird species can survive for any length of time in the high latitude polar regions. Vuilleumier (1996) reported that up to 29 different species have been sighted between the high Arctic Ocean and the North Pole over a number of years. However, the number of species declined with increasing latitude, as did their absolute numbers. The most common species north of 83° N were the kittiwake (*Rissa tridactyla*), Ross's gull (*Rhodostethia rosea*), the ivory gull (*Pagophila eburnea*) and the fulmar (*Fulmarus glacialis*). In the Northern Hemisphere, breeding sites for birds are restricted to circumpolar land masses that extend typically up to around 80° N, of which the best known are the Canadian, Alaskan, Scandinavian and Russian Arctic areas and the islands of the North Atlantic Ocean. In the southern hemisphere, there are even fewer species, with no resident land birds at all on the continent of Antarctica. South of the polar front there are 38 species of breeding birds including seven species of penguins, four albatrosses and 20 smaller petrels. The two most prominent species on continental Antarctica are the emperor penguin *Aptenodytes forsteri* and the Adélie penguin *Pygoscelis adeliae* (Stonehouse, 1989).

In this chapter, it is not possible to provide a comprehensive review of the current state of knowledge regarding the physiology of all the bird species that live in these areas. The intention, therefore, is to focus on the body mass changes, growth, field energetics, cardiovascular physiology and thermoregulation of a few representative examples. The polar regions present a challenging environment for any animal attempting to live or breed there. First, there is the severe cold for which the regions are renowned and second, there are the changes in photoperiod which, above the polar circles, can vary between 24 h of daylight and 24 h of darkness. How birds have evolved to cope with these factors will form the basis of this review.

Regulation of body mass

In general, birds are well adapted to exploit the temporary abundance of food and good conditions present in the polar regions during the summer months as they can fly to more hospitable climes during the autumn and return again in the spring. Thus, the ability to fly long distances and to migrate successfully to suitable winter habitats is a dominant feature among birds breeding at high latitudes.

Not all birds migrate to avoid the harsh winter climates in the polar regions. Some polar species, such as the Svalbard ptarmigan (*Lagopus mutus*), remain throughout the winter months. The biology of the ptarmigan illustrates some important environmental factors influencing the physiology of birds living in Arctic regions. During the short breeding season, parents and offspring are exposed, for much of the time, to continuous daylight, with rapid changes in daylength at the beginning and end of the season. These changes in daylength can have a direct effect on the birds' behaviour and physiology. For example, body mass regulation during the year is the result of a complex interaction of photoperiod, ambient temperature, activity levels, food intake and reproductive condition (Stokkan, Lingard & Reierth, 1995; Lindgard, Nasland & Stokken, 1995).

During the Autumn, ptarmigans increase their food intake and fatten for the winter, but body mass begins to decline again during the early winter months as the temperature falls. This decline in winter body mass can be prevented by maintaining ptarmigan at higher than ambient temperatures, although body mass eventually decreases on exposure to the lengthening days of the approaching spring. Those birds kept at a higher than ambient temperature and short daylengths did not show a decline in body mass the following spring (Stokkan *et al.*, 1995). A confounding variable in this study was the effect of photoperiod on the activity cycles of the birds. The autumnal increase in body mass was associated with both hyperphagia and a very low level of locomotor activity, while exposure to continuous daylight resulted in a lower level of fattening and a higher, though intermediate, level of activity (Lindgard *et al.*, 1995). During the following spring, exposure to increasing daylengths resulted in a four-fold increase in activity levels that was correlated with a reduction in body mass. This contrasts with the situation in birds maintained on continuously long days that maintained a moderate amount of activity. Overlaying these energetic responses to either changes in activity levels or to changes in thermoregulatory requirements was evidence that there was a sliding set point for body mass that was affected by the change in photoperiod during the spring. When ptarmigans, exposed to a 'natural' photoperiod but kept at a higher than ambient temperature, were starved for a period during the winter, their body mass declined but recovered briefly upon refeeding before declining along with that of control birds during spring (Stokkan *et al.*, 1995).

Thus, seasonal regulation of body mass is affected by a number of environmental factors and, in turn, has important influences on many behavioural, ecological and energetic traits (reviewed by Witter & Cuthill, 1993). For example, increased body mass could adversely affect the aerobic flight performance (Hedenstrom & Alerstam, 1992) or take-off performance (Lee *et al.*, 1996) of birds and consequently increase vulnerability to cursorial and/or aerial predation. Cotter and Gratto (1995) illustrated the subtle importance of even relatively small changes in body mass for the mortality of birds. They showed that ptarmigans fitted with radio transmitters that were 3.6% of body mass had a significantly reduced survival rate, compared with the non-significant reduction when carrying radio transmitters representing 2.3% of body mass. However, for relatively sedentary species, such as the Svalbard ptarmigan, storing sufficient fat during the autumn is vital to prevent death by starvation during the winter months.

Generally, during the initial stages of starvation (or phase I), fatty acid concentration rises in the blood and helps to reduce both the rate of glucose oxidation and the utilisation of amino acids (from protein degradation) for gluconeogenesis. After a relatively short period (approximately 24 h), an intermediate stage is reached (phase II). During this stage, the level of protein degradation has stabilised at a relatively low level and fatty acid metabolism is high and includes the production of ketone bodies by the liver further to reduce the requirement for glucose by certain tissues such as the heart. The final stage (phase III) is reached after many days of starvation during which the fatty acid, protein and glucose 'stores' have been severely depleted. Essential proteinatious tissues then start to be catabolised to provide substrates for gluconeogenesis. Death may quickly follow once this stage is reached.

When fat Svalbard ptarmigan were starved for 15 days, fatty acid use gradually increased and protein catabolism initially fell in the classic phase I response to starvation (Lindgard *et al.*, 1992). Subsequently, protein catabolism accounted for around only 9% of total energy demand during phase II, while two individuals eventually showed an increase in protein catabolism as they moved into the potentially fatal phase III of starvation. By contrast, lean Svalbard ptarmigan starved for only five days, showed an immediate increase in the level of protein catabolism (typical of phase III), with protein providing 22% of total energy demand on day two and up to 41% by day five of starvation.

In many species of ducks and geese, the female incubates the eggs almost continuously for between 24 and 30 days without significant amounts of refeeding. The body mass loss during this self imposed starvation can be greater than 30% of initial body mass (Korschgen, 1977; Le Maho *et al.*, 1981; Gabrielsen *et al.*, 1991; Boismenu, Gautheir & Larochelle, 1992). Greater snow geese (*Anser caerulescens atlantica*), which breed in the Canadian Arctic, show this ability to survive prolonged starvation during incubation (Boismenu *et al.*, 1992) and also exhibit the three phases regarding protein catabolism.

In the Antarctic, several species also undergo long periods of food deprivation, probably the most famous being the winter fast of the male emperor penguins. The duration of this fast is approximately 120 days during which the body mass decreases by 41%. However, unlike the geese and Svalbard ptarmigan, free-fatty acids do not increase substantially during the fast and reflects a possible limitation to ketogenesis (Groscolas, 1986). It may be that the decrease in plasma free-fatty acid and β-hydoxybutyrate seen during prolonged fasting provide the cues for the bird to depart to sea (Groscolas, 1986).

Of course, incubating birds can reduce their field metabolic rate (FMR) to a minimum through excellent body insulation and a reduction in locomotor activity (Stokkan, 1992). Even the overwintering Svalbard ptarmigan has a relatively low FMR, and fasting emperor penguins can reduce their energy requirements through huddling (Ancel *et al.*, 1997). By contrast, birds undertaking migratory flights, must maintain extremely high levels of energy expenditure. However, the underlying requirement, to maximise fatty acid metabolism and minimise protein catabolism, may be similar to that for incubation and overwintering birds. Indeed, the relative mass losses are similar between the various behaviours. In all three cases, one of the critical issues is to what degree can protein metabolism be reduced so that phase III of starvation can be delayed for as long as possible. Thus the ability to fatten and survive periods of food deprivation is a common theme among birds living at high latitudes and is employed during various behaviours throughout the year.

Chick growth

The survival strategy of Arctic breeding species that forage primarily on the land, such as wading birds and geese, is dependent on the young birds being able to develop quickly so that they can accompany their parents on the autumnal, southerly migration. An example of such a species is the barnacle goose (*Branta leucopsis*). This species undertakes an annual migration, of approximately 2500 km, between the breeding areas on the Svalbard archipelago and the overwintering sites on the south west coast of Scotland (Owen & Gullestad, 1984). Due to the relatively short Arctic summers, the goslings have only around 12 weeks after hatching in which to mature and prepare for this long flight (Owen & Black, 1989). Bishop *et al.* (1995, 1996, 1997) studied the biochemical and morphological development of the locomotor and cardiac muscles of these goslings, from 1 week of age through to the moment the birds left Svalbard on the migration. These results were compared with those from wild adults and captive geese.

Biochemical development

Mass-specific activity of four different mitochondrial enzymes, including citrate synthase (CS) and 3-hydroxyacyl-CoA dehydrogenase (HAD) showed broadly similar changes in activity during development (Bishop *et al.*, 1995). Throughout most of development, the cardiac muscle had the highest relative capacity for aerobic metabolic flux (as indicated by the activity of CS; Bishop *et al.*, 1995) and fatty acid oxidation (as indicated by the activity of HAD, Fig. 1a).The activity of CS and HAD in the pectoralis muscle was initially very low, but increased very rapidly between five and seven weeks of age (when the birds first begin to fly) and reached a peak in the pre-migratory birds. The development of various glycolytic enzymes, lactate dehydrogenase (LDH, Fig. 1b), hexokinase (HEX, Fig. 1c) and glycogen phosphorylase (G.PHOS, Fig. 1d), was more tissue-specific compared to that of the mitochondrial enzymes (Bishop *et al.*, 1995). Development of anaerobic capacity (i.e. activity of LDH) and ability to utilise intramuscular glycogen stores (as indicated by activity of G.PHOS) in the pectoralis muscle shows a similar exponential increase in activity as that for CS and HAD, but values for glucose utilisation (i.e. activity of HEX) remain very low at all ages. Overall, the results suggest that the relative use of circulating plasma glucose by the pectoralis muscle during flight is low. The intramuscular stores of glycogen are of primary importance during short-burst activity but these are only of intermediate importance during longer-term activity. Fatty acid oxidation is the main fuel during long-distance flight, as is typical for all long-distance migrants studied (Blem, 1976). In general, moderate activity levels were detected in the semimembranosus muscle for all the enzymes measured (Bishop *et al.*, 1995), suggesting that the leg muscles are capable of utilising circulating glucose, intra-cellular glycogen and fatty acids in almost equal amounts.

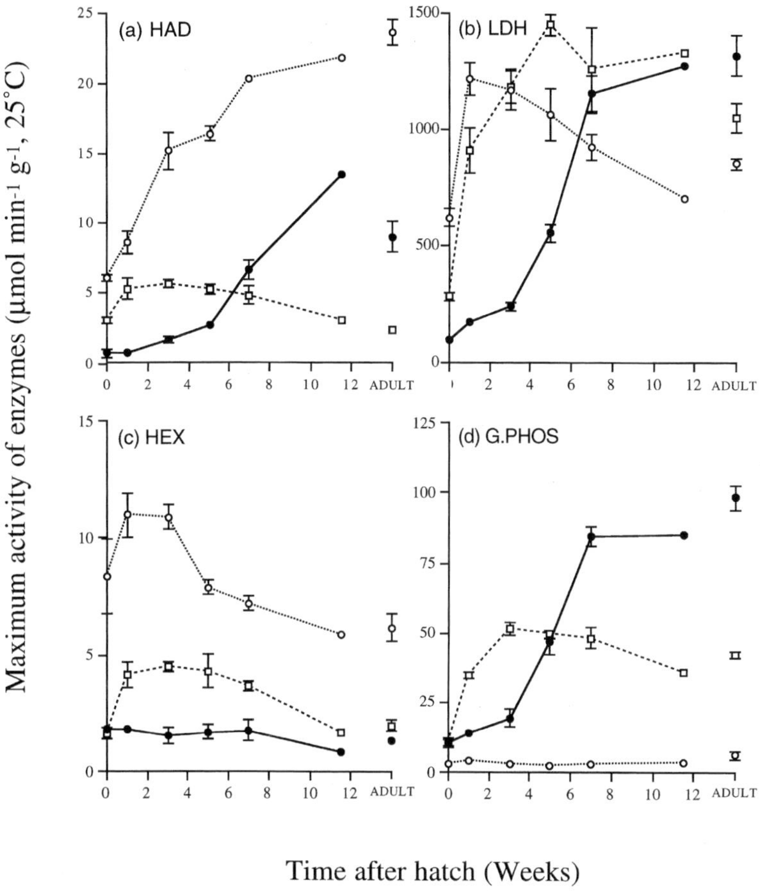

Fig. 1. Mean ± SE of maximum activities of aerobic and glycolytic enzymes (μmol min^{-1} g^{-1} wet wt, 25 °C) in pectoralis (filled dots, solid line), heart (open dots, dotted line) and semimembranosus (open squares, dashed line) muscles plotted against age of wild barnacle geese. (a) 3-hydroxyacyl-CoA dehydrogenase (HAD), (b) Lactate dehydrogenase (LDH), (c) Hexokinase (HEX) and (d) Glycogen phosphorylase (G.PHOS). n=2 to 8, except for a single gosling at 11.5 weeks of age. Adult geese and 11.5 week old gosling are pre-migratory (see text).

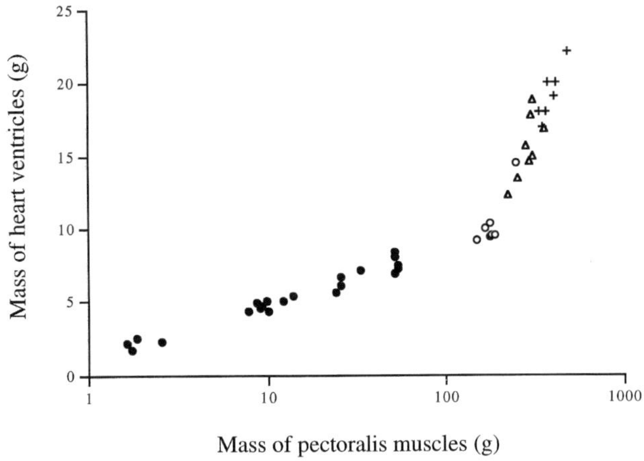

Fig. 2. Scatter diagram of the masses (g) of the heart ventricles plotted against the mass of the pectoralis muscles of wild barnacle geese. Goslings of 1–5 weeks of age (filled dots), fledgling goslings of 7 weeks of age (open dots), adult geese captured at 7 weeks post-hatch (time elapsed since population hatch date; open triangles) and pre-migratory adult geese (crosses).

However, the cardiac muscle had extremely low levels of G.PHOS indicating that it can only utilise the aerobic pathways and that the metabolic fuel must be primarily supplied by the blood. LDH values were relatively high in the heart, while pyruvate kinase levels were very low (Bishop *et al.*, 1995), indicating that the heart is capable of using circulating lactate as a fuel.

Morphological development

Bishop *et al.* (1996) studied the morphological development of the barnacle geese. They showed that the pectoralis, cardiac and leg muscles express different patterns of development to each other both before and after fledging. In goslings between 1 and 5 weeks of age, the growth of the ventricles of the heart is almost linear, as is the growth of the leg muscles, whereas the flight muscles show an exponential pattern of development. When the goslings first start flying, the leg muscles stop growing, while the heart and flight muscles show an accelerated growth pattern. During this period, there is a complete transition in the physiological control of the patterns of growth of these muscles. Figure 2 shows that the masses of the ventricular and pectoralis muscles are related semi-logarithmically to each other throughout both the pre- and post-flight development.

Control factors

Bishop *et al.* (1997) compared the development of a captive, sedentary population of barnacle geese (kept in an outside aviary at Birmingham, UK) with that of the wild migratory population from Svalbard. This study investigated to what extent the migratory specialisations of the cardiac and locomotory muscles might be determined by (a) developmental processes, and (b) differences in relative levels of activity. Both pre- and post-fledging development of the goslings were very similar between the two populations. In particular, there was no significant difference in the slope of the regressions relating pre-migratory increases in the masses of the pectoralis muscles to changes in body mass, for both wild and captive geese (Fig. 3(a) and (b)). There was, however, a small (90–95%), but significant ($P<0.001$), reduction in the average mass of the pectoralis muscles of captive adults compared to wild adults, during this period. Thus, the flight-restricted, captive birds still exhibited a flight muscle hypertrophy (Bishop *et al.*, 1997). This is in contrast to the conclusion of Marsh and Storer (1981) who suggested that the correlation between flight muscle mass and body mass in Cooper's hawk (*Accipiter cooperii*) is a natural analogue of 'power' training during flight. The results of the barnacle goose study suggest a couple of possible mechanisms. First, that the pectoralis muscle mass may be determined predominantly by endogenous processes during pre-migratory fattening. Secondly, that the flight muscles of the captive, flight-restricted geese may still be able to detect and respond appropriately to the increase in wing loading during this period. Perhaps the infrequent bouts of wing-flapping, and occasional take-offs, exhibited by captive birds provided a sufficient mechanistic link by which the birds could maintain the relationship between the mass of the flight muscles and the body mass.

Bishop *et al.* (1997) also measured the mass-specific aerobic capacity of the flight muscles of barnacle geese by measuring the maximum activity of CS. Figures 3c and 3d show the change in activity of CS in the pectoral muscles of the same birds that the data on pectoralis muscle mass were obtained. There was little difference between the two populations up to seven weeks of age. Subsequently, captive goslings older than around 11 weeks post-hatch and captive adults had lower levels of CS than those of wild, pre-migratory geese. Thus, the activity of CS in the pectoralis muscle of adult, captive geese, and captive goslings over 11 weeks of age, is only around 60% of that measured in wild geese. In addition, CS activity in the pectoralis muscles of a group of post-moulting adult geese were similar to those of long-term captive birds. Thus, it is suggested that the rise in the activity of CS in the wild pre-migratory birds may be a reaction to the increase in flight activity *per se*, while the rise in CS activity during the development of the goslings may be primarily under endogenous control.

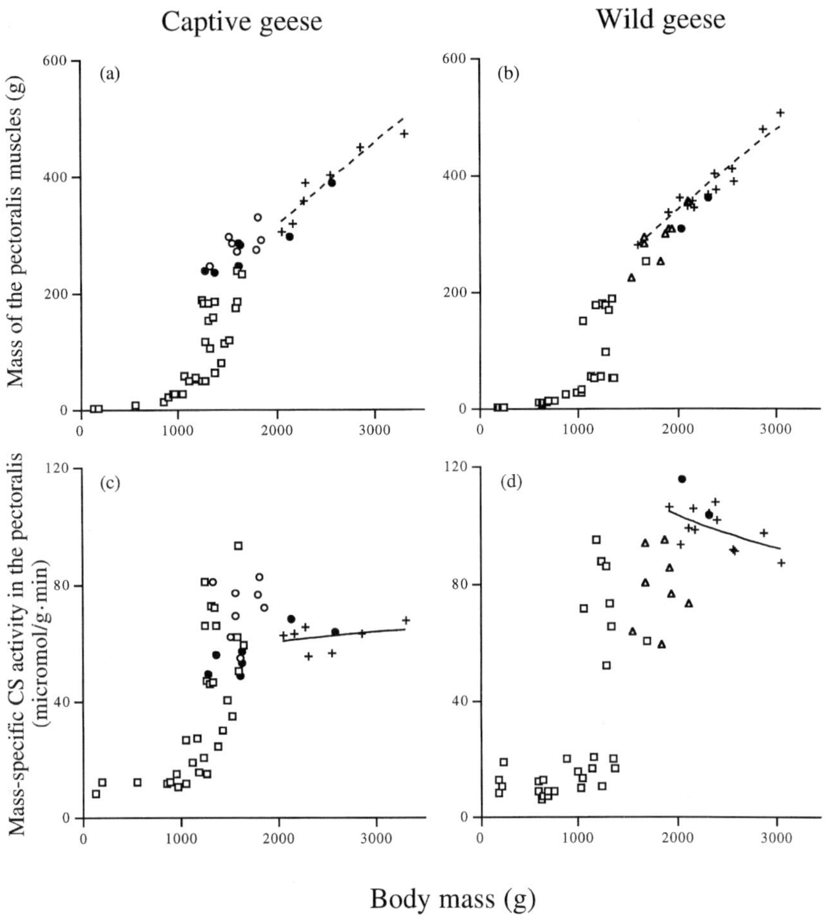

Fig. 3. Scatter diagram of the masses of the pectoralis muscles (g) and the mass-specific citrate synthase (CS) activity in the pectoralis muscles (μmol g^{-1} min^{-1}) of (a and c) captive and (b & d) wild barnacle geese, plotted against body mass (g). Goslings of 1–7 weeks of age (open squares), goslings of 12–20 weeks of age (filled circles), adults at 7 weeks post-hatch (open triangle) and pre-migratory adults at >10 weeks post-hatch (crosses).

Energy expenditure

Effect of latitude

The resting metabolic rate (RMR) of eight species of Arctic breeding seabirds, ranging from the little auk (0.165 kg) to the eider duck (1.6 kg), have been studied by Gabrielsen (1994). Field metabolic rate (FMR) was also measured in four of these species. The results confirmed the general view that seabirds breeding at high latitudes (either northern or southern) have a higher RMR than that of seabirds at lower latitudes. However, a component of this increase may be a result of the continuous day lengths experienced by these birds during the breeding season. RMRs of birds are usually lower at night than during the day (Aschoff & Pohl, 1970), while Gabrielsen (1994) found that the RMR of Arctic seabirds showed no diurnal rhythm when exposed to constant light. However, the fulmar (*Fulmarus gracilis*) had a relatively low RMR that is typical of Antarctic procellariiformes (Adams & Brown, 1984). The fulmar also had a significantly lower body temperature than other Arctic breeding seabirds, in keeping with its phylogeny (Gabrielsen, Mehlum & Karlsen, 1988).

The FMR towards the end of the chick-rearing period consists of the costs of a combination of behavioural and physiological activities. These include locomotion, the specific dynamic action of feeding and thermoregulation. It is to be expected that FMR will be at its highest during this period of the breeding season. The ratio of FMR/RMR for four species of Arctic seabirds (Gabrielsen, 1994) was 2.6 (kittiwake), 3.0 (black guillemot), 3.8 (common guillemot) and 3.6 (little auk). The exact ratio is largely a reflection of the costs of foraging for the different species. For example kittiwakes forage over a large distance but have a relatively low cost of transport due to their high aspect ratio wings and low wing loading. Black guillemots, on the other hand, forage over short distances but have a relatively high cost of flight and must expend additional energy during diving. The FMR/RMR ratios measured by Gabrielsen (1994) do not exceed the proposed maximum sustainable FMR value of 4 times the RMR (Drent & Daan, 1980; Drent & Klaasen, 1989). The structural argument for the association of a high RMR with a high FMR is based on the assumption that the muscular and cardiovascular adaptations required to provide a higher level of sustainable power output may be more costly to maintain even at rest. The functional argument is based on the idea that high levels of foraging activity will 'carry-over' into high levels of RMR during the recovery period. However, neither of these hypotheses justifies a 'fixed' ratio between FMR and RMR. Indeed, Gabrielsen (1994) showed that the FMR/RMR ratio of common guillemots can vary on a daily basis between 1.7 and 6.8.

Effect of locomotion

Walking

The penguins have evolved a body shape that is ideal for an aquatic existence, yet they can still walk or toboggan more than 100 km inland. Several studies have investigated the energy cost of walking in the penguin (Pinshow, Fedak & Schmidt-Nielsen, 1977; Le Maho & Dewasmes, 1984; Bevan *et al.*, 1995*a*). These studies have revealed that it is a very inefficient mode of transport. The inefficiency is primarily due to the kinetic energy changes associated with a short leg length that requires a higher stride frequency than that seen in more cursorially adapted birds (Pinshow *et al.*, 1977). The specific cost of transport of the emperor penguin (*Aptenodytes forsteri*) was found to peak when birds weighed 23 kg. This is the approximate weight of the penguins when they leave the colony to trek back to the sea (Le Maho & Dewasmes, 1984). At lower body mass, their cost of transport reduces so while they lose mass during the trek, their cost of transport will also decrease. Penguins have one other mode of terrestrial travel: tobogganing. This may be energetically less demanding but at the cost of damage to the feathers (Wilson *et al.*, 1991). Adélie penguins (*Pygoscelis adeliae*) therefore balance the lower energy costs of tobogganing with the time and energy expended on plumage care (Wilson *et al.*, 1991).

The albatrosses also have a 'waddling' gait when walking. Only one study to date has measured the energy cost of walking in this group of birds (Bevan *et al.*, 1994). These authors found that any form of leg exercise in the black-browed albatross (*Diomedea melanophrys*) caused a substantial increase in energy expenditure. Simply standing caused the oxygen consumption to increase four-fold above the resting level. The high rate of oxygen consumption by the leg muscles when the bird is walking is probably an adaptation to the feeding behaviour of black-browed albatrosses at sea rather than to any land-based activity. At sea, the feeding black-browed albatross regularly dives, and also engages in feeding frenzies where the birds perform a lot of swimming. Both activities are foot propelled and energetically expensive.

Swimming and diving

For birds using the cold polar seas, there will be an increased energy cost due to the high heat capacity and conductance of water. Simply sitting on the water invariably causes an elevation in metabolic rate (Barré & Roussel, 1986; Jenssen, Ekker & Bech, 1989; Bevan & Butler, 1992*a*; Bevan *et al.* 1994, 1997). Even the gentoo (*Pygoscelis papua*) and king penguins (*Aptenodytes patagonica*) with their adaptations to an aquatic existence have a nearly twofold elevation in their metabolic rates when resting in water (Bevan *et al.*, 1995*a*; Culik *et al.*, 1996; Butler, Bevan & Woakes, 1997). However, penguins are morphologically adapted for locomotion underwater and several studies have examined the cost, in terms of energy expenditure, of swimming in the penguins (e.g. Hui, 1988; Kooyman & Ponganis 1994; Culik, Wilson & Bannasch, 1994; Bevan *et al.*, 1995*a*; Culik *et al.*, 1996). Table 1 provides data on several species of penguin. Obtaining empirical data on the energy cost of swimming in penguins is vital for estimating the energy expenditure of these birds while diving. If the energy expended during diving and the usable oxygen stores within the body of the bird are known, it is possible to determine the proportion of the energy, if any, that is provided by anaerobic metabolism. At present there is insufficient accurate information on either of these to assess accurately the contribution that anaerobiosis makes.

It must be remembered that all the studies mentioned (except that of Hui, 1988) made measurements on penguins in static or dynamic water channels. When at sea, many other factors could affect the bird's energetic needs. For example, the air trapped between the feathers and within the extensive respiratory system can become compressed when diving to depth, thus reducing the buoyancy and the energy needed to descend. Another factor that could affect the metabolic rate is regional body temperature and this is discussed later.

Effect of foraging

The common guillemot (*Uria aalge*) spends 4–10 hours per day foraging at sea, and much of its time must be spent just covering its own requirements (Gabrielsen, 1994; Brekke & Gabrielsen, 1994). In addition, the high wing-loading and the high cost of flight in this species, mean that only around 1.5% body mass can be carried as food for the chick. Thus, only 7% of the total energy gathered by adult common guillemots is provided to the chick (Gabrielsen, 1994). In contrast, the smaller black guillemot (*Cepphus grylle*), which forages over shorter distances and has a lower wing-loading, can supply around 38% of its gathered energy to chick rearing. Thus, while the black guillemot can raise two chicks which fledge at around 100% of adult body mass, the common guillemot can only raise one chick which must leave the nest at less than 30% of adult body mass. Gabrielsen (1994) also shows that little auks (*Alle alle*) are only able to deliver around 15% of their gathered energy to raise a single chick to around 65% of adult mass. It would appear, therefore, that the growth of common guillemot and little auk chicks is constrained by the high foraging costs of the adult birds, and results in both these altricial species having an early fledging strategy. Once fledged, subsequent survival of most Arctic seabirds depends on their ability to forage and avoid predation while spending most of their time at sea.

A detailed knowledge of the energy expenditure of seabirds is thus essential if we are to understand their role in the nutrient fluxes within the marine system and also the limitations that foraging methods place on the breeding ecology of birds. However, the behaviour of birds foraging at sea is impossible to observe continuously in order to determine their activity and energy

Table 1. *Energy expenditure of swimming penguins*

Species	Mass (kg)	T_W (°C)[e]	Rest in water[f] ($W\ kg^{-1}$)	min COT speed[g] ($m\ s^{-1}$)	min COT[h]	
					($J\ kg^{-1}\ m^{-1}$)	($W\ kg^{-1}$)
HP[a]	3.8	19.2	7.20	0.5 – 1.25	13.4	6.7 – 16.8
CP[b]	3.8	4	8.75	2.4[i]	3.7	8.9
AP[b]	4.0	4	8.40	2.2[i]	4.9	10.8
GP[b]	5.5	4	8.20	1.8[i]	7.6	13.7
KP[c]	11.5	9.1	4.65	2.2	4.7	10.3
EP[d]	20.8	3.3	2.19	–	–	–

Notes:

Species code: HP=Humboldt penguin, AP=Adélie penguin, CP=chinstrap penguin (*P.antarctica*), GP=gentoo penguin, KP=king penguin, EP=emperor penguin.

References: [a] Hui (1988), [b] Culik *et al.* (1994), [c] Culik *et al* (1996), [d] Kooyman & Ponganis (1994).

[e] Water temperature.

[f] Energy expended while resting in water.

[g] swim speed at which the minimum cost of transport (COT) occurs.

[h] minimum COT.

[i] preferred swimming speed in the wild.

budgets. Several studies have investigated the field metabolic rate of polar birds both in the Arctic and the Antarctic. The doubly labelled water (DLW) technique has been used extensively to determine the energy budgets of sea-birds (Adams, Brown & Nagy, 1986; Costa & Prince, 1987; Obst, Nagy & Ricklefs, 1987; Birt-Friesen *et al.* 1989; Davis, Croxall & O'Connell, 1989; Nagy, Siegfried & Wilson, 1984; Gabrielsen, 1994). However, the technique has its practical limitations. It can only be used over relatively short periods and is, as a consequence, restricted to specific phases of the reproductive cycle (Costa & Prince, 1987). The technique also provides only a mean rate of energy expenditure over the entire monitoring period (for a review of the technique, see Tatner & Bryant, 1989). A number of studies have tried to solve these problems by monitoring the time spent at sea and the time spent at the nest to partition the energy usage (Adams *et al.*, 1986; Davis *et al.*, 1989; Kooyman *et al.* 1992). Other studies have used the proportion of time a bird spends on the water (Prince & Francis, 1984) to determine the costs of specific behaviours such as flight (Adams *et al.*, 1986; Costa & Prince, 1987). However, to understand fully the complexities of the energy budgets of pelagic sea-birds, a method is required that can both measure the energy expenditure over long periods and be used at a much finer time resolution. One such method is the use of heart rate (Butler, 1993).

Recording the heart rate and body temperature of free-ranging birds has implications not only for the study of the cardiovascular physiology of birds and how it varies with behaviour, but also for the study of the bird's energy expenditure. The latter is possible, as heart rate is related to oxygen consumption (and therefore to energy expenditure). This relationship is described by the Fick equation: oxygen consumption = heart rate $\times$ cardiac stroke volume $\times$ tissue oxygen extraction. If stroke volume and tissue oxygen extraction remain constant or vary systematically then the heart rate of a bird will be directly related to its oxygen consumption. Heart rate can therefore be used to estimate the energy expenditure of a bird provided that it is first calibrated (Butler, 1993). Calibrations of heart rate against oxygen consumption have been made for many birds (e.g. Culik *et al.*, 1990; Nolet *et al.*, 1992; Bevan *et al.*, 1994, 1995*a*; Butler *et al.*, 1997) and all have shown a describable relationship between the two variables. Validation experiments in a number of birds have shown the heart rate technique to be as robust as DLW for estimating metabolic rate (Nolet *et al.*, 1992; Bevan *et al.*, 1994, 1995*a*).

A study on the black-browed albatross illustrates the use of heart rate for determining the energy expenditure of free-ranging birds (Bevan *et al.*, 1994, 1995*b*). Eighteen black-browed albatrosses were implanted with heart rate data loggers (Woakes, Butler & Bevan, 1995). Five birds were also equipped with salt-water switches (to determine when and for how long the birds spent on the water) and another two birds with satellite transmitters (Fig. 4a and

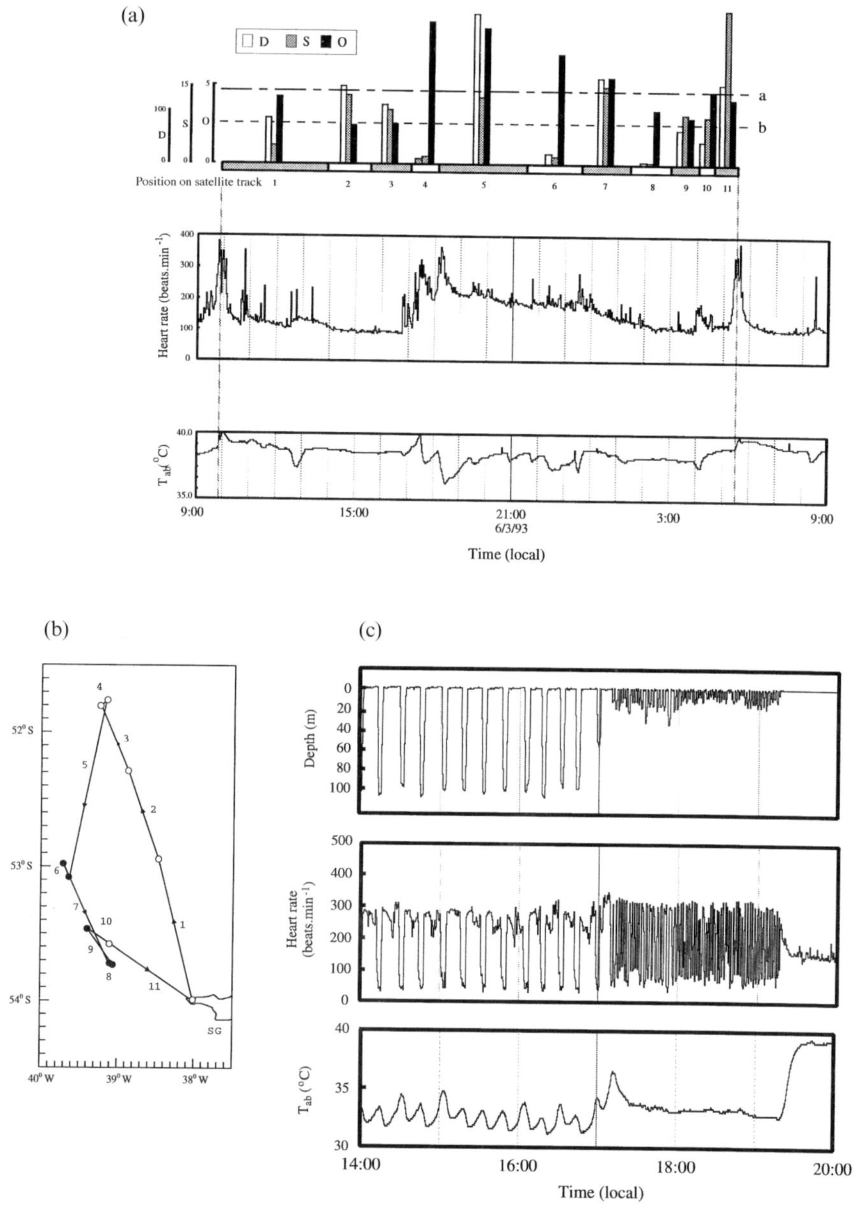

Fig. 4. Data obtained from a free-ranging black-browed albatross and South Georgian shag. (a) Heart rate (middle trace) and abdominal temperature (bottom trace) obtained from a black-browed albatross on a single foraging trip. Upper trace represents estimates of distance travelled (D, km),

b). The energy expenditure during incubation was 2.22 W kg^{-1} and was not significantly different, at 2.42 W kg^{-1}, during brooding. When foraging at sea during these phases, the energy expended by the birds increased to 4.72 W kg^{-1} and 4.63 W kg^{-1}, respectively. The overall rates of energy expenditure during incubation and brooding (3.63 W kg^{-1} and 3.54 W kg^{-1}, respectively) were significantly lower than the costs during the chick-rearing phase (5.80 W kg^{-1}). Using the data from the salt-water switches, it was possible to estimate the energy expended by the birds during flight (6.21 W kg^{-1}) and when on the water (5.77 W kg^{-1}). These rates were not significantly different. The cost of flight was, however, considerably lower than had been estimated by doubly labelled water (Costa & Prince, 1987). The mean cost of flight was only two times the basal metabolic rate and it was clear from the birds with satellite transmitters, that the energy expended during flight could be even lower (Fig. 4b).

Cardiovascular physiology

Most birds in the Antarctic and sub-Antarctic use the sea to obtain their food and employ various foraging methods with which to exploit this resource. For instance, both the wandering albatross (*Diomedea exulans*) and Wilson's storm petrel (*Oceanides oceanus*) forage at the sea surface. However, the most common foraging behaviour used by birds in the polar region is diving, e.g. the penguins, the shags, the diving petrels and even the mollymawk albatrosses (Prince, Huin & Weimerskirch, 1994). The cardiovascular adjustments to breathhold diving have been well documented (for a review see Butler & Jones, 1997) so this chapter will concentrate on those concerning free-ranging Antarctic birds.

Forced submergence causes a profound bradycardia, a reduction in cardiac output, a peripheral vasoconstriction and an increase in the energy supplied by anaerobic metabolism (Irving, 1939). This is known as the 'classic' dive-response. The bradycardia is taken to be indicative that the other adjustments are also occurring. In 1973, a pioneering study by Millard, Johansen and Milsom measured blood flows in gentoo and Adélie penguins both in the laboratory and when the birds were freely-diving. From their results they deduced that, in the penguin, the cardiovascular adjustments

speed (S, m s^{-1}) and energy expenditure (O, W kg^{-1}) over each leg of the flight path. The horizontal lines a and b represent the mean ground speed of flying black-browed albatrosses and the mean rate of energy expenditure of incubating birds, respectively. (From Bevan *et al.*, 1995*b*.) (b) Flight path of the same bird using satellite telemetry. (c) Traces of depth (top trace), heart rate (middle trace) and abdominal temperature (bottom trace). (Adapted from Bevan *et al.*, 1997.)

were 'a composite of diving and exercise response' (Millard *et al.*, 1973). This has since been seen to be true for other species of diving birds (Bevan & Butler, 1992*b*). However, although not being forced to dive, the penguins studied by Millard *et al.* (1973) were still restrained to a given area. A data logger has recently been developed to record heart rates from free-ranging birds (Woakes *et al.*, 1995). Heart rate is chosen for a number of reasons (i) it is relatively easy to monitor, (ii) it can provide an insight into other cardiovascular adjustments that may be occurring, e.g. by monitoring the heart rate during diving, it may be possible to infer whether any changes are occurring in the peripheral blood flow, (iii) it can be used to determine the metabolic rate of a bird if calibrations are first made between heart rate and energy expenditure (see p. 445) and (iv) it can be used to determine the psychological state of birds (Jungius & Hirsch, 1979).

Heart rate data loggers have been deployed in free-ranging gentoo, king and emperor penguins, the black-browed albatross and the South Georgian shag *Phalacrocorax georgianus* (Kooyman *et al.*, 1992; Bevan *et al.*, 1997; Butler *et al.*, 1997). Figure 4c shows heart rate and abdominal temperature data obtained from a South Georgian shag using an implanted heart rate and temperature data logger. Behavioural information was provided by a time-depth recorder attached to the feathers of the birds. Table 2 provides data on the heart rates during diving from a number of different studies. The resting heart rates of birds, taken while at the nest, were 99.4 ± 10.5 beats min^{-1}, 76.3 ± 4.4 beats min^{-1} and 104.0 ± 13.1 beats min^{-1} for the gentoo penguin, king penguin and South Georgian shag, respectively. During diving the mean minimum heart rates of these species were 94.4 ± 4.8 beats min^{-1}, 66.2 ± 0.9 beats min^{-1} and 64.8 ± 5.8 beats min^{-1}, respectively. The mean heart rate of the gentoo penguin during diving was 155 ± 9 beats min^{-1} while in the South Georgian shag it was 104 ± 14 beats min^{-1}.

In the South Georgian shag, mean heart rate during diving was not significantly different from the resting measurements, but they are probably lower than the heart rates exhibited by these birds when they are exercising at the surface (cf. Woakes & Butler, 1983). The mean heart rate measured over the complete dive cycle (dive + surface interval) was significantly higher than the resting level, suggesting that these birds have an increased metabolic rate during diving. The minimum heart rate of the South Georgian shag during diving was, however, lower than the resting rate. This lower than resting heart rate can therefore be perceived as a true 'diving bradycardia' which suggests a switch to anaerobic metabolism and/or a lower than resting aerobic metabolism. A gradual reduction in the energy provided by aerobic metabolism is also indicated by the progressive reduction in heart rate during a dive (Fig. 5a). In the gentoo penguin the mean heart rate during diving was elevated above the resting level implying an increased metabolic rate. However, the

Table 2. *Heart rates (beats min^{-1}) obtained from several species of penguin and from the South Georgian shag for different activities*

Species	N[a]	Rest (air)[b]	Rest (water)[c]	Min[d]	Dive[e]	Cycle[f]	Reference
EP*	5	72			63.3		Kooyman *et al.*, 1992
AP	2[g]	67, 77	89, –		107, –		Culik, 1992
AP	?	90			123		Millard *et al.*, 1973
HP	3	121	139		134		Butler & Woakes, 1984
GP*	13	83.8					Nimon, Schroter & Stonehouse, 1995
GP*	8	99		94	155	176	Butler *et al.*, 1997
GP	7	87	144			177	Bevan *et al.*, 1995*a*
KP*	7	76		66			Butler *et al.*, 1997
SGS*	4	104		65	104	213	Bevan *et al.*, 1997

Notes:

Species code: GP=gentoo penguin, EP=emperor penguin, AP=Adélie penguin, HP=Humboldt penguin, KP=king penguin, SGS=South Georgian shag.

* denotes heart rates recorded from free-ranging birds.

[a] The number of individual birds recorded from.

[b] Rest (air)=heart rates from birds resting in air.

[c] Rest (water)=heart rates from birds resting in water.

[d] Min=minimum heart rate of birds recorded when submerged.

[e] Dive=mean heart rate of birds recorded when submerged.

[f] Cycle=mean heart rate over the dive cycle (dive+surface interval).

[g] Values are from the individual birds.

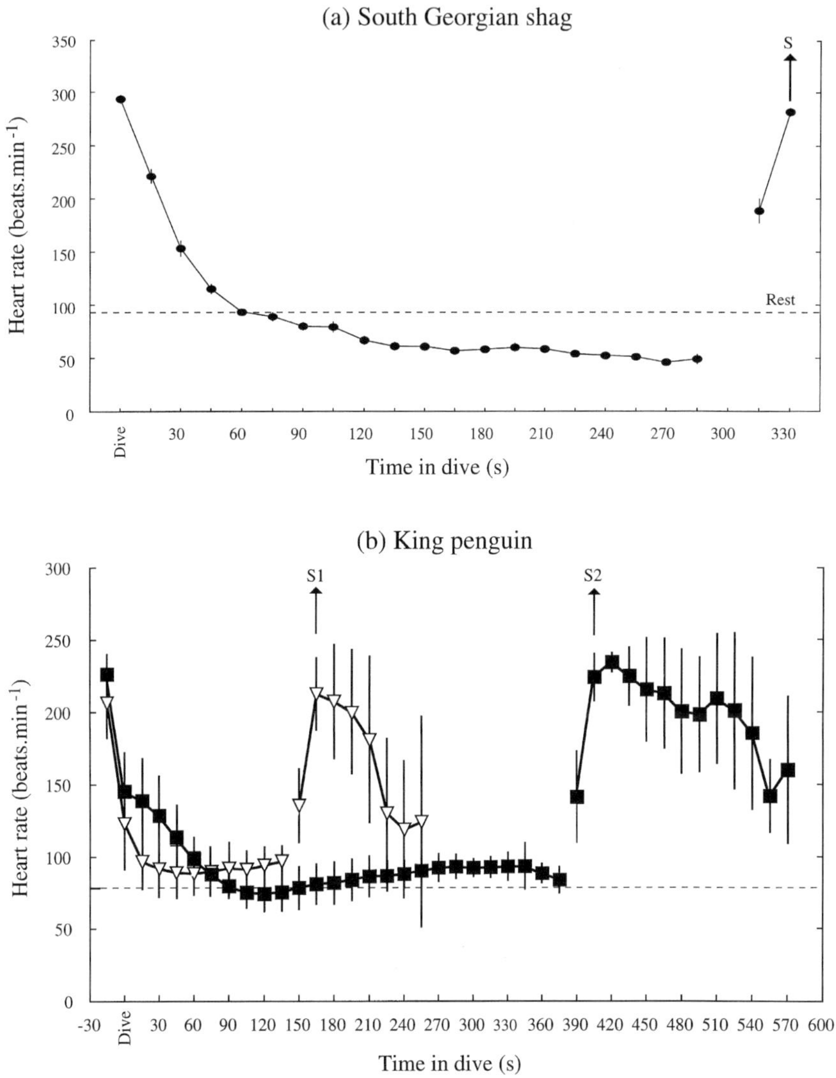

Fig. 5. Heart rate during diving from (a) a male South Georgian shag and (b) a male king penguin. Symbols are the mean heart rate taken as a dive progresses and the vertical lines are the standard error of the mean. When not present, the SE lines lie within the symbol. The dashed horizontal line represents the mean resting heart rate of that bird. (a) S marks the heart rate when the bird surfaces and the point before is the value obtained 15 s before surfacing (Adapted from Bevan *et al* 1997.) (b) ∇ are the mean heart rates from dives <180 s and ■ are the means obtained from dives >180 s. S1 and S2 denote the heart rates obtained when the bird surfaces after dives <180 s and dives >180s, respectively.

rate was lower than when swimming in a water channel (Table 2). This supports the argument of Millard *et al.* (1973) that diving is a balance between an exercise and dive response.

The king penguin dives deeper than the gentoo penguin and for much longer durations. Figure 5b shows the heart rates of a king penguin during diving for short (<180 s) and long (>180 s) periods. Note that heart rate during the longer dives falls at a slower rate but to a lower level than during the shorter ones. Heart rates during submergence in both the long and short dives are at approximately resting levels and suggest that the metabolic rate of these exercising birds is also at the resting level. The slower fall in heart rate over the initial 90 s of the longer dives probably reflects the extra work and time needed to travel to greater depths.

Thermoregulation

Polar birds have had to evolve mechanisms to cope with the rigorous thermal environment in which they live. Some of the earliest physiological experiments on penguins were designed to investigate thermoregulation in these birds. Sladen, Boyd and Pedersen (1966) monitored the body temperature (T_b) of emperor and Adélie penguins via telemetry when the birds were in their rookeries. T_b ranged from 37.6 °C to 39.7 °C in the emperor penguin and from 37.7 °C to 39.6 °C in the Adélie penguin; the latter was monitored for 14 days. Another probe placed in the proventriculus registered a temperature that was 0.3 °C higher than that in the abdomen. The authors also found that the abdominal temperature of an Adélie penguin rose by 1.0 °C after displaying for 15 min. Groscolas (1986) found that the body temperature of the emperor penguin changed with time during the breeding fast. When the birds arrived at the colony, the mean body temperature was 38.9 ± 0.1 °C, but within three days it had dropped significantly to 37.8 ± 0.2 °C. Groscolas (1986) surmised that the elevation in T_b on arrival was due to the exertions of getting to the colony. T_b remained constant during the fast with an average value of 37.1 ± 0.1 °C. However, T_b was found to increase by 1.2 °C at the end of the fast, probably owing to an increase in activity associated with chick feeding and the interaction with the returning female. It may be that the reduction in T_b is an adaptation to reduce the overall energy costs of the birds during the fast.

Chappell *et al.* (1989) found that external air temperatures and wind had little effect on T_b (measured in the proventriculus) of Adélie penguins and blue-eyed shags *Phalacrocorax atriceps*. Only in small Adélie penguin chicks that were still downy, did a combination of cold and wind cause a reduction in T_b. However, the oxygen consumption of all age classes of both species had to be elevated, in some cases by nearly threefold, to maintain T_b in cold windy conditions. It would therefore appear that, apart from severe weather conditions, Antarctic birds can easily maintain body temperature while on land but that a combination of wind and cold can place a high burden on the energy budget of the birds.

The biggest thermoregulatory problem that seabirds face is that of the cold water. In juvenile king penguins, the first cold water immersion caused a 4.5-fold increase in metabolic rate and yet T_b was still reduced by 2–3 °C (Barré & Roussel, 1986). Birds acclimatised to being immersed had a consistently lower metabolic rate than those that had never been immersed. After 4 h of immersion in water at 7 °C, T_b had fallen in non-acclimatised king penguins by approximately 1 °C, whereas acclimatised birds showed no such decrease. In the Macaroni penguin (*Eudyptes chrysolophus*), the T_b of non-acclimatised birds had fallen to 33.8 °C in water at 17 °C but acclimatised birds maintained their body temperature at 40.6 °C when in water at 11 °C. Interestingly, non-acclimatised Macaroni penguins could maintain body temperature when they were allowed to swim in water at 6 °C, whereas the T_b of inactive birds fell by 2.2 °C after only 10 min in the water (Barré & Roussel, 1986). Kooyman *et al.* (1976) investigating the effect of pressure, found that the T_b of Adélie penguins was 38.9 °C when the birds were resting in air and only slightly lower at 38.3 °C when immersed in water at 5 °C at a simulated depth of 20 m. It is not known what effect higher pressures will have on the thermoregulation of diving birds. At a depth of 100 m, all air spaces will be compressed to an eleventh of their original volume and any insulation from the air trapped in the feathers will be lost.

When birds dive, they are surrounded by the cold medium, thereby increasing its capacity as a heat sink. Movement through the water may further increase the heat loss by disturbing the boundary layer. Abdominal temperature (T_{ab}) has been recorded from a number of species when foraging (diving) at sea (e.g. South Georgian shag, gentoo penguin, king penguin; Bevan *et al.*, 1997; Butler *et al.*, 1997; Handrich *et al.*, 1997). In all species, there is a progressive drop in T_{ab} related to diving activity (Fig. 6). Figure 6 shows how the fall in T_{ab} is related to time within a dive bout. In the South Georgian shag, T_{ab} falls by 4.9 °C from 40.0 ± 0.5 °C at the start of a dive bout to 35.1 ± 1.7 °C at the end. In one bird T_{ab} fell by an average of 9.4 °C. Similarly, in the gentoo penguin, T_{ab} fell from 38.8 ± 0.1 °C to 36.2 ± 0.6 °C at the start and end of a dive bout respectively. The minimum T_{ab} recorded from all birds was between 25.3 °C and 36.9 °C (mean ± SE=33.6 ± 1.5 °C). King penguins show the greatest fall in T_{ab} after diving from 38.5 ± 0.04 °C to a mean (± SE) daily minimum T_{ab} of 33.1 ± 1.2 °C. The minimum T_{ab} recorded during each foraging trip from seven free-ranging king penguins ranged from 22.7 °C to 34.4 °C (mean ± SE=29.7 ± 1.4 °C).

The reasons behind the decreases in T_{ab} associated with diving are a matter

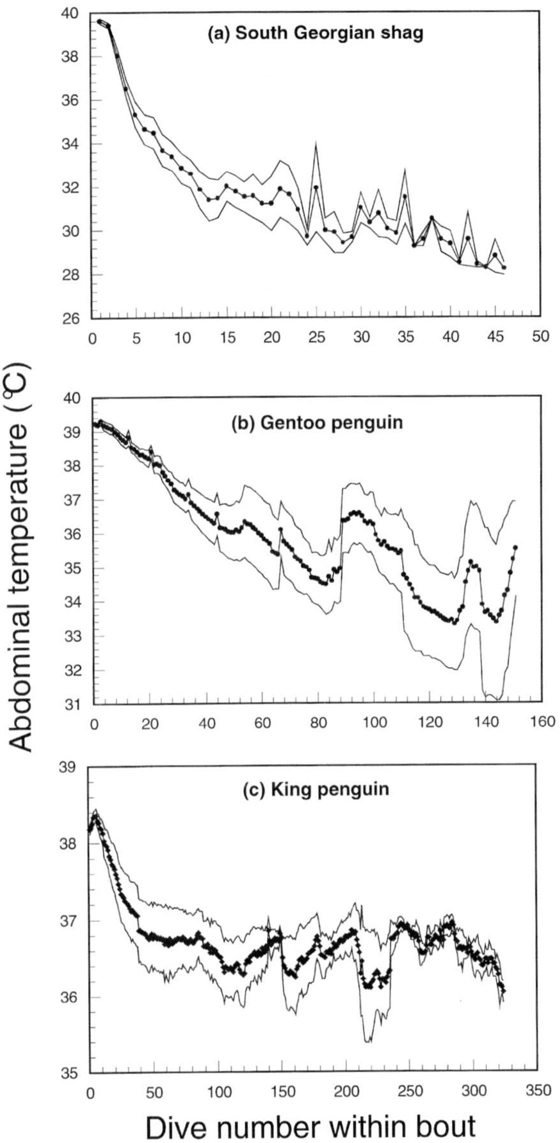

Fig. 6. Abdominal temperature (°C) as a function of dive number within a bout for (a) a South Georgian shag, (b) a gentoo penguin and (c) a king penguin. Each point represents the mean value of abdominal temperature taken at the start of the dive for that dive number. The lines above and below the mean values represent the SE of the mean.

of conjecture. It may be due to the ingestion of cold prey items (Grémillet & Plös, 1994) or to the physical act of diving in a cold medium. In the king penguin, T_{ab} is often lower than stomach temperature suggesting that the cooling effect is coming from the exterior (Handrich *et al.*, 1997). Whatever the reason, a puzzle remains. Given that active penguins seem able to maintain T_{ab} and that diving birds are active when submerged, why is it that diving penguins and shags become hypothermic in at least some regions of the body? One possible explanation is connected to the conservation of oxygen stores during submergence. If T_{ab} reflects a temperature change that is occurring in other tissues, then the metabolic rate of these tissues will also fall, thus reducing the overall metabolic rate (Hill *et al.*, 1987). In other words, the birds may allow their body temperature to fall to enable them to extend their diving capabilities. This would help explain the apparent discrepancy between the estimated aerobic dive limit (ADL) and the observed dive durations (Butler *et al.*, 1995; Culik *et al.*, 1996; Butler *et al.*, 1997; Bevan *et al.*, 1997). Another piece of evidence comes from the observation that, in the king penguin, gentoo penguin and South Georgian shag, there is a small but significant rise in T_{ab} prior to diving. This suggests that the birds are capable of elevating their metabolic rate to maintain body temperature (Butler *et al.*, 1997; Bevan *et al.*, 1997).

Having to ingest prey that is at the same temperature as the surrounding water is another source of heat loss for foraging birds. Cold prey has to be heated up to body temperature placing a drain on valuable energy supplies. Wilson and Culik (1991) calculated that as much as 13% of the daily energy expenditure of Adélie penguins may be used in heating the ingested prey. The stomach temperature of many species has been monitored in the field, primarily to determine when they feed, as a sharp drop in stomach temperature implies that an ingestion event has taken place (Wilson *et al.*, 1995). The extent of the temperature change has even been used to calculate the actual amount of food ingested (Wilson *et al.*, 1995). However, this technique makes the assumption that the body temperature is at the normal resting level and that temperature decreases are confined to the stomach. This may hold true for non-diving or surface feeding birds but, as has been illustrated, all deep diving birds studied to date show a reduction in abdominal temperature that is related to diving *per se*. The difference between the temperature registered in the stomach and the external temperature will therefore be less and the amount ingested will be overestimated. In surface feeding birds such as the albatrosses, though they too dive (Prince *et al.*, 1994), the stomach temperature probably gives a true reflection of ingestion events. T_{ab} also falls in the black-browed albatross and these events are probably associated with ingestion as they all occur when the bird is on the water (Bevan *et al.*, 1995*b*; Wilson *et al.*, 1995).

In spite of the hostile thermal environment, polar birds are well adapted to contend with the extreme conditions that they encounter. Sustained bad weather will, however, place a high energetic burden on adult birds that are incubating or brooding and on chicks that are left unattended. The act of diving probably puts the greatest burden on the thermoregulatory system but it is possible that birds use this to their advantage to conserve oxygen stores during submergence.

Summary and conclusions

Birds living and breeding in the polar regions have evolved to take advantage of the brief periods of high productivity in these areas. Some species are permanent residents of the polar regions whereas others migrate before the harsh winter conditions prevail. The ability to increase body mass and to survive periods of starvation is a common theme amongst the polar birds. This is true whether the birds are incubating, overwintering or migrating. In birds that migrate, young birds that have been raised during the summer have to be developed sufficiently to undertake the long migrations before the winter arrives. A number of different factors are involved in the control of the development of the skeletomuscular system. There is a tendency for basal metabolic rate to increase with latitude. Differences in the ratio of field:basal metabolic rate (usually less than four) between species probably reflect the different flight costs of each species. In some species, such as the albatrosses, the energy costs of flight can be as low as when the birds are incubating. Many polar birds use the marine environment to obtain their food. Although they use a wide range of foraging techniques, for many species the predominant one is diving. There are a wide range of adaptations to diving and although these adaptations are not exclusive to polar birds, many of the most aquatically evolved, e.g. the penguins, are primarily polar species. The penguins, although adapted for an aquatic existence, can walk considerable distances but it is a very inefficient form of locomotion. The cost of swimming, on the other hand, can be very low. Free-ranging penguins and shags have minimum heart rates during diving that are as low as, or even lower than, resting rates. This suggests that the birds during submergence may be using anaerobic metabolism and/or reducing their aerobic metabolism. A possible mechanism for the latter is that reduced body temperatures have been recorded in several species during submergence and these may indicate a reduction in metabolism.

Acknowledgements

This work was supported by grants from the NERC and the BBSRC, UK.

References

Adams, N.J. & Brown, C.R. (1984). Metabolic rates of sub-Antarctic Procellariiformes: a comparative study. *Comparative Biochemistry and Physiology*, **77A**, 169–73.

Adams, N.J., Brown, C.R. & Nagy, K.A. (1986). Energy expenditure of free-ranging wandering albatrosses *Diomedea exulans*. *Physiological Zoology*, **59**, 583–91.

Ancel, A., Visser, H., Handrich, Y., Masman, D. & Le Maho, Y. (1997). Energy saving in huddling penguins. *Nature*, **385**, 304–5.

Aschoff, J. & Pohl, H. (1970). Der Ruheumsatz von Vogeln als Funktion der Tageszeit und der Korpergrosse. *Journal of Ornithology*, **111**, 38–47.

Barré, H. & Roussel, B. (1986). Thermal and metabolic adaptation to first cold-water immersion in juvenile penguins. *American Journal of Physiology*, **251**, R456–62.

Bevan, R.M. & Butler, P.J. (1992*a*). The effects of temperature on the oxygen consumption, heart rate and deep body temperature during diving in the tufted duck *Aythya fuligula*. *Journal of Experimental Biology*, **163**, 139–51.

Bevan,R.M. & Butler, P.J. (1992*b*). Cardiac output and blood flow distribution during swimming and voluntary diving of the tufted duck (*Aythya fuligula*). *Journal of Experimental Biology*, **168**, 199–217.

Bevan, R.M., Woakes, A.J., Butler, P.J. & Boyd, I.L. (1994). The use of heart rate in estimating the energetics of black-browed albatrosses, *Diomedea melanophrys*. *Journal of Experimental Biology*, **193**, 119–38.

Bevan, R.M., Woakes, A.J., Butler, P.J. & Croxall, J.P. (1995*a*). Heart rates and oxygen consumptions of exercising gentoo penguins *Physiological Zoology*, **68**, 855–77.

Bevan, R.M., Woakes, A.J., Butler, P.J.,& Prince, P.A. (1995*b*). The energetics of free-living albatrosses. *Philosophical Transactions of the Royal Society B*, **350**, 119–31.

Bevan, R.M., Boyd, I.L., Butler, P.J., Reid, K., Woakes, A.J. & Croxall, J.P. (1997). Heart rates and abdominal temperatures of free-ranging South Georgian shags, *Phalacrocorax georgianus*. *Journal of Experimental Bio*logy, **200**, 661–75.

Birt-Friesen, V.L., Montevecchi, W.A., Cairns, D.K. & Macko, S.A. (1989). Activity specific metabolic rates of free-living northern gannets and other seabirds. *Ecology*, **70**, 357–67.

Bishop, C.M., Butler, P.J., Egginton, S., El Haj, A.J. & Gabrielsen, G.W. (1995). Development of metabolic enzyme activity in locomotor and cardiac muscles of the migratory barnacle goose. *American Journal of Physiology*, **269**, R64–72.

Bishop, C.M., Butler, P.J., El Haj, A.J., Egginton, S. & Loonen, M.J.J.E. (1996). The morphological development of the locomotor and cardiac muscles of the migratory barnacle goose (*Branta leucopsis*). *Journal of Zoology*, **239**, 1–15.

Bishop, C.M., Butler, P.J., El Haj, A.J. & Egginton, S. (1997). Comparative development of captive and migratory populations of the barnacle goose. *Physiological Zoology*, (in press).

Blem, C.R. (1976). Patterns of lipid storage and utilization in birds. *American Zoologist*, **16**, 671–84.

Boismenu, C., Gautheir, G. & Larochelle, J. (1992). Physiology of prolonged fasting in greater snow geese (*Anser caerulescens atlantic*). *Auk*, **109**, 511–21.

Brekke, B. & Gabrielsen, G.W. (1994). Assimilation efficiency of Kittiwakes and Brunnichs guillemots fed caplin and arctic cod. *Polar Biology*, **14**, 71–83.

Butler, P.J. (1993). To what extent can heart rate be used as an indicator of metabolic rate in free-living marine mammals? *Symposia of the Zoological Society London*, **66**, 317–32.

Butler, P.J. & Jones, D.R. (1997). The physiology of diving of birds and mammals. *Physiological Reviews* (in press).

Butler, P.J. & Woakes, A.J. (1984). Heart rate and aerobic metabolism in Humboldt penguins, *Spheniscus humboldti*, during voluntary dives. *Journal of Experimental Biology*, **108**, 419–28.

Butler, P.J., Bevan, R.M., Woakes, A.J., Croxall, J.P. & Boyd, I.L. (1995). The use of data loggers to determine the energetics and physiology of aquatic birds and mammals. *Brazilian Journal of Medical and Biological Research*, **28**, 1307–17.

Butler, P.J., Bevan, R.M. & Woakes, A.J. (1997). The energetics of free-ranging penguins using the heart rate technique. *Proceedings of the Vth European Conference Wildlife Telemonitoring*, Strasbourg.

Chappell, M.A., Morgan, K.R., Souza, S.L. & Bucher, T.L. (1989). Convection and thermoregulation in two Antarctic seabirds. *Journal of Comparative Physiology*, **159B**, 313–22.

Costa, D.P. & Prince, P.A. (1987). Foraging energetics of grey-headed albatrosses *Diomedea chrystoma* at Bird Island, South Georgia. *Ibis*, **129**, 149–58.

Cotter, R.C. & Gratto, C.J. (1995). Effects of nest and breed visits and radio transmitters on rock ptarmigan. *Journal of Wildlife Management*, **59**, 93–8.

Culik, B.M. (1992). Diving heart rates in Adélie penguins *(Pygoscelis adeliae)*. *Comparative Biochemistry and Physiology*, **102A**, 487–90.

Culik, B.M., Woakes, A.J., Adelung, D., Wilson, R.P., Coria, N.R. & Spairani, H.J. (1990). Energy requirements of Adélie penguin (*Pygoscelis adeliae*) chicks. *Journal of Comparative Physiology*, **160B**, 61–70.

Culik, B.M., Wilson, R.P. & Bannasch, R. (1994). Underwater swimming at low energetic cost by pysgoscelid penguins. *Journal of Experimental Biology*, **197**, 65–78.

Culik, B.M., Putz, K., Wilson, R.P., Allers, D., Lage, J., Bost, C.A. & Le Maho, Y. (1996). Diving energetics in king penguins *(Aptenodytes patagonica)*. *Journal of Experimental Biology*, **199**, 973–83.

Davis, R.W., Croxall, J.P. & O'Connell, M.J. (1989). The reproductive energetics of gentoo (*Pygoscelis papua*) and macaroni (*Eudyptes chysolophus*) penguins at South Georgia. *Journal of Animal Ecology*, **58**, 59–74.

Drent, R.H. & Daan, S. (1980). The prudent parent: energetic adjustments in avian breeding. *Ardea*, **68**, 225–52.

Drent, R.H. & Klaasen, M. (1989). Energetics of avian growth: the causal link with BMR and metabolic scope. In *Physiology of Cold Adaptation in Birds*, eds. C. Bech & R.E. Reinertsen. pp. 349–59. New York: Plenum Press.

Gabrielsen, G.W. (1994). Energy expenditure in Arctic seabirds. PhD thesis. University of Tromsø, Tromsø, Norway.

Gabrielsen, G.W., Mehlum, F. & Karlsen, H.E. (1988). Thermoregulation in four species of arctic seabirds. *Journal of Comparative Physiology*, **157B**, 713–8.

Gabrielsen, G.W., Mehlum, F., Karlsen, H.E., Andresen,O. & Parker, H. (1991). Energy cost during incubation and thermoregulation in female common eider, *Somateria mollissima. Norsk Polarinstitutt Skrifter*, **195**, 51–62.

Grémillet, D. & Plös, A.L. (1994). The use of stomach temperature records for the calculation of daily food intake in cormorants. *Journal of Experimental Biology*, **189**, 105–15.

Groscolas, R. (1986). Changes in body mass, body temperature and plasma fuel levels during the natural breeding fast in male and female emperor penguins *Aptenodytes forsteri. Journal of Comparative Physiology*, **156B**, 521–7.

Handrich, Y., Bevan, R.M., Charrasin, J-B., Pütz, K., Butler, P.J., Lage, J., Le Maho, Y. & Woakes, A.J. (1997). Diving hypothermia in foraging king penguins. *Nature*, **388**, 64–7.

Hedenstrom, A. & Alerstam, T. (1992). Climbing performance of migrating birds as a basis for estimating limits for fuel-carrying capacity and muscle work. *Journal of Experimental Biology*, **164**, 19–38.

Hill, R.D., Schneider, R.C., Liggins, G.C., Schuette, A.H., Elliott, R.L., Guppy, M., Hochachka, P.W., Qvist, J., Falke, K. & Zapol, W.M. (1987). Heart rate and body temperature during free diving of Weddell seals. *American Journal of Physiology*, **253,** R344–51.

Hui, C.A. (1988). Penguin swimming. II. Energetics and behaviour. *Physiological Zoology*, **61**, 344–50.

Irving, L. (1939). Respiration in diving mammals. *Physiological Reviews*, **19**, 112–34.

Jenssen,B.M., Ekker,M. & Bech,C. (1989). Thermoregulation in winter-acclimatized common eiders (*Somateria mollissima*) in air and water. *Canadian Journal of Zoology*, **67**, 669–73.

Jungius, V.M. & Hirsch, U. (1979). Herzfrequenzänderungen bei Brutvögeln in Galapagos als Folge von Störungen durch Besucher. *Journal of Ornithology*, **120**, 299–310.

Kooyman, G.L. & Ponganis, P.J. (1994). Emperor penguin oxygen consumption, heart rate and plasma lactate levels during graded swimming exercise. *Journal of Experimental Biology*, **195**, 199–209.

Kooyman, G.L., Gentry, R.L., Bergman, W.P. & Hammel, H.T. (1976). Heat loss in penguins during immersion and compression. *Comparative Biochemistry and Physiology*, **54A**, 75–80.

Kooyman, G.L., Ponganis, P.J., Castellini, M.A., Ponganis, E.P., Ponganis, K.V., Thorson, P.H., Eckert, S.A. & Le Maho, Y. (1992). Heart rates and swim speeds of emperor penguins diving under sea ice. *Journal of Experimental Biology*, **165**, 161–80.

Korschgen, C.E. (1977). Breeding stress of female eiders in Maine. *Journal of Wildlife Management*, **41**, 360–73.

Le Maho, Y. & Dewasmes,G. (1984). Energetics of walking in penguins. In *Seabird Energetics*, ed. G.C. Whittow & H. Rahn, pp. 235–43. New York and London: Plenum Publishing Corp.

Le Maho, Y., Vu Van Kha, H., Koubi, H., Dewasmes, G., Girard, J., Ferre, P. & Cagnard, M. (1981). Body composition, energy expenditure and plasma metabolites in long-term fasting geese. *American Journal of Physiology*, **241**, E342–54.

Lee, S.J., Witter, M.S., Cuthill, I.C. & Goldsmith, A.R. (1996). Reduction in escape performance as a cost of reproduction in gravid starlings, *Sturnus vulgaris. Proceedings of the Royal Society of London B*, **263**, 619–23.

Lindgard, L., Stokkan, K.A., Le Maho, Y. & Groscolas, R. (1992). Protein utilization during starvation in fat and lean Svalbard ptarmigan (*Lagopus mutus hyperboreus*). *Journal of Comparative Physiology*, **162B**, 607–13.

Lindgard, K., Naslund, S. & Stokkan, K.A. (1995). Annual changes in body mass in captive Svalbard ptarmigan – role of changes in locomotor activity and food intake. *Journal of Comparative Physiology*, **165B**, 445–9.

Marsh, R.L. & Storer, R.W. (1981). Correlation of flight muscle size and body mass in Cooper's Hawks: a natural analogue of power training. *Journal of Experimental Biology*, **91**, 363–8.

Millard, R.W., Johansen, K. & Milsom, W.K. (1973). Radiotelemetry of cardiovascular responses to exercise and diving in penguins. *Comparative Biochemistry and Physiology*, **46A**, 227–40.

Nagy, K.A., Siegfried, W.R. & Wilson, R.P. (1984). Energy utilization by free-ranging jackass penguins, *Spheniscus demersus*. *Journal of Ecology*, **65**, 1648–55.

Nimon, A.J., Schroter, R.C. & Stonehouse, B. (1995). Heart-rate of disturbed penguins. *Nature*, **374**, 415.

Nolet, B.A., Butler, P.J., Masman, D. & Woakes, A.J. (1992). Estimation of daily energy-expenditure from heart-rate and doubly labeled water in exercising geese. *Physiological Zoology*, **65**, 1188–216.

Obst, B.S., Nagy, K.A. & Ricklefs, R.E. (1987). Energy utilization by Wilson's storm petrel (*Oceanites oceanicus*). *Physiological Zoology*, **60**, 200–10.

Owen, M. & Gullestad, N. (1984). Migration routes of Svalbard barnacle geese Branta leucopsis with a preliminary report on the importance of the Bjornoya staging area. *Norsk Polarinstitutt Skrifter*, **181**, 67–77.

Owen, M. & Black, J.J. (1989). Factors affecting survival of barnacle geese on migration from the breeding grounds. *Journal of Animal Ecology*, **58**, 603–17.

Pinshow, B., Fedak, M.A., Schmidt-Nielsen, K. (1977). Terrestrial locomotion in penguins: it costs more to waddle. *Science*, **195**, 592–4.

Prince, P.A. & Francis, M.D. (1984). Activity budgets of foraging grey-headed albatrosses. *Condor*, **86**, 297–300.

Prince, P.A., Huin, N. & Weimerskirch, H. (1994). Diving depths of albatrosses. *Antarctic Science*, **6**, 353–4.

Sladen, W.J.L., Boyd, J.C. & Pedersen, J.M. (1966). Biotelemetry studies on penguin body temperatures. *Antarctic Journal of US*, **1**, 142–3.

Stokkan, K. (1992). Energetics and adaptations to cold in ptarmigan in winter. *Ornis Scandinavica*, **23**, 366–70.

Stokkan, K.A., Lindgard, K. & Reierth, E. (1995). Photoperiodic and ambient temperature control of the annual body mass cycle in Svalbard ptarmigan. *Journal of Comparative Physiology*, **165B**, 359–65.

Stonehouse, B. (1989). *Polar Ecology*. Blackie and Son Ltd.

Tatner, P. & Bryant, D.M. (1989). Doubly-labelled water technique for measuring energy expenditure. In *Techniques in Comparative Respiratory Physiology: An Experimental Approach*, ed. C.R.Bridges & P.J.Butler, pp. 77–112. Cambridge: Cambridge University Press.

Vuilleumier, F. (1996). Birds observed in the Arctic Ocean to the North Pole. *Arctic and Alpine Research*, **28**, 118–22.

Wilson, R.W. & Culik, B.M. (1991). The cost of a hot meal: faculative specific dynamic action may ensure temperature homeostasis in post-digestive endotherms. *Comparative Biochemistry and Physiology*, **100A**, 151–4.

Wilson, R.P., Culik, B.M., Adelung, D., Coria, N.R. & Spairani, H.J. (1991). To slide or stride: when should Adélie penguins (*Pygoscelis adeliae*) toboggan. *Canadian Journal of Zoology*, **69**, 221–5.

Wilson, R.W., Putz, K., Grémillet, D., Culik, B.M., Kierspel, M., Regel, J., Bost, C.A., Lage, J. & Cooper, J. (1995). Reliability of stomach temperature changes in determining feeding characteristics of seabirds. *Journal of Experimental Biology*, **198**, 1115–35.

Witter, M.S. & Cuthill, I.C. (1993). The ecological costs of avian fat storage. *Philosophical Transactions of the Royal Society of London B.*, **340**, 73–92.

Woakes, A.J. & Butler, P.J. (1983). Swimming and diving in tufted ducks, *Aythya fuligula*, with particular reference to heart rate and gas exchange. *Journal of Experimental Biology*, **107**, 311–29.

Woakes, A.J., Butler, P.J. & Bevan,R.M. (1995). An implantable data logger system for heart rate and body temperature: its application to the estimation of field metabolic rates in Antarctic predators. *Medical and Biological Engineering and Computing*, **33**, 145–51.

PART V Applied approaches

R.G. BOUTILIER

Physiological ecology in cold ocean fisheries: a case study in Atlantic cod

Despite a long and distinguished history of scientific investigation in marine fisheries research, there remain a large number of gaps in our knowledge of the processes that determine both the numbers and sizes of the animals produced. The major problem is one of the enormous domain within which the production system occurs, with physical and biological coupling mechanisms operating over temporal and spatial scales of several orders of magnitude. In practical terms, these varying scales represent different subdisciplines in marine fisheries science, e.g. ranging from the physics of large oceanic structures to the physical structure of the genetic material of the biological components. When working in isolation, each subdiscipline can offer descriptive interpretations appropriate to its own scale of study. But when combined within the same theoretical and experimental paradigm, they can consolidate to offer mechanistic insight into the dynamic processes regulating population quality and quantity.

Figure 1 outlines the marine fish production process and can, with appropriate differences in emphasis, represent either an aquaculture-based or harvest-based fishery, the latter being the main emphasis of this report. As with any biological system, the main building blocks are the genetic materials present in the gene pools of the populations in question. These genetic elements interact in both time and space with physical–biological coupling processes to affect both the numerical yield (i.e. through natural mortality), as well as the size and/or quality of each numerical component (i.e. through the imposition of bioenergetic costs). Through a presumed combination of stochastic and selective processes, the physiological and genetic attributes of the surviving phenotypes are returned to the gene pool of the population when each of the individuals eventually reach reproductive maturity. In the context of fisheries management, the fishery itself acts as a selective predator (Fig. 1) of variable magnitude, sometimes to the detriment of fish stocks and to the socio-economic aims of providing a sustainable resource. Various attempts

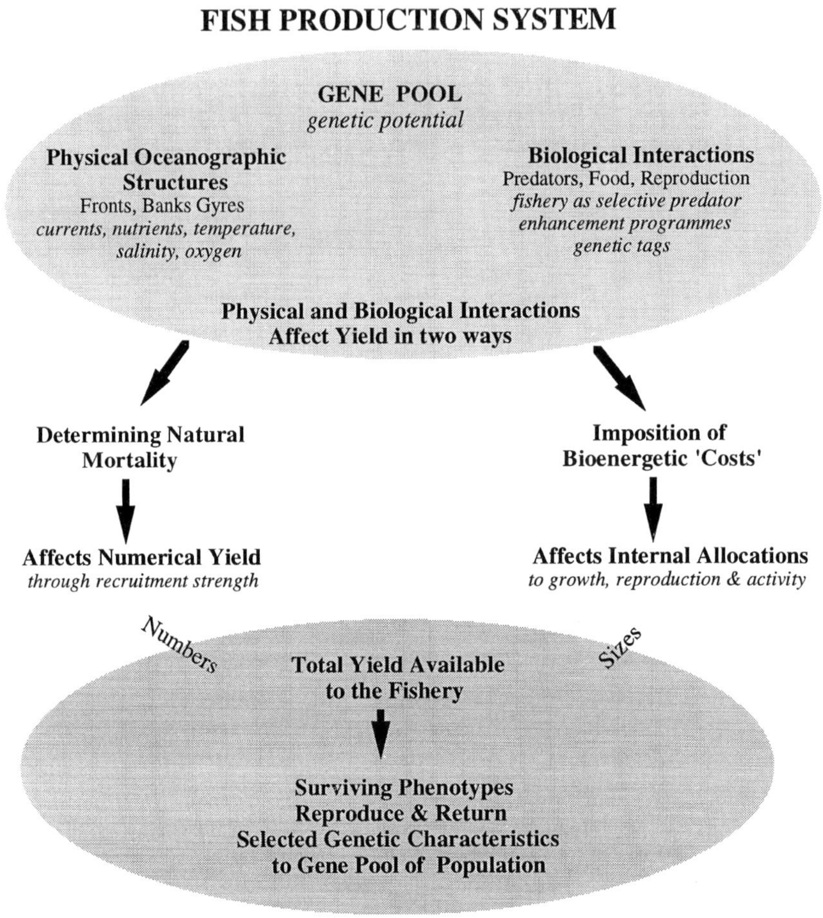

Fig. 1. An outline of the physical and biological coupling processes involved in marine fish production.

can also be made to revive depleted stocks through fish enhancement programmes.

Law (1991) argues that the way in which a fishery is exploited may exert a strong influence on the way in which a population of fish evolves and asks whether it might be possible to manage such evolution, for example, through the selective harvesting of individuals at different stages of development? Clearly, the framework in Fig.1 is as applicable to fundamental studies of physiological ecology and evolution, as it is to fisheries management. Multidisciplinary approaches that tell us more about which genetic characters will respond to the selective effects of fishing may help determine not only the paths of evolution (Law, 1991) but also the intervention strategies that need to be implemented in order to provide for sensible exploitation of our fishery resources. It is within this context of population biology and evolution, that the tools of physiological ecology have been applied to studies of the heavily exploited cod fishery of the North Atlantic ocean.

The Atlantic cod, *Gadus morhua*, is a cold-ocean species of considerable commercial importance in itself as well as being a representative or 'model' of a key component of northern marine ecosystems. The approach has been multidisciplinary, examining the production system of *Gadus morhua* from ecological, physiological and genetical perspectives. Physiological ecology is in itself a multidisciplinary field embracing areas of biochemistry, physiology, behavioural ecology, population genetics and evolutionary biology. Its application to fish production systems (Fig. 1) can therefore occur at many levels and, when undertaken in collaboration with physical and biological oceanographers working at the appropriate scales, it has the potential to offer powerful insights into fisheries management.

A fisheries management approach

It is possible to identify at least three major concerns in fisheries management that fall within the realm of the physiological ecologist. The first is whether there is a need for fisheries managers to undertake a multispecies (i.e. ecosystem) approach. Proponents of this approach would argue that the abundances of a population are regulated by food-chain interactions, and that a fishery for one commercially important species may have considerable impact on the predator–prey interactions and competition for food resources or space, of co-occurring commercially viable species. Ecological theory predicts that the major competition for resources should occur at the intraspecific level, since individuals of the same species will have the most similar ecological needs (Ridley, 1993). Knowledge of interspecific competition is also vital since the next most ecologically similar species will exert the next greatest negative influence, and so on throughout the food web. Obviously, the relative importance of these intra- and interspecific interactions need to be identified in relation to the animals' growth and behaviour in nature.

A second major concern is whether it is possible to predict the geographic movements and oceanic distributions of cod in relation to seasonal or yearly changes in physical and chemical 'structures' in the environment. Gradients of temperature, water flow, salinity and dissolved oxygen may constitute 'natural fences' which assemblages of cod must either manoeuvre around or swim through as they proceed along their migratory routes. Increased knowledge of the animals' physiological (e.g. metabolic) and behavioural (e.g. avoidance) responses to such physical and chemical 'barriers' may facilitate the construction of oceanographic-based models which can be used to predict how the environment dictates the distributions and movements of cod in both time and space.

A third set of fundamental questions involve population-level regulation which, from the point of view of the fishery, ultimately determines biomass quantity and quality. Of foremost interest is whether the species in question can be subdivided into geographically discrete spawning units, and if so, how do the units remain distinct and what minimum spawning stock is required to ensure that the unit will remain self-sustaining (Sinclair & Page, 1995). Furthermore, very little is known about geographic-based differences in physiological traits (see review by Garland & Adolph, 1991), and whether such differences affect the allocation of internal energy resources to biological processes such as growth, reproduction and activity metabolism.

A brief life history of cod

Larval survivorship

Fisheries science is based on systematic study of the dynamic interactions between physical and biological processes, on the one hand, and the abundance and distribution of fishes on the other. The primary objective is to improve the ability to predict the causes of year-to-year changes in the numbers of animals that 'recruit' to a population. In the ecological and evolutionary literature, 'recruitment' normally refers to the numbers of individuals that survive and enter the reproductive population, whereas recruitment in fisheries science can have the more limited definition of being the numbers that survive to harvestable size, which does not always mean reproductive maturity (Frank & Leggett, 1994). Interannual variability in recruitment within a commercial fishery can result in large expansions and contractions of harvestable biomass, with attendant economic and social impacts. The numbers of individuals that recruit to the fishery within a given year define the year-class strength.

For many species of marine fishes including cod, interannual variability in recruitment is thought to be determined largely by differences in mortality during the planktonic stage of eggs and larvae. Cod are one of the most fecund of all vertebrates, with many adult females laying between two and five million eggs in a breeding season (May, 1967); and, in exceptionally large individuals (140 cm length), up to 12 million eggs (Powles, 1958). The fertilised eggs float to surface-waters and remain there while they are hatching (Dahl *et al.*, 1984). During this time, the eggs and newly hatched larvae are 'passive drifters', whose retention near the spawning grounds, or wide-scale dispersal away from them, depends on various hydrodynamic features such as large oceanic gyres and current-flow patterns. Interannual variations in the physical environment can therefore have a strong influence on whether cod eggs and larvae are transported to suitable nursery grounds or not (Davidson & deYoung, 1995), as can any attendant changes in biological interactions such as predation pressure and food availability (Bakun *et al.*, 1982). Taken together, these physical and biological factors at the earliest life history stages are important determinants of survivorship and overall spawning success of cod. Indeed, mortalities of eggs and larvae are appallingly high, with only about one egg in a million surviving to become a mature cod.

Though the vast majority of past research has focused on the more than 99% of pelagic eggs and larvae that do not recruit to the juvenile stage, this approach has failed to demonstrate unequivocally, that recruitment variability and subsequent year-class strength are correlated with marine larval abundance and/or mortality (Taggart & Frank, 1990). On the other hand, less attention has been paid to the ecology and physiology of young-of-year cod (i.e. 0+ cod), even though there have been suggestions that the abundance of demersal 0+ cod may be a better indicator of year-class strength than larval abundance (deLafontaine *et al.*, 1994).

Juvenile cod settlement, growth and post-settlement mortality

Cod differ from many other demersal marine fishes in that their planktonic larvae transform into a pelagic juvenile phase at 2–3 months post-hatch at sizes >20 mm (Lough *et al.*, 1989). There is then a gradual transition from a pelagic to demersal life stage as the fish grow to 30–100 mm (Lough &Potter, 1993; Tupper & Boutilier, 1995*a*,*b*). Most studies of the ecology of juvenile cod have been carried out by trawl census in open ocean environments, such as George's Bank (Lough *et al.*, 1989; Lough & Potter, 1993). However, a good many juvenile cod recruit to warmer inshore environments along the Northeast Atlantic where they settle in a variety of habitat types including reefs or rock ledges, sand, gravel or cobble bottoms, and seagrass beds (Gotceitas & Brown, 1993; Tupper & Boutilier, 1994, 1995*a*,*b*). Although initial settlement strengths of 0+ cod do not appear to differ between habitat types, post-settlement survival rates and subsequent juvenile densities are certainly much higher in the more structurally complex habitats, owing to the greater availability of shelter and decreased predator efficiencies.

Using visual census techniques, we have studied over a 6-month period, the settlement, growth and postsettlement mortality of 0+ cod (Tupper & Boutilier, 1995*a*), from their appearance in warm-water inshore habitats in July to their disappearance in December, by which time the habitats have become much cooler and in some cases frozen-over. By marking newly

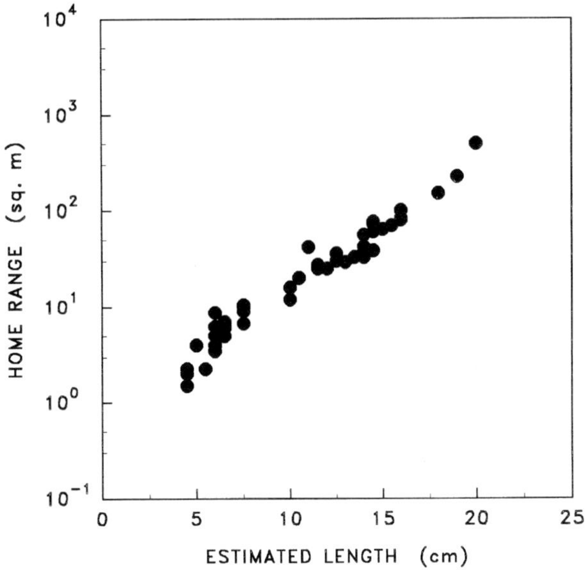

Fig. 2. The relationship of home range size (in square metres) to length of 0+ Atlantic cod (*Gadus morhua*) at Back Cove Reef, St. Margaret's Bay, Nova Scotia, from June to December, 1991. The perimeter of each home range was outlined on a 1:1000 scale drawing of the experimental field site. The area covered by the home range was then extrapolated from the area measured on the scale drawing. (From Tupper & Boutilier, 1995*b*.)

settled juveniles with an acrylic dye, it has been possible to make biweekly measurements of behavioural activities within each habitat type, as well as length/growth estimates of individual animals inhabiting rocky reef inshore environments (1–2 m depth, 5–30 m from shoreline). The newly settled 0+ cod become site-attached and compete for shelter sites by establishing territories (Fig. 2) which they defend vigorously against all intruders, not unlike many tropical reef fish. Growth and territory size are determined by size at settlement within the framework of a size-dependent social hierarchy, i.e. fish that settle earliest/largest grow more quickly and defend larger territories than those settling smaller/later. Given that early settlement and large size at settlement confer distinct competitive advantages to juveniles, it is necessary to know more about whether size at settlement is predetermined by some measurable factor (genetic or physiological), as it might then become possible to predict population- or stock-specific postsettlement survival, and subsequent recruitment, of cod.

Juvenile cod appear to congregate in inshore environments on the rising arm of a temperature curve (Fig. 3). Settlement at this time may confer physio-

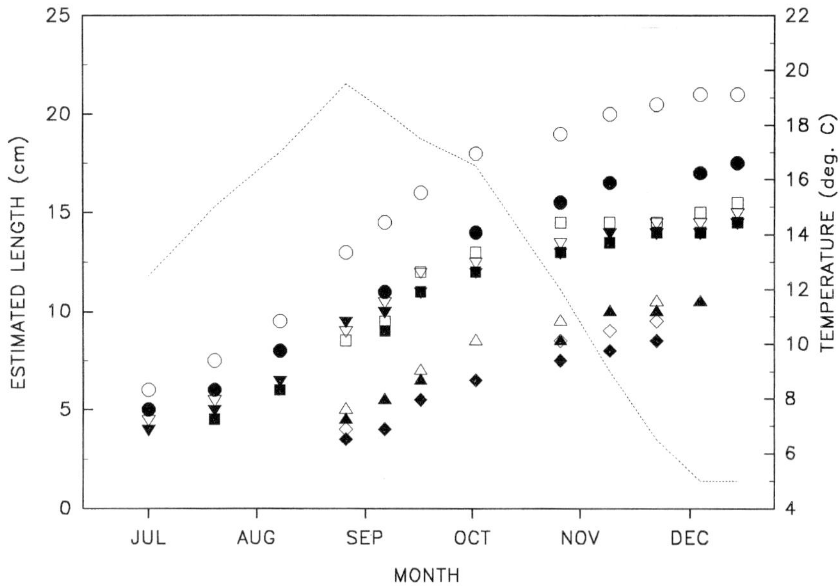

Fig. 3. Temporal change in water temperature and in estimated total length of 10 individual 0+ Atlantic cod (*Gadus morhua*) at Back Cove Reef, St. Margaret's Bay, Nova Scotia, from June to December, 1991. Symbols represent individual fish; first appearance of a symbol indicates 'settlement' of that individual. Measurements of habitat temperature taken at each sampling time are illustrated by the dotted line. (From Tupper & Boutilier, 1995*b*.)

logical advantages to young-of-year fish that recruit to warm-water inshore, as compared to cold-water offshore, environments. Settlement of fish during a period of enhanced growth potential may not only reduce the amount of time spent in smaller and presumably more vulnerable early juvenile stages (Walsh, 1987) but may also facilitate pre-winter storage of lipids, since survival is generally higher in large individuals (Henderson, Holmes & Bamber, 1988). Nevertheless, behavioural interactions can have equal or greater importance than physiological processes *per se* in determining growth rates of 0+ cod. For example, in the populations studied, the presence of resident juvenile cod with established territories appeared to depress the growth rates of the late settlers. Thus, both size at first settlement, and the timing of settlement, may determine the competitive success and growth potential of individual 0+ cod. Although normally considered to be a cold-water species, those individuals that move to warmer inshore habitats appear to be acting as thermal opportunists in that these 'nursery areas' probably serve to maximise the animals' growth, survival, and eventual recruitment to the commercial fishery.

The fact that 0+ cod compete for shelter sites indicates that shelter from predation is a potentially limiting resource and that adult population size of cod is not determined solely by reproductive output and/or survival of the planktonic stage. High mortalities of juveniles can result from density-dependent postsettlement interactions, mainly competition for shelter. Tupper and Boutilier (1995*a,b*) found that growth rates of juvenile cod were highest in seagrass beds, while the efficiency of predators was lowest and survival highest on rocky reefs and cobble bottoms. Thus, trade-offs occur between energy gain and predation risk. The population structure of Atlantic cod may be less influenced by the patterns of larval supply than by post-settlement processes such as habitat-specific growth and mortality.

Bioenergetic models

Variations in the overall biomass available to a fishery can be attributed to the interaction of a number of physical and biological factors which affect both the numerical yields (i.e. year class strength) as well as the size-at-age (Fig.1). Knowledge of the growth rates of the individuals within a fishery are therefore of major concern because variations in size-at-age will influence the overall biomass directly, as well as determine the amount of surplus energy available for reproduction. Changes in the availability of Atlantic cod to the fishery are known to be greatly influenced by size-selective fishing pressure (e.g. Hutchings & Myers, 1994), as well as interannual changes in temperature (Rose *et al.*, 1994), and prey availability (Lilly, 1987). Both water temperature and food consumption will have obvious direct effects on the way in which energy is taken in, assimilated, and allocated both to growth and reproductive effort. There is therefore a need to develop robust bioenergetic models to improve understanding of the physiological mechanisms underlying the fish production system.

Bioenergetic models for the analysis of fish production systems (Fig. 1) are quite advanced (e.g. Kerr, 1982; Kitchell, 1983; Brett, 1986), but their effective application to marine cold-water fisheries has been hampered by our limited knowledge of the physiology and bioenergetics of individual species. The models take a general form whereby growth rates are estimated by subtracting energy expenditures and losses (i.e. activity, reproduction, feeding-induced thermogenesis, excretion) from the total energy intake. Because data for temperature and prey availability can be obtained from historical databases of catch statistics and oceanographic surveys, the realised (i.e. observed) growth rates can be compared with potential growth rates (i.e. those generated from models that assume maximum consumption rates). By varying the inputs to such models, and assessing the differences between realised and potential growth rates, estimates can be made of the effects that variations in temperature and prey abundance might have had on the growth rates of selected stocks over several years (Krohn, Reidy & Kerr, 1997). In addition, if we were to know the amount of energy being used by the individuals of a given stock, and the size and abundance of that stock, it would then become possible to predict the annual amount of energy required to sustain the population.

Energetics of activity

The metabolic cost of swimming is a major component of the energy budget of actively predatory fish, and bioenergetic models are very sensitive to this parameter (Kerr, 1982; Boisclair & Leggett, 1989). Nevertheless, there is a surprising lack of research on the energetics of activity in most of the commercially important cold-ocean species. For example, even very recent bioenergetic studies on cod have had to rely on estimates of activity metabolism taken from other species, like haddock (Tytler, 1969, in Björnsson, 1993). Others have been forced to make large extrapolations of existing data on standard metabolism of cod (Saunders, 1963), based on theoretical assumptions that the swimming cost of fish searching for prey should be one to two times the standard metabolism (Weihs, 1975; Ware, 1975) or as high as two to three times the standard metabolism (Ware, 1978; Kerr, 1982), depending on temperature. Thus, even though the extensive studies of Brett and his coworkers on Pacific salmon (e.g. Brett, 1964; Brett & Glass, 1973; Brett & Groves, 1979) have pointed the way to the development of robust bioenergetic models (Brett, 1986), similar foundation studies on commercially important cold-ocean species, at appropriate temperatures, are only beginning to emerge (Nelson, Tang & Boutilier 1994, 1996; Steffenson, 1993; Blaikie & Kerr, 1996; Fig. 4).

Metabolic costs of food consumption

Another energetic cost about which very little is known is that associated with the ingestion of food. The so-called specific dynamic action or SDA, is thought to reflect the costs of the nutritive process, including the energetic requirement for ingestion, digestion and absorption of foodstuffs (for review see Jobling, 1981). It is observed characteristically as a demonstrable posprandial increase in the oxygen consumption of adult cod (i.e. 1.6 times the routine metabolic rate; Saunders, 1963), and juvenile cod (up to 98% of the metabolic scope; Soofiani & Hawkins, 1982), and is considered an inescapable cost of growth (Jobling, 1981). In terms of the parameterisation of bioenergetic models, there is again very little information on how SDA might vary with temperature, activity, or feeding rates to affect overall growth in many marine fishes.

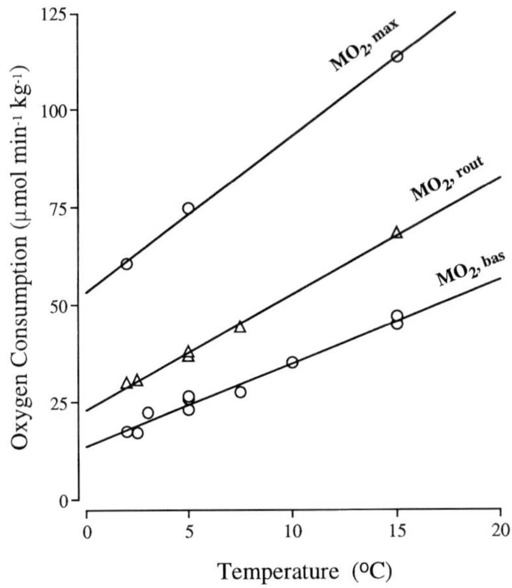

Fig. 4. Relationship between acclimation temperature and whole animal oxygen consumption of adult Atlantic cod at rest ($\dot{M}o_{2,rout}$) and at maximal swimming activity in a swim-tunnel ($\dot{M}o_{2,max}$). Basal metabolic rates ($\dot{M}o_{2,bas}$) were taken at the y-intercept of the measured relationship between swimming speed and $\dot{M}o_2$. (Data are compiled from Saunders, 1963; Tang *et al.*, 1994; Nelson *et al.*, 1994, 1996; Claireaux *et al.* 1995*a*; and unpublished measurements of Y. Tang, J. A. Nelson and R.G. Boutilier for cod at 15 °C.)

Cod migration: physiology, behaviour and oceanography

Although the possibility of being able to fully parameterise the advanced bioenergetic models of fisheries biology is still a long way off, recent laboratory experiments on swimming energetics of cod as a function of temperature (Nelson *et al.*, 1994, 1996; Fig. 4) and food consumption (Björnsson, 1993; Blaikie & Kerr, 1996), together with oceanographic studies of foraging activity (DeBlois & Rose, 1995) and migration of cod (Rose, 1993; Rose, de Young & Colbourne, 1995), can be used to illustrate simplified energy relationships (Fig. 5). During their seasonal migrations, Atlantic cod form large cohesive aggregations (>10 km in diameter; Rose, 1993) which are thought to both improve foraging efficiency and predator avoidance (DeBlois & Rose, 1995). Mean swimming speeds of these large shoals (corrected for prevailing water current speeds) range from 0.08–0.28 m/s, which are well below the maximum sustainable swimming speeds found in laboratory studies (0.4–0.8 m/s;

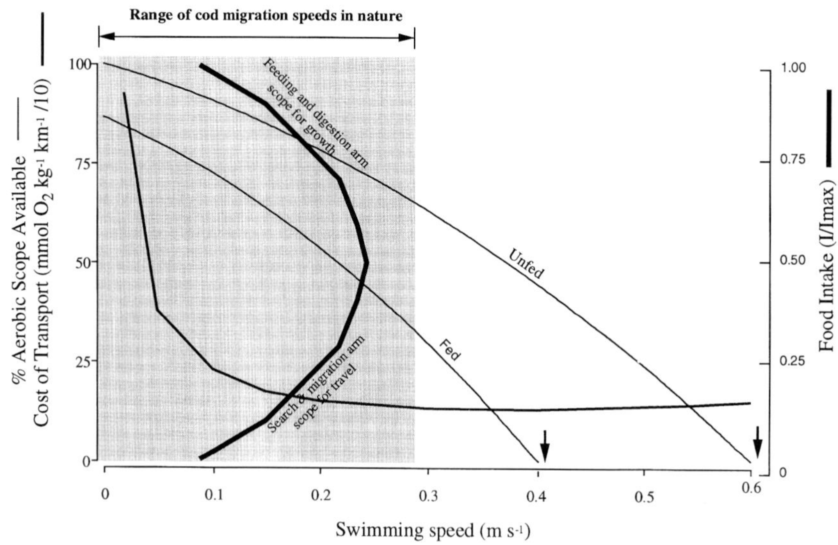

Fig. 5. Model of the energetics of foraging activity in migrating cod. Relationship between swimming speed and fractional food intake (intake/maximal intake) from Björnsson (1993). The percentage of aerobic scope available for activity at each speed, and the costs of transport, were estimated from exponential relationships between swimming speed and oxygen consumption in fed and unfed fish (Blaikie & Kerr, 1996; D. Webber, S.R. Kerr and R.G. Boutilier, unpublished data). Aerobic scope was calculated as the difference between maximal oxygen consumption ($\dot{M}o_{2,max}$) and basal oxygen consumption ($\dot{M}o_{2,bas}$) and the percentage aerobic scope available was the fraction of total aerobic scope remaining unused at each given swimming speed. Arrows at 0.4 and 0.6 m/s swimming speeds equal $\dot{M}o_{2,max}$ for unfed and fed fish, respectively.

Wardle, 1977; Tytler, 1978; Soofiani and Priede, 1985; Tang *et al.*, 1994; Nelson *et al.*, 1994, 1996). Even so, individuals within such shoals may swim at much higher speeds (e.g. Arnold *et al.*, 1994). Cod displacements of up to 59 km have been observed in a single day (Rose *et al.*, 1995), representing an average ground speed of 0.68 m/s, though it is unknown in that case to what extent the cod were exploiting transport advantages such as drifting with currents or swimming in advancing flows (cf. Metcalfe, Arnold & Webb, 1990).

Activity metabolism in relation to food availability

Cod migration occurs as a series of stops and starts (mean daily displacements over weeks to months range from 7–24 km), as the animals search for, and find, prey (typically capelin, *Mallotus villosus*, or shrimp, *Pandalus borealis*). The highest swimming speeds occur either when food is absent or when prey are initially detected (Jones, 1978). Upon entering abundant food patches, however, shoal speeds of migrating cod decrease by a factor of 3–5 (from 20 to 4–7 km/day; Deblois & Rose, 1995). Indeed, Björnsson (1993) showed that the mean swimming speeds of cod tested in large mesocosms increased from 0.0 to 0.08 m/s when search for prey was initiated, became highest (0.20–0.24 m/s) when food density was one-half maximal intake, and then decreased when food was either unlimited or entirely absent. This dome-shaped relationship between food intake and swimming speed (Björnsson, 1993) is illustrated in Fig. 5 where food intake is presented as a fraction of maximum food intake (i.e. I/I_{max}). Also shown is the percentage of aerobic scope available to an idealised animal, calculated from aerobic scope measurements in starved animals swimming in a swim-tunnel (Nelson *et al.*, 1994, 1996) and in animals fed prior to the swimming performance tests (Blaikie & Kerr, 1996). Note that unfed animals at basal metabolic rate (i.e. zero swimming speed) have 100% of their aerobic scope available, whereas when the same animal is fed, its aerobic scope at zero swimming speed decreases by ~15%. Maximum sustainable swimming speeds, and maximal oxygen consumptions, of animals fed a ration equivalent to 2.5 % of their own body weight were 0.40 m/s (Blaikie & Kerr, 1996) as opposed to 0.60 m/s, at maximal O_2 consumption in unfed animals (Fig. 5).

As starved animals begin to search for and encounter prey, their aerobic scope is maximal, i.e. for fuelling sustained bouts of chasing down prey or repaying costs of anaerobically fuelled 'bursts' of prey capture (Fig. 5). The search for food becomes most intense as intake rates reach one-half maximal intake, which is still well within aerobically sustainable swimming speeds. Presumably, the decrease in swimming speeds at high food ingestion rates signals a shift in aerobic scope towards growth. For animals migrating in the wild, it may be more cost-effective to stop and rest after encountering abundant prey, channelling energy into growth rather than locomotion. Migrating at very slow speeds would in any case be energetically extravagant in terms of fuel efficiency due to the correspondingly high costs of transport (see Fig. 5). However, when increased swimming speeds become necessary as hungry animals resume their search and/or migration for food, they appear to do so at speeds of 0.20 m/s and above, where costs of transport are minimal and aerobic scopes for activity are highest.

Physiology and behaviour of free-swimming cod

Although physiologists are well practised at gathering information about the activity energetics of fish under controlled laboratory-based experiments, relatively little is known about the bioenergetic responses of animals behaving naturally in the wild. The lack of information arises because of the necessity of tracking physiological indices of metabolism in animals moving about freely in environments whose physical and chemical 'structure' is either unknown or unpredictable. Even though it is now possible to track certain physiological variables through the use of telemetric devices, interpretation of the data requires contemporaneous knowledge of both ocean structure as well as any ongoing intra- and inter-specific behavioural interactions (Fig. 1). All of these criteria can rarely be met, at least for animals patrolling large volumes of water in open ocean environments. As an initial step towards the ultimate goal of open ocean tracking, cod were fitted with ultrasonic telemeters that provided continuous records of the heart rate and position (depth) of animals released into a large 125m^3 mesocosm. This 10.5 m deep and 4 m diameter 'Tower Tank' can be chemically and thermally stratified (i.e. controlled 'structure') and observation windows enable video surveillance of the animals behavioural interactions.

These experiments were designed to complement, as far as possible, a number of large-scale oceanographic studies in which the migratory movements of Northern Newfoundland cod were tracked, using high-resolution sonar techniques, as the animals moved over hundreds of kilometres through defined thermal fields (e.g. Rose, 1993). Our interest was twofold. First, was to estimate the energy expenditure and cost of transport of cod, over the range of cruising ground speeds recorded during migrations (Fig. 5). Secondly, was to examine the behavioural responses of cod to changing temperature fields (Claireaux *et al.*, 1995*a*) since the migrating animals appear to track a thermally stable corridor or 'cod highway' (Rose, 1993), as they move from their offshore breeding grounds to inshore areas of feeding. The animals travel this 'warm' 2–2.5 °C highway with great precision, only veering onto thermally heterogeneous 'offroads' (−1.7–3.6 °C) when encountering prey (DeBlois & Rose, 1995). We therefore tested the physiological and behavioural responses of adult cod, fitted with depth and heart rate telemeters, to small perturbations in their thermal environment. After animals had been acclimated for weeks to a stable temperature in the mesocosm, they were found to be extremely sensitive to any small perturbations in their thermal environment, e.g. imposed temperature increases of 2–3 °C led to a doubling of heart rate (Fig. 6) and an estimated 30% rise in metabolic rate (Claireaux *et al.*, 1995*a*). This marked metabolic effect in response to relatively small changes in ambient temperatures indicated that cod on their migratory highway (Rose, 1993) may be exquisitely sensitive to changing temperature fields, and might be avoiding them whenever possible, on energetic grounds. To test this further, we acclimated animals for 1 month to a stable 5 °C environment, and then presented the cod with a thermally heterogeneous environment, in which a 1 metre layer of 5 °C could be

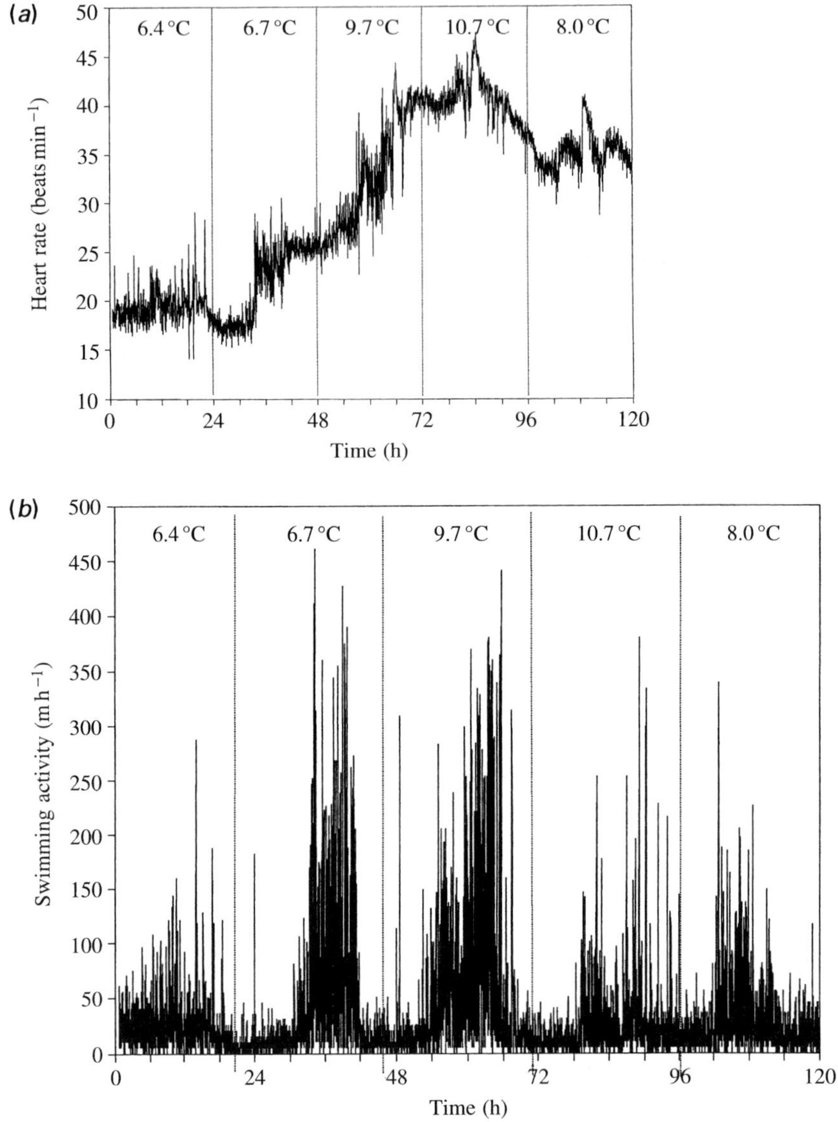

Fig. 6. Telemetered heart rate measurements (*a*) and estimates of spontaneous swimming activity (*b*) for an individual cod free to move about in a 10 metre deep, 4 metre diameter 'Tower Tank'. Heart rate reponses to changes in mean daily temperatures are shown over a period of five days. By summing the vertical displacements of the fish (*b*), measured by an ultrasonic depth telemeter, we estimated total distances travelled as: day 1, 1.94 km; day 2, 3.69 km; day 3, 4.52 km; day 4, 2.14 km; day 5, 2.36 km. Records

experimentally manipulated in the vertical column. Cod tracked the 5 °C water with an accuracy of ±0.3 °C, thus providing us with the evidence to conclude that temperature gradients in the natural environment of less than 1 °C could play an important role in determining patterns of movement of cod in the wild.

Cod may display similar degrees of circumspection when defining migratory routes through waters of heterogeneous salinity or oxygen, since mesocosm experiments suggest that they will avoid movement into such areas, unless provoked by the opportunity for a meal (Fig. 7; Claireaux *et al.*, 1995*a*). Just as was seen with temperature, there are energetic costs of movement into and out of zones of reduced salinity (e.g. 30‰ to 26‰, typifying the range found in the Scotian Shelf for example), and animals are sensitive in detecting and avoiding exposure to salinities outside those to which they have been acclimated (Fig. 7*a*). The effects of movement into zones of low oxygen are perhaps more apparent from an energetics point of view since it is obvious that this will impose a direct stress on the oxygen uptake and transport systems (Claireaux & Dutil, 1992). Indeed, cod avoid movement into hypoxic zones unless presented with the opportunity to feed, when they will make short 'dives' to capture the prey and then quickly retreat to normoxic zones to digest and assimilate the meal (Fig. 7*b*).

Eutrophication-induced hypoxia is a growing problem, particularly in inshore environments which often serve as 'nursery areas' for juvenile cod (e.g. Tupper & Boutilier, 1995a,b) and as feeding and spawning grounds for adults. Many such areas are found in and around shallow-water seas such as the Kattegat and Baltic, and are characterised by strong vertical stratifications in salinity (i.e. haloclines), which reduce the convective mixing of surface and bottom waters (Pihl, 1989; Bagge, Steffensen & Bay, 1991). In temporal studies of the Kattegat, it has been found that at bottom-water PO_2 levels of 32–36 mmHg and above, variations in fish biomass are quite small. However, severe hypoxic conditions ($PO_2 \leq 10$ mmHg) that arise during certain seasons, cause massive emigration of many demersal species from the affected areas, reducing the total biomass of demersal fish species

in (*b*) show the existence of a daily cycle in the vertical distance travelled by the fish, activity being low at night and highest during the daylight hours. Note that the swimming activity increased as the mean daily temperaures were increased gradually from 6.4 °C at day 1 to 9.7 °C at day 3, and thereafter returned to reference levels as temperatures were decreased gradually from 10.7 °C on day 4 to 8.0 °C on day 5. Movements of the individual cod tracked the movements of the school of 10 cod with which it was associated. (From Claireaux *et al.*, 1995*a*.)

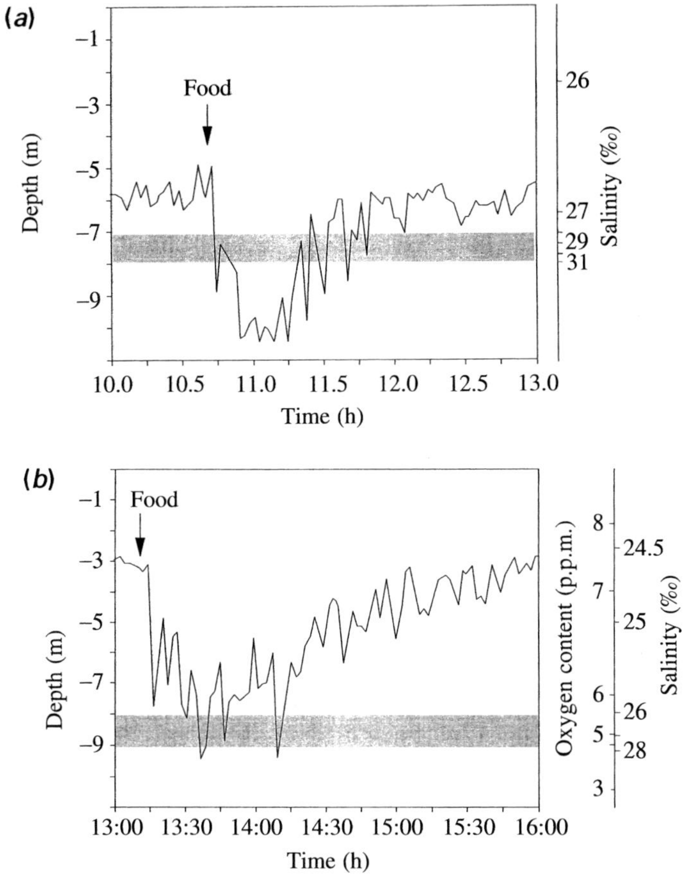

Fig. 7 (*a*) Telemetered depth records from an individual cod free to move about in a 10 metre deep, 4 metre diameter 'Tower Tank' stratified for salinity at ~7 metres depth. Cod acclimated to 26‰ salinity avoided the higher salinity water in the bottom 3 metres of the tank until presented with the opportunity to feed there, whereupon they immediately moved into the higher salinity water and remained there for as long as food was available (45 min). When no more food was available, the fish returned above the 'halocline' symbolised by the shaded area. (*b*) When the bottom layer was stratified for salinity as in (*a*) but also made hypoxic (note O_2 scale), fish avoided the hypoxic region until presented with the opportunity to feed. In this case, however, rather than swimming into the hypoxic zone and remaining there to feed, the fish moved from their initial position down to the vicinity of the 'chemocline' and from there made a number of short 'dives' to capture food, after which they immediately returned to the normoxic waters above the chemocline. Note that the data in (*a*) and (*b*) are estimates of the average depth over 30 s, taken at 2 min intervals. In the case of hypoxia, 'round trips' below the transition zone were often so fast that we could not identify every one of them. (From Claireaux *et al.*, 1995*b*.)

by up to 80% (Pihl, 1989). Cod in particular are often one of the first species to leave (Pihl, 1989), since even at the low temperatures of the bottom waters of the Baltic (<5 °C; Bagge *et al.*, 1991), these levels of aquatic hypoxia are known to be lethal to the Atlantic cod populations of these regions (Steffensen, 1993). The increasing incidence of hypoxia in areas of strong salinity stratification (e.g. Falkowski, Hopkins & Walsh, 1980; Rossignol-Strick, 1985), and perhaps in other areas yet to be identified, may displace demersal species like cod away from historical spawning sites and change migratory routes. Such displacements may bear important consequences to the distributions of cod in both time and space, and alter biomass yields at the population level, through changes in recruitment strength. Although O_2-limiting conditions are well known to occur in brackish-water, shallow-bottom seas, hypoxic regions along certain banks and shelfs in more open-ocean environments have also been recorded. Since cod appear to actively avoid such regions (Fig. 7*b*), there is reason to encourage more temporal and spatial definition of 'oxygen fields', particularly along migration routes and in identifiable spawning and recruitment sites.

The problems associated with predicting the abundance, distribution and movements of migrating marine fish are indeed very challenging. Their resolution requires detailed information not only on the dynamic interactions between biological and physical processes already mentioned, but also on whether there is a genetic basis for such interaction.

Population genetics of Atlantic cod

cDNA markers for cod

Much current knowledge concerning both the levels and patterns of genetic variation in natural populations has been determined by polymorphisms at soluble protein, or allozyme, loci. Advances in the field of recombinant DNA technology have enabled genetic polymorphism to be scored at the level of DNA to a much higher degree than would be possible by allozymes (Pogson & Zouros, 1994). The ability to score substantial numbers of polymorphic sites throughout a species' genome allows key insights to be made into two fundamental areas of research: (i) the causative basis of correlations between an individual's level of heterozygosity and fitness-related traits (such as growth or viability) in single populations, and (ii) the extent of population subdivision that occurs throughout the species' range and the evolutionary processes that may be responsible for this substructure. These questions have become tractable in cod owing to the development of a suite of polymorphic DNA markers for the Atlantic cod *Gadus morhua* (Pogson, 1994).

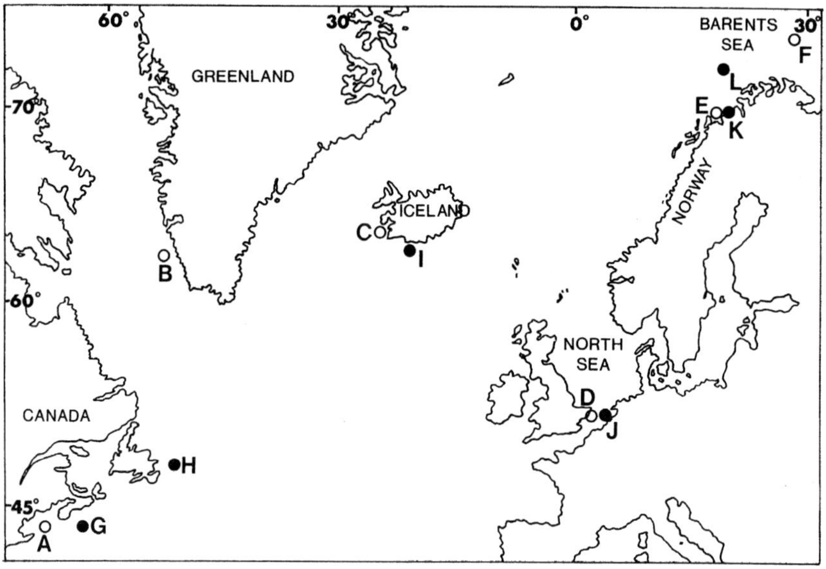

Fig. 8. Geographic location of cod (*Gadus morhua*) populations sampled throughout the North Atlantic. Open circles are samples characterised at the protein level by Mork *et al.* (1985); closed circles are the locations of samples scored for nuclear RFLP loci by Pogson *et al.* (1995). A, Gulf of Maine, B, Greenland, C, Iceland, D, North Sea, E, Malangen, F, Barents Sea, G, Nova Scotia, H, Newfoundland, I, Iceland, J, North Sea, K, Balsfjord and L, Barents Sea. The mean difference separating five of the six samples taken from geographically similar regions is estimated to be less than 350 km; the remaining pair (Newfoundland and Greenland) differ in location by ~1600 km. (From Pogson *et al.* 1995.)

Genetic differentiation among geographically isolated cod populations

High levels of gene flow can be inferred from a species' potential for dispersal since species with high dispersal potential often exhibit low levels of population differentiation (e.g. Waples, 1987), whereas those with more restricted means of dispersal do not (Larson, Wake & Yanev, 1984). As gene flow is a powerful homogenising force that restricts genetic divergence among populations, it follows that animals with wide dispersal patterns, such as many marine fish species (cf. Ward, Woodwark & Skibinski, 1994), might be expected to show low levels of population subdivision. Atlantic cod are almost certainly one of the most widely distributed species in the north Atlantic, occupying a geographic range that extends from the Gulf of Maine to the Barents Sea (Fig. 8), some 8000 km. Mature cod are also known to aggregate in scales of tens of thousands of fish (Rose, 1993), and to undergo large-scale migrations (e.g. Rasmussen, 1959; Gulland & Williamson, 1962; Templeman, 1974). Thus, Atlantic cod not only meet the apparent preconditions for limited population structure, but evidence of homogeneous patterns of allozyme variation and of low levels of mtDNA polymorphism among widely distributed populations (Mork *et al.*, 1985; Smith *et al.*, 1989; Carr & Marshall, 1991; Arnason & Rand, 1992), supports the notion of there being low levels of population differentiation in this species.

However, not all genetic markers support this claim. For example, Norwegian coastal and Arctic cod stocks exhibit highly significant differences in haemoglobin 1 (Hb-1) and transferrin (Tf) blood protein loci (Frydenberg *et al.*, 1965; Sick, 1965; Moller, 1968; Dahle & Jorstad, 1993). More recently, the patterns of variation observed in the ten polymorphic allozyme loci sampled by Mork *et al.* (1985) were compared with 17 cDNA-based nuclear RFLPs (restriction fragment length polymorphisms) scored for six populations (Pogson, Mesa & Boutilier, 1995) sampled from the same geographic regions as in Mork *et al.* (1985) (Fig. 8). The highly significant correlations between genetic and geographic distance exhibited by the nuclear RFLP loci effectively eliminates gene flow as the reason for the rather uniform allozyme patterns (Fig. 9), since gene flow is a process that affects all loci equally. The data therefore implicate a role for selection as the causative agent in protein variation in this, and perhaps other species of marine fishes for which homogeneous patterns of allozyme variation have been observed (Pogson *et al.*, 1995). The high degree of genetic differentiation seen in Fig. 9, also suggests that biological data collected on animals sampled from a geographically discrete region, may not be as readily transferable to other populations as was previously thought possible.

Geographic variation in physiological traits

The activity metabolism of cod has been studied from three distinct geographic regions: the Scotian Shelf, coastal Newfoundland, and the brackish-water Bras d'Or Lakes in Cape Breton, Nova Scotia. The Newfoundland population experiences very low and narrow ranges of temperatures (−1.5 to 2 °C) compared to Scotian Shelf populations (2 to 8 °C), whereas the Bras d'Or population occupies a habitat with broad temperature variation (−1 to 20 °C) and salinities (21 to 24.5‰) much lower than those of either the Newfoundland or Scotian populations (31.3 to 33.5‰).

Work in this area has focused on the identification of cod populations inhabiting different physico-chemical environments, and evaluating whether performance (or the physiological support of that performance) varies between these environments. Activity capacities are of general interest to

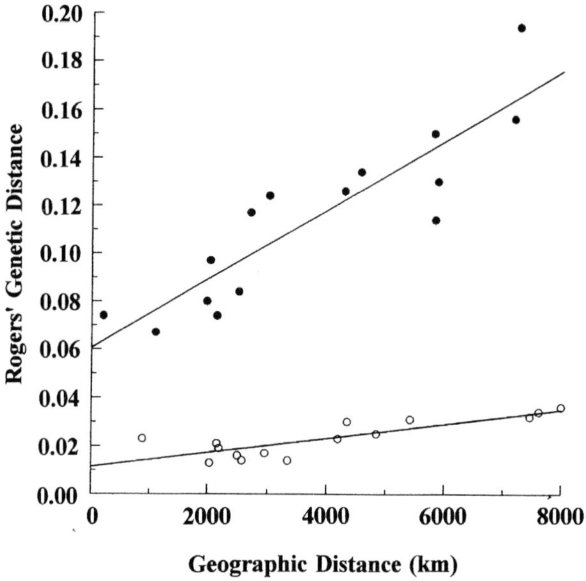

Fig. 9. Correlations between Rogers' genetic distance and geographic distance for samples collected from Atlantic cod (*Gadus morhua*) populations at the various sites in Fig. 9 and scored for allozyme loci (open circles; Mork *et al.*, 1985), or nuclear RFLP loci (closed circles). (From Pogson *et al.*, 1995.)

physiological ecologists because they can help to identify the structural and functional boundaries within which a suite of behavioural activities take place (Bennett 1991), be they predator–prey encounters (burst and escape activity), fighting oncoming currents, chasing down prey or reproducing (maximum exertion), or cruising at sustained speeds during large-scale migrations (endurance exercise). And, behaviours such as these have long interested evolutionary biologists and fisheries biologists alike (Frank & Leggett, 1994) because small intraspecific differences in activity capacities may mean the difference between predator avoidance or prey capture, social status and growth, and reproductive success or failure (Kerr, 1982). Yet, despite such obvious links between activity capacities and fitness, the field of fish physiology has for many years focused itself almost entirely at the inter-species level. Although there has been occasional interest in whether certain activity-related traits can be attributed to population origin and/or the environmental history of individual species of fish (Thomas & Donahoo, 1977; Taylor & McPhail, 1986; Giles, 1991; Nelson, 1990; Nelson & Mitchell, 1992), only a handful of studies have been similarly focused on cold-water, commercial fish species (Nelson *et al.*, 1994, 1996).

Discrete differences in activity capacities between populations of Atlantic cod inhabiting brackish inshore waters, and their fully marine offshore counterparts have recently been described (Nelson *et al.*, 1994). The intraspecific differences in activity metabolism observed (aerobic v. anaerobic scope, cost of transport) cannot be attributed simply to the salinity differences between habitats, since the differences do not disappear when individuals from each geographic region are acclimated for several months to the native salinity of the comparison population (Nelson *et al.*, 1996). The importance of these findings in terms of fisheries management and bioenergetic modelling, is the simple fact that the metabolic costs for activity are unlikely to be uniform across populations (i.e. stocks).

Conclusion

Behavioural, physiological and molecular genetic techniques have been applied to improve our understanding of the ecology of Atlantic cod. Many juvenile cod avoid cold open-ocean temperatures by recruiting to warmer inshore 'nursery areas' which probably serve to maximise the animals' growth, survival and eventual recruitment to the commercial fishery. We have ultrasonic telemetry to monitor physiological correlates of migratory decision making processes of adult cod in response to heterogeneous low temperature environments. Temperature gradients are by far the most important 'natural fences' that migrating cod encounter in the open ocean and knowledge of the animals' corresponding behaviour and metabolic physiology is providing important clues as to how the thermal environment determines the distributions of cod in both time and space. Application of a new suite of molecular markers has revealed that cod maintain discrete genetic stocks throughout the North Atlantic and our physiological studies suggest that these geographically and genetically isolated populations exhibit different capacities for metabolic performance. While a major aim of the studies has been to show the degree of intraspecific variation in physiological traits that relate to evolutionary fitness, they are also important for the management strategy adopted for a fishery. At present, many cod populations are managed as 'stock complexes', which fails to recognise the existence of genetically distinct groups, either within, or between, more arbitrarily-set management zones. Our findings negate a basic premise in current procedures for assessing cod production: namely, the assumption that all codfish are alike. Cod abundance alone is not a sufficient indicator of potential reproductive output, and therefore of potential yield, because of the wide intraspecific variation that cod of different sizes, ages, condition and genotype can proportionately contribute to gamete production and therefore to future year class strength.

Acknowledgements

This paper reviews work done in collaboration with several colleagues to whom I am greatly indebted: Guy Claireaux, Ione Hunt von Herbing, Steve Kerr, Kate Mesa, Jay Nelson, Grant Pogson, Yong Tang and Dale Webber. Thanks are also extended to Hollie Blaikie and Martha Krohn for access to unpublished data. This paper was presented in a symposium entitled 'The Integration of Zoological Sciences into Fisheries Management' (funded by The Canadian Centre for Fisheries Innovation, the Department of Fisheries and Oceans Canada and the Newfoundland Government Department of Fisheries and Agriculture), at the Canadian Society of Zoologists Annual Meeting, held at Memorial University, Newfoundland, in May 1996. I thank Derek Burton and George Rose for their invitation. The research in this paper was funded by grants from the Ocean Production Enhancement Network (NSERC Canada) to S.R. Kerr and RGB.

References

Arnason, E. & Rand, D.M. (1992). Heteroplasmy of short tandem repeats in mitochondrial DNA of Atlantic cod, *Gadus morhua*. *Genetics*, **132**, 211–20.

Arnold, G.P., Greer Walker, M., Emerson, L.S. & Holford, B.H. (1994). Movements of cod (*Gadus morhua* L.) in relation to the tidal streams in the southern North Sea. *ICES Journal of Marine Science*, **51**, 207–32.

Bagge, O., Steffensen, E. & Bay, J. (1991). The fluctuations in abundance of the stock of cod compared to environmental changes and the fishery. *North Atlantic Fisheries Organization Scientific Council Meeting* – September 1991, NAFO SCR Doc. 91/97, Serial No. N1986.

Bakun, A., Beyer, J., Pauly, D., Pope, J.G. & Sharp, G.D. (1982). Ocean sciences in relation to living resources. *Canadian Journal of Fisheries and Aquatic Sciences*, **39**, 1059–70.

Bennett, A.F. (1991). The evolution of activity capacity. *Journal of Experimental Biology*, **160**, 1–23.

Björnsson, B. (1993). Swimming speed and swimming metabolism of Atlantic cod (*Gadus morhua*) in relation to available food: a laboratory study. *Canadian Journal of Fisheries and Aquatic Sciences*, **50**, 2542–51.

Blaikie, H.D. & Kerr, S.R. (1996). Effect of activity level on apparent heat increment in Atlantic cod, *Gadus morhua*. *Canadian Journal of Fisheries and Aquatic Science*, **53**, 2093–9.

Boisclair, D. & Leggett, W.C. (1989). The importance of activity in bioenergetic models applied to actively predatory fishes. *Canadian Journal of Fisheries and Aquatic Sciences*, **46**, 1859–67.

Brett, J.R. (1964). The respiratory metabolism and swimming performance of young sockeye salmon. *Journal of the Fisheries Research Board of Canada* **21**, 1183–226.

Brett, J.R. (1986). Production energetics of a population of sockeye salmon, *Oncorhynchus nerka*. *Canadian Journal of Zoology*, **64**, 555–64.

Brett, J.R. & Glass, N.R. (1973). Metabolic rates and critical swimming speeds of sockeye salmon (*Oncorhynchus nerka*) in relation to size and temperature. *Journal of the Fisheries Research Board of Canada*, **30**, 379–87.

Brett, J.R. & Groves, T.D.D. (1979). Physiological energetics. In *Fish Physiology* Vol VIII, ed. W.S. Hoar, D.J. Randall & J.R. Brett, pp. 279–352. New York: Academic Press.

Carr, S.M. & Marshall, H.D. (1991). Detection of intraspecific DNA sequences variation in the mitochondrial cytochrome b gene of Atlantic cod (Gadus morhua) by the polymerase chain reaction. *Canadian Journal of Fisheries and Aquatic Sciences*, **48**, 48–52.

Claireaux, G. & Dutil, J.D. (1992). Physiological response of the Atlantic cod (*Gadus morhua*) to hypoxia at various environmental salinities. *Journal of Experimental Biology*, **163**, 97–118.

Claireaux, G., Webber, D.M., Kerr, S.R. & Boutilier, R.G. (1995*a*). Physiology and behaviour of free swimming Atlantic cod (*Gadus morhua*) facing fluctuating temperature conditions. *Journal of Experimental Biology*, **198**, 49–60.

Claireaux, G., Webber, D.M., Kerr, S.R. & Boutilier, R.G. (1995*b*). Physiology and behaviour of free swimming Atlantic cod (*Gadus morhua*) facing fluctuating salinity and oxygenation conditions. *Journal of Experimental Biology*, **198**, 61–9.

Dahl, E., Danielssen, D.S., Moksness, E. & Solemdal, P. (1984). *Flodevigan Rapportser*, Part 1: The Propagation of cod *Gadus morhua* L., pp. 1–10.

Dahle, G. & Jorstad, K.E. (1993). Haemoglobin variation in cod – a reliable marker for Arctic cod (*Gadus morhua* L.). *Fisheries Research*, **16**, 301–11.

Davidson, F.J.M. & deYoung, B. (1995). Modelling advection of cod eggs and larvae on the Newfoundland Shelf. *Fisheries Oceanography*, **4**, 33–51.

DeBlois, E.M. & Rose, G.A. (1995). Effect of foraging activity on the shoal structure of cod (*Gadus morhua*). *Canadian Journal of Fisheries and Aquatic Sciences*, **52**, 2377–87.

deLafontaine, Y., Lambert, T, Gilly, G.R., McKone, W.D. & Miller, R.J. *(eds.)*. (1994). Juvenile stages: the missing link in fisheries research. *Canadian Technical Reports in Fisheries and Aquatic Sciences*, 1890.

Falkowski, P.G., Hopkins, T.S. & Walsh, J.J. (1980). An analysis of factors affecting oxygen depletion in the New York Bight. *Journal of Marine Research*, **38**, 479–506.

Frank, K.T. & Leggett, W.C. (1994). Fisheries ecology in the context of ecological and evolutionary theory. *Annual Review of Ecology and Systematics*, **25**, 401–22.

Frydenberg, O., Moller, D., Naevdal, G. & Sick, K. (1965). Haemoglobin polymorphism in Norwegian cod populations. *Hereditas*, **53**, 257–71.

Garland, T. Jr. & Adolph, S.C. (1991) Physiological differentiation of vertebrate populations. *Annual Review of Ecology and Systematics* **22,** 193–228.

Giles, M.A. (1991). Strain differences in hemoglobin polymorphism, oxygen consumption, and blood oxygen equilibria in three hatchery broodstocks of arctic charr, *Salvelinus alpinus. Fish Physiology and Biochemistry*, **9**, 291–301.

Gotceitas, V. & Brown, J.A. (1993). Substrate selection by juvenile Atlantic cod (*Gadus morhua*): effects of predation risk. *Oecologia*, **93**, 31–7.

Gulland, J.A. & Williamson, G.R. (1962). Transatlantic journey of tagged cod. *Nature* **195**, 921.

Henderson, P.A., Holmes, R.H.A. & Bamber, R.N. (1988). Size-selective overwintering mortality in the sand smelt, *Atherina boyeri* Risso, and its role in population regulation. *Journal of Fish Biology*, **33**, 221–33.

Hutchings, J.A. & Myers, R.A. (1994). What can be learned from the collapse of a renewable resource? Atlantic cod, *Gadus morhua*, of Newfoundland and Labrador. *Canadian Journal of Fisheries and Aquatic Sciences*, **51**, 2126–46.

Jobling, M. (1981). The influences of feeding on the metabolic rate of fishes: a short review. *Journal of Fish Biology*, **18**, 385–400.

Jones, R. (1978). Estimates of the food consumption of haddock (*Melanogrammus aeglefinus*) and cod (*Gadus morhua*). *Journal du Conseil International pour l'Exploration de la Mer*, **38**, 18–27.

Kerr, S.R. (1982). Estimating the energy budgets of actively predatory fishes. *Canadian Journal of Fisheries and Aquatic Sciences*, **39**, 371–9.

Kitchell, J.F. (1983). Energetics. In *Fish Biomechanics*, ed. P.W. Webb & D. Weihs, pp. 312–38. New York: Praegar Publishers.

Krohn, M.M. & Boisclair, D. (1994). Use of a stero-video system to estimate the energy expenditure of free-swimming fish. *Canadian Journal of Fisheries and Aquatic Sciences*, **51**, 1119–27.

Krohn, M., Reidy, S. & Kerr, S.R. (1997). Bioenergetic analysis of the effects of temperature and prey availability on growth and condition of northern cod (*Gadus morhua*). *Canadian Journal of Fisheries and Aquatic Sciences*, **54** (Suppl. 2), 113–21.

Larson, A., Wake, D.B. & Yanev, K.P. (1984). Measuring gene flow among populations having high levels of genetic fragmentation. *Genetics*, **106**, 293–308.

Law, R. (1991). Fishing in evolutionary waters. *New Scientist*, **2** March 1991, 35–7.

Lilly, G.R. (1987). Interactions between Atlantic cod (*Gadus morhua*) and capelin (*Mallotus villosus*) off Labrador and eastern Newfoundland: a review. *Canadian Technical Reports in Fisheries and Aquatic Sciences*, **1567**, 37pp.

Lough, G.R. & Potter, D.C. (1993). Vertical distribution patterns and diel migrations of larval and juvenile haddock *Melanogrammus aeglefinus* and Atlantic cod *Gadus morhua* on Georges Bank. *Fisheries Bulletin*, **91**, 281–303.

Lough, G.R., Valentine, P.C., Potter, D.C., Auditore, P.J., Bolz, G.R., Nielson, J.D. & Perry, R.I. (1989). Ecology and distribution of juvenile cod and haddock in relation to sediment type and bottom currents on eastern Georges Bank. *Marine Ecology Progress Series*, **56**, 1–12.

May, A.W. (1967). Fecundity of Atlantic cod. *Journal of the Fisheries Research Board of Canada*, **24**,1531–51.

Metcalfe, J.D., Arnold, G.P. & Webb, P.W. (1990). The energetics of migration by selective tidal stream transport: an analysis for plaice tracked in the southern North Sea. *Journal of the Marine Biological Association of the UK*, **70**, 149–62.

Moller, D. (1968). Genetic diversity in cod. *Hereditas*, **60**, 1–32.

Mork, J., Ryman, N., Stahl, G., Utter, F. & Sundnes, G. (1985). Genetic variation in Atlantic cod (*Gadus morhua*) throughout its range. *Canadian Journal of Fisheries and Aquatic Sciences*, **42**, 1580–7.

Nelson, J.A. (1990). Muscle metabolite response to exercise and recovery in yellow perch *(Perca flavescens)*: comparison of populations from naturally acidic and neutral waters. *Physiological Zoology*, **63**, 886–908.

Nelson, J.A. & Mitchell, G.S. (1992). Blood chemistry response to acid exposure in yellow perch *Perca flavescens*: comparison of populations from naturally acidic and neutral environments. *Physiological Zoology*, **65**, 493–514.

Nelson, J.A., Tang, Y. & Boutilier, R.G. (1994). Differences in exercise physiology between two Atlantic cod *(Gadus morhua)* populations from different environments. *Physiological Zoology*, **67**, 330–54.

Nelson, J.A., Tang, Y. & Boutilier, R.G. (1996). The effects of salinity change on the exercise performance of two Atlantic cod (*Gadus morhua*) populations from different environments. *Journal of Experimental Biology*, 199, 1295–309.

Pihl, L. (1989). Effects of oxygen depletion on demersal fish in coastal areas of the south-east Kattegat. In *Reproduction, Genetics and Distributions of Marine Organisms.* Proceedings of 23rd European Marine Biology Symposium, pp.431–9. Denmark: Olsen & Olsen.

Pogson, G.H. (1994). Contrasting patterns of DNA polymorphism in the deep-sea scallop (*Placopecten magellanicus*) and the Atlantic cod (*Gadus morhua*) revealed by random cDNA clones. In *Genetics and Evolution of Aquatic Organisms*, ed. A.R. Beaumont, pp. 207–17. London: Chapman & Hall.

Pogson, G.H. & Zouros, E. (1994). Allozyme and RFLP heterozygosities as correlates of growth rate in the scallop Placopecten magellanicus: a test of the associative overdominance hypothesis. *Genetics*, **137**, 221–31.

Pogson, G.H, Mesa, K.A. & Boutilier, R.G. (1995). Genetic population structure and gene flow in the Atlantic cod *Gadus morhua*: a comparison of allozyme and RFLP loci. *Genetics*, **139**, 375–85.

Powles, P.M. (1958). Studies of reproduction and feeding in Atlantic cod (*Gadus callarius* L.) in the southwestern Gulf of St. Lawrence. *Journal of the Fisheries Reseaarch Board of Canada*, **15**, 1382–402.

Rasmussen, B. (1959). On the migration pattern of the West Greenland stock of cod. *Annals in Biology, Copenhagen*, **4**, 123–4.

Ridley, M. (1993). *Evolution*. Oxford: Blackwell Scientific Publications. 670 pp.

Rose, G.A. (1993). Cod spawning on a migration highway in the north-west Atlantic. *Nature*, **366**, 458–61.

Rose, G.A., Atkinson, B.A., Baird, J., Bishop, C.A. & Kulka, D.W. (1994). Changes in the distribution of Atlantic cod and thermal variations in Newfoundland waters, 1980–1992. *ICES Marine Science Symposium* **198**, 542–52.

Rose, G.A., deYoung, B. & Colbourne, E.B. (1995). Cod (*Gadus morhua* L.) migration speeds and transport relative to currents on the north-east Newfoundland Shelf. *ICES Journal of Marine Science*, **52**, 903–13.

Rossignol-Strick, M. (1985). A marine anoxic event on the Brittany coast, July 1982. *Journal of Coastal Research*, **1**, 11–20.

Saunders, R.L. (1963). Respiration of the Atlantic cod. *Journal of the Fisheries Research Board of Canada*, **20**, 373–86.

Sick, K. (1965). Haemoglobin polymorphism of cod in the North Sea and the north Atlantic Ocean. *Hereditas*, **54**, 49–69.

Sinclair, M. & Page, F. (1995). Cod fishery collapses and North Atlantic GLOBEC. *US GLOBEC News*, **no. 8**, March 1995, University of California Berkeley.

Smith, P.J., Birley, A.J., Jamieson, A. & Bishop, C.A. (1989). Mitochondrial DNA in the Atlantic cod, *Gadus morhua*: lack of genetic divergence between eastern and western populations. *Journal of Fish Biology*, **34**, 369–73.

Soofiani, N. & Hawkins, A.D. (1982). Energetic costs at different levels of feeding in juvenile cod, *Gadus morhua* L. *Journal of Fish Biology*, **21**, 577–92.

Soofiani, N. & Priede, I.G. (1985). Aerobic metabolic scope and swimming performance in juvenile cod, *Gadus morhua* L. *Journal of Fish Biology*, **26**, 127–38.

Steffensen, J.F. (1993). Ventilatory and respiratory responses in fish: adaptations to the environment. In *The Vertebrate Gas Transport Cascade: Adaptations to Environment and Mode of Life*, ed. E. Bicudo, pp. 60–71. Boca Raton, London, Tokyo: CRC Press.

Taggart, C.T. & Frank, K.T. (1990). Perspectives on larval fish ecology and recruitment processes: probing the scales of relationships. In *Large Marine Ecosystems*, ed. by K. Sherman, L.M. Alexander & B.D. Gold, pp. 151–64. Washington, DC, USA: American Association for the Advancement of Science.

Tang, Y., Nelson, J.A., Reidy, S., Kerr, S.R. & Boutilier, R.G. (1994). A reappraisal of activity metabolism in the Atlantic cod *(Gadus morhua)*: *Journal of Fish Biology*, **44**, 1–10.

Taylor, E.B. & McPhail, J.D. (1986). Prolonged and burst swimming in anadromous and freshwater threespine stickleback, *Gasterosteus aculeatus*. *Canadian Journal of Zoology*, **64**, 416–20.

Templeman, W. (1974). Migrations and intermingling of Atlantic cod (*Gadus morhua*) stocks of the Newfoundland area. *Journal of the Fisheries Research Board of Canada*, **31**, 1073–92.

Thomas, A.E. & Donahoo, M.J. (1977). Differences in swimming performance among strains of rainbow trout (*Salmo gairdneri*). *Journal of the Fisheries Research Board of Canada*, **34**, 304–6.

Tupper, M. & Boutilier, R.G. (1995*a*). Effects of habitat on settlement, growth and post-settlement survival of age 0+ Atlantic cod (*Gadus morhua*). *Canadian Journal of Fisheries and Aquatic Sciences*, **52**, 1834–41.

Tupper, M. & Boutilier, R.G. (1995*b*). Size and priority at settlement affect growth and competitive success of newly settled Atlantic cod. *Marine Ecology Progress Series*, **118**, 295–300.

Tupper, M., and Boutilier, R.G. (1994). Settlement and growth of age 0+ cod (*Gadus morhua*) in St. Margaret's Bay, Nova Scotia. In *Juvenile Stages: The Missing Link in Fisheries Research*, ed. Y. deLafontaine, T. Lambert, G.R. Gilly, W.D. McKone & R.J. Miller. *Canadian Technical Reports in Fisheries and Aquatic Sciences*, **1890**, 89–90.

Tytler, P. (1969). Relationship between oxygen consumption and swimming speed in the haddock *Melanogrammus aeglefinus*. *Nature*, **221**, 274–5.

Tytler, P. (1978). The influence of swimming performance on the metabolic rate of gadoid fish. In *Physiology and Behaviour of Marine Organisms*, ed. D.S. McLusky & A.J. Berry, Oxford: Pergamon Press.

Walsh, W.J. (1987). Patterns of recruitment and spawning in Hawaiian reef fishes. *Environmental Biology of Fishes*, **18**, 257–76.

Waples, R.S. (1987). A multispecies approach to the analysis of gene flow in marine shore fishes. *Evolution*, **41**, 385–400.

Ward, R.D., Woodwark, M. & Skibinski, D.O.F. (1994). A comparison of genetic diversity levels in marine, freshwater, and anadromous fishes. *Journal of Fish Biology*, **44**, 213–32.

Wardle, C.S. (1977). Effects of size on the swimming speeds of fish. In *Scale Effects in Animal Locomotion*, ed. T.J. Pedley, pp. 299–313. London: Academic Press.

Ware, D.M. (1975). Growth, metabolism and optimal swimming speed of a pelagic fish. *Journal of the Fisheries Research Board of Canada*, **32**, 33–41.

Ware, D.M. (1978). Bioenergetics of pelagic fish: theoretical change in swimming speed and ratio with body size. *Journal of the Fisheries Research Board of Canada*, **35**, 220–8.

Weihs, D. (1975). An optimum swimming speed of fish based on feeding efficiency. *Israel Journal of Technology*, **13**, 163–7.

Index